Guide to
Residential Carpentry

JOHN L. FEIRER

Head, Industrial Technology and Education Department
Western Michigan University
Kalamazoo, Michigan

GILBERT R. HUTCHINGS

Professor, Industrial Technology and Education Department
Western Michigan University
Kalamazoo, Michigan

Guide to
Residential Carpentry

GLENCOE PUBLISHING COMPANY

BENNETT & McKNIGHT DIVISION

Distributed to the book trade by Macmillan Publishing Company, New York.

Macmillan Publishing Company
866 Third Avenue, New York, NY 10022
Collier Macmillan Canada, Inc.

ISBN 0-02-000490-7

Macmillan books are available at special discounts for bulk purchases for sales promotions, premiums, fund-raising, or educational use. For details contact:

Special Sales Director
Macmillan Publishing Company
866 Third Avenue
New York, NY 10022

10 9 8 7 6 5 4 3 2 1

Printed in the United States of America

Cover: "Super Square" framing square courtesy of Orem Research, Inc., Hinsdale, IL
Other tools and materials courtesy of Carver Lumber Co., Peoria, IL.

Preface

RESIDENTIAL CARPENTRY has been designed as a basic text for students in industrial education classes at the high school, vocational school, and the community junior college levels. It has been written to prepare the learner to do quality work on the job in building construction. This book will also be helpful in apprenticeship training and for do-it-yourselfers in building and remodeling.

Wood is a remarkable space-age material. It possesses the same fundamental composite structure as some of the best manufactured materials developed for space flights, yet it is a renewable material, one that this country need never deplete. Wood can be easily worked, with relatively simple equipment. With such advantages, it is natural that wood remains the basic material for seven of every ten homes built in the United States.

Craftspeople in the building trades represent the largest single group of skilled workers in our nation's labor force, comprising about three of every ten skilled workers. Carpenters alone number nearly one-third of all building craftspeople. Carpenters are actually in short supply and will continue to be in the years ahead. Therefore, it is important for schools at secondary level and above to consider adding a program in carpentry and building construction.

In preparing this text, the authors have reviewed courses of study from schools, trade unions, and public agencies to determine what should be included.

They have also carefully reviewed technical materials from companies and other organizations involved in building construction. Based on this material, they have composed a text that is current and technically correct. Each area contains concise descriptions of materials and techniques. The book is generously illustrated with drawings and photographs to give the reader a thorough grasp of materials and techniques, technical know-how and related information.

This book can be used in many ways. It is an ideal text for anyone who is building a home on site. It can also serve as a course of instruction for students building sectional homes or modules in a shop or laboratory.

Students who follow the instructions in this text should achieve the following behavioral objectives:
- To know the career opportunities in the building trades.
- To understand the basic materials used in residential home construction.
- To be able to read prints and technical material necessary to build a home.
- To demonstrate competency in the use of hand and machine tools used in house construction.
- To do acceptable workmanship in building construction, based on recognized building codes.

The many companies and individuals who have contributed to this publication are recognized in the list of acknowledgments.

Acknowledgments

PERSONAL

Susan M. Hutchings
Daniel W. Irvin
John Polderman
William K. Purdy
Walter C. Schwersinske

COMMERCIAL, PROFESSIONAL, AND GOVERNMENTAL

The Abitibi Corporation
Acorn Products Company
Air Vent, Inc.
Allied Chemical, Barrett Division
Alsco, Incorporated
Aluminum Company of America
American Concrete Institute
American Forest Products Industries, Incorporated
American Hardboard Association
American Ladder Institute
American Olean
American Optical Corporation
American Plywood Association
Amerigo
Andersen Corporation
Armstrong Cork Company
Atlas
Automated Building Components, Incorporated
Baker-Roos, Incorporated
Benjamin Moore and Company
Berry Industries, Incorporated
Bethlehem Steel
Billings
Bird and Son, Incorporated
The Black & Decker Manufacturing Company
Boice-Crane Company
Bostitch, Incorporated
Boston Chamber of Commerce
Bowater Board Company
Bruce Oak Flooring
Buck Brothers, Incorporated
California Redwood Association
The Celotex Corporation
Clark Equipment
Clausing Corporation
Condon-King Company, Incorporated
Consumers Power Company
Continental Homes
Copper Development Association
Crawford Door Company
Curtis Companies, Incorporated

E. I. dupont de Nemours and Company, Incorporated
Delta
Desa Industries, Incorporated
De Walt
Dexter Lock Company
Disston Saw Company
The Donley Brothers Company
Dow Chemical Company
Feather-Lite Manufacturing Company
Federal Pocket Door Frame Company
The Flintkote Company
Formica Corporation
General Electric Company
Georgia-Pacific Corporation
Grand Rapids Sash and Door Company
Greenberg-May Prod., Inc.
Greenlee Brothers and Company
Harr, Hedrich-Blessing
H C Products Company
Heart Truss and Engineering
Homeway Corporation
Independent Nail and Packing Company
Inland Steel Company
International Conference of Building Officials
International Harvester Company
Irwin
The I-XL Furniture Company, Incorporated
Johns-Manville
Johnson-Howard Lumber Company
Kemper Brothers, Incorporated
Koppers
Lennox Industries, Incorporated
Los Angeles Building and Construction Trades Council
Macklanburg-Duncan Co.
Manor House Cupolas
Manpower Magazine
Maple Flooring Manufacturers Association
Marvin Windows
Masonite Corporation
McKee Door Company
Memphis Moldings, Incorporated
Metric Monitor
Miller Lumber Company
Millers Falls Company
Millwork Supply Company
Minnesota & Ontario Paper Company
Modernfold Industries
Morgan Company
Mueller Brass Company
Mutschler Brothers Company

National Association of Home Builders Research Foundation, Incorporated
National Manufacturing Company
National Oak Flooring Manufacturers Association
National Research Council
National Woodwork Manufacturers Association
Nautilus Industries, Incorporated
Nord
Nordahl Sliding Door Pockets
Nu-Wood Sheathing
Oliver Machinery Company
Overhead Door Corporation
Owens Corning Fiberglas Corp.
Paslode Company
Patent Scaffolding Company
Frank Paxton Lumber Company
Pease Company
Pittsburgh Plate Glass Company
Porta-Table Corporation
Porter-Cable Machine Company
Portland Cement Association
Powermatic, Incorporated
Raynor Manufacturing Company
Red Cedar Shingle & Handsplit Shake Bureau
Redman Industries, Incorporated
Remington Arms Company, Incorporated
Republic Steel Corporation
Research Products Corporation
Rock Island Millwork
Rockwell Manufacturing Company
Rolscreen Company
Russell Jennings
Safway Steel Products, Inc.
St. Regis Paper Company, Panelyte Division
Sanford Roof Trusses
H. J. Scheirich Company
Scholz Homes, Incorporated
Scovill, Caradco Division
Sears Roebuck and Company
Senco Products, Incorporated

Shakertown Corporation
Shopmate
Simonds Saw and Steel Company
Simplex Products Group
Simpson Timber Company
Spotnail, Incorporated
Stanley Iron Works, Incorporated
Stanley Tools
Stanley Door Systems, Inc.
Sterling Homes, Incorporated
Structural Clay Products Institute
Textron Company
Timber Engineering Company
Trusswall Systems Corporation
United States Department of Agriculture
United States Department of Commerce, National Bureau of Standards
United States Forest Service, Forest Products Laboratory
United States Gypsum Company
United States Plywood Corporation
United States Savings and Loan League
United States Steel Corporation
The Upson Company
Van Mark Products Corporation
Vega Industries, Incorporated
Weather Shield Manufacturing, Inc.
Weiser Company
Western Electric Company
Western Red Cedar Lumber Association
Western Wood Moulding and Millwork Producers
Western Wood Products Association
Westinghouse
Weyerhaeuser Company
Woodbridge Ornamental Iron Company
Wood Conversion Company
Wood-Mode Kitchens
Wood-Mosaic Corporation
Zinc Institute, Inc.

Table of Contents

TABLE OF CONTENTS

SECTION 1

Introduction

Carpentry in Housing

1

Wood is perhaps the most important of all building materials. Four out of every five housing units are wood frame construction, Fig. 1-1. Wood has long been preferred for homes and other housing units because of its high strength, resistance to wear, and ability to hold fasteners. Even in housing units built of other materials, a great deal of wood is needed to finish the interior.

Wood is one of our most popular and widely used materials, universally loved for its beauty, its warmth, and its naturalness. Yet wood has many other qualities which are generally unknown—its incredible strength; its durability; its ease of maintenance; its adaptability; its ability to insulate; and its renewability. Pound for pound, wood can be stronger than steel. For example, a wood block just 1″ square and 2¼″ long can support five tons. Wood is also elastic. That explains why wood floors are less tiring to walk on than concrete. Wood can bend while under the stress of high winds or earthquakes. A dramatic example of wood's strength is laminated beams that can span wide distances well over 200 feet.

Wood is a natural insulator. A well designed and built wood home can save energy and reduce utility bills. Heat will move through 1″ of wood fifteen times slower than concrete and almost 2,000 times slower than through aluminum. A roof of wood shingles and shakes can help contain heat and keep cold out. For example, wood shingles have two times greater insulation value than asphalt shingles.

Wood can last many lifetimes. Houses built in New England over three hundred years ago are still standing. Wood exposed to the ele-

1-1. *Many houses are almost all wood in construction.*

ments wears away very little. Woods can be made more resistant to decay by using treated material. Thousands of houses have foundations constructed of preservative-treated wood. Wood is beautiful finished or natural. Chalets of wood have weathered for centuries without paint or stain. Wood is indeed the ideal building material.

People are dependent on trees and wood for housing and for other commodities such as furniture, boats, sporting equipment, and many other items. By the year 2000 the population of the United States will exceed 300 million. Wood products must double by then to house Americans adequately and to meet our other material needs.

In the housing industry carpentry is divided into two categories, "rough" and "finished," Fig. 1-2. Examples of "rough" carpentry include wood forms for concrete, wood framing and all exterior work on a home, except trim. "Finished" carpentry includes building stairs, cabinets, installing wood paneling, doing trim work, and many other details. This finish work is sometimes called *cabinetmaking* because this is a term usually used to describe fine quality woodworking.

Housing is a multi-billion dollar industry that exceeds four percent of our gross national product. Carpentry in housing deals with both *new construction* and *remodeling*, Fig. 1-3. In recent years the dollar volume of remodeling has equaled the amount of money spent on new construction. In the years ahead it is estimated that remodeling will become an even larger segment of this multi-billion dollar industry. There are unlimited possibilities for home improvements. This includes not only enlarging existing older homes but making many kinds of interior and exterior improvements. Most remodeling jobs require a broad

1-2a. Framing is a part of rough carpentry.

knowledge of carpentry. Carpentry has been a most important craft for centuries. Today carpenters not only work with wood and wood products such as plywood, hardboard, and particleboard, but many other materials such as plastics, metals and glass. Their skills are important and versatile, making them the largest group of people in the building trades.

Before you begin any project, whether it is new or old construction, make sure that you plan carefully. For new construction a complete set of house plans, lists of materials and specifications are essential. In remodeling, a detailed sketch or drawing of the job to be

1-2b. Interiors such as trim and cabinets are part of finish carpentry.

done must be available before plans can be made.

Carpenters must be skilled in the use of hand and power equipment, including portable power tools. Carpenters must also know how to read and sometimes make drawings. They must be familiar with a wide range of building materials and how to order them. They must also know all of the basic techniques of construction, including several different ways of doing the same job. Carpenters must make decisions on how to complete a project. Carpenters must also be able to deal with lumber yard and building suppliers so that they can get the correct kind and amount of material for the job. Carpentry is an exciting kind of work because each job varies so greatly. The units that follow will help you grasp the basic fundamentals involved in residential carpentry.

The construction industry directly or indirectly employs about 15% of our working population. This means that the chances are one in six or seven that you will work in some area of building construction, which is one of America's largest industries. Your career may

1-3. *Here's a typical remodeling job. The lower illustration shows the original Cape Cod. The upper illustration shows what it looks like after a face-lift.*

be that of a carpenter, architect, drafter, a lumberyard salesperson or any of the hundreds of other building construction opportunities.

In construction there are three career levels that require special training and education: craft, technical, and professional. See *Occupational Outlook Handbook* for details.

QUESTIONS

1. What percent of all housing has wood frame construction?

2. Name some of the qualities of wood that are generally unknown.

3. How does wood compare to concrete and steel as an insulator? Explain.

4. Describe the two categories of carpentry.

5. How does remodeling compare to new construction as to volume of business?

6. Give a brief description of what a carpenter must know and be able to do.

Reading Prints

2

The ability to read and understand drawings, prints, and plans is basic to all construction. Sketches, drawings, and prints are a kind of language. They tell you everything you need to know to build something, including the materials needed. Fig. 2-1. By means of *lines*, *symbols*, and *dimensions*, the ideas of the designer or architect are conveyed to you. To be a good builder you must be able to interpret correctly the sketch, drawing, or print so you can visu-

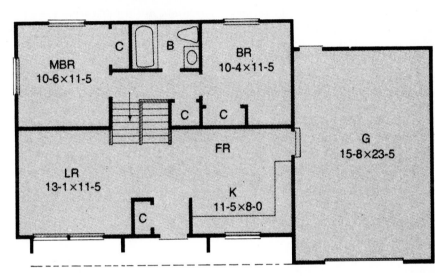

2-1. *Floor plan for a two bedroom home.*

(Right hand plan)

alize the size and shape of the product to be built. It would not be possible to convey this information in any other way.

A *sketch* or *drawing* of something to be built is the original idea put on paper. A *print* is an exact copy of the drawing. In the building industry most prints are called *blueprints;* the paper has a blue background with white lines. Blueprints are commonly used in house construction and other building trades because they do not fade when exposed to sunlight. They are made on chemically treated paper which shows the drawing in white against a blue background.

An architect, designer, or drafter usually has the responsibility for making the original drawing. Many carpenters also have the ability to make a good sketch. They must often take measurements "on the job" and then make sketches which are sometimes used to do the building or remodeling. At other times the sketches are reviewed, refined, and then made into a set of drawings and prints. Pictorial sketches are sometimes used.

METHODS OF MEASUREMENT

Two common systems of measurement are used worldwide. The United States currently uses the customary (English) system of measurement while all of the other industrial nations in the world use the metric system. The United States is moving toward the metric system; so you should be acquainted with it. The three common units of measure used are those for *length, liquid measure,* and *weight.* Fig. 2-2.

In the customary (English) system, lengths are given in inches, feet, yards, and miles. In the metric system, lengths are given in millimetres, centimetres, metres, and kilometres. Fig. 2-3. A metre, which is the basic unit of length, is slightly longer than a yard (39.37″). Since the entire metric system is based on units of ten, the millimetre is equal to 1/1000 of a metre and a centimetre is 1/100 of a metre. A kilometre is 1000 times a metre. Actually 1″ is equal to 25.4 millimetres. The two common length measures used in the customary system for buildings are the inch and the foot, while the only two length measurements used in the metric system are the millimetre and the metre. It is easy to convert from one to the other as you can see in Fig. 2-4a. In building construction

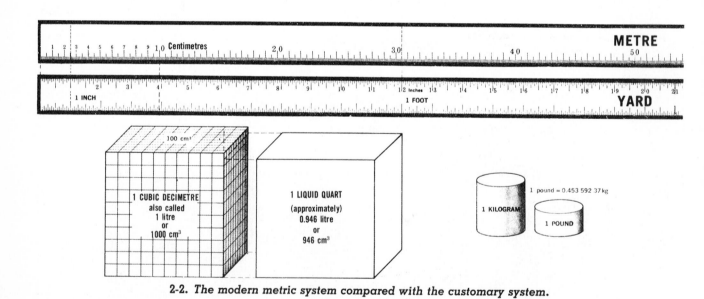

2-2. *The modern metric system compared with the customary system.*

2-3. *Note that one inch is approximately 25 mm.*

Metric Units for Common Use			
		Conversion Factors (approximate)	
Quantity	Metric unit and symbol	Customary to Metric Units	Metric to Customary Units
Length	millimetre (mm) centimetre (cm) metre (m) kilometre (km)	1 in = 25.4 mm* 1 in = 2.54 cm* 1 ft. = 30.5 cm 1 yd = 0.914 m 1 mi = 1.61 km	1 mm = 0.0394 in 1 cm = 0.394 in 1 m = 3.28 ft 1 km = 0.62 mi
Weight (Mass)	gram (g) kilogram (kg) metric ton or tonne (t)	1 oz = 28.3 g 1 lb = 454 g 1 long ton = 1.02†	1 g = 0.0353 oz 1 kg = 2.2 lb 1 t = 0.98 long ton
Area	square centimetre (cm²) square metre (m²) hectare (ha) square kilometre (km²)	1 in² = 6.45 cm² 1 ft² = 929 cm² 1 yd² = 0.836 m² 1 acre = 0.405 ha 1 mi² = 2.59 km²	1 cm² = 0.155 in² 1 m² = 10.8 ft² 1 m² = 1.2 yd² 1 ha = 2.47 acres 1 km² = 0.386 mi²
Volume	cubic centimetre (cm³) cubic metre (m³)	1 in³ = 16.4 cm³ 1 ft³ = 28 300 cm³ 1 yd³ = 0.765 m³	1 cm³ = 0.061 in³ 1 m³ = 35.3 ft³ 1 m³ = 1.31 yd³
Volume (liquids and gases)	millilitre (ml) litre (L or l) kilolitre (kl)	1 fl oz = 29.6 ml 1 liq qt = 0.946 litre 1 gal = 3.79 litres	1 ml = 0.033 8 fl oz 1 litre = 1.057 liq qt 1 kl = 264 gal
Time Interval	second (s) minute (min) hour (h)		
Speed	metre per minute (m/min)	1 ft/min = 0.305 m/min	1 m/min = 3.28 ft/min
Pressure	kilopascal (kPa)	1 psi = 6.89 kPa	1 kPa = 0.145 psi
Energy	kilojoule (kJ)	1 Btu = 1.06 kJ	1 kJ = 0.948 Btu
Power	kilowatt (kW)	1 hp = 0.746 kW	1 kW = 1.34 hp
Temperature	degree Celsius (°C)	°C = 5/9 (°F-32)	°F = 9/5°C + 32

*Accurate

2-4a. Metric units and their conversion factors.

all millimetre measurements are rounded off to the closest full measurement. For example, 1″ is equal to 25 millimetres. Fig. 2-4b.

The *liquid* measure in the customary system is in quarts and gallons, while in the metric system it is in litres. A litre is about 5% more than a quart. Liquid measure for finishing materials including paints is normally given in litres, half-litres and quarter-litres. *Weight* measure in the customary system is given in pounds, while in the metric system it is in kilograms. A kilogram is approximately 2.2 pounds.

In metric countries, particularly Britain, a standard module is 300 millimetres, which is very close to one foot. In architectural drawings, all building measurements are given in millimetres and all site measurements in metres and fractions of a metre. Figs. 2-5a & b.

To get better acquainted with both systems of measurement, it is

Customary (English)	Metric				
	Actual	Accurate Woodworkers' Language	Tool Sizes	Lumber Sizes	
				Thickness	Width
1/32 in	0.8 mm	1 mm bare			
1/16 in	1.6 mm	1.5 mm			
1/8 in	3.2 mm	3 mm full	3 mm		
3/16 in	4.8 mm	5 mm bare	5 mm		
1/4 in	6.4 mm	6.5 mm	6 mm		
5/16 in	7.9 mm	8 mm bare	8 mm		
3/8 in	9.5 mm	9.5 mm	10 mm		
7/16 in	11.1 mm	11 mm full	11 mm		
1/2 in	12.7 mm	12.5 mm full	13 mm	12 mm	
9/16	14.3 mm	14.5 mm bare	14 mm		
5/8 in	15.9 mm	16 mm bare	16 mm	16 mm	
11/16 in	17.5 mm	17.5 mm	17 mm		
3/4 in	19.1 mm	19 mm full	19 mm	19 mm	
13/16 in	20.6 mm	20.5 mm	21 mm		
7/8 in	22.2 mm	22 mm full	22 mm	22 mm	
15/16 in	23.8 mm	24 mm bare	24 mm		
1 in	25.4 mm	25.5 mm	25 mm	25 mm	
1¼ in	31.8 mm	32 mm bare	32 mm	32 mm	
1⅜ in	34.9 mm	35 mm bare	36 mm	36 mm	
1½ in	38.1 mm	38 mm full	38 mm	38 mm(or 40 mm)	
1¾ in	44.5 mm	44.5 mm	44 mm	44 mm	
2 in	50.8 mm	51 mm bare	50 mm	50 mm	
2½ in	63.5 mm	63.5 mm	64 mm	64 mm	
3 in	76.2 mm	76 mm full		75 mm	75 mm
4 in	101.6 mm	101.5 mm		100 mm	100 mm
5 in	127.0 mm	127 mm			125 mm
6 in	152.4 mm	152.5 mm			150 mm
7 in	177.8 mm	178 mm bare			
8 in	203.2 mm	203 mm full			200 mm
9 in	228.6 mm	228.5 mm			
10 in	254.0 mm	254 mm			250 mm
11 in	279.4 mm	279.5 mm			
12 in	304.8 mm	305 mm bare			300 mm
18 in	457.2 mm	457 mm full	460 mm		
24 in	609.6 mm	609.5 mm			
36 in	914.4 mm	914.5 mm		Panel Stock Sizes	
48 in—4'	1219.2 mm	1220 mm or 1.22 m		1220 mm or 1.22 m width	
96 in—8'	2438.4 mm	2440 mm or 2.44 m		2440 mm or 2.44 m width	

2-4b. *A conversion table can be used in woodworking.*

a good idea to use measuring tools marked both in inches and feet and in millimetres, centimetres, and metres. Some rules are numbered in centimetres, with ten small divisions (millimetres) between each one. However, most rules show measurements in millimetres only. Whenever possible, use a rule that is numbered in millimetres since these are easier to read. There will be less chance of error.

In architectural layouts, the general practice is to show all residential drawings in millimetres. Site plans are dimensioned in metres.

Building materials come in standard sizes, and most homes are built according to certain dimensional design standards. When the building industry converts to metrics, new standards will replace the customary ones. The design module for building materials will be 100 mm, which is slightly less than four inches.

Softwoods will not change in size, except that dimensions will be given in millimetres. For example, a standard 2 × 4, which measures 1½″ × 3½″, will be listed as 38 × 89 mm. This is a soft conversion. But, all panel stock will undergo a hard conversion, and the standard size will be 1200 × 2400 mm.

SCALE

Drawings must often be reduced from actual size so they will fit on a piece of paper. Care is taken to make such drawings according to *scale;* that is, exactly

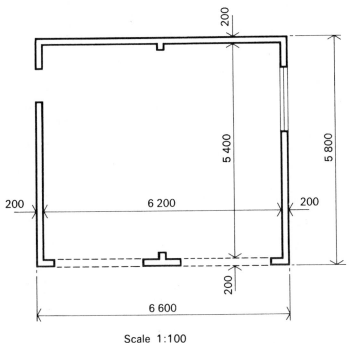

Scale 1:100

Note: Dimensions in
millimetres

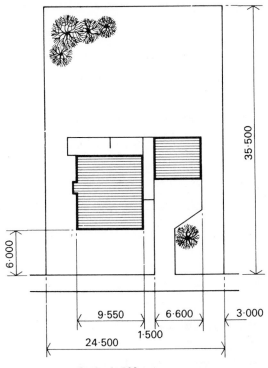

Scale 1:500
Note: Dimensions in metres

2-5b. Example of metric site plan.

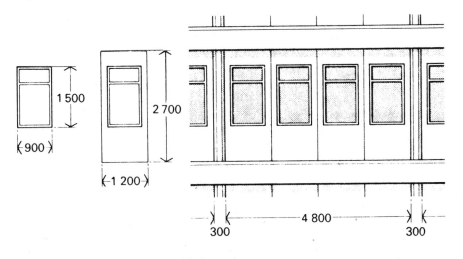

All dimensions are
in millimetres

2-5a. Example from metric house plans.

most scale drawings are, it will
probably be drawn to one of the
following common scales.

Customary Scales

6"   equals 1' (read "six inches
       equals one foot"): half size.

3"   equals 1': one-fourth size.

1½"   equals 1': one-eighth size.

1"   equals 1': one-twelfth size.

¾"   equals 1': one-sixteenth size.

½"   equals 1': one twenty-fourth
       size.

⅜"   equals 1': one thirty-second
       size.

¼"   equals 1': one forty-eighth
       size.

³⁄₁₆" equals 1': one sixty-fourth
       size.

⅛"   equals 1': one ninety-sixth
       size.

A scale of ¼" equals 1' is often
used for drawing buildings and
rooms. Detail drawings, which
show how parts of a product are

in proportion to full size. Fig. 2-6.
For example, an architect can
represent any size of building on
a single piece of paper by drawing
it to a certain scale. The scale is
not a unit of measurement but
represents the ratio between the
size of the object as drawn and its
actual size. If the drawing is exactly the same size as the object
itself, it is called a full-size or full-
scale drawing. If it is reduced, as

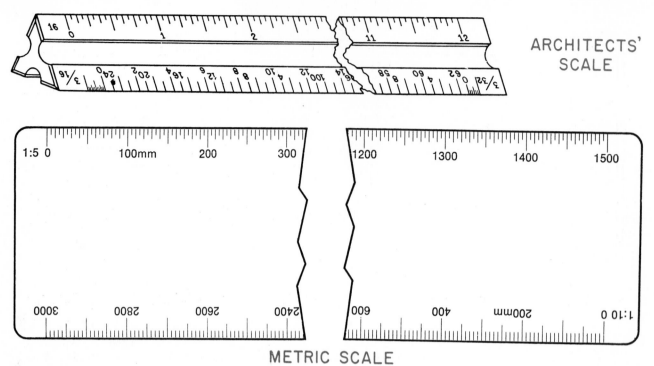

ARCHITECTS' SCALE

METRIC SCALE

2-6. *Two types of tools for making scale drawings—the architect's scale (for customary measurements) and the metric scale.*

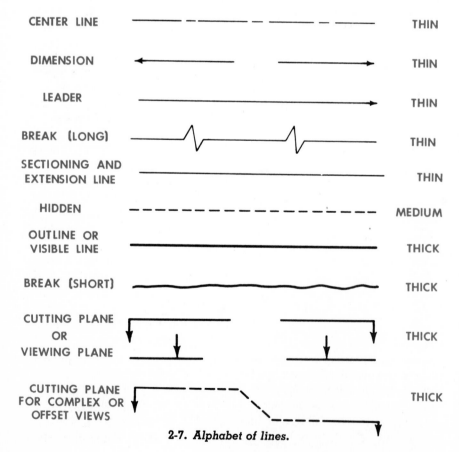

CENTER LINE		THIN
DIMENSION		THIN
LEADER		THIN
BREAK (LONG)		THIN
SECTIONING AND EXTENSION LINE		THIN
HIDDEN		MEDIUM
OUTLINE OR VISIBLE LINE		THICK
BREAK (SHORT)		THICK
CUTTING PLANE OR VIEWING PLANE		THICK
CUTTING PLANE FOR COMPLEX OR OFFSET VIEWS		THICK

2-7. *Alphabet of lines.*

made, are prepared to scales of ⅜″, ½″, ¾″, or 1¼″ equal 1′.

Metric Scales

The preferred metric scales are as follows:

1 equals 1: full size   1 equals 1250
1 equals 2: half size 1 equals 2500
1 equals 5
1 equals 10
1 equals 20
1 equals 50

ELEMENTS OF DRAWING

A drawing consists of lines, dimensions, symbols, and notes. *Lines* show the shape of a product and include many details of construction. Fig. 2-7. *Dimensions* are numbers that tell the sizes of each part as well as overall sizes. The craftspeople must follow these dimensions in making the materials list and the layout. *Symbols* are used to represent things that would be impractical to show by drawing, such as doors, windows, electrical circuits, and plumbing and heating equipment. Fig. 2-8. Some drawings also contain *notes* or written information to explain something not otherwise shown. Frequently in these notes *abbreviations* are given for common words.

Lines

The lines described below are used for all drawings.

Centerlines. These are composed of long and short dashes, alternately and evenly spaced with a long dash at each end; at intersections the short dashes cross. Very short centerlines may be broken if there is no confusion with other lines.

Dimension Lines. Dimension lines terminate in arrowheads at each end. On construction drawings they are unbroken. On production drawings they are broken only where space is required for the dimension.

Leader Lines. These lines are used to indicate a part or portion

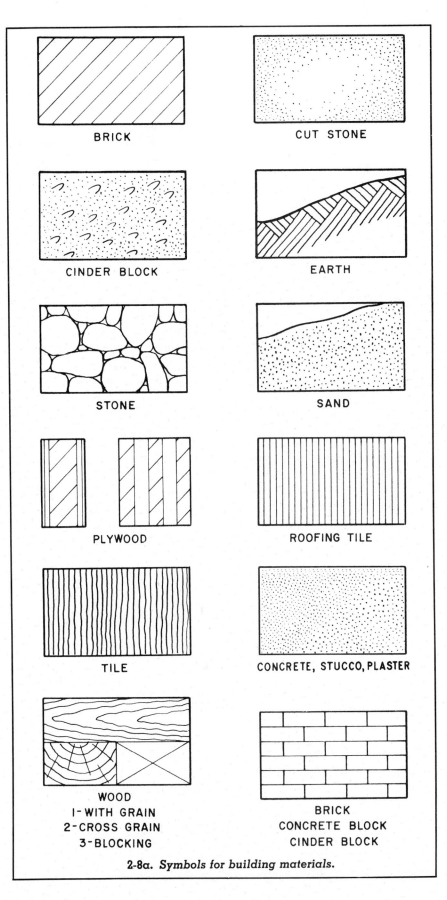

2-8a. *Symbols for building materials.*

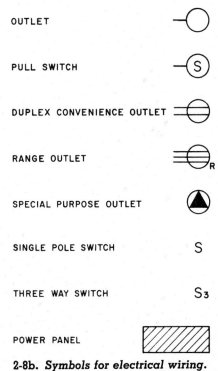

2-8b. *Symbols for electrical wiring.*

21

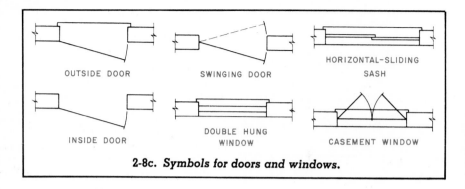

2-8c. *Symbols for doors and windows.*

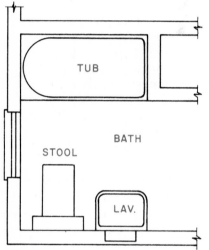

2-8d. *Symbols for plumbing fixtures.*

to which a note or other reference applies. They terminate in an arrowhead or a dot. Arrowheads should always terminate at a line; dots should be within the outline of an object. Leaders should terminate at any suitable portion of the note, reference, or dimension.

Break Lines. Short breaks are indicated by solid, freehand lines. Full, ruled lines with freehand zigzags are used for long breaks.

Sectioning Lines. Sectioning lines indicate the exposed surfaces of an object in a sectional view. They are generally full, thin lines, but they may vary with the kind of material shown.

Extension Lines. Extension lines indicate the extent of a dimension and should not touch the outline.

Hidden Lines. Hidden lines consist of short dashes evenly spaced and are used to show the hidden features of a part. They always begin with a dash in contact with the line from which they start, except when such a dash would form the continuation of a full line. Dashes touch at corners. Arcs start with dashes at the *tangent* points (where they touch each other).

Outline or Visible Lines. The outline or visible line represents those lines of the object which can actually be seen.

Cutting Plane Lines. These lines show where a section has been taken from the building drawings for detail representation.

ARCHITECTURAL WORKING DRAWINGS

Architectural drawings, Fig. 2-9, are prepared as *presentation drawings* or as *working drawings*. Presentation drawings require techniques of pictorial drawing, such as perspective (showing depth) and shading. Construction drafters are

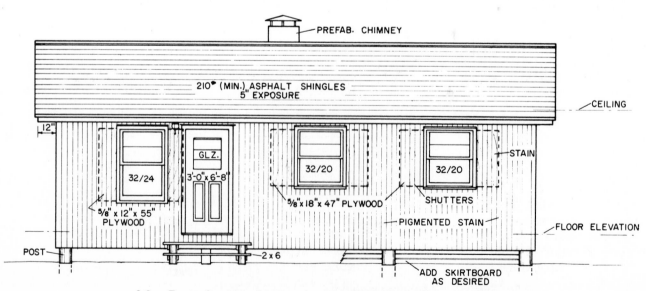

2-9a. *Front elevation drawing for a small home. Scale—¼" = 1'-0".*

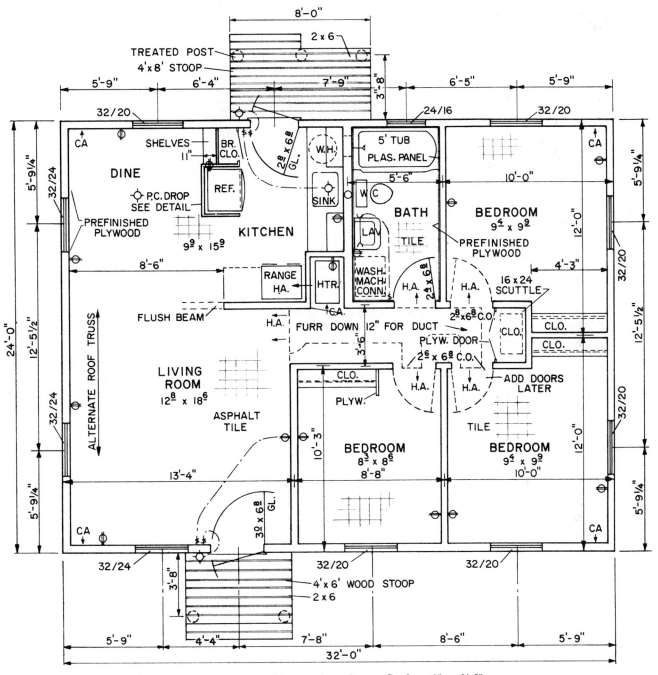

2-9b. *Floor plan drawing for a home. Scale—¼″ = 1′-0″.*

not concerned with presentation drawings. They prepare architectural working drawings consisting of plans, elevations (which show heights), sections and details (close-up views), and isometric views. (Isometrics are constructed around three basic lines that form 120° angles.)

Structural Members

In working with architectural drawings, you will find the following structural members referred to often. Therefore you will need to be familiar with them. Here they are classified according to use.

Footings. Footings rest on soil material and transmit their received load onto the soil. The natural material on which a footing

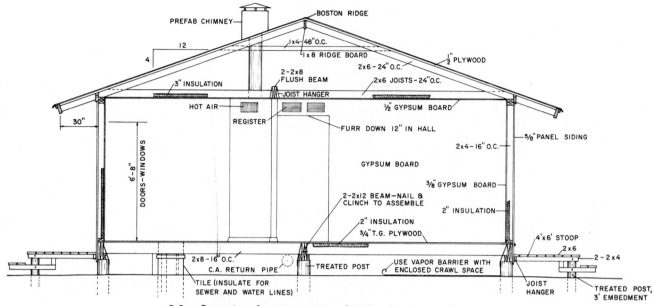

2-9c. *A section drawing giving details of rafters and joists.*

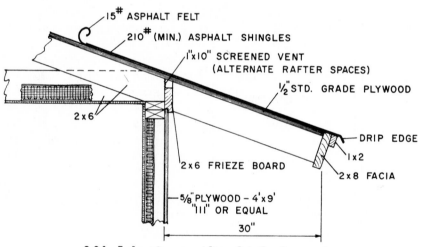

2-9d. *A drawing providing details of a cornice.*

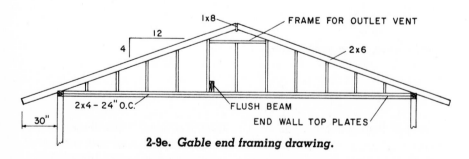

2-9e. *Gable end framing drawing.*

rests is called the *foundation bed.* Footings support columns, piers, pilasters, walls, and similar loads.

Usually, footings are made of concrete, although wood or timber may be used.

Vertical Members. Vertical members are in compression; that is, they support loads acting downward at the top. Columns, posts, studs, and piers are those most often encountered.

• **Columns.** Columns may be steel, timber, or concrete. They rest on footings and are the principal load-carrying vertical members.

• **Piers.** Piers are of concrete, timber, or masonry construction. They rest on footings and support horizontal or vertical members. In bridge construction, a pier is an intermediate support for the adjacent ends of two bridge spans.

• **Studs and Posts.** Studs are vertical members used in wood-frame construction, spaced close together in walls. Posts are heavier vertical members used in wood-frame construction, usually at corners.

Horizontal Members. Those most frequently encountered include the following: joists, beams, girders, and lintels.

• **Joists.** These are lightweight beams spaced four feet or less from the center of one to the center of the next. Joists take the load directly.

- **Beams.** Beams, like joists, take the load of the floor directly, but they are spaced wider than four feet on center.
- **Girders.** Girders take the load of either joists or beams and are generally the heaviest horizontal members in a structure.
- **Lintels.** Lintels are beams which span door or window openings and carry the structure above those openings.

Roof Members. The following roof members are most common:

- **Common rafters.** These members that run square with top plate and extend to the ridge board.
- **Hip rafters.** Those that extend from the outside angle of the plates toward the apex of the roof.
- **Jack rafters.** Those that are square with the top plate and intersect a hip rafter.
- **Valley rafters.** Those that extend from an inside angle of the plates toward the ridge.
- **Cripple rafters.** Those that cut between valley and hip rafters.
- **Purlin.** A timber that supports several rafters at one or more points, or one that supports the roof sheathing directly.
- **Trusses.** Structural members that connect together to span the space between the walls of a building. They support the roof and floor loads.

Flooring

Subflooring is laid atop joists or trusses. Building paper is put between subflooring and finished flooring where required.

Sheathing, Siding, and Roofing

Structural members are covered with suitable materials to form the outside walls and the roof. Insulation is placed between sheathing and interior materials.

Utilities

Heating, air conditioning, wiring, and plumbing are the utilities or mechanical systems of a building. They are represented by drawings.

Finishing and Painting

Glazing, plastering, finish trim, and painting complete the building.

PRINCIPLES OF CONSTRUCTION DRAWING

Construction drawings are based on the same general principles as are all other technical drawings. The shape of a structure is described in *orthographic* (multiview) drawings, made to scale. Its size is described by *figured dimensions*, whose extent is indicated by dimension lines, arrowheads, and extension lines. Overall relationships are shown in *general drawings* similar to assembly drawings. Important specific features are shown in *detail drawings* usually drawn to a larger scale than the general drawings. Additional information about size and material is furnished in the specific and general *notes*. If you are familiar with other types of working drawings, you will find obvious similarities in construction drawings. However, there are certain terms and uses of drawings that are found only in the construction field. Chiefly these are related to the materials and methods of construction and the conventional practices of construction drawing.

Views in Construction Drawings

The views of a structure are presented in general and detail drawings. General drawings consist of plans and elevations; detail drawings are made up of sectional and specific detail views.

PLANS

A plan is a top view—a projection on a horizontal plane. Several types of plan views are used for specific purposes, such as site plans, foundation plans, and floor plans.

Site Plan. A site plan shows the building site with boundaries, contours, existing roads, utilities, and other physical details such as trees and buildings. Site plans are drawn from notes and sketches based upon a survey. The layout of the structure is superimposed on the contour drawings, and corners of the structure are located by reference to established natural objects or other buildings.

Foundation Plans. A foundation plan is a top view of the footing or foundation walls, showing their area and location by (1) distances between centerlines and (2) distances from reference lines or boundary lines. Foundation walls are located by dimensions from the corner of the building to the wall itself. All openings in foundation walls are shown.

Floor Plan. Floor plans, commonly referred to as *plan views*, are cross-section views of a building. The horizontal cutting plane is placed so that it includes all doors and window openings. A floor plan shows the outside shape of the home; the arrangement, size, and shape of rooms; types of materials; thickness of walls and partitions; and the types, sizes, and locations of doors and windows for each story. A plan may also include details of framework and structure, although these features are usually shown on separate drawings called framing plans.

ELEVATIONS

Elevations are external views of a structure; they may be drawn to show views of the front, rear, and right or left side. They correspond to front, rear, and side views in orthographic projections on vertical planes. An elevation is a picturelike view of a building that shows exterior materials and the height of windows, doors, and rooms. It may also show the ground level surrounding the structure, called the *grade*.

FRAMING PLANS

Framing plans show the size, number, and location of the structural members constituting the building framework. Separate framing plans may be drawn for the floors, walls, and roof. The floor framing plan must specify the sizes and spacing of joists, girders, and columns used to support the floor. Detail drawings are added, if necessary, to show the methods of anchoring joists and girders to the columns and foundation walls or footings. Wall framing plans show the location and method of framing openings and ceiling heights so that studs and posts can be cut. Roof framing plans show the construction of the rafters used to span the building and support the roof. Size, spacing, roof slope, and all necessary details are shown.

Floor Framing. Framing plans for floors are basically plan views of the girders and joists. The unbroken double-line symbol is used to indicate joists which are drawn in the positions they will occupy in the completed building. Double framing around openings and beneath bathroom fixtures is shown where used.

Wall Framing. Wall framing plans are *detail drawings* showing the locations of studs, plates, sills, and bracing. They show one wall at a time. Usually they are elevation views.

Roof Framing. Framing plans for roofs are drawn in the same manner as floor framing plans. A drafter should imagine that he or she is looking down on the roof before any of the roofing material (sheathing) has been added. Rafters are shown in the same manner as joists.

SECTIONAL VIEWS

Sectional views, or sections, provide important information as to height, materials, fastening and support systems, and concealed features. They show how a structure looks when cut vertically by a plane. The cutting plane is not necessarily continuous but, as with the horizontal cutting plane in building plans, may be staggered to include as much construction information as possible. Like elevations, sectional views are vertical projections. Being detail drawings, they are drawn to large scale. This facilitates reading and provides information that cannot be given on elevation or plan views. Sections may be classified as *typical* and *specific*.

Typical Sections. Typical sections are used to show construction features that are repeated many times throughout a structure.

Specific Sections. When a particular construction feature occurs only once and is not shown clearly in the general drawing, a cutting plane is passed through that portion. The cutting plane is indicated by lines on the general drawing. These lines, which sometimes have arrowheads, have a letter at each end. The same letter is used at each end of the line. These letters then become part of the title of the section drawing. Thus the cutting-plane lines show the relationship between the general drawing and the section drawing. They show the location of that portion of the general drawing which is represented in the section drawing.

DETAILS

Details are large-scale drawings showing the builders of a structure how its various parts are to be connected and placed. Details do not use a cutting-plane indication, but they are closely related to section drawing because sections are often used as parts of detail drawings. The construction at doors, windows, and eaves is customarily shown in detail drawings. Such drawings are also used whenever the information provided in elevations, plans, and sections is not clear enough for the mechanics on the job. They are usually grouped so that references may be made easily from general drawings.

Dimensioning Construction Drawings

Plan views are dimensioned both outside and inside the building lines. Outside dimensions describe changes and openings in the exterior wall in addition to overall dimension. Inside measurements locate partitions relative to each other and to exterior walls. All horizontal dimensions are shown in a plan view.

Notes in Construction Drawings

Notes in a set of construction drawings are clear, direct statements regarding such matters as materials, construction, and finish. They are included wherever necessary to provide information not clearly indicated by the dimensions. There are two kinds of notes—specific and general.

All notes on the drawings themselves are *specific*. Such notes may add to the dimensioning information or they may explain a procedure or material standard. When more than one line of explanatory notes is placed on a drawing, lower case lettering is used. Titles and subtitles are always in upper case letters. Many terms frequently used on construction drawings are abbreviated to save space.

General notes are usually grouped according to material of construction in a tabular form called a *schedule*.

The notes with a set of construction drawings are so extensive that they cannot all be placed on the drawings themselves. All of the general notes and many specific notes are made into a separate list called the *specifications*. These notes tell the manner in which work will be performed, designate what materials and finishes are to be used, and establish the responsibility of the unit performing the work.

Although it is not a drafter's or a carpenter's job to prepare specifications, these workers should be familiar with such notes. This is because the specifications give detailed instructions regarding materials and methods of work and are therefore an important source of information related to the drawings.

A drafter experienced in preparing construction drawings can be of assistance to the specifications writer. Specifications should be written clearly and briefly.

Bill of Materials in Construction Drawings

A bill of materials is a table of information that tells the requirements for a given project. It shows the item number, name, description, quantity, kind of material, stock size, and sometimes the weight of each piece.

QUESTIONS

1. Why is it important to learn to read prints?
2. What is a blueprint?
3. Define scale.
4. Describe the basic elements of drawing.
5. Name some of the common lines.
6. Name the common views needed in a construction drawing.
7. What is an elevation?
8. What are framing plans?
9. Why are sectional views shown?
10. What are detail drawings and what is their purpose?
11. What are specifications?
12. Define a bill of materials.

SECTION II

Materials, Tools, Machines, and Equipment

Building Materials

3

A wide variety of materials—wood, metal, plastics, ceramics and combinations of these materials—are needed for residential construction. Fig. 3-1. Of major interest are the following:

LUMBER AND WOOD PRODUCTS

Hardwoods and Softwoods

The terms *hardwood* and *softwood* identify woods according to two main types of trees; they do not indicate actual softness or hardness. Some hardwoods are actually softer than certain of the softwood species. Softwoods are those that come from evergreens *(conifers)* which are cone-bearing or needle-bearing trees. Common examples are pine, fir, cedar, and redwood. Hardwoods are cut from the broadleaf, *deciduous* trees. A deciduous tree is one that sheds its leaves annually. Some common hardwoods are walnut, mahogany, maple, birch, cherry, and oak.

Grades of Lumber

So that the purchaser can know the quality of the wood to buy, lumber is classified in grades, according to the defects in it. In the grading process, a lumber piece must meet the lowest requirements for its grade. Wide differences in quality are found in the same grade because some lumber is much better than the minimum for its grade, but not quite good enough for a higher classification.

HARDWOOD LUMBER GRADING

Hardwoods are available in three common grades, *firsts* and *seconds* (FAS), *select*, and *No. 1 common.* Generally, firsts and seconds are used for built-ins and paneling.

Grading rules for hardwood lumber have been established by the National Hardwood Lumber Association. Each different kind of hardwood lumber has a slightly different grading standard.

SOFTWOOD GRADING

The National Bureau of Standards of the Department of Commerce has established American Lumber Standards for softwood lumber. These standards are intended as guides for the different associations of lumber producers, each of which has its own grading rules and specifications. Most of the major associations, such as the Western Wood Products Association, Redwood Inspection Service, Southern Pine Inspection Bureau, and others participated in developing these grading rules. If you are working with just a few kinds of lumber most of the time, you

3-1. Lumber, plywood, rigid plastic sheathing, steel and concrete products are all being used on this construction site.

BOARDS

Appearance Grades	Selects	B & Better *(IWP—Supreme) C Select (IWP—Choice) D Select (IWP—Quality)	Select and Finish grades are widely used for interior walls, moldings, cabinets, siding, architectural woodwork, interior and exterior trim, and other applications.	
	Finish	Superior Prime E		
	Paneling	Clear (Any Select or Finish Grade) No. 2 Common—Selected for Knotty Paneling No. 3 Common—Selected for Knotty Paneling		
	Siding (Bevel, Bungalow)	Superior Prime	(Other siding grades are shown for western red cedar and boards.)	
	Western Red Cedar	Finish Paneling and Ceiling	Clear Heart A B	Bevel Siding	Clear—V.G. Heart A—Bevel Siding B—Bevel Siding C—Bevel Siding
	Boards (Commons) Siding, Paneling, Shelving, Sheathing, and Form Lumber	No. 1 Common (IWP—Colonial) No. 2 Common (IWP—Sterling) No. 3 Common (IWP—Standard)	No. 1 Common boards are the ultimate in small-knot material for appearance uses, but less expensive No. 2 and No. 3 Commons are most often used in housing for paneling, siding and shelving.	

*Idaho White Pine

DIMENSION LUMBER/ALL SPECIES

Light Framing	Construction Standard Utility	This category for use where high strength values are **not** required such as studs, plates, sills, cripples, blocking, etc.
Studs	Stud	A popular grade for load and non-load bearing walls. Limited to 10-ft. and shorter.
Structural Light Framing Joists and Planks	Select Structural No. 1 No. 2 No. 3	These grades fit engineering applications where higher strength is needed, for uses such as trusses, joists, rafters, and general framing.

3-2. A grade selector chart of recommended grades following the grading rules used by the Western Wood Products Association.

should obtain grading rules from the associations involved and become acquainted with them. For example, if most of your building lumber comes from the western states, then the grading rules published by the Western Wood Products Association should be used. These are available in a reference entitled *Western Woods Use Book.*

One of the main ideas of these standards is to divide all softwood lumber for *grading purposes* into two groups:
● Dry lumber that is seasoned or dried to a moisture content of 19 percent or less.
● Green lumber, which has a moisture content in excess of 19 percent.

DEFECTS IN STANDARD GRADES

Defects in lumber are the faults which detract from the quality of the piece, either in appearance or utility. Some of the more common terms and definitions used by the woodworker are:

Check. Lengthwise grain separation, usually occurring through the growth rings as a result of seasoning.

Decay. Disintegration of wood substance due to action of wood-destroying fungi.

Knot. Branch or limb embedded in the tree and cut through in the process of lumber manufacture; classified according to size, quality, and occurrence. To determine the size of a knot, average the maximum length and maximum width unless otherwise specified.

Pitch. Accumulation of resin in the wood cells in a more or less irregular patch.

Pitch-pocket. An opening between growth rings which usually contains or has contained resin, bark, or both.

Shake. A lengthwise grain separation between or through the growth rings. May be further classified as ring shake or pitch shake.

Split. Lengthwise separation of the wood extending from one surface through the piece to the opposite or an adjoining surface.

Stain. Discoloration on or in lumber other than its natural color.

Summerwood. Denser outer portion of each annual ring, usually without easily visible pores, formed late in the growing period, not necessarily in summer.

Torn Grain. Part of the wood torn out in dressing.

Wane. This is bark or lack of wood from any cause on the edge or corner of a piece.

Warp. Any variation from a true or plane surface; includes bow, crook, cup, or any combination thereof.

R = Recommended   D = Dark
G = Good   M = Medium
B = Best   L = Light

Uses and Characteristics of Western Wood Species	Western Cedars	Douglas Fir	Ponderosa Pine	Sugar Pine	Lodgepole Pine	Idaho White Pine	Hem-Fir	Western Larch	Engelmann Spruce
USES									
Joists, Rafters, Trusses (span controls species and size)	R	R	R	R	R	R	R	R	R
Studs	R	R	R	R	R	R	R	R	R
Plates	R	R	R	R	R	R	R	R	R
Wall & Roof Sheathing	R	R	R	R	R	R	R	R	R
Subflooring	R	R	R	R	R	R	R	R	R
Siding	B	G	G	G	G	B	G	G	B
Shelving	G	G	G	G	G	G	G	G	G
Fence Posts	B	G	G	G	G	G	G	G	G
Fence Boards, Gates	G	G	G	G	G	G	G	G	G
Interior Paneling	G	G	G	G	G	G	G	G	G
Heartwood Paneling Color	D	M	L	L	L	L	L	M	L
CHARACTERISTICS									
Works Easily	G	G	G	B	G	B	G	G	G
Dimensionally Stable	B	B	G	G	G	G	G	G	G
Fastener Holding Power	G	B	G	G	G	G	B	G	G
Glues Well	G	B	B	B	B	B	B	G	B
Mills Smoothly	G	G	B	B	B	B	G	B	B
Naturally Durable	B	G	G	G	G	G	G	G	G
Resists Splintering	G	G	B	B	B	B	B	B	B
Paintability	B	B	B	B	B	B	G	G	B

3-3. *Recommended species of wood produced by the Western Wood Products Association.*

• *Bow.* Deviation flatwise from a straight line from end to end of a piece, measured at the point of greatest distance from the straight line.

• *Crook.* Deviation edgewise from a straight line from end to end of a piece, measured at the point of greatest distance from the straight line.

• *Cup.* Deviation flatwise from a straight line across the width of a piece, measured at the point of greatest distance from the line; classified as slight, medium, and deep.

BASIC SELECTION FACTORS

Attention to the following points and familiarity with common lumber abbreviations will simplify the selection and specifications of softwood lumber.

Product Classification. Identify product names for clarity. Examples: paneling, structural decking, joists, rafters, studding, beams, and siding. Fig. 3-2.

Species. Include all suitable species. With more species to choose from, you may be able to lower your costs. Check with your local supplier. Fig. 3-3.

When wood color, grain, durability or other special characteristics are important, select and specify the species accordingly.

Grade. Specify standard grades as described in the official grading rules. Consider all grades suitable for the intended use. For economy, it is recommended that the lowest suitable grade be specified.

Stress Rating. When strength is a factor, specify the stress rating requirements without reference to grades.

Stress-grade lumber is structural lumber that has been scientifically graded with electronic devices and stamped with information to indicate the specific load it will support.

Size. For standard products such as boards and framing, specify the nominal size by thickness and width in full inches. Example: 1 × 6, 1 × 8, 2 × 4, 2 × 6. Fig. 3-4. (Nominal sizes are larger than actual or dressed sizes. For example, a nominal 2 by 4 is 1½ by 3½ dressed.)

Surface Texture. Indicate whether lumber is to be smooth surface (surfaced) or rough surface (rough).

Seasoning. Specify "dry" lumber to assure long-range product quality, stability, increased nail-holding power, improved paintability, and workability.

Grade Stamps. Specify grade-stamped framing lumber, sheathing, and other construction items. Finish lumber and decking may also be grade-stamped on ends or backs where the stamp will not be visible in use and may be so specified if desired. Fig. 3-5.

How Lumber is Sold

Most lumber is sold by the board foot. A board foot is a piece 1″ thick by 12″ wide by 12″ long. Stock that is less than 1″ thick is figured as 1″. There are three common ways of figuring board feet:

1. Board feet equals thickness in inches times width in inches times length in feet divided by 12. For example, a board that measures 2″ by 4″ by 12′ contains 8 board feet. Using the formula shown above:

$$Bd.\ ft. = \frac{T\ (in.) \times W\ (in.) \times L\ (ft.)}{12}$$

$$Bd.\ ft. = \frac{2 \times 4 \times 12}{12} = 8$$

2. Board feet equals thickness in inches times width in feet times length in feet. For example, a piece of white pine 2″ thick by 12″ wide by 6′ long would be 12 board feet.

$$Bd.\ ft. = T\ (in.) \times W\ (ft.) \times L\ (ft.)$$
$$Bd.\ ft. = 2 \times 1 \times 6 = 12$$

3. Board feet equals thickness in inches times width in inches times length in inches divided by 144. For example, a board measuring 2″ × 4″ × 7.5′ would contain 5 board feet.

$$Bd.\ ft. = \frac{T \times W \times L}{144}\ (in.)$$

$$Bd.\ ft. = \frac{2 \times 4 \times 90}{144}$$

$$= \frac{720}{144} = 5$$

Lumber is sold by the board foot, by 100 board feet, or by 1000 board feet (M). For example, if lumber sells at $700 per (M), it cost $70 for 100 board feet and $.75 for one board foot.

Lumber for basement and outdoor construction such as porches and decks should be pressure treated with chemicals so that it will withstand rot and insects.

Plywood

Softwood plywoods are produced in accordance with PS-1-74, the standard for the manufacture of *construction and industrial* plywood. Fig. 3-6. The *American Plywood Association Product Guide* should be consulted for specific details on grades for use in residential constructions. Plywoods are available with several different types of cores. Fig. 3-7.

Thickness (Inches)			Face Width (Inches)		
			Minimum Dressed		
Nominal	Dry	Green (Unseasoned)	Nominal	Dry	Green (Unseasoned)
1	3/4	25/32	2	1 1/2	1 9/16
1 1/4	1	1 1/32	3	2 1/2	2 9/16
1 1/2	1 1/4	1 9/32	4	3 1/2	3 9/16
2	1 1/2	1 9/16	5	4 1/2	4 5/8
2 1/2	2	2 1/16	6	5 1/2	5 5/8
3	2 1/2	2 9/16	7	6 1/2	6 5/8
3 1/2	3	3 1/16	8	7 1/4	7 1/2
4	3 1/2	3 9/16	9	8 1/4	8 1/2
			10	9 1/4	9 1/2
			11	10 1/4	10 1/2
			12	11 1/4	11 1/2

3-4. Nominal and actual dry sizes of softwood lumber.

3-5. Grade stamp: (a) The official Western Wood Products Association mark on a piece of lumber is assurance of its assigned grade. Grading practices of Western Wood Products Association member mills are supervised to assure uniformity. (b) Each mill is assigned a permanent number for grade stamp purposes. (c) An example of an official grade name abbreviation. The official grade name, as defined by the Association, gives positive identification to graded lumber. (See grade selector chart.) (d) This mark identifies wood species. (e) Symbol denotes moisture content of lumber when manufactured. S-Dry indicates seasoned lumber. S-Green indicates unseasoned or "green" lumber. It is recommended that S-Dry lumber be used for all enclosed framing.

Over 70 different wood species including softwoods and native and imported hardwoods are used in manufacturing plywood. These woods are divided into five groups. Fig. 3-8. Group 1 is the strongest and stiffest; Group 5 is the weakest. Group 1 is made up largely of Douglas Fir and Southern Pine.

Most softwood plywoods are of veneer-core construction, Fig. 3-9. The top quality veneer used is N.

The quality of the others range from A to D with D having lowest quality.

Plywoods are made in two basic types: *exterior*, made with 100 percent water-proof glue and *interior*, made with highly moisture resistant glue. *Engineered* or *performance rated* plywoods that are used for subflooring, sheathing, forms, and other special purposes have slightly different grade marks. Fig. 3-10. A double number such as 32/16 appears on the grade mark. The number at the left shows the maximum space for supports when used for roof decking and the number at the right tells the maximum spacing of floor joist or trusses for subflooring. Also sometimes the note EXP (exposure) 1 or 2 appears. Use EXP 2 when construction delays are anticipated.

ENGINEERED BOARDS

Engineered boards include hardboard, particleboard, and oriented strand board (OSB). Fig. 3-11.

Hardboard. Hardboard is a smooth panel without knots or grain, yet it is almost entirely natural wood. It contains cellulose for strength and lignin for bonding power the same as the tree itself. The difference is that the fibers are rearranged to provide special properties. Hardboard is tough and

Softwood Plywood

CONDENSED GUIDE TO PLYWOOD GRADES AND USES/ EXTERIOR TYPE

Grade Designation	Description and Most Common Uses	Typical Grade-Trademarks	Veneer Grade			Most Common Thicknesses (Inch)				
			Face	Inner Plies	Back					
APA C-C EXT	Unsanded grade with waterproof bond for subflooring and roof decking, siding on service and farm buildings, crating, pallets, pallet bins, cable reels, treated-wood foundations	C-C 42/20 APA EXTERIOR PS 1 74 000	C	C	C	5/16	3/8	1/2	5/8	3/4
APA Structural I C-C EXT and APA Structural II C-C EXT	For engineered applications in construction and industry where full Exterior type panels are required. Unsanded. See species group requirements	STRUCTURAL I C-C 32/16 APA EXTERIOR PS 1 74 000	C	C	C	5/16	3/8	1/2	5/8	3/4
APA Sturd-I-Floor EXT	For combination subfloor-underlayment under resilient floor coverings where severe moisture conditions may be present, as in balcony decks. Possesses high concentrated- and impact-load resistance during construction and occupancy. Touch-sanded. Available square edge or tongue-and-groove	STURD-I-FLOOR 20oc 5/8 INCH EXTERIOR APA 000 NRB-108/FHA-UM-66	C plugged	C	C				19/32 5/8	23/32 3/4
APA underlayment C-C plugged EXT	For application over structural subfloor. Provides smooth surface for application of resilient floor coverings where severe moisture conditions may be present. Touch-sanded	APA UNDERLAYMENT C-C PLUGGED GROUP 2 EXTERIOR 000 PS 1 74	C plugged	C	C		3/8	1/2	19/32 5/8	23/32 3/4
APA C-C plugged EXT	For use as tile backing where severe moisture conditions exist. For refrigerated or controlled atmosphere rooms, pallet fruit bins, tanks, box car and truck floors and linings, open soffits. Touch-sanded	APA C-C PLUGGED GROUP 2 EXTERIOR 000 PS 1 74	C plugged	C	C		3/8	1/2	19/32 5/8	23/32 3/4
APA B-B Plyform Class I and Class II EXT	Concrete form grades with high reuse factor. Sanded both sides. Mill-oiled unless otherwise specified. Special restrictions on species. Available in HDO and Structural I. Class I most commonly available	APA PLYFORM B-B CLASS I EXTERIOR 000 PS 1 74	B	C	B				5/8	3/4
APA A-B EXT	Use where the appearance of only one side is less important	A-B G-1 EXT-APA PS1-74 000	A	C	B	1/4	3/8	1/2	5/8	3/4
APA A-C EXT	Use where the appearance of only one side is important. Soffits, fences, structural uses, boxcar and truck lining. Tanks, trays, commercial refrigerators	APA A-C GROUP 1 EXTERIOR 000 PS 1 74	A	C	C	1/4	3/8	1/2	5/8	3/4
APA B-B EXT	Utility panel with solid faces	B-B G-2 EXT-APA PS1-74 000	B	C	B	1/4	3/8	1/2	5/8	3/4
APA B-C EXT	Utility panel for farm service and work buildings, boxcar and truck lining, containers, tanks, agricultural equipment. Also as base for exterior coatings for walls, roofs	APA B-C GROUP 1 EXTERIOR 000 PS 1 74	B	C	C	1/4	3/8	1/2	5/8	3/4
APA HDO EXT	High Density Overlay. Has a hard, semi-opaque resin-fiber overlay both faces. Abrasion resistant. For concrete forms, cabinets, counter tops, signs, tanks. Also available with skid-resistant screen-grid surface	HDO A-A G-1 EXT-APA PS1-74 000	A or B	C or C plgd	A or B		3/8	1/2	5/8	3/4
APA MDO EXT	Medium Density Overlay. Smooth, opaque, resin-fiber overlay one or both faces. Ideal base for paint, both indoors and outdoors	MDO B-B G-2 EXT-APA PS1-74 000	B	C	B or C		3/8	1/2	5/8	3/4

3-6a. A condensed guide for Exterior type plywood grades and uses.

dense. It is available in panel sizes up to 4' wide and 8' long, in thicknesses from 1/8" to 1/4".

There are three basic types of hardboard. *Service* is for miscellaneous interior uses such as storage areas, closet liners, shelving, cabinet backs, and drawer bottoms. *Standard* is for interior paneling,

Softwood Plywood

CONDENSED GUIDE TO PLYWOOD GRADES AND USES/ INTERIOR TYPE

Grade Designation	Description and Most Common Uses	Typical Grade-Trademarks	Veneer Grade			Most Common Thicknesses (inch)				
			Face	Inner Plies	Back					
APA C-D INT	For wall and roof sheathing, subflooring, industrial uses such as pallets. Most commonly available with exterior glue (CDX). Specify exterior glue where construction delays are anticipated and for treated-wood foundations	C-D 32/16 (APA) INTERIOR PS 1-74 000 / C-D 24/0 (APA) INTERIOR PS 1-74 000 EXTERIOR GLUE	C	D	D	5/16	3/8	1/2	5/8	3/4
APA Structural I C-D INT and APA Structural II C-D INT	Unsanded structural grades where plywood strength properties are of maximum importance: structural diaphragms, box beams, gusset plates, stressed-skin panels, containers, pallet bins. Made only with exterior glue. See species group requirements. Structural I more commonly available	STRUCTURAL I C-D 24/0 (APA) INTERIOR PS 1-74 000 EXTERIOR GLUE	C	D	D	5/16	3/8	1/2	5/8	3/4
APA Sturd-I-Floor INT	For combination subfloor-underlayment. Provides smooth surface for application of resilient floor covering. Possesses high concentrated- and impact-load resistance during construction and occupancy. Manufactured with exterior glue only. Touch-sanded. Available square edge or tongue-and-groove	STURD-I-FLOOR 24oc T&G 23 32 INCH (APA) INTERIOR 000 EXTERIOR GLUE NRB 108/FHA-UM-66	C plugged		D				19/32 5/8	23/32 3/4
APA Sturd-I-Floor 48 O.C. (2-4-1) INT	For combination subfloor-underlayment on 32- and 48-inch spans. Provides smooth surface for application of resilient floor coverings. Possesses high concentrated- and impact-load resistance during construction and occupancy. Manufactured with exterior glue only. Unsanded or touch-sanded. Available square edge or tongue-and-groove	STURD-I-FLOOR 48oc 2-4-1 T&G 1 1 8 INCH (APA) INTERIOR 000 EXTERIOR GLUE NRB 108/FHA-UM-66	C plugged	C & D	D	1-1/8				
APA underlayment INT	For application over structural subfloor. Provides smooth surface for application of resilient floor coverings. Touch-sanded. Also available with exterior glue	APA UNDERLAYMENT GROUP 1 INTERIOR 000 PS 1-74 EXTERIOR GLUE	C plugged	C & D	D		3/8	1/2	19/32 5/8	23/32 3/4
APA C-D plugged INT	For built-ins, wall and ceiling tile backing, cable reels, walkways, separator boards. Not a substitute for Underlayment or Sturd-I-Floor as it lacks their indentation resistance. Touch-sanded. Also made with exterior glue	APA C-D PLUGGED GROUP 2 INTERIOR 000 PS 1-74 EXTERIOR GLUE	C plugged	D	D		3/8	1/2	19/32 5/8	23/32 3/4
APA A-B INT	Use where appearance of one side is less important but where two solid surfaces are necessary	A-B G-1 INT-APA PS1-74 000	A	D	B	1/4	3/8	1/2	5/8	3/4
APA A-D INT	Use where appearance of only one side is important. Paneling, built-ins, shelving, partitions, flow racks	APA A-D GROUP 1 INTERIOR 000 PS 1-74	A	D	D	1/4	3/8	1/2	5/8	3/4
APA B-B INT	Utility panel with two solid sides. Permits circular plugs	B-B G-2 INT-APA PS1-74 000	B	D	B	1/4	3/8	1/2	5/8	3/4
APA B-D INT	Utility panel with one solid side. Good for backing, sides of built-ins, industry shelving, slip sheets, separator boards, bins	APA B-D GROUP 2 INTERIOR 000 PS 1-74	B	D	D	1/4	3/8	1/2	5/8	3/4
APA Decorative panels	Rough-sawn, brushed, grooved, or striated faces. For paneling, interior accent walls, built-ins, counter facing, displays, exhibits	APA DECORATIVE GROUP 4 INTERIOR 000	C or btr.	D	D		5/16	3/8	1/2	5/8

3-6b. A condensed guide for Interior type plywood grades and uses.

underlayment, screens, and wardrobe doors. Perforated hardboard is available for storage areas and garage liners where items can be hung from small hooks and brackets inserted in the holes. Special prefinished decorative panels are also available in a variety of patterns and textures. *Tempered* is produced in a manufacturing process which introduces oil into the board which is permanently "set"

Three-ply Veneer Core Construction

Five-ply Veneer Core Construction

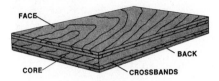

Multiply Veneer Core Construction

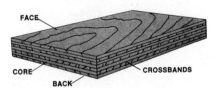

Five-ply Lumber Core Construction

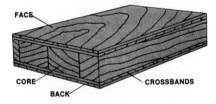

Five-ply Particleboard Core Construction

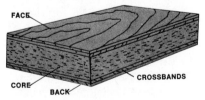

3-7. A variety of cores are used in plywood construction.

Group 1	American Birch, Douglas Fir, Sugar Maple, and Southern Pine
Group 2	Cypress, White Fir, Western Hemlock, Philippine Mahogany, Black Maple, Red Pine, and Yellow Poplar
Group 3	Red Alder, Alaska Cedar, Eastern Hemlock, Jack Pine, Redwood, and Black Spruce
Group 4	Quaking Aspen, Western Red Cedar, Eastern Cottonwood, and Sugar Pine
Group 5	Basswood, Balsam Fir, and Balsam Poplar

3-8. Wood species used in manufacturing plywood. Group 1 is the strongest; Group 5 is the weakest.

N	Intended for natural finish. Selected all heartwood or all sapwood. Free of open defects. Allows some repairs.
A	Smooth and paintable. Neatly made repairs permissible. Also used for natural finish in less demanding applications.
B	Solid surface veneer. Repair plugs and tight knots permitted. Can be painted.
C	Sanding defects permitted that will not impair the strength or serviceability of the panel. Knotholes to 1½" and splits to ½" permitted under certain conditions.
C plugged	Improved C veneer with closer limits on knotholes and splits. C plugged face veneers are fully sanded.
D	Used only in interior type for inner plies and backs. Permits knots and knotholes to 2½" in maximum dimension and ½" larger under certain specified limits. Limited splits permitted.

3-9. Quality ratings for veneer-core plywood. The top quality veneer rating is N, and D has the lowest quality.

3-10. A grade mark used for Engineered or Performance-rated plywoods.

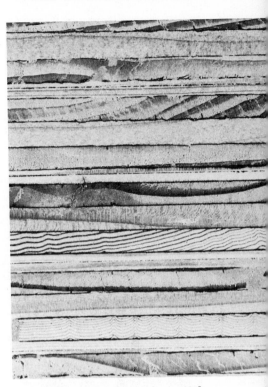

3-11. Edges of plywood particleboard and hardboard.

with a heat process. This gives the board greater strength, abrasion resistance, and moisture resistance. Tempered hardboard is used for exterior applications such as soffits, shutters, fencing, flower boxes, windbreaks, and garden sheds. Specially treated hardboards are available in exterior siding patterns, both lap and panel.

Particleboard. Particleboard is another wood panel made of wood residues from lumber and plywood manufacturing. The product is noted for its surface smoothness, strong internal bond, good screw- and nail-holding power, and dimensional stability. Like hardboard, it is without knots or grain, but it is less dense and comes in thicker panels.

Particleboard may be used for shelving, drawers, bookcases, cabinets, countertops, and the like. Thicknesses generally available are ¼" to 1½". It has a very smooth surface as a corestock material and laminates can be glued easily. It also accepts stain, varnish, paint or lacquer like any wood product and for this purpose can be primed or filled. To minimize warping, both sides of each panel should be finished the same.

3-12. *Attractive use of vertical siding.*

3-13. *4' × 8' paneling with plank design.*

Oriented Strand Board. Oriented strand board (OSB) is a panel composed of wood wafers bonded together with plastic resin. It is commonly used for floor systems and sheathing. It differs from particleboard in that the pieces of wood that make up the panel are much larger and arranged in layers to improve the structural strength of the panel.

Siding

Siding is produced in a variety of shapes and patterns of plywood, hardboard, lumber, plastics, and metal (principally aluminum). Fig. 3-12. Plywood sidings come in a wide variety of attractive patterns and textures. Hardboard sidings come in several patterns and are also available with textured surfaces. There are a number of types of lumber siding available. Sidings are suitable for exterior or interior use. Edge sealing is recommended for all edges on both panel and lap sidings.

Paneling

Factory finished interior wall paneling is plywood or hardboard with an almost unlimited variety of face treatments, most of which are designed to bring the look of real wood into the home. Fig. 3-13. Many paneling faces are, in fact, hardwood or softwood veneers finished in a number of ways to enhance their visual appeal. Other faces are printed or otherwise finished to imitate wood grain.

OTHER BUILDING MATERIALS

Gypsum Board

Gypsum board is a building panel consisting of gypsum, fibers, and other ingredients. It is finished on both sides with special paper to provide smooth surfaces and panel reinforcement. The properties that make gypsum board important are gypsum's inherent resistance to fire and to the transmission of sound. It is commonly used for interior *dry wall* construction. Fig. 3-14.

Masonry Products

Masonry products include such items as poured concrete, concrete block, brick, stone, and ceramic tile.

Roofing

Roofing materials may be of wood (shingles or shakes), metal, or asphalt (rolls and/or shingles). Fig. 3-15.

Metal Products

Metal products (other than siding and roofing) include nails, dry wall screws, reinforcing bars,

3-14. *Installing dry wall.*

3-16. *Preformed metal for the fascia of a roof.*

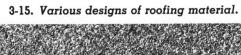

3-15. *Various designs of roofing material.*

3-17. *Various types of solid insulation.*

welded wire fabric, I-beams, truss connectors, hardware, doors, and many other products. Fig. 3-16.

Insulation

Insulation products are available as rigid or liquid foams, sheathing, fiberglass, loose bag, and a variety of other types. Fig. 3-17. Most insulation is made of plastics or a combination of plastics and other materials.

Plastics

Plastics are used in many products including insulation, flooring, counter tops, tile, siding, hardware, and finishing.

All materials used in construction should be specified in the prints and bills of materials. If a choice must be made between various products, the builder must consider such factors as cost, labor, availability, and design.

QUESTIONS

1. How is softwood lumber ordered?
2. Define the term "board foot".
3. How should lumber be treated for outdoor construction?
4. Explain how exterior-type plywood differs from interior plywood.
5. Name two species of wood from the strongest group for plywood construction.

6. Describe three kinds of engineered boards.
7. Name four types of materials used for siding.
8. What are the two principal materials used for paneling?
9. Describe gypsum board.

Safety

4

In building construction, safety is regarded as an important part of the total operation for at least two reasons. One is the natural concern for people's welfare. Another is financial. Building contractors know that injuries to employees are costly since they reduce the efficiency of the working force and may result in expensive medical bills and law suits. Therefore many contractors have safety programs which are intended to protect employees.

Safety is of prime importance in the operation of the power tools you will be using in building construction. Therefore you should not just read but *really learn* the safety rules for any job you do, and put the rules into practice. Do this for each piece of equipment you use. Learn to make safety a habit as you develop your skills.

This unit contains information on how to work safely in the building construction trades.

The important thing to remember is that serious accidents will not occur if workers are safety con-scious and follow recommended precautions. The following information should be valuable in aiding you to develop the correct attitudes toward safety and acquaint you with certain safety precautions.

DRESS

• Wear safety glasses whenever your work involves a threat to your eyes. (Sometimes this is required by law.) Fig. 4-1.
• Clothing should suit your individual needs and be suitable for the

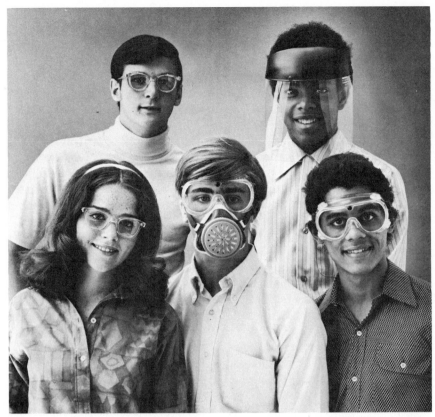

4-1. *Always wear adequate eye protection for the job to be performed. Sometimes it is also necessary to wear a mask for operations such as spray painting.*

4-2. *Wear long sleeves, gloves, and safety glasses when installing insulation.*

prevailing weather conditions. Fig. 4-2. Avoid loose fitting clothing that would restrict freedom of movement.

● In a shop or on a job site, you will probably work harder and safer in clothes you are not afraid to get dirty. If at all possible, as you enter the shop each day you should exchange your regular clothing for coveralls or other accepted working attire.

● Avoid wearing pants or overalls that are too long. Cuffs made by turning up the legs of pants tend to catch heels, causing falls.

● To avoid catching on nails, keep the sleeves of shirts or jackets buttoned.

● Keep hair cut short or keep long hair in place with a visor or hair net.

● To protect feet from protruding nails, wear shoes with thick, sturdy soles.

● To protect feet from falling objects, wear safety shoes or boots with steel toe caps.

● Wear a hard hat when exposed to overhead work or whenever there is danger from falling objects. Fig. 4-3.

● Remove neckties; also rings, wristwatches, neck chains, and other jewelry.

MATERIAL HANDLING

● Long pieces of material should be carried by two persons.

● Use the muscles in your legs and knees to lift heavy objects. Do not lift with your back muscles. To do so could result in painful back injuries.

● Observe caution when carrying planks or other objects across frozen, wet, or otherwise slippery footing.

AVOID FALLS

● Learn to watch your footing; avoid objects that could trip you.

● Check scaffolding and temporary walkways before walking on them.

4-3. *Supervisory and sales person-nel should never enter the job site without wearing a hard hat and eye protection.*

Be sure the supports are strong and secure.

• Use only ladders which are in good condition and set up properly.

PRACTICE GOOD HOUSEKEEPING

• Materials and equipment should be stacked straight and neat.

• Keep aisles and walkways clear of tools, materials, and debris.

• To prevent fires and reduce haz-ards which cause accidents, dispose of scraps and rubbish daily.

• Whenever you see protruding nails, remove them or bend them down immediately.

• When working above other peo-ple, place tools and materials where they will not fall and cause injuries.

GENERAL SAFETY

The following is a list of general safety rules to be used in the shop or on the job site as you work on and around machines and construc-tion. These rules will help to pro-tect you and others who are near you while you work.

• Always walk—do not run.

• Never talk to or interrupt any-one who is working on a machine.

• Remove power plug or turn off power supply to a machine when changing cutters or blades.

• Never leave tools or pieces of stock lying on the table surface of a machine being used.

• When finished with a machine, turn off the power and wait until the blade or cutter has come to a complete stop before leaving.

• Always carefully check stock for knots, splits, metal objects, and other defects before machining.

• Do not use a machine until you understand it thoroughly. Any tool with a sharp cutting edge can cause serious injury if mishandled.

• Use guards on power equipment. It should be understood that using guards does not necessarily pre-vent accidents. Guards must be used correctly if they are to pro-vide fullest protection. Also, it is impossible to do some operations, especially on the circular saw, with the regular guard in place. There-fore there are times when special guards should be used.

• Always keep your fingers away from the moving cutting edges. The most common accident is caused by trying to run too small a piece through a machine.

• Keep the floor around the ma-chine clean. The danger from fall-ing or slipping is always great.

• Make all adjustments with the power off and the machine at a dead stop.

• Always use a brush to clean the table surface.

• Always keep your eyes focused on where the cutting action is tak-ing place.

• Always use sharp tools.

• When using tools for set-up work on a machine: (1) Select the right tool for the job. (2) Keep it in safe condition. (3) Keep it in a safe place.

• Report strange noises or faulty operation of machines to the in-structor.

• Follow the suggestions for each machine given in this book.

The Congress of the United States in April of 1971 made the Federal Occupational Safety and Health Act (OSHA) an official part of the national labor law. The pur-pose of this law is " . . . to assure so far as possible every working man and woman in the nation safe and healthful working conditions and to preserve our human re-sources." This law affects all em-ployees who are working in the building trades where one or more workers are employed. As an indi-vidual employed in the building trades it is just as important to de-velop safe work attitudes and hab-its as outlined by this law as it is to develop the skills of your trade. Building trades employers will be looking for men and women with these traits for their benefit and welfare as well as yours as a skilled worker. Thus it is important to know and follow safety rules.

GENERAL SAFETY RULES FOR PORTABLE POWER TOOLS

• Never use portable power tools in contact with water, including rain, or if any part of your body is

4-4. *Whenever this drawing of a guard appears with an illustration, a guard must be used for the opera-tion shown.*

in contact with moisture. Be sure the power plug is removed before making any adjustments.

• Portable power tools should be properly grounded with a three-prong grounded plug. If a grounded receptacle is not available, use a three-to-two prong adapter plug which has been properly grounded.

• Always wear approved eye protection.

• Always disconnect the power plug when the work is completed.

• Be sure the switch is in the "off" position before connecting the power plug.

• Always use the recommended extension cord size.

NOTE: In the following units on tools and machines, many illustrations show dangerous operations being performed on machines without guards. The guards have been removed so that the photographs will show the operations more clearly. Whenever a drawing of a guard appears with an illustration, **a guard must be used** for the operation that is shown. Fig. 4-4.

QUESTIONS

1. Why is safety so important to the building contractor?

2. Why should students wear some form of work clothes for building construction?

3. What features should you look for when buying shoes to be used for working on a job site?

4. Do guards prevent accidents on power machines? Explain.

5. Why should portable power tools be grounded?

6. Explain why it is important to wear proper eye protection at all times, even when you are not operating equipment.

Scaffolds and Ladders · 5

SCAFFOLDS

A scaffold is a temporary or movable platform to stand on when working at a height above the floor or ground. The scaffold must also support the weight of the worker's tools and materials. Scaffolds make it possible to work safely, in a comfortable and convenient position, with both hands free.

Scaffolding is of two general types:

• Wood scaffolding, constructed on the job.

• Manufactured scaffolding.

Wood Scaffolding

When constructing wood scaffolding, select clear straight lumber for maximum strength. Fig. 5-1. Use adequate bracing and nail securely using a duplex head nail. Fig. 5-2. These nails can be driven in tightly and still be easily pulled when dismantling.

Manufactured Scaffolding

Manufactured scaffolding is designed to be readily assembled or dismantled. Fig. 5-3. The scaffold planks may be set at various

heights for comfort and safety. For interior use, casters are installed for easy movement. The end frames may be assembled in a staggered position, making it possible to work off a stairway. Where additional height is necessary, the units may be stacked.

Scaffolding Safety

• All scaffolding should be plumb and level. Use adjusting screws, not blocks, to adjust to uneven grading conditions.

• Adequate support should be pro-

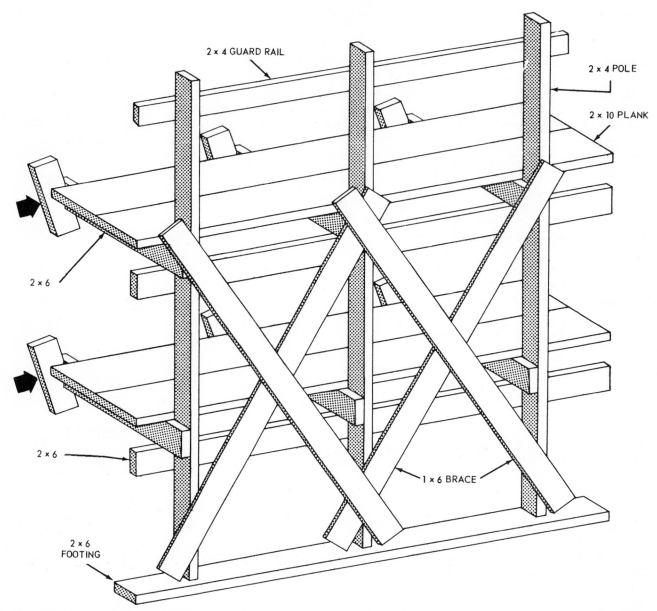

2 × 4 GUARD RAIL

2 × 4 POLE

2 × 10 PLANK

2 × 6

2 × 6

1 × 6 BRACE

2 × 6
FOOTING

5-1. *A scaffold must be safe. Use adequate supports and bracing. Note the blocks at the large arrows. These blocks are attached to the building. They are notched to receive the 2″ × 6″ boards.*

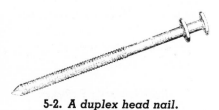

5-2. *A duplex head nail.*

vided. Use base plates, making sure that they rest firmly on the ground.

• All braces should be fastened securely.
• Cross braces should never be climbed. Access to scaffolds should be by stairs or fixed ladders only.
• Wall scaffolds should be securely anchored.
• Free-standing scaffold towers must be secured by guying (attaching guy ropes or wires) or other means.
• Proper guard rails should be pro-

vided. Add toe boards when required on planked or staged areas.
• Ladders or makeshift equipment should never be used on top of the scaffold.
• A scaffold should never be overloaded. Inspect the scaffolding assembly regularly.
• Lumber used for scaffold planks must be properly inspected and graded for that purpose. Both ends of planks must be cleated to pre-

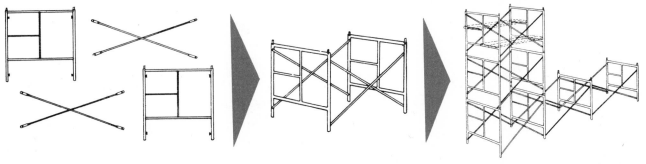

5-3. *Manufactured scaffolding components may be assembled in a variety of sizes and shapes, depending on the job requirements.*

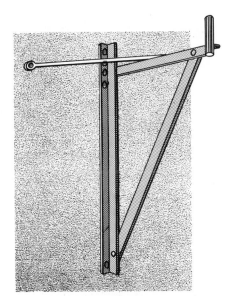

5-4a. *A nail-attached wall scaffold bracket.*

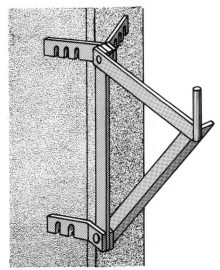

5-4b. *A nail-attached corner scaffold bracket.*

vent planks from sliding off supports. If planking is to be continuous, it should have at least a 12″ overlap, and should extend at least 6″ beyond the center of the support. Also do not extend the plank too far beyond the supports because such planks tend to be unstable.

• Whenever necessary for stability, planks should be nailed or clamped to the scaffold.

BRACKETS, JACKS, AND TRESTLES

Brackets

Scaffold planks may be supported by special brackets which are available for sidewall and roof installations. There are various styles of *sidewall* and *corner brackets*. Some are nailed to the studs while others are bolted or hooked directly around the studding. Fig. 5-4. The nail-attached wall and corner brackets are secured to the wall with 20d nails driven at an angle into the wall stud through the tapered holes in the bracket. The brackets may be easily removed without pulling the nails. Any nails remaining after brackets are removed are driven flush.

Many styles of *roofing brackets* are available for various applications. Roofing brackets are attached with nails through the roof sheathing and into the rafters. They can be removed without pull-

ing the nails. When they are removed, the nails are driven flush with the roof, and the shingles cover the nails that had held the bracket in place. One style holds a 2″ × 4″ or 2″ × 6″ flat against the roof. Others will hold the 2″ × 4″ or 2″ × 6″ on edge at right angles to the roof. A third type positions a 2″ × 6″ so that it provides a level walkway on a roof. Fig. 5-5.

Ladder Jacks

A ladder jack is a device for hanging a scaffold plank from a ladder. A jack can be used over or under any ladder that has rungs rather than steps. It has two hooks at the top and two at the bottom which fit close to the ladder siderails, preventing excessive loads on the ladder rungs. The ladder jack adjusts to the pitch of the ladder and can be hooked over the rungs with one hand. Fig. 5-6. The scaffold plank is then placed onto the horizontal projection to provide a convenient work platform so you don't have to move the ladder frequently. Fig. 5-7.

Trestles

A trestle consists of two jacks as shown in Fig. 5-8. Trestles are available in a wide range of sizes, from 16″ to 12′, each with height adjustments at approximately 3″ intervals. Timber is used as a *ledger* (a stationary support) between the jacks to support the scaffold. This

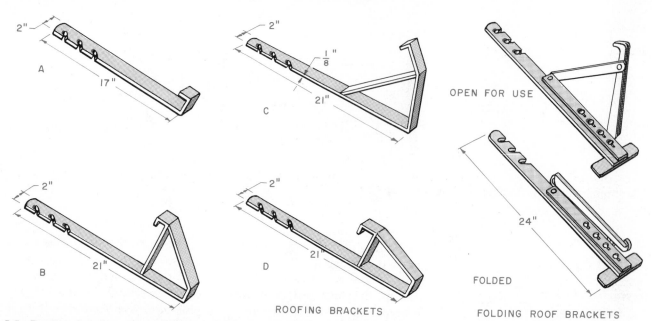

ROOFING BRACKETS   FOLDING ROOF BRACKETS

5-5. *Roofing brackets. The folding roof bracket adjusts to various roof pitches, from 90° to level walkways. It is ideal for use on steep roofs and will handle planks as large as 2″ × 10″. A. This roofing bracket holds a 2″ × 4″ or 2″ × 6″ flat against the roof. B. This bracket supports a 2″ × 6″ at a right angle to the roof. C. This bracket positions a 2″ × 6″ so that it provides a level walkway. D. This bracket supports a 2″ × 4″ at a right angle to the roof.*

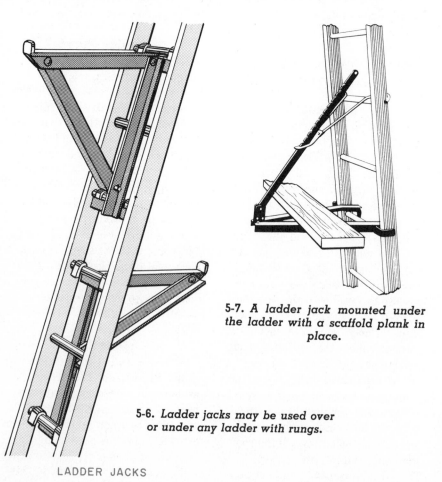

5-7. *A ladder jack mounted under the ladder with a scaffold plank in place.*

5-6. *Ladder jacks may be used over or under any ladder with rungs.*

LADDER JACKS

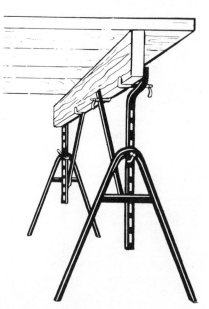

5-8. *Trestles used to support scaffolding.*

type of scaffold is sometimes used by plasterers and gypsum dry-wall applicators for working on ceilings. The scaffolding is set up over the entire floor area, on trestles about 18″ to 20″ high. The workers then

Folding

Straight

Extension

5-9. *The three types of ladders: folding, straight, and extension.*

can work continuously without stopping to take the scaffolding apart, move it, and set it up again.

LADDERS

Carpenters often must use ladders for high work. Ladders are usually made of wood or aluminum, and they come in many sizes. Commonly they are made in lengths from 3' to 50', with special three-section ladders available for reaching greater distances.

There are three basic types of ladders (Fig. 5-9):
- Folding (stepladder).
- Straight.
- Extension.

To set up a straight or extension ladder, place the lower end against a base so it cannot slide. Then grasp a rung at the upper end with both hands. Raise the top end and walk forward under the ladder, moving the hands to grasp other rungs as you proceed. When the ladder is erect, lean it forward to the desired position. Check the angle, height, and stability at top and bottom. Fig. 5-10.

When using a stepladder, always be certain that the four legs are firmly supported and that the spreaders are straight and level. Never stand on the top step of the ladder. The ledge on the back of the ladder is for holding tools and materials. Do not use it as a step.

When working from a ladder, set

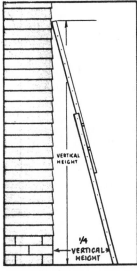

VERTICAL HEIGHT

¼ VERTICAL HEIGHT

5-10. *For safety on a ladder, the pitch or angle should be such that the horizontal distance at the bottom is one-fourth the working length of the ladder.*

it where the work can be reached with ease. Never lean out far to one side. Relocate the ladders as necessary so the work area can be reached without much leaning.

When going up or down, grip the ladder firmly and place your feet squarely on the rungs. Make certain your shoes and the rungs are free of mud and grease. While working, it is recommended that one leg be wrapped around a rung. When using a ladder for access to a roof, the ladder should extend above the edge of the roof by at least 3'.

Ladder Safety

- Inspect ladders carefully. Keep nuts, bolts, and other fastenings tight. Oil moving metal parts frequently. Do not allow makeshift repairs. Never straighten or use a bent metal ladder.
- Ladders must stand on a firm, level surface. Always use safety feet with nonslip bases.
- Face the ladder when climbing up or down.
- Always place the ladder close

enough to the work to avoid dangerous overreaching.
- Keep your weight centered between both side rails.
- Keep steps and rungs free of oil, grease, paint, or other slippery substances.
- Be sure that stepladders are fully open and the spreader straight.
- Never stand or climb on the top, pail rest, or rear rungs of a stepladder.
- Never place ladders in front of doors or openings unless appropriate precautions are taken.
- Always insure that the working length of the ladder will reach the support height required. It should be lashed or otherwise secured at the top to prevent slipping and should extend at least 3' above a roof or other elevated platform. Never stand on the top three rungs.
- Position the ladder so the horizontal distance of the ladder foot from the top support is one-fourth the working length of the ladder (75° angle). (*Working length* is the distance from the ground to the top support.) Always make sure that both side rails are fully supported top and bottom.
- Overlap extension-ladder sections by the following amounts: 3' for total extended lengths up to 36'; 4' for total lengths of 36' to 48'; and 5' for total lengths of 48' to 60'.
- When using a two-section extension ladder, place it so the upper section is outermost.
- Be sure all locks on extension ladders are securely hooked over rungs before climbing. Adjust the height of an extension ladder only when standing at the base of the ladder.
- Metal and water conduct electricity. *Do not use metal, metal reinforced, or wet ladders where direct contact with a live power source is possible.* Provide for tem-

porary insulation of any exposed electrical conductors near the place of work.

• A ladder is intended to carry only one person at a time. Do not overload.

• Never use ladders in a horizontal position.
• Store ladders in dry, cool, ventilated places.
• Never use ladders after prolonged immersion in water or expo-

sure to fire, chemicals, or fumes which could affect their strength.

QUESTIONS

1. What is a scaffold?
2. What are the two types of scaffolds?
3. What is a duplex head nail?
4. List several kinds of scaffold brackets.
5. What is a ladder jack?
6. List the three basic types of ladders.

7. Describe the method of setting up an extension ladder.

8. At what angle should an extension ladder be set against a building?

9. How much should the sections of an extension ladder overlap when working at a height of 39′?

Hand Tools 6

LAYOUT, MEASURING, AND CHECKING DEVICES

Tool	Description	Uses
Bench Rule Fig. 6-1.	A 12″ or 1′ (or 300-mm) rule. One side is divided into eighths, the other into sixteenths. A metric rule is divided into centimetres or millimetres.	1. To make simple measurements. 2. To adjust dividers. *Caution:* Never use as a straightedge.

6-1a.

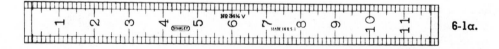

6-1b.

Zig-Zag Rule
Fig. 6-2.

A folding rule of 6′ or 8′ (or 2-m) length.

1. To measure distances greater than 2′ (600 mm), place the rule flat on the stock.
2. To measure less than 2′ (600 mm) it is better to use the rule on edge. (This instrument is good for inside measurement, since the reading on the brass extension can be added to the length of the rule itself.)

Flexible Tape Rules
Fig. 6-3.

A flexible tape that slides into a metal case. Comes in lengths of 6′, 8′, 10′, 12′, 50′, and 100′ (2 m to 50 m). The steel tape has a hook on the end that adjusts to true zero.

1. To measure irregular as well as regular shapes.
2. To make accurate inside measurements. (Measurement is read by adding 2″ (50 mm) to the reading on the blade.)

Try Square
Fig. 6-4.

A squaring, measuring, and testing tool with a metal blade and a wood or metal handle.

1. To test a surface for levelness.
2. To check adjacent surfaces for squareness.
3. To make lines across the face or edge of stock.

Combination Square
Fig. 6-5

Consists of a blade and handle. The blade slides along in the handle or head. There is a level and a scriber in the handle.

1. To test a level or plumb surface.
2. To check squareness—either inside or outside.
3. To mark and test a 45-degree miter.
4. To gauge-mark a line with a pencil.

Sliding T Bevel
Fig. 6-6.

A blade that can be set at any angle to the handle. Set with a framing square or protractor.

1. To measure or transfer an angle between 0 and 180 degrees.
2. To check or test a miter cut.

Dividers
Fig. 6-7.

A tool with two metal legs. One metal leg can be removed and replaced with a pencil. To set the dividers, hold both points on the measuring lines of the rule.

1. To lay out an arc or circle.
2. To step off measurements.
3. To divide distances along a straight line.

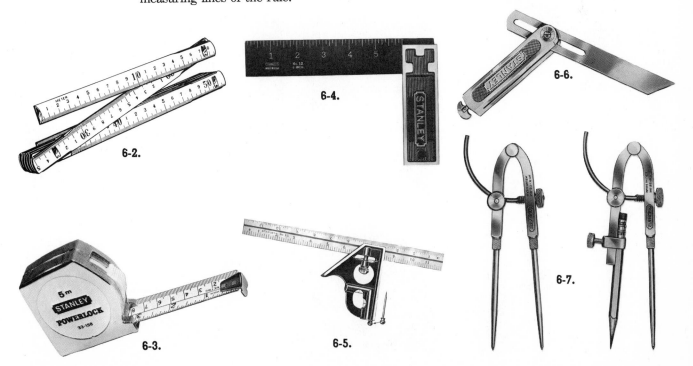

6-6.

6-4.

6-2.

6-3.

6-5.

6-7.

Framing or Rafter Square Fig. 6-8.

A large steel square consisting of a blade, or body, and a tongue.

1. To check for squareness.
2. To mark a line across a board.
3. To lay out rafters and stairs.

Carpenter's Level Fig. 6-9.

A rectangular metal or wood frame with several level glasses.

To check whether a surface is level or plumb.

Scratch Awl Fig. 6-10.

A pointed metal tool with handle.

1. To locate a point of measurement.
2. To scribe a line accurately.

Trammel Points Fig. 6-11.

Two metal pointers that can be fastened to a long bar of wood or metal.

1. To lay out distances between two points.
2. To scribe arcs and circles, larger than those made with dividers.

Plumb Bob and Line Fig. 6-12

A metal weight with a pointed end. The opposite end has a hole for attaching the cord.

1. To determine the corners of buildings.
2. To establish a vertical line.

SAWING TOOLS

Tool	Description	Uses
Back Saw Fig. 6-13.	A fine-tooth crosscut saw with a heavy metal band across the back to strengthen the thin blade.	1. To make fine cuts for joinery. 2. To use in a miter box.
Crosscut Saw Fig. 6-14.	A hand saw in lengths from 20″ to 26″ with from 4 to 12 points per inch. A 22″, 10-point saw is a good one for general purpose work.	1. To cut across grain. 2. Can be used to cut with the grain. *Caution:* Never cut into nails or screws. Never twist off strips of waste stock.
Rip Saw Fig. 6-15.	A hand saw in lengths from 20″ to 28″. A 26″, 5½-point saw is good for general use.	To cut with the grain. *Caution:* Support the waste stock. Never allow end of saw to strike the floor.

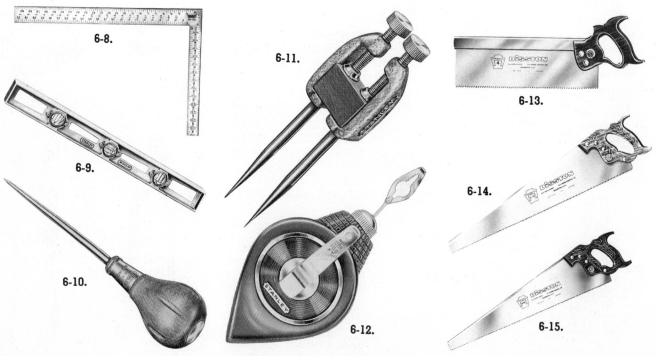

6-8.

6-11.

6-13.

6-9.

6-14.

6-10.

6-12.

6-15.

Compass Saw Fig. 6-16.	A 12″ or 14″ taper blade saw.	1. To cut gentle curves. 2. To cut inside curves.
Keyhole Saw Fig. 6-17.	A 10″ or 12″ narrow taper saw with fine teeth.	To cut small openings and fine work.
Miter Box Saw Fig. 6-18.	A longer back saw (24″ to 28″).	Used in a homemade or commercial miter box for cutting miters or square ends.
Coping Saw Fig. 6-19.	A U-shaped saw frame permitting 4⅝″ or 6½″ deep cuts. Uses standard 6½″ pin-end blades.	1. To cut curves. 2. To shape the ends of molding for joints. 3. For scroll work.

EDGE-CUTTING TOOLS

Tool	Description	Uses
Jack Plane Fig. 6-20.	A 14″ or 15″ plane.	1. Ideal for rough surface where chip should be coarse. 2. Also used to obtain a smooth, flat surface.
Jointer Plane Fig. 6-21.	A 22″ or 24″ plane.	1. To smooth and flatten edges for making a close-fitting joint. 2. For planing long boards such as the edges of doors.
Block Plane Fig. 6-22.	A small plane with a single, low-angle cutter with the bevel up.	1. To plane end grain. 2. For small pieces. 3. For planing the ends of molding, trim, and siding.
Chisels Fig. 6-23.	A set usually includes blade widths from ⅛″ to 2″.	To trim and shape wood.
Surform Tool Fig. 6-24.	Available in plane file type. Also round, or block-plane types. A blade with 45-degree cutting teeth.	For all types of cutting and trimming.

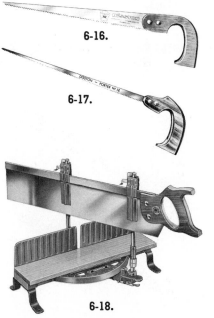

6-16.

6-17.

6-18.

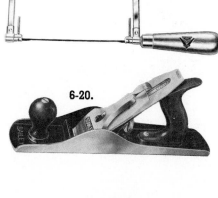

6-19.

6-20.

6-21.

6-22.

6-23.

6-24.

| Hatchet Fig. 6-25. | A cutting tool with a curved edge on one side and a hammer head on the other. Has hammer-length handle. | To trim pieces to fit in building construction. For nailing flooring. |

| Utility Knife Fig. 6-26. | An all-purpose knife with retractable blade. | 1. To cut and trim wood, veneer, hardboard, and particleboard. 2. To make accurate layouts. |

DRILLING AND BORING TOOLS

Tool	Description	Uses
Auger Bit Fig. 6-27.	May be either single-twist or double-twist bit. Comes in sizes from No. 4 (¼″) to No. 16 (1″).	1. To bore holes ¼″ or larger. 2. Single twist bit is better for boring deep holes.
Expansion Bit Fig. 6-28.	A bit that holds cutters of different sizes. Sometimes this tool is called an expansive bit.	1. To bore a hole larger than 1″. 2. One cutter will bore holes in the 1″ to 2″ range. 3. A second cutter will bore holes in the 2″ to 3″ range.
Brace Fig. 6-29.	Two common types—the plain for a full swing, and the ratchet for close corners.	To hold and operate bits.
Foerstner Bit Fig. 6-30.	A bit with a flat cutting surface on the end.	1. To bore a shallow hole with a flat bottom. 2. To bore a hole in thin stock. 3. To bore a hole in end grain. 4. To enlarge an existing hole.
Bit or Depth Gauges Fig. 6-31.	Two types—one is a solid clamp, the other a spring type.	To limit the depth of a hole.
Twist Drill (a) or Bit Stock Drill (b) Fig. 6-32.	A fractional-sized set from ⅟₆₄″ to ½″ is best.	To drill small holes for nails, screws, etc.

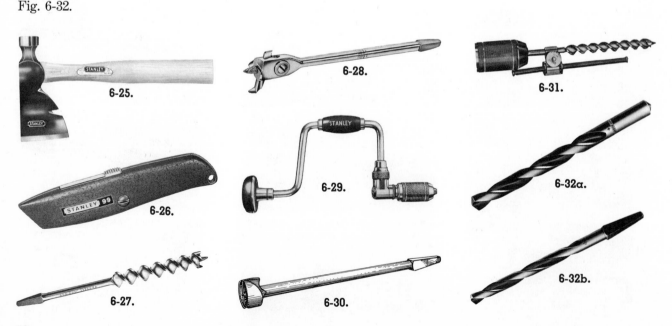

6-25. 6-28. 6-31. 6-26. 6-29. 6-32a. 6-27. 6-30. 6-32b.

| Hand Drill Fig. 6-33. | A tool with a 3-jaw chuck. | To hold twist-drills for drilling small holes. |

METALWORKING TOOLS*

Tool	Description	Uses
Hacksaw Fig. 6-34.	A U-shaped frame with handle. Uses replaceable metal-cutting blades.	To cut all types of metal fasteners, hardware, and metal parts.
Cold Chisel Fig. 6-35.	A tool-steel chisel with cutting edge especially hardened and tempered for cutting metal. Angle between bevel surface is about 60 degrees.	1. To cut off a rivet or nail. 2. To get a tight or rusted nut started.
Adjustable Wrench Fig. 6-36.	An extra-strong, lightweight, thin-jawed tool with one adjustable jaw. Wrench develops greatest strength when hand pressure is applied to the side that has the fixed jaw.	1. To make adjustments on machines, when there is plenty of clearance. 2. To install and replace knives and blades.
Open-end Wrench Fig. 6-37.	A non-adjustable wrench with accurately machined openings on either end. Sizes of openings are stamped on the tool. For variety of work, a complete set is needed.	1. To make adjustments on machines where there is plenty of clearance. 2. To install and replace knives and blades.
Box Wrench Fig. 6-38.	A metal wrench with two enclosed ends. Heads are offset from 15 to 45 degrees.	To make adjustments where there is limited space for movement.
Vise-grip Wrench Fig. 6-39.	An all-purpose tool with double-lever action that locks the jaws on the work.	Used as a substitute for a vise, clamp, pipe wrench, fixed wrench, or adjustable wrench.
Combination Pliers Fig. 6-40.	An all-purpose, slip-joint adjustable pliers.	To hold and turn round pieces. Never used on heads of nuts or bolts.

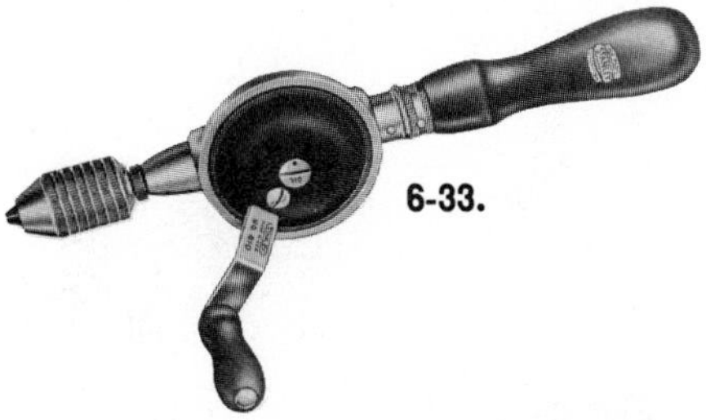

6-33.

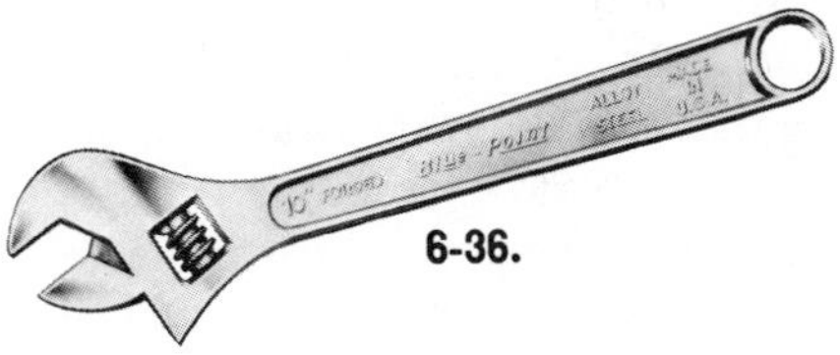

6-36.

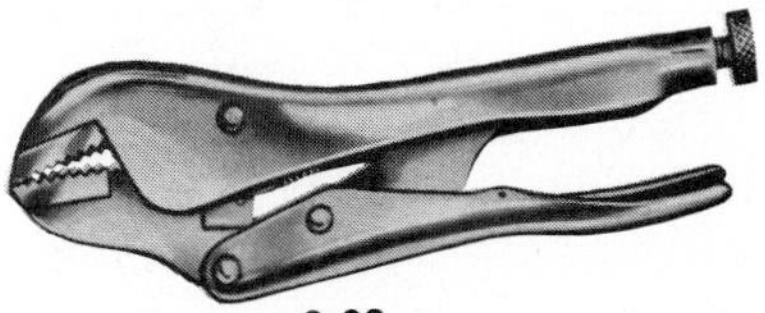

6-39.

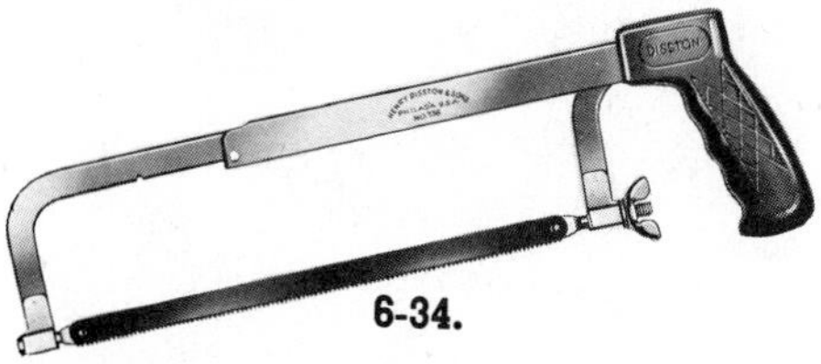

6-34.

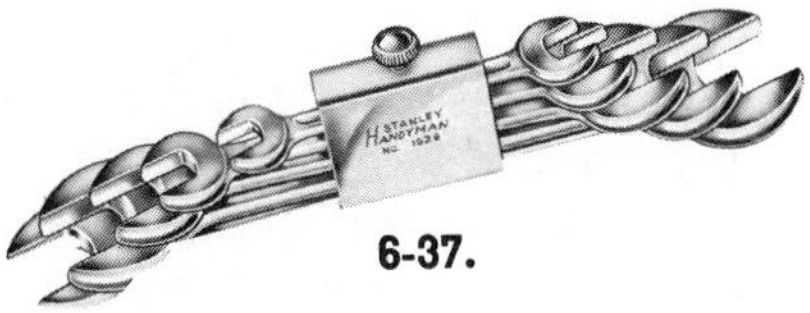

6-37.

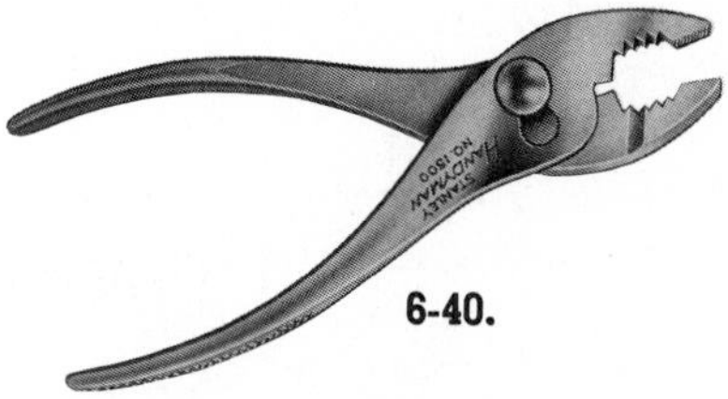

6-40.

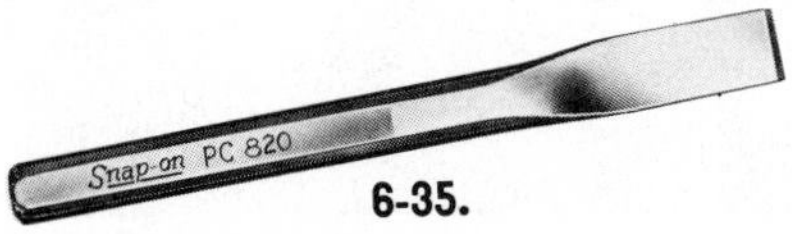

6-35.

6-38.

*In carpentry, many metalworking tools are needed to set up and adjust machinery and to work with metal hardware and fasteners.

QUESTIONS

1. Name several kinds of common rules.
2. What are four uses for the combination square?
3. Why are flexible tapes useful measuring tools?
4. What is the difference between a crosscut saw and a ripsaw? Describe their teeth.
5. Name the hand saws that are used for cutting irregular curves.
6. Why are planes made in different lengths?

7. How does a bit differ from a drill?
8. Name the tool used for operating bits. For operating drills?
9. Describe several uses for the Foerstner bit and the two devices that are used to limit the depth of a hole.
10. Why are metalworking tools needed in a wood shop?

Radial-Arm Saw 7

SAFETY

• Always keep the safety guard and the anti-kickback device in position.
• Make sure the clamps and locking handles are tight.
• When crosscutting, adjust the anti-kickback device (sometimes called "fingers") to clear the top of the work by about ⅛". This acts as a guard to prevent your fingers from coming near the revolving saw.
• Make sure the stock to be cut is held tightly against the fence.
• For crosscutting, dadoing, and similar operations, pull the saw into the work.
• Return the saw to the rear of the table after each cut.
• For ripping, make sure that the blade is rotating upwards toward you. Use the anti-kickback device

to hold work firmly against the table. Feed the stock from the end opposite the anti-kickback device.
• Keep your hands away from the danger area—that is, the path of the saw blade.
• Be sure the power is off and the saw is *not* rotating before making any adjustments.
• Always use a sharp saw or cutter.
• Allow the saw to reach full speed before making a cut.
• Hold the saw to prevent it from coming forward, before turning on the power.
• This saw tends to feed itself into the work. Therefore it is necessary to regulate the rate of cutting by holding back the saw. Otherwise it will feed faster than it can cut, causing the motor to stall.

• Use a brush or stick to keep the table clear of all scraps and sawdust.

RADIAL-ARM CONTROLS
RIGHT SIDE

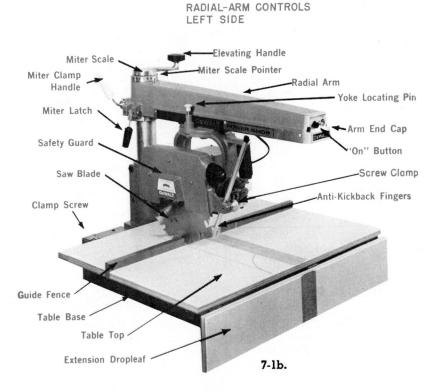

Rip Pointer
Rip Scale
"Off" Button
Key Switch
Dust Spout
Yoke Handle
Yoke
Bevel Locating Pin

Rip Lock
Line Cord
Column
Yoke Clamp Handle
Motor Restart
Motor
Right-Hand Motor Arbor

Space Boards
Bevel Scale
Bevel Clamp Handle

7-1a. *Study the names of the parts and controls. You must know them to follow directions for making adjustments and cuts. See Fig. 7-1b.*

RADIAL-ARM CONTROLS
LEFT SIDE

Miter Scale
Miter Clamp Handle
Miter Latch
Safety Guard
Saw Blade
Clamp Screw
Guide Fence
Table Base
Table Top
Extension Dropleaf

Elevating Handle
Miter Scale Pointer
Radial Arm
Yoke Locating Pin
Arm End Cap
"On" Button
Screw Clamp
Anti-Kickback Fingers

7-1b.

The radial-arm saw, Fig. 7-1, is a very versatile machine. It can be used for ripping, dadoing, grooving, and various combinations of these cuts. Many of these operations can be performed more easily on the radial-arm saw than on any other machine. For instance, a long board can be cut into shorter lengths easily because the board remains stationary on the table while the saw is pulled through the stock. Another advantage is that the saw blade is on top of the work so that when dadoes, grooves, and stop cuts are made, the cut is always in sight.

INSTALLING THE SAW BLADE

Remove the guard by removing the wing nut on top of the motor housing. Fig. 7-2. Raise the blade so it will clear the table top when it is removed.

To remove the arbor nut, hold the arbor with one wrench and turn the nut clockwise with a second wrench. Fig. 7-3. *Do not attempt to hold the blade with a block of wood while you loosen the nut.* If you do, the saw will climb onto the block and will thus be forced out of alignment.

Place the blade on the arbor. Make certain the teeth at the bottom are pointing away from you and toward the column. Replace the collar with the recessed side against the blade. Replace and securely tighten the nut. Then replace the guard. Fig. 7-4.

CROSSCUTTING

1. Mount a crosscutting or combination saw blade on the arbor.

2. Adjust the radial arm to zero (at right angles to the guide fence) and set the motor so that the blade will be at right angles to the table top. Lock the radial arm with the miter clamp handle.

3. Turn the elevating handle down until the teeth are about $\frac{1}{16}''$

53

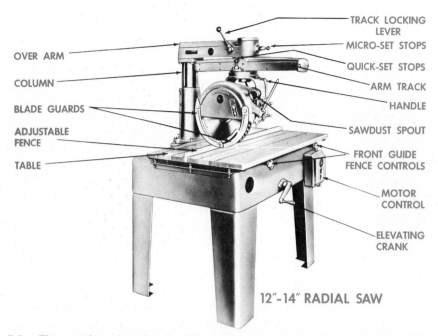

OVER ARM
COLUMN
BLADE GUARDS
ADJUSTABLE FENCE
TABLE

TRACK LOCKING LEVER
MICRO-SET STOPS
QUICK-SET STOPS
ARM TRACK
HANDLE
SAWDUST SPOUT
FRONT GUIDE FENCE CONTROLS
MOTOR CONTROL
ELEVATING CRANK

12"-14" RADIAL SAW

7-1c. This machine has the double arm with the arm track pivoting from the upper arm directly over the work area. This places the saw cut nearer the center of the table on both the left- and right-hand miters.

7-2. The guard is held firmly on the motor housing by a wing nut.

below the surface of the wood table. (The blade should follow the saw kerf already cut in the table.)

4. Adjust the anti-kickback fin-

gers about ⅛" above the work surface.

5. With one hand hold the work on the table firmly against

the guide fence. The layout line should be in line with the path of the saw.

6. Turn on the power and allow

7-3. The arbor is held in place with a hex wrench. On some machines the arbor is held by using an open-end wrench on a flat area between the blade and motor housing. Use the correct size wrench to turn the arbor nut.

7-4. The teeth on the bottom of the saw blade point away from the operator and toward the column. Not all blades will have an arrow to show the direction of rotation.

7-5. When cutting stock to length, place the workpiece against the fence and slowly pull the saw into the stock. (Note the guard printed with this caption. This means a guard must be used for this operation.)

7-6. When ripping, adjust the "fingers" on the anti-kickback device to project about 1/8" below the surface of the workpiece. As the work is fed, the "fingers" will ride up on the surface of the work. Always use a push stick to feed the workpiece past the saw blade as the cut is finished.

7-7. For your safety and the safety of others working around you, always set the guards properly for maximum protection.

the saw to come to full speed. Grasp the motor yoke handle and pull the saw firmly but slowly through the work. Fig. 7-5.

7. When the cut is completed, return the saw behind the guide fence. Then turn off the power.

RIPPING

1. Mount a combination or ripping blade. Pull the entire motor carriage to the front of the arm. Pull up on the locating pin above the yoke. Rotate the yoke 90 degrees *clockwise* until the blade is parallel to the guide fence. The motor should be "outboard" (that is, away from the column) so it will not obstruct the cutting. Fig. 7-6. When ripping wide panels it is necessary to rotate the yoke counterclockwise so the motor will be "inboard" (that is, toward the column). This will increase the ripping capacity.

2. Move the motor assembly along the radial arm until the correct width is shown on the rip scale. Tighten the *rip clamp* (on opposite side of radial arm from locating pin). Lower the saw until the blade just touches the wood table.

3. Adjust the guard so that the infeed end clears the work slightly (about 1/8"). Adjust the anti-kickback device so that the points are 1/8" below the surface of the workpiece. Fig. 7-7.

4. Turn on the power. Make sure the saw is rotating upwards toward you. Hold the work against the guide fence and feed it into the blade. Never feed the work from the anti-kickback end. Use a push stick to complete the cut.

QUESTIONS

1. Describe the main safety precautions to follow in operating a radial-arm saw.

2. Tell how to do crosscutting on a radial-arm saw.

3. In ripping stock on a radial-arm saw, does the work or the saw move? Explain the action.

Jointer

SAFETY

- *Always keep the knives of the jointer sharp.* Dull knives tend to cause kickback and also result in poor planing.
- *The fence should be tight.* Never adjust the fence while the jointer is running.
- Adjust the depth of cut before the jointer is turned on.
- *See that the guard is in place and operating easily.* If the regular guard is removed, a special guard must be provided.
- *Always allow the machine to come to full speed before using it.*
- Check the stock for knots, splits, metal particles, and other imperfections before jointing. Defective stock may break up or be thrown from the jointer.

- *Keep the left hand back from the front end of the board when feeding.*
- *Stand to the side of the jointer, never directly behind it.* In case of kickback you will be out of the way.
- *Cut with the grain, never against it.*
- *Always use a push stick or push block.*
- *Do not try to take too heavy a cut.*
- *Use common sense about when stock is too thin or too short to joint safely.*
- Never apply pressure to the board with your hand directly over the cutterhead.
- Use a brush to clean shavings off the table. Never use your hand.

Although the jointer is not used for a great variety of operations, it is one of the most frequently used machines in a typical shop. Common uses of the jointer are for surfacing a board and for planing an edge or an end. It can also be used for cutting a rabbet, bevel, chamfer, or taper.

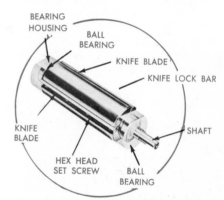

8-2. A 6″ jointer head with the parts named.

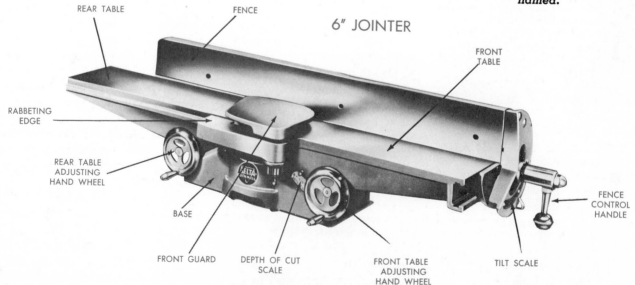

6″ JOINTER

8-1. A 6″ jointer with the parts named.

The jointer has a circular cutterhead which usually has three or four blades (or knives). The blades rotate, shearing off small chips of wood, thus producing a smooth surface on the workpiece.

SIZE

The size of a jointer is indicated by the length of the knives. Since most jointing operations are performed on the edge of stock, a 6″ or 8″ jointer is most common.

The length of the bed also affects

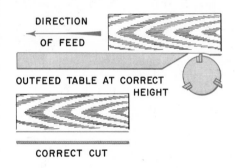

DIRECTION OF FEED

OUTFEED TABLE AT CORRECT HEIGHT

CORRECT CUT

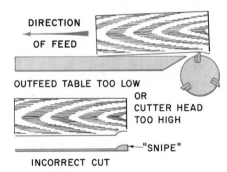

DIRECTION OF FEED

OUTFEED TABLE TOO LOW
OR
CUTTER HEAD TOO HIGH

"SNIPE"

INCORRECT CUT

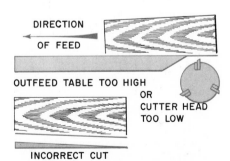

DIRECTION OF FEED

OUTFEED TABLE TOO HIGH
OR
CUTTER HEAD TOO LOW

INCORRECT CUT

8-3. The jointer must be adjusted so the outfeed table is at exactly the same height as the cutterhead knife at its highest point. Otherwise a taper or a recess will be cut. (A recess is sometimes called a "snipe.")

the usefulness of the jointer, since a longer bed provides better support for jointing longer pieces.

PARTS

The *frame or base* of the jointer has two tables—the *front or infeed table* and the *rear or outfeed table.* Fig. 8-1. On most machines, both of these tables are adjustable, although there are some on which only the infeed table can be raised or lowered.

The *cutterhead* is the heart of the jointer. As mentioned, it consists of the head itself and three or more knives. This assembly usually operates on two roller bearings. Fig. 8-2.

The *fence* provides support for the work while it is fed, on edge or on end, through the machine. The fence can be adjusted to various angles, usually up to 45 degrees both ways from the vertical position.

The *guard* is a protective device covering the cutterhead. It either swings out of the way or lifts up. Most operations, except rabbeting on some jointers, and certain tapering, should be done with the guard in place.

The jointer usually operates at about 4000 rpm.

ADJUSTMENTS

Aligning and Adjusting the Outfeed Table

The top of the outfeed table must be at exactly the same height as the knife blades at their highest point of revolution. If the table is too low, the board will drop down onto the knives as it leaves the infeed table. This will cause a recess to be cut at the end of the board. If the table is too high, the board will be slightly tapered. Figure 8-3 shows correct and incorrect cuts.

To align the knives with the table as just mentioned, turn the cut-

terhead until one blade is at its highest point. Release the table locking screw on the side of the jointer. Lower the outfeed table (rear) until it is below the blade; then place a straightedge on the outfeed table with one end projecting over the blade. Fig. 8-4. Turn the table up slowly until it is in line with the knife at the highest point. Turn the cutterhead over slowly by hand until there is very light contact between the knives and the bottom of the straightedge. Tighten the lock nut. Once the outfeed table is set, it does not require changing except for certain cuts such as stop chamfers and bevels, and recess cuts. If the outfeed table is the fixed kind, raise or lower the cutterhead until the knives are even with the outfeed table.

Adjusting the Infeed Table

The distance the infeed (front) table is below the knives determines the depth of cut. The depth of cut to be taken will depend on:
• The width of the surface being jointed.
• The kind of wood and grain pattern.
• Whether you are making a rough or finish cut.

Loosen the lock on the side of the infeed table, then turn the handle beneath the table to raise or lower it. There is a pointer and scale, indicating the depth of cut, which must be checked periodically for accuracy. Fig. 8-5.

Adjusting the Position of the Fence

For most operations it is desirable to have the fence at an exact right angle to the table. To adjust the fence, loosen the knob or lever that holds it in position; then set the fence at a 90-degree angle to the table. To check that the angle is correct, hold a square against the table and fence. Fig. 8-6. The

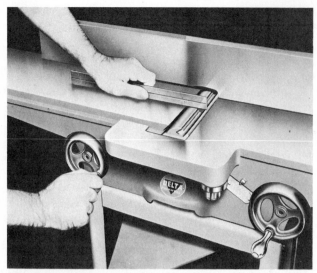

8-4. *Adjusting the outfeed table. Raise the table slowly until the straightedge rests evenly on the table and the knife. Always replace the guard after making this adjustment.*

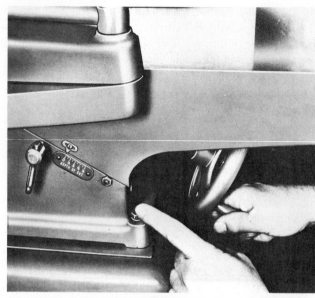

8-5. *Always check the depth of cut before making a cut on the jointer.*

fence can be moved in or out. When cutting, never expose any more of the blade than necessary.

The fence can also be *tilted* 45 degrees to right or left. This can be set on the tilt scale and checked with a protractor head of a combination square set or a sliding T bevel. There is a pointer and scale to indicate the tilt.

BASIC PROCEDURES

1. Check the fence for squareness and the infeed table for depth

of cut before turning on the machine. If the jointer has been used for some other operation, make a trial cut after resetting it.

2. Adjust the *depth of cut* with these things in mind:
• The amount of stock to be removed. Take a light cut for such operations as face planing or end planing and a slightly heavier cut for edge planing.
• The kind of wood. A light or heavy cut may be made on soft woods; a light cut is best on hard woods.

• The kind of planed surface. Take a heavier cut for removing stock and a lighter cut for finishing.

3. Change the position of the fence periodically to distribute the wear on the jointer knives.

4. When duplicate parts are needed, do the jointing operations first; then cut the stock into the desired smaller pieces.

5. If you are right-handed, stand to the left of the jointer with your left foot forward and right foot back and beneath the infeed table. Move your body along as you do the planing operation.

6. Always check a board for warp and wind first. Fig. 8-7. Place a concave surface down for the first cuts. If the board has twist, balance it on the high corners to take the first cuts.

Planing a Surface

1. Check the board for warp and for direction of grain. Be certain the jointer is correctly adjusted.

2. Hold the board firmly on the infeed table with your left hand toward the front of the board and your right hand on the push block.

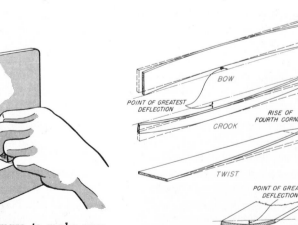

8-6. *Use a try square to make sure the fence is set at right angles to the table.*

8-7. *Common kinds of warp that can be removed on the jointer.*

BOW

POINT OF GREATEST DEFLECTION

RISE OF FOURTH CORNER

CROOK

TWIST

POINT OF GREATEST DEFLECTION

8-8. *Using the one-handed push block.*

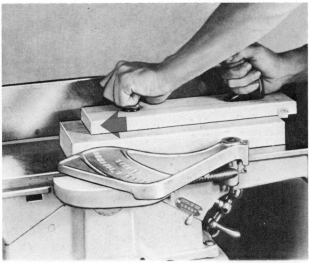

8-9. *Using a push block to do facing on short stock. Note the use of the push block or holddown. The knob is held in the left hand and the handle in the right.*

Fig. 8-8. The push block is hooked on the end of the board over the infeed table. Apply equal pressure with both hands. Fig. 8-9.

3. Turn on the machine and allow it to come to full speed.

4. Move the stock forward, keeping your left hand back of the cutterhead. When about half to two-thirds of the board has passed the cutterhead, move the left hand to the board over the outfeed table. Fig. 8-10.

5. After most of the board has passed over the cutter, move the right hand to the portion of the board over the outfeed table to finish the cut. *Never place your hand directly over the cutterhead. Fig. 8-11.*

Planing an Edge

The most common use for the jointer is planing or jointing an edge. An edge is said to be jointed when it is at right angles to the face of the board and is true along its entire length. Fig. 8-12.

1. Check the fence for squareness. Generally, for safest operation, it is best to set the fence as close as possible to the left side of the machine.

2. Select the best edge and determine the grain direction.

3. Adjust for proper depth of

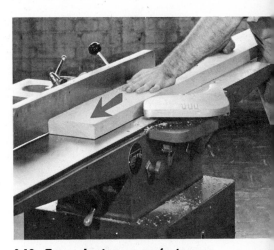

8-10. *Face planing or surfacing on an 8" jointer. Note how the left hand is kept back from the front edge of the board.*

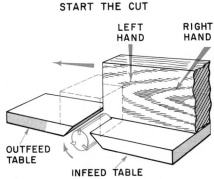
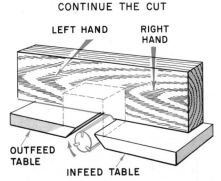
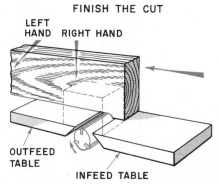

8-11. *Correct method of feeding when the hands are moved as stock passes over the cutterhead. The danger area is shown in color.*

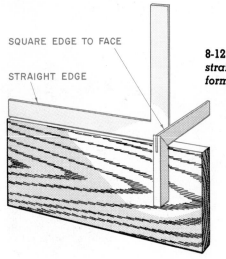

SQUARE EDGE TO FACE

STRAIGHT EDGE

8-12. A properly jointed edge is straight along its entire length and forms a 90° angle with the working face.

cut. To insure parallel edges on the stock, rip to width and allow just enough extra stock to joint off the sawn edges.

4. Hold the stock firmly against the infeed table and the fence. The jointed or planed surface of the board should be against the fence.

5. For the right-handed person, the left hand is a guide and the right hand pushes the stock across the cutterhead. Move the left hand along with the board until the major portion of the board is over the outfeed table; then move the right hand to the other side of the cutterhead, to the stock over the outfeed table. Do not push the board too fast, as this will make a rippled edge.

QUESTIONS

1. What function do rotary cutters on a jointer perform?

2. How is the size of a jointer indicated?

3. How is the depth of cut adjusted on a jointer?

4. What happens if the outfeed table of a jointer is too low?

5. What happens if the outfeed table of a jointer is too high?

6. List five safety rules for operating a jointer.

7. What three factors should be considered in adjusting for depth of cut?

8. When should a push block be used?

9. Indicate the most common use for a jointer.

Circular Saw

SAFETY

• Make all adjustments when the power is off and the blade has stopped revolving.

• Always adjust the saw blade so it protrudes just enough above the stock to cut completely through.

• Never reach over a revolving saw; instead, bring the cut piece back around the side of the machine.

• Keep your fingers away from the saw blade at all times.

• Always keep the guard and splitter in place unless this is impossible for the kind of cut you are making.

• If the cut you are making doesn't permit use of the regular guard, use a feather board or a special guard.

• When crosscutting with the miter gauge, never use the fence for a stop unless a clearance block is used.

- Always push the stock through with a push stick when ripping stock that cannot be fed safely by hand.
- Never stand directly behind the blade.
- Always use a sharp blade.
- When ripping, place the jointed edge against the fence.
- Keep the saw table clean. Remove all scraps with a brush or push stick, *never with your fingers*.

- Remove rings, watches and other items that might catch in the saw. Wear garments with short or tight sleeves.
- Use the proper saw blade for the operation being performed.
- Always hold the stock firmly against the miter gauge when crosscutting and against the ripping fence when ripping.
- Be certain the fence is clamped securely.

- When someone assists you, they should not *pull* the stock. They only *support* the stock.
- Do not saw warped material on the circular saw.
- If stock must be lowered onto the revolving blade for certain cuts, use stops and guards. Never have your hands in line with the blade.

Many fundamental woodworking operations can be done with the circular saw. It can be used not only for cutting stock to size but also for cutting many joints.

SIZE

The size of the circular saw is indicated by the diameter of blade recommended for its use. Typical sizes are the 8″ or 10″.

BLADES

There are ten kinds of saw blades. Fig. 9-1. In selecting a blade, make sure that you secure one with the correct diameter arbor hole size. Never attempt to install a blade that has too large a hole.

PARTS

Study the parts of the circular saw, as shown in Fig. 9-2. Notice that this is a tilting arbor saw. The table top has two grooves cut in it into which the miter gauge fits. These are parallel to the saw blade. The miter gauge comes equipped with a stop rod which can be adjusted in length for cutting duplicate parts. A fence clamps to the table for all ripping operations. Also available are table extensions which can be fastened to the sides of the table top and are especially convenient when cutting long or large stock such as a sheet of plywood. An opening in the center of the table is covered by a throat

plate. A guard, which drops over the blade, is always fastened to the back or side of the table. This should be kept in place whenever possible. (There are some operations for which regular guards cannot be used; a special guard or a feather board should then be used.) There is also a splitter which is usually a part of the back of the guard. This fits directly back of the saw blade and is slightly thicker than the blade. It keeps the saw kerf open as the cutting is done.

ADJUSTMENTS

Installing or Removing a Saw Blade

1. Remove the throat plate. This usually snaps in and out of position.

2. Select a wrench to fit the arbor nut. On some saws the arbor has a left-hand thread and must be turned clockwise to loosen. However, some saws have a right-hand thread. If so, you must turn it counterclockwise to remove. Always check the thread before loosening. A good rule to remember is that the nut always loosens by turning it in the direction the teeth are pointing (direction of blade rotation). If the nut doesn't come off easily, hold a piece of scrap wood against the blade to keep the arbor from turning. Fig. 9-3.

3. Remove the nut and the collar, and take off the old blade.

4. Replace the blade in the correct position, with the saw teeth pointing in the direction of blade rotation. Replace the collar and nut. Tighten it firmly, but not too tight. The nut tightens against the rotation and will not come off. Replace the throat plate.

Raising the Saw Blade

There is a wheel or lever on the front of the machine to raise or lower the blade. Often, in addition, there is a lock that must be loosened when making this adjustment. To raise the blade to the proper position, hold the workpiece against the side of the blade and carefully turn the blade until the top tooth is at the correct height. For most cutting, the top of the blade should extend no more than ⅛″ above the stock. On many joint cuts, however, the blade must be set for the exact depth of cut.

Tilting the Saw Blade

A lever or handle on the side of the machine tilts the blade. A pointer or scale on the front indicates the degree of tilt. There is usually a lock to hold the blade in position when it is tilted.

Adjusting the Fence

A ripping fence is fastened to the table for all ripping operations and for many other cutting jobs. It is usually placed to the right of the blade. To adjust the fence to the

Cut-Off Blade
Flat ground. Fine teeth are set for fast smooth cutting across the grains of hard or soft wood, sheeting, flooring and wood molding. Easily resharpened.

Rip Blade
Flat ground. Designed for cutting hard and soft woods with the grain. Teeth set for clearance. Easily resharpened.

All Purpose Combination
Flat ground. A blade for all general purpose work. Smooth cutting in any direction through all types of wood. Easily resharpened.

Chisel Tooth Combination
Flat ground. For fast rugged general purpose cutting. Rips, crosscuts, mitres hard and soft wood, wallboard, and heavy construction gauge plywood. Easily resharpened.

Planer Combination
Flat ground. Economical blade for smooth crosscutting, ripping, and mitring. Fast cutting teeth are set for clearance. Resharpenable.

Plywood Blade
Flat ground. Smooth economical cutting of paneling, plywood, and laminates. Also used where, occasional nails are found, such as old flooring. Easily resharpened.

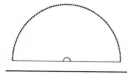

Planer Combination-Hollow Ground
Premium blade offers fast, smooth crosscutting, ripping and mitring. Fast cutting teeth, hollow ground for clearance. Excellent blade for all furniture and cabinet work where exceptionally smooth cutting is required. Resharpenable.

Plywood Blade-Hollow Ground
A premium blade hollow ground for clearance designed for virtually splinter free cutting of plywood, paneling, and laminated woods. Easily resharpened.

Carbide Tipped Blade
Long lasting premium blade for smooth and fast cutting of wood, hardboard, plastics, laminates, thin brass and aluminum. Longer lasting than conventional blades.

Multi-Tooth Carbide Tipped Blade
Long lasting premium blade for fast rugged general purpose cutting of woods, wallboard, and heavy construction gauge plywood. Longer lasting than conventional blades.

9-1. *Kinds of saw blades.*

pointer and scale on the miter gauge for setting it to any degree right or left. Most gauges have automatic stop positions at 30, 45, 60, and 90 degrees.

RIPPING
Install a ripping, combination, or easy-cut blade for these operations:

Cutting Wide Stock to Width
1. When the width of the board to be cut is 6″ or more, this is considered a wide cut. Adjust the fence and blade accordingly.

2. Turn on the machine. Place the board over the table. Apply pressure against the fence with the left hand, and push the board forward with the right. If the board is longer than 6′ or 8′, have a helper stand behind the saw to hold the piece up after it passes the blade. If a helper is not available, use a roller stand as shown in Fig. 9-4.

3. Feed the stock at an even speed into the blade about as fast as it will cut. Be careful not to overload the saw. Hold your right hand close to the fence as you push the end of the board through the saw. Fig. 9-5.

4. If extremely thick or hard wood is being cut, it is often necessary to cut partway through the board, then turn the board over and complete the cut.

Cutting Narrower Stock to Width
1. When cutting stock narrower than 6″, observe the same general practices as in starting the cut on wide stock.

2. As the end of the board reaches the front of the table, use a push stick to do the work you began with your right hand, guiding the board between the blade and

correct position, first move it to an approximate location. Hold a rule or try square at right angles to the fence and carefully measure the distance from the fence to one tooth bent toward the fence.

Adjusting the Miter Gauge
The miter gauge, which is used for crosscutting operations, can be used in either groove of the table but usually is placed in the groove to the left of the blade. There is a

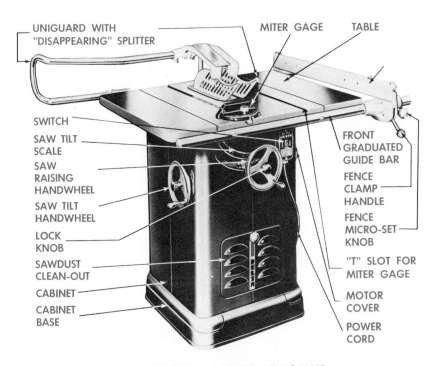

UNIGUARD WITH "DISAPPEARING" SPLITTER
MITER GAGE
TABLE
SWITCH
SAW TILT SCALE
SAW RAISING HANDWHEEL
SAW TILT HANDWHEEL
LOCK KNOB
SAWDUST CLEAN-OUT
CABINET
CABINET BASE
FRONT GRADUATED GUIDE BAR
FENCE CLAMP HANDLE
FENCE MICRO-SET KNOB
"T" SLOT FOR MITER GAGE
MOTOR COVER
POWER CORD

10″ TILTING ARBOR UNISAW®

9-2. *Parts of a 10″ tilting-arbor circular saw.*

9-5. *Ripping on the circular saw.*

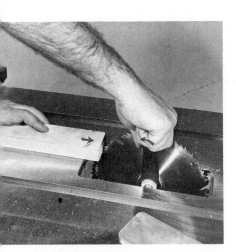

9-3. *Hold the blade with a piece of scrap wood. The nut will loosen if turned in the direction the teeth are pointing.*

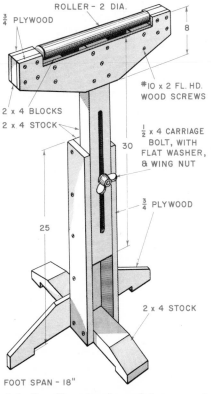

ROLLER – 2 DIA.
¾ PLYWOOD
8
#10 x 2 FL. HD. WOOD SCREWS
2 x 4 BLOCKS
2 x 4 STOCK
½ x 4 CARRIAGE BOLT, WITH FLAT WASHER, & WING NUT
30
¾ PLYWOOD
25
2 x 4 STOCK
FOOT SPAN – 18″

9-4. *A roller stand used to support long stock when ripping.*

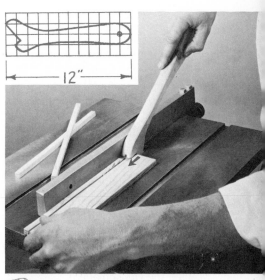

12″

9-6. *Using a push stick in ripping narrow stock.*

the fence. Fig. 9-6. *Never under any circumstances cut narrow stock without a push stick.* It is good practice to hang the push stick conveniently at the side of the

saw so that you don't take a chance and cut without it.

3. If very narrow stock is being cut, it may be a good idea to cut half the length of the stock, pull it back out, reverse it, and complete the cut from the other end.

CROSSCUTTING

Install a crosscutting, hollow-ground, or combination blade. Use the miter gauge for all crosscutting

9-7. *Crosscutting narrow pieces.*

Cutting Short Boards

Place the gauge in the groove in the side toward the longest portion of the board. Fig. 9-7. Hold the stock firmly against the gauge and advance it slowly into the blade. Never drag the cut edge back across the blade.

Cutting Long Pieces

If the board is longer than 6′, have a helper support the other end.

Cutting Plywood

Because of its construction and often because of its size, plywood presents special cutting problems. Since grain directions of alternating plies are at right angles to each other, there is a tendency to split out the ends of cross-grain layers, no matter what the direction of the cut. The glue lines are also a problem in that they dull the blade. Finally, since plywood is glued up in large sheets, the workpiece is often too large to fit conveniently on the table of a circular saw.

To reduce these problems to a minimum, adjust the blade so it will barely clear the top of the plywood, and place the stock with the good side up. Fig. 9-8. Then use one of the three following methods to guide the stock:
- The miter gauge can be reversed in the groove when the cut is

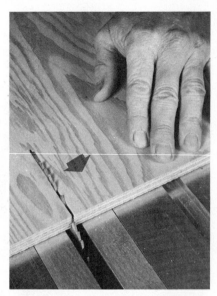

9-8. *Plywood should be cut with the good face up. Use a combination, crosscut, or plywood blade.*

operations. For added support of the workpiece some operators like to fasten permanently a long support board to the miter gauge. Always remove the ripping fence for crosscutting. Carefully mark a line on the edge of the stock nearest the blade. (You must be able to see the mark easily, so you can begin the cut accurately.) Be sure the miter gauge is set to cut the correct angle. It is also a good practice to use the stop rod as an aid to prevent the stock from moving while the cut is made.

started, to guide the stock for as long a cut as possible. Then the gauge can be removed and slipped into its regular position to complete the cut.
- Another suggestion for sawing plywood is to clamp a straight-edged board on the underside of the plywood. This will act as a guide against the edge of the table.
- The ripping fence can be used as a guide in cutting plywood to size.

QUESTIONS

1. Name five kinds of blades that can be used on a circular saw.

2. Name the two devices that are used to guide stock when cutting on a circular saw.

3. Tell how to remove a saw blade.

4. Explain two ways of supporting long stock for ripping.

5. In ripping narrow stock, what safety device should be used?

6. What is meant by resawing?

7. When cutting plywood, should the good side be up or down?

Portable Power Tools

10

Portable power tools are widely used in carpentry to increase productivity and improve quality. Portable power tools are operated either with electricity or air (pneumatic). In selecting these tools, make sure the words *industrial*, *commercial* or *builders* is included in the description to indicate a tool designed for heavy and continuous use. Fig. 10-1.

ELECTRICAL TOOLS

In selecting a tool, make sure that there is a symbol on each tool that indicates it meets the standards of an independent testing lab, such as the Underwriters' Laboratory. The tool should protect you from electrical shock either by one of two safety systems: (1) A tool with external grounding has a wire that runs from the housing through the power cord to a third prong on the power plug. When this prong is connected to a grounded three-hole electrical outlet, the grounding wire will carry any current that leaks past the electrical insulation of the tool. (2) A double insulated tool has an extra layer of electrical insulation which eliminates the need for a three-prong plug and grounded outlet.

The quantity of electric current a tool normally draws during operation is listed on the name plate. As a general rule, the larger the power tool, the greater the amperage rating. Heavy-duty power tools draw 12 to 13 amperes, and

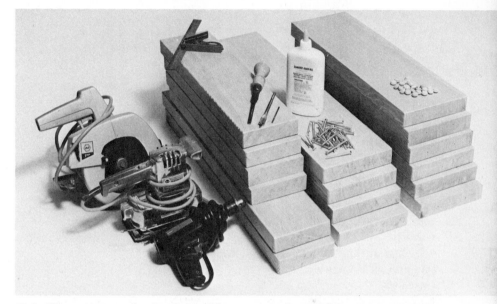

10-1. *Three common kinds of portable power tools used in carpentry include the electric drill, pad sander, and power saw.*

these should never be connected to an electrical circuit on which other tools are used. Always choose a recommended extension cord size based on the name plate amperage and the cord length.

Portable Electric Drill

The portable electric drill is used for drilling, boring, installing wood screws, and many other uses. Fig. 10-2. Most drills have a pistol grip handle. Some models include side handles. Fig. 10-3. The size of the drill is indicated by the largest diameter drill or bit it will hold, such as ¼″. Many drills have either two speeds or variable speeds used to drill some materials at lower speeds. The drill with both variable

speed and reverse is good for driving and removing screws.

The bits or other accessories are inserted in the part of the drill called a chuck. Fig. 10-4. The three-jaw gear type is most common. To install the drill or bit, turn the drill chuck by hand until the jaws are opened wide enough to take the desired size. Insert the bit shank in the chuck as far as possible and then close the jaws by hand. Tighten the chuck with the key.

OPERATING INSTRUCTIONS
• Disconnect drill from power source before removing or installing drill bit.
• Wear goggles or a face shield.

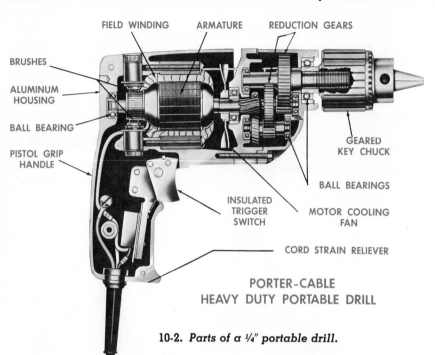

FIELD WINDING | ARMATURE | REDUCTION GEARS

BRUSHES

ALUMINUM HOUSING

BALL BEARING

PISTOL GRIP HANDLE

GEARED KEY CHUCK

BALL BEARINGS

INSULATED TRIGGER SWITCH

MOTOR COOLING FAN

CORD STRAIN RELIEVER

PORTER-CABLE
HEAVY DUTY PORTABLE DRILL

10-2. *Parts of a ¼″ portable drill.*

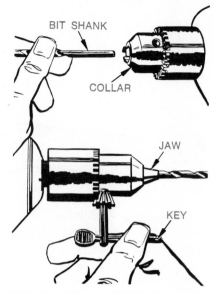

BIT SHANK

COLLAR

JAW

KEY

10-4. *Installing a drill in the chuck.*

• Before connecting to power source, make sure switch is in off position.

• Before starting drill, make certain drill bit is securely gripped in the chuck.

• Check to see that key has been removed from chuck before starting drill.

• Locate exact point of desired penetration and mark with center punch or awl.

• Drill with even, steady pressure and let the drill do the work.

• When work is completed, disconnect drill from power source and remove drill bit.

SIDE HANDLE

PISTOL GRIP

10-3. *Using a drill with a side handle.*

Portable Power Saw

A portable power saw, sometimes called a *cut-off*, *electric*, or *circular hand saw*, is widely used by the carpenter to do all types of cutting. Fig. 10-5. An excellent use of this saw is to cut material that has been attached to a building. For example, roof boards or plywood can be trimmed off after they have been installed.

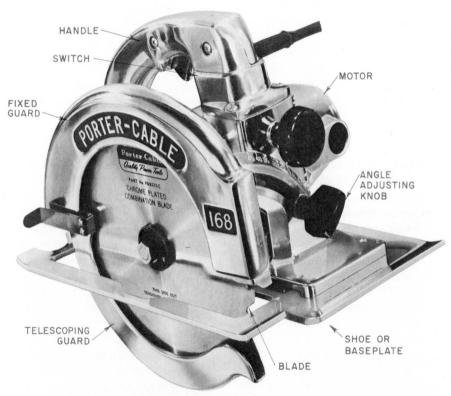

HANDLE

SWITCH

FIXED GUARD

MOTOR

ANGLE ADJUSTING KNOB

TELESCOPING GUARD

SHOE OR BASEPLATE

BLADE

10-5. *Parts of a portable power saw.*

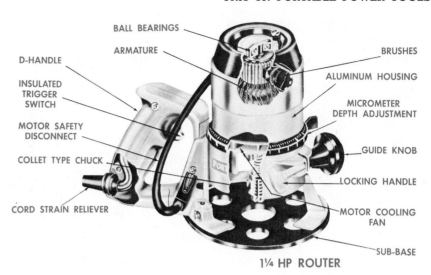

BALL BEARINGS
ARMATURE
BRUSHES
D-HANDLE
ALUMINUM HOUSING
INSULATED TRIGGER SWITCH
MICROMETER DEPTH ADJUSTMENT
MOTOR SAFETY DISCONNECT
GUIDE KNOB
COLLET TYPE CHUCK
LOCKING HANDLE
CORD STRAIN RELIEVER
MOTOR COOLING FAN
SUB-BASE

1¼ HP ROUTER

10-8. Portable router with parts named.

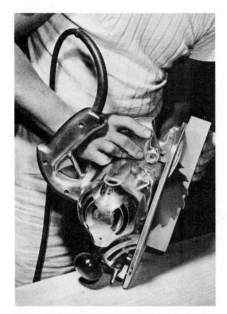

10-6. Adjusting the saw for the depth of cut.

10-7. Straight cutting. Notice that the guard covers nearly all of the blade that is not in contact with the workpiece.

The work capacity of the portable power saw is determined by the depth of vertical cut that can be made. Fig. 10-6. Most saws have a slip clutch or special washers where the blade fastens to the drive shaft. The slip clutch and washers prevent motor burn-out if the saw sticks and also reduce the likelihood of kickback and loss of control.

The blade guard includes a stationary upper guard that covers the front, top, and back of the saw blade. The lower guard covers the blade bottom when the saw is not in use. As you push a running saw into the workpiece, the lower guard moves backward and upward into or outside the upper guard. Fig. 10-7.

The shoe consists of one or more skids that rest on the workpiece and hold the saw upright when it is operating. The more substantial the shoe is, the more stable the saw. A rip guard is a scale side extension to the shoe that guides the saw blade parallel to the edge of the work. The depth adjustment moves the shoe up and down. The tilt angle adjusts the tilt of the shoe as much as 45 degrees. The saw has a pistol grip for the right hand and, in some models, a front auxiliary handle for the left.

An adaptation of this tool is called a *power miter box*. It is a power saw mounted in a miter box and is an important labor-saving device for the carpenter. It is useful in cutting miters for such operations as fitting and installing molding.

OPERATING INSTRUCTIONS
- Wear goggles or a face shield.
- With saw disconnected from power source, make sure blade is the proper type for the work to be done and that the arbor nut is tight.
- Check to see that the telescoping blade guard is functioning properly before connecting saw to power source.
- Make certain that work to be cut is firmly supported and free of any obstructions.
- Bring the saw blade up to the desired point of cut, back up slightly and start the saw motor. When full speed is reached, advance the saw through the work.
- When through the cut, release the switch, apply brake or wait until blade stops before placing on bench.
- When work is completed, disconnect saw from power source.

Portable Router
A portable router is used for shaping the surfaces and edges of stock. Fig. 10-8. While it is considered primarily a cabinetmaking tool, it has wide uses in carpentry. For example, when paneling is installed around a door or window, a small amount of the material might be left. The router is then used for accurately trimming the paneling to the frame of the door or window. It is also used for trimming

10-9. *Using a saber saw to cut an opening for an electrical outlet.*

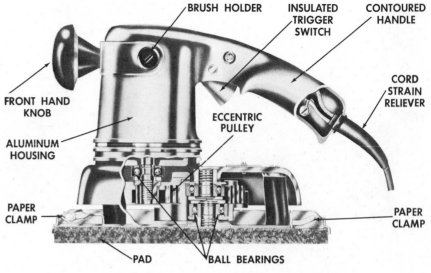

10-11. *Parts of a half-pad finishing sander.*

plastic laminate, hinged butt routing, and other uses.

OPERATING INSTRUCTIONS

- Wear goggles or face shield.
- **CAUTION!** While inserting router bits, making adjustments and when router is not in actual use, *disconnect from power source.*
- Select proper router bit for work to be done, insert shank in collet chuck and tighten collet nut.
- Be sure switch is *off* before inserting plug into power source.
- Make certain that workpiece is rigidly held in proper position and free of obstructions.

- Make a trial cut on a piece of scrap lumber.
- Keep cutting pressure constant; do not overload router.
- When work is completed, disconnect router from power source and remove router bit.

Other Portable Power Tools

The saber saw is used for on-the-job cutting of irregular shapes and openings. Fig. 10-9. Portable belt and pad sanders are used for

sanding cabinet work and trim. Figs. 10-10 and 10-11. The portable electric plane is used instead of a hand plane for trimming the edge of stock, fitting doors and cabinets and many other uses. Fig. 10-12. The power miter box is a portable power saw mounted in a miter box. It is used for cutting miters and for fitting and installing moldings. Fig. 10-13.

PNEUMATIC (AIR POWERED) TOOLS

Pneumatic means that the device operates with air pressure. There are two major types of air-powered tools, namely, *twisting* or *rotating tools* such as drills, wrenches, sanders and grinders; and *percussion* (striking) *tools* such as nailers and staplers. The most commonly used air-operated tools in carpentry are the nailer and stapler. Fig. 10-14. A portable air compressor is needed to operate these tools on the job.

Operating Instructions

- Always wear safety glasses and a hard hat when operating air-powered tools.
- Make sure that all the pipe fit-

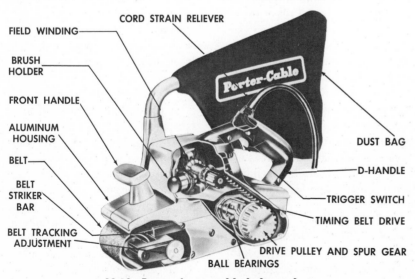

10-10. *Parts of a portable belt sander.*

10-12. *A portable electric plane.*

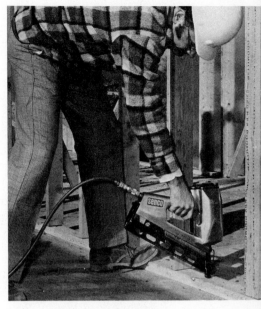

10-14. *Using a nailer to fasten wall framing to flooring.*

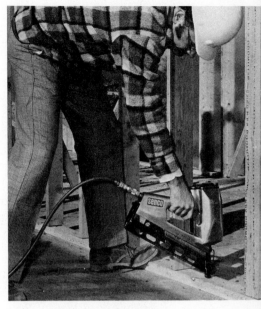

10-13. *A power miter box.*

tings are tight and secure before opening the valve in the air line. Use the recommended air pressure from 90 to 100 psi.

• Make sure that the compressed air is free from dust and excessive moisture.

• When using on-the-job nailers and staplers, keep the portable air compressor as close to the work as possible.

• Air hoses should not be too long. They should be placed so that they will not be a hazard to workers moving around the job. Hoses laid across walkways or curled underfoot can be tripped over. When hoses must be left lying on the ground or on the floor of the building, a board should be laid along each side of the hose.

• Make sure all the safety standards established by the manufacturer are followed.

• Use the right size and kind of nail or staple for the particular operation. Manufacturers of nailers produce various kinds of staples and nails with recommended uses. Different types and sizes of nails and staples are designed for different kinds of jobs. For example, nails up to and including 16d may be used for framing, while a

smaller size might be satisfactory for applying some types of roofing materials.

Using Nailers and Staplers

• Get acquainted with the tool you are using to determine how it is loaded.

• Use the correct size and kind of nail or staple for the job.

• Make sure the tool is properly connected to the air hose and to the air compressor.

• Hold the tool firmly on the surface before pulling the switch.

QUESTIONS

1. Describe the two methods of protecting the operator when using portable electrical power tools.

2. List three safety rules that should be observed when using the portable electric drill.

3. How far should the drill bit be inserted in the chuck?

4. List three safety rules for the portable power saw.

5. What is the portable power saw normally used for in carpentry?

6. What is the principal use for the router in carpentry?

7. Name three other portable power electric tools.

8. What does the term "pneumatic" mean?

9. Name two types of pneumatic tools.

SECTION III

Foundations

Footings and Foundations 11

Before the carpenter can frame up a house, the lot must be surveyed, the land cleared, the top soil pushed to one side, the house staked out, the basement and/or footings dug, and the foundation completed. While it is not the job of carpenters to install the footings and foundations (except for all-weather wood foundations), they must know how it is done so that the building is started on a solid base. Carpenters do, however, construct the forms for most poured concrete. Fig. 11-1. A masonry basement with poured concrete footings and concrete block walls is the most common foundation for a wood house.

LAYING OUT THE FOUNDATION FOR A MASONRY BASEMENT

Laying out a foundation is the important beginning in house construction. It is a simple but extremely important process. If the foundation is square and level, you will find all later jobs easier, from rough to finish construction.

1. Be sure your house location on the lot complies with local regulations. If property lines are in question, verify location of lot-line corners by city, county, or private surveyor. Once property lines are established, it is vital that you review city, county, or state requirements on locating the house with respect to property lines. Most regulations require that a house be set back from the street a specified

distance (often 25' to 30') and that sides of the house be set back from adjoining property lines (often 5' to 10').

2. Decide on house location, based on required setbacks and other factors such as the natural drainage pattern of the lot; then level or at least rough-clear the site.

3. Lay out the outside foundation lines. Fig. 11-2. Locate each outside corner of the house and drive small stakes into the ground. Drive tacks into tops of these stakes to indicate the outside line of the foundation wall (not footings). Check squareness of house by measuring the diagonals, corner to corner, to see that they are equal. You can check squareness of any corner by measuring along one side for 6', along the other for 8'.

The length of the diagonal line across these two end points should measure exactly 10'. Fig. 11-3.

4. After corners are located and squared, drive three 2" × 4" stakes at each corner. Fig. 11-4. Locate these stakes 3' to 4' outside the actual foundation line. Nail 1" × 6" batter boards horizontally so that their top edges are all level and at the same grade. Hold a string-line across tops of opposite batter boards at two corners and, with a plumb bob, adjust so that it is exactly over the tacks in the two corner stakes. Cut saw kerfs the same depth since the string-line not only shows the outside edges of the foundation but will provide a reference line to ensure uniform depth of footing excavation. Making similar cuts in all eight batter boards and stringing the four lines in posi-

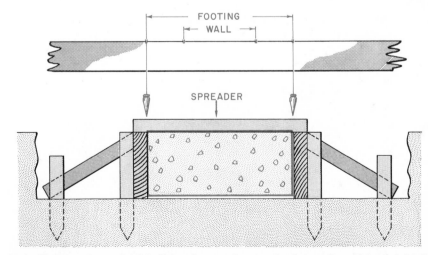

11-1. *Footing form detail. Note the notches in the batter board from which the plumb bobs are hung, to designate the footing width and wall thickness.*

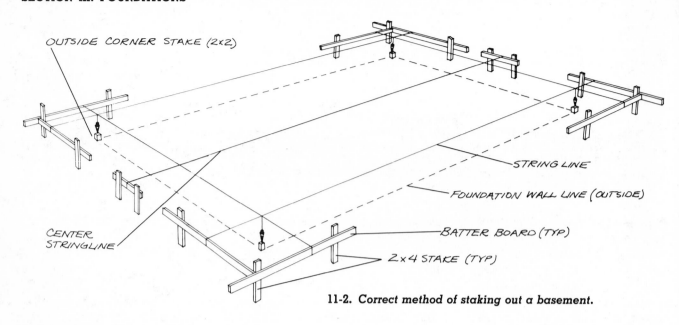

OUTSIDE CORNER STAKE (2x2)

STRING LINE

FOUNDATION WALL LINE (OUTSIDE)

CENTER STRINGLINE

BATTER BOARD (TYP)

2 x 4 STAKE (TYP)

11-2. *Correct method of staking out a basement.*

10'

6'

90°

8'

CORNER TACK

OUTSIDE CORNER STAKE (2x2)

11-3. *Use this technique to make sure the corner is square.*

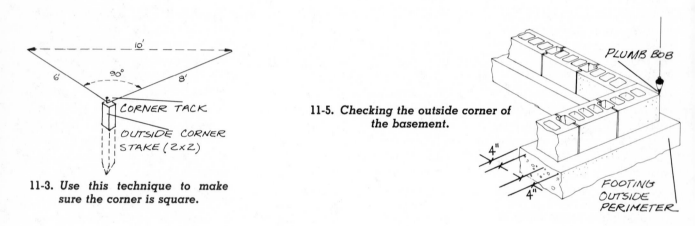

11-5. **Checking the outside corner of the basement.**

PLUMB BOB

4"

4"

FOOTING OUTSIDE PERIMETER

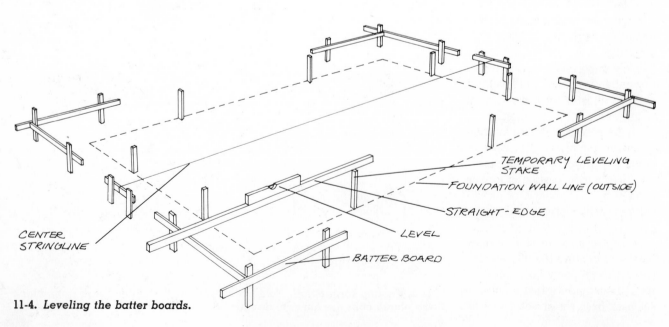

TEMPORARY LEVELING STAKE

FOUNDATION WALL LINE (OUTSIDE)

STRAIGHT-EDGE

LEVEL

BATTER BOARD

CENTER STRINGLINE

11-4. *Leveling the batter boards.*

tion establishes the lines for the outside of the foundation. Foundation lines are then accurately established.

5. Locate the lengthwise girder location, usually on the centerline of the house. Double check the house plans for exact position since sometimes a girder will be slightly off centerline to support an interior bearing wall. To find the line, measure the correct distance from corners. Then install batter boards and locate string-line.

6. Check for foundation levelness. The top of the foundation must be level around the entire outside of the house. The most accurate way to check this is to use a surveyor's level.

The next best approach is to make sure that batter boards and the string-lines are all absolutely level. Do this with a piece of straight lumber 10' to 14' long (judge its straightness by sighting along the surface). Using this straightedge together with a carpenter or masonry level, drive temporary stakes around the outside of the house. These stakes should be spaced a distance apart not exceeding the length of the straightedge lumber. Fig. 11-4. Place one end of the straightedge on a batter board, check for exact levelness, and drive another stake to the same height. Each time a stake is driven, the straightedge

and level should be reversed end-for-end. This will ensure accuracy in establishing the height of each stake with reference to the batter board. Make final check on overall levelness. Level the last stake with the first batter board where you started. If the straightedge is level here, then you have a level foundation base line.

During foundation excavation, the corner stakes and temporary leveling stakes will be removed.

INSTALLING FOUNDATION AND POST FOOTING FOR CONCRETE BLOCK FOUNDATIONS AND POST FOOTINGS

When the foundation layout is completed, study the house drawing details and construction notes.

Poured concrete footings are most commonly used. Footings, properly sized and constructed, prevent settling or cracking of walls. Footings must be completely level and must extend at least 12" below the frost line and at least 6" into undisturbed soil. These requirements dictate the depth at which the footings are placed. Do not place foundations on black top soil.

A row of post footings will be located along the centerline of the house. These support posts and girders, in turn, support the floor joists or trusses. Minimum height for post footings should be 8" above finish grade in crawl-space foundations.

Footings

1. Preliminary excavation may vary, depending on soil conditions. To allow adequate working space,

excavate about 2' beyond the outside of the wall.

2. Using a plumb bob from the foundation, lay out a string-line and locate the outside corners of the block foundation. Measure about 4" beyond corner points to establish footing edge lines. Fig. 11-5. Outline both outside and inside perimeters of the footing with string. Dig the footing trench to the required depth. Install forms of ½" waterproof plywood or 2 × 8s supported by 2" × 4" stakes driven into the soil about 2' apart. Fig. 11-6. These stakes should be driven below the top of the form-boards to help in leveling the concrete.

3. Using the lengthwise centerline string, mark locations for floor-supporting posts. Post spacing is specified on the house plan, as is post footing size. Post footings are generally about 20" square, but may vary in size depending on allowable soil pressure and post spacing. Some require a steel rod extending out the top of the post footing to secure the post and keep it in place. The bottom of the footing should be at the same level as the perimeter footing. Height should be a minimum of 8" above ground in a crawl space house. Build forms of ½" exterior plywood and 2 × 4s. End-post footings may be poured at the same time as perimeter footings.

4. Order and place concrete with at least 2000 psi, 28-day strength. Place concrete in forms in thin layers. Spade and tamp concrete carefully between pours to prevent air pockets. The top of the footing must be smooth and level all around.

5. Cure concrete and strip forms. In warm weather, leave forms in place for three days, sprinkling daily with water so concrete will not dry too quickly. In cool weather, leave forms in place for 7 days.

2'

2"x 8" OR PLYWOOD FORMBOARD

4" 8" 4"
TYP

2x4 STAKE

11-6. *Here you see the forms with the concrete poured and the first two layers of concrete block installed.*

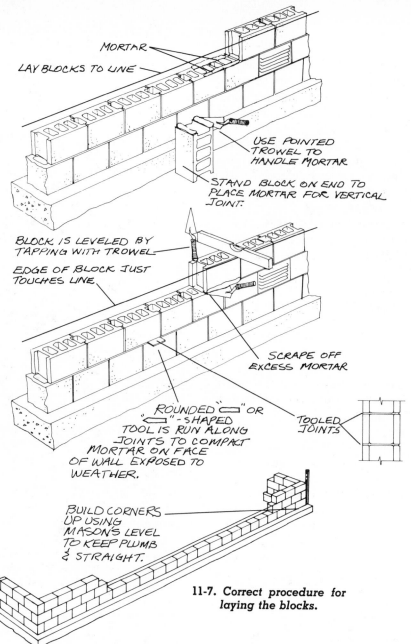

MORTAR

LAY BLOCKS TO LINE

USE POINTED TROWEL TO HANDLE MORTAR

STAND BLOCK ON END TO PLACE MORTAR FOR VERTICAL JOINT.

BLOCK IS LEVELED BY TAPPING WITH TROWEL

EDGE OF BLOCK JUST TOUCHES LINE

SCRAPE OFF EXCESS MORTAR

ROUNDED "⌐" OR "⌐"-SHAPED TOOL IS RUN ALONG JOINTS TO COMPACT MORTAR ON FACE OF WALL EXPOSED TO WEATHER.

TOOLED JOINTS

BUILD CORNERS UP USING MASON'S LEVEL TO KEEP PLUMB & STRAIGHT.

11-7. Correct procedure for laying the blocks.

lowable, provided the average for the entire course is no more than ½"). Mark each joint on the foundation. Check the house plan for any required openings for crawl space vents, drains, utilities, etc.

7. Prepare the mortar. Mix 2 parts masonry cement (or 1 part each of Portland cement and hydrated lime) with 4 to 6 parts of damp mortar sand. Add just enough water to make a plastic mortar that clings to the trowel and block but is not so soft that it squeezes down too much when laying block. After mixing, place mortar on a wet mortar board near where blocks will be laid.

8. Lay blocks as shown in Fig. 11-7. The finished height of the foundation wall should be approximately 12" above finish grade level. Build each corner up to full height to establish required thickness of joints. Use corner blocks with one flat end at the corner. Build corners up using a mason's level to keep blocks plumb and level. Then stretch the line between corners to guide laying additional blocks. For the first course of blocks, place mortar on footing for the full width of the block. For succeeding courses, apply mortar on face shells only.

Concrete Block Foundation Wall

6. Accurately locate outside wall corners on footings, using a plumb bob. (Block walls should be centered on the footing.) String a cord tightly between the corners to outline the outside of the block wall. Mark with chalk, or use a chalk line. Lay the first course of blocks without mortar all around the perimeter to determine joint spacing and whether you will have to cut any blocks. Space blocks ⅜" apart (a ¾" maximum joint is al-

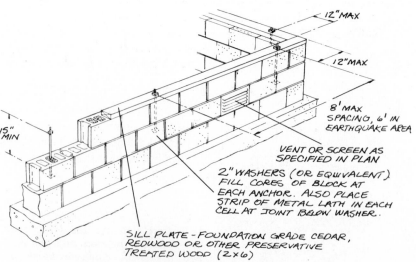

12" MAX

12" MAX

8' MAX SPACING, 6' IN EARTHQUAKE AREA

VENT OR SCREEN AS SPECIFIED IN PLAN

2" WASHERS (OR EQUIVALENT) FILL CORES OF BLOCK AT EACH ANCHOR. ALSO PLACE STRIP OF METAL LATH IN EACH CELL AT JOINT BELOW WASHER.

15" MIN

SILL PLATE—FOUNDATION GRADE CEDAR, REDWOOD OR OTHER PRESERVATIVE TREATED WOOD (2x6)

11-8. Installing the sill plate.

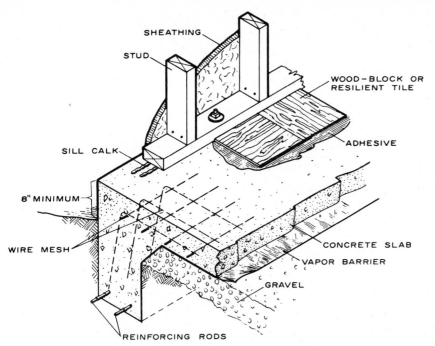

11-9. *A concrete floor and foundation combination (thickened-edge slab).*

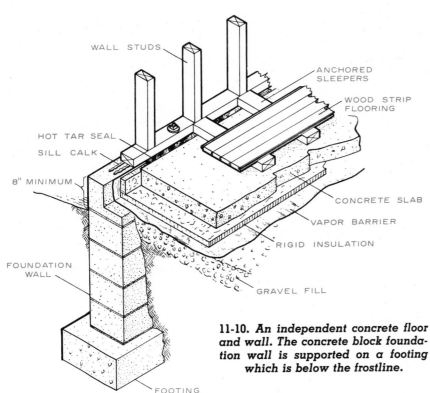

11-10. *An independent concrete floor and wall. The concrete block foundation wall is supported on a footing which is below the frostline.*

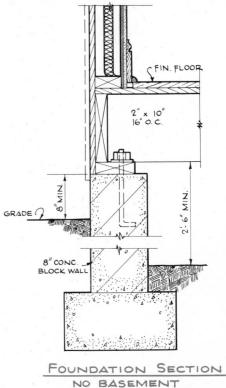

FOUNDATION SECTION
NO BASEMENT

11-11. *Footing for a house with a crawl space. The footings are of poured concrete and the walls are of concrete block.*

9. Set anchor bolts for the sill plate. Before laying the last two courses, locate and position anchor bolts as shown on plans. Fig. 11-8. Be sure to provide at least two bolts per individual sill plate. Fill all cells in the top course, or cover with 4″ solid masonry units, or use a wood sill plate wide enough to bear on both the inner and outer shells of the blocks. Install fiberglass sill or metal termite shield between the foundation and wood sill plate.

10. After the block wall is completed, wait at least 7 days. Do not backfill until at least floor sheathing is installed. If a drain is provided, slope soil in the crawl space toward the drain.

OTHER METHODS OF BUILDING A FOUNDATION

Houses without basements can be built on:
● Combined slab floor and foundation. Fig. 11-9.
● Independent concrete floor and foundation wall. Fig. 11-10.
● Footings (concrete) with poured concrete foundation wall and with crawl space.
● Footings (concrete) with concrete block wall and with crawl space. Fig. 11-11.
● Footings (concrete) with wood posts or piers. Fig. 11-12.

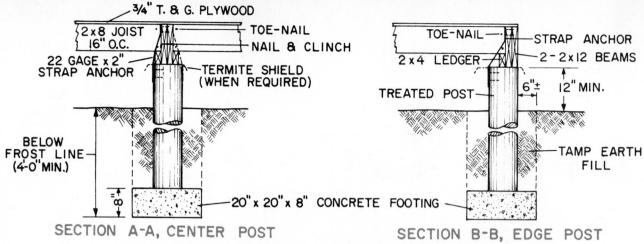

SECTION A-A, CENTER POST

3/4" T. & G. PLYWOOD
2x8 JOIST 16" O.C.
22 GAGE x 2" STRAP ANCHOR
TOE-NAIL
NAIL & CLINCH
TERMITE SHIELD (WHEN REQUIRED)
BELOW FROST LINE (4'-0" MIN.)
8"
20" x 20" x 8" CONCRETE FOOTING

SECTION B-B, EDGE POST

TOE-NAIL
2x4 LEDGER
TREATED POST
STRAP ANCHOR
2-2x12 BEAMS
6"±
12" MIN.
TAMP EARTH FILL

11-12. Foundation details for concrete footings and treated posts.

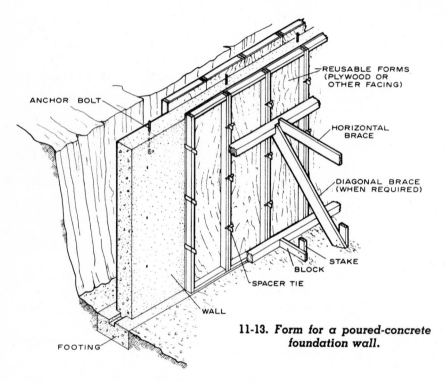

ANCHOR BOLT
REUSABLE FORMS (PLYWOOD OR OTHER FACING)
HORIZONTAL BRACE
DIAGONAL BRACE (WHEN REQUIRED)
STAKE
BLOCK
SPACER TIE
WALL
FOOTING

11-13. Form for a poured-concrete foundation wall.

• Footings (gravel, coarse stone or crushed stone) with pressure-treated, all-weather wood foundation panels. All-weather wood foundations are built using chemically preserved lumber and plywood. The system consists of a set of pre-fabricated pressure-treated plywood sheath stud walls set below grade. The typical panel is made from either 2″ × 4″ or 2″ × 6″ studs placed 16″ on center and sheathed with plywood. Panels can be any width and as high as 8′ when measured from the footing plate to the foundation sill. All fasteners must be silicon-bronze, copper, or heavy dipped zinc-coated steel. The framing anchors, if used, must also be zinc-coated steel. The footing plate is set in on one end and extends out at the other so that each adjoining panel will overlap. This design permits an interlocking of panels at the corners. These panels can be placed on footings of gravel, coarse stone, or crushed stone.

Houses with basements can be built on:
• Footings (concrete) with concrete block walls assembled with exterior bonding material of a cement-fiberglass mix.
• Footings (concrete) with poured concrete walls. Fig. 11-13.

QUESTIONS

1. Describe the most common way of constructing a foundation.
2. Can a foundation be checked for levelness without a surveyor's level? Explain.
3. Describe three methods of providing a foundation for houses without basements.

4. Name three ways of providing a foundation for houses with basements, using concrete materials.
5. Which type of foundation does the carpenter have to construct?

SECTION IV

Framing

Framing Methods

12

Most homes in the United States and Canada are of wood-frame construction. Many are covered with wood siding; other common coverings include wood shingles, composition shingles or siding, brick veneer, and stucco.

Wood-frame houses have several important advantages. In general, frame construction costs less than other types. It provides more house for a given price and better

insulation, thereby increasing comfort to the occupants and reducing heating and air-conditioning costs.

Wood is easily worked and is suitable for use with a wide variety of exteriors. This flexibility allows architects and builders to produce nearly any architectural style.

A well-built wood-frame home is very durable. Some of the oldest existing buildings in North America are Paul Revere's house in Bos-

ton, built in 1677, and the "House of Seven Gables" in Salem, Mass., constructed in 1668. Fig. 12-1.

TYPES OF WOOD FRAMING

Buildings framed of lumber usually belong in one of two main classes:

- *Multiple-member assemblies,* often called *conventional framing.*
- *Plank-and-beam framing,* which consists of heavier members, more widely spaced.

Each of these framing methods has its advantages, which will be discussed.

Conventional Framing

Conventional framing consists of multiple small members (joists, studs, and rafters) so joined that they act together and share the loads in supporting the structure. When assembled and sheathed, these members form complete floor, wall, and roof surfaces. Two common types of conventional framing are *platform* and *balloon.*

PLATFORM-FRAME CONSTRUCTION

As the name indicates, in this type of construction the floors are complete platforms, independent of the walls. The subfloor extends to the outside edges of the building and provides a platform upon which exterior walls and interior partitions are erected. Platform construction is generally used for one-story houses. It is also used

12-1. *Paul Revere House. This is the oldest home in Boston, built around 1677. Revere left from here on his historic ride to Lexington. The house is still open to the public as a museum.*

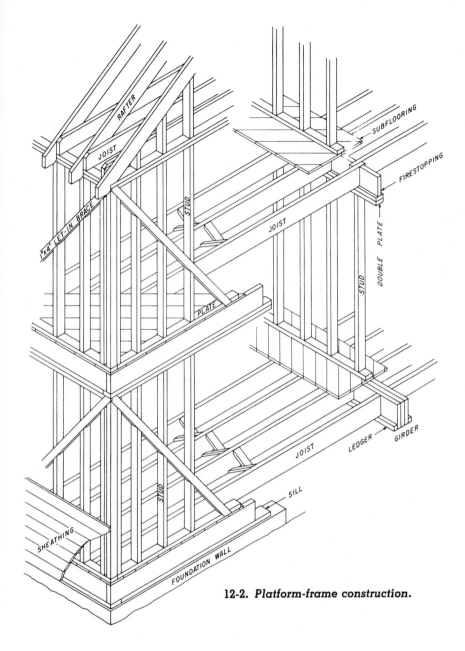

12-2. *Platform-frame construction.*

type of construction is that the studs are continuous from sill to top plate. Fig. 12-3. Studs and first-floor joists rest on the anchored sill. Second-floor joists bear on $1'' \times 4''$ ribbon strips which have been let into the inside edges of the studs.

In this type of construction there is less cross-grain wood framing than in conventional construction. Wood expands and contracts across the grain but is relatively stable with the grain. Therefore balloon-frame construction is less likely to be affected by expansion and contraction than conventional construction. This is an advantage for certain types of buildings. Specifically, balloon framing is good for two-story buildings on which the exterior covering is of brick veneer, stone veneer, or stucco. With such buildings, movement of the wood framing under the masonry veneer can be a serious problem.

If exterior walls are of solid masonry, it is also desirable to use balloon framing for interior bearing partitions. Again, this is because there is relatively little cross-grain wood framing. This minimizes dimensional changes in the walls and reduces variations in settlement which may occur between exterior walls and interior supports.

In balloon framing, blocks are placed between the joists to serve the dual purpose of solid bridging and fire stopping. Solid bridging holds the joists' ends in line; fire stopping prevents the vertical and horizontal spaces from acting as flues in the event of fire. Fig. 12-4.

Balloon-frame construction is rarely used today. The longer framing members required for this type of construction are not readily available, and they cost more than the materials used in platform-frame construction. Therefore balloon framing construction techniques will not be discussed in detail in this book.

alone or in combination with balloon construction for two-story structures. Building techniques in most parts of the United States have been developed almost entirely around the platform system. Fig. 12-2. This book will therefore concentrate on platform-frame construction.

Compared with balloon framing, platform construction is easier to erect because at each floor level it provides a flat surface on which to work. It is also easily adapted to various methods of prefabrication. Each level of a two-story house is constructed separately. With a platform-framing system, it is common practice to assemble the wall framing on the floor and then tilt the entire unit into place.

BALLOON-FRAME CONSTRUCTION

The feature which identifies this

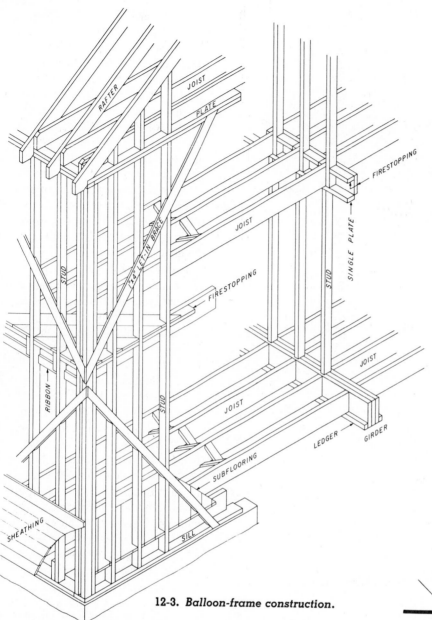

12-3. *Balloon-frame construction.*

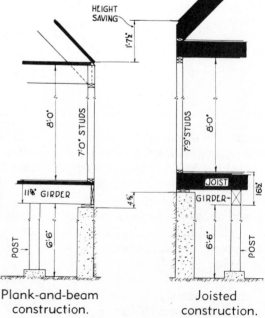

12-4. *First-floor framing at exterior wall—balloon-frame construction.*

framing and its covering also provide lateral bracing for the building.

Conventional framing uses joists, rafters, and studs spaced 12″ to 24″ on centers; the plank-and-beam method calls for fewer and larger-sized pieces, spaced farther apart.

There are many advantages to be gained through the use of the plank-and-beam system of framing. One of the best points is the architectural effect provided by the ex-

Plank-and-Beam Construction

In the plank-and-beam method of framing, plank subfloors or roofs, usually a 2″ nominal thickness, are supported on beams spaced up to 8′ apart. The ends of the beams are supported on posts or piers. Wall spaces between posts are provided with supplementary framing as needed for attachment of exterior and interior finish. This extra

12-5a. *Comparison of height of plank-and-beam house with conventionally framed house.*

Plank-and-beam construction.

Joisted construction.

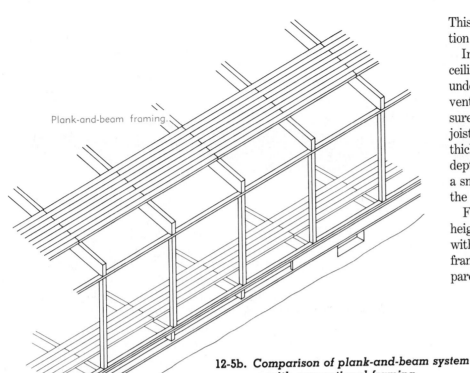

Plank-and-beam framing.

This results in a substantial reduction of labor on the job site.

In plank-and-beam framing, the ceiling height is measured to the underside of the plank, but in conventional construction it is measured to the underside of the joists. The difference between the thickness of the plank and the depth of the joist gives the building a smaller volume and also reduces the height of the interior walls.

Figure 12-5a compares the height of a plank-and-beam house with that of a conventionally framed house. Figure 12-5b compares the two framing methods.

12-5b. *Comparison of plank-and-beam system with conventional framing.*

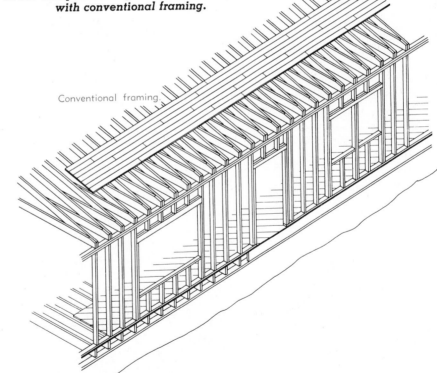

Conventional framing.

posed plank-and-beam ceiling. In this type of construction, the roof plank serves as the ceiling, which provides added height to the living area at no additional cost. Generally, the planks are selected for appearance; therefore no further ceiling treatment is required except the application of a finish.

In a well-planned plank-and-beam framed structure, there are important savings in labor. As mentioned, the pieces are larger, and there are fewer of them than in conventional framing. The cross-bridging of joists is eliminated, and larger but fewer nails are required.

QUESTIONS

1. List several advantages of wood-framed houses.

2. How does platform-frame construction differ from balloon construction?

3. How does plank-and-beam construction differ from platform-frame construction?

Floor Framing

13

Floor framing consists of posts, girders, sills, joists or floor trusses, and subflooring. Fig. 13-1. All members are tied together to support the load on the floor and to give support to the exterior walls.

POSTS

A post is a wooden or steel member which supports girders. A wood post must be solid and not less than 6″ × 6″ in size. It should rest on the top of a masonry pedes-

tal that is at least 3″ above the floor. Fig. 13-2a. Steel posts may be H-Sectional, I-Sectional, or round. They have steel bearing plates at each end. Fig. 13-2b.

Wood posts should be square at both ends and securely fastened to the girder. Fig. 13-3. When necessary, a bearing plate should be placed between the post and the girder. Fig. 13-4. Posts are generally spaced 8′ to 10′ on center depending on the size and strength of the girder in relation to the load it must support.

GIRDERS

Girders are large principal beams used to support the floor joists. They may be of wood or steel.

For steel girders, I-beams are commonly used. An advantage of

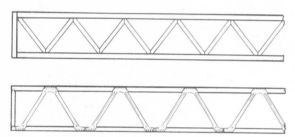

13-1a. Floor trusses of various designs are often used instead of solid wood joists.

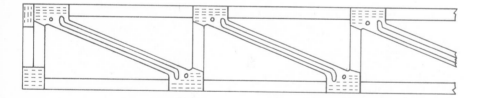

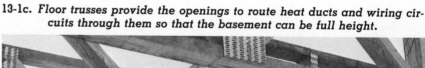

13-1c. Floor trusses provide the openings to route heat ducts and wiring circuits through them so that the basement can be full height.

13-1b. Metal hangers that fit over the sill are often used with floor trusses.

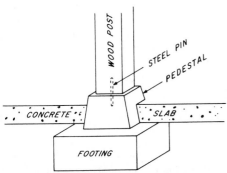

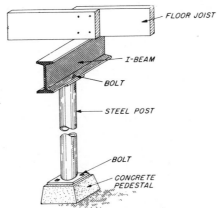

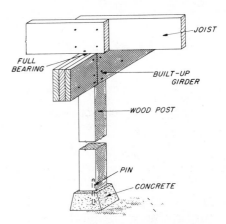

13-2a. *The foundation and pedestal for a basement or cellar post. If the post is wood, a steel pin must be set into the pedestal to secure the post.*

13-2b. *A steel post used with a steel I-beam. A wood beam may also be used with a steel post. Notice the flanges, welded onto each end of the post, for bolting the post into position.*

13-3a. *A wood post under a built-up girder. Notice that the joint in the girder is over the support.*

POST TO BEAM CONNECTION METHODS

POST TO FOOTING CONNECTION METHODS

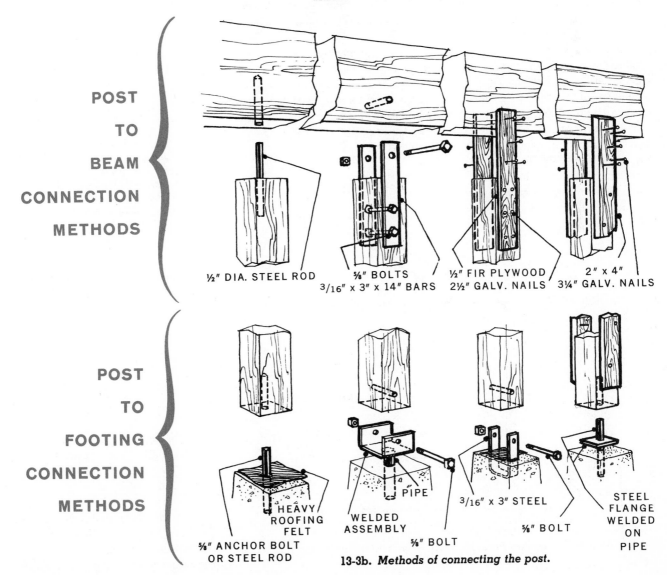

13-3b. *Methods of connecting the post.*

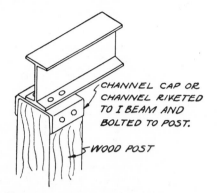

13-4. *If a wood post is used with a steel girder, a cap should be provided for bolting the post in position.*

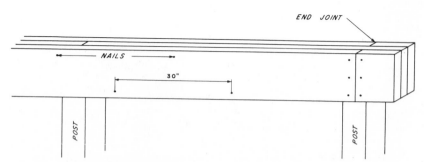

13-5a. *A built-up wood beam. The nails in the nailing pattern are 20d.*

steel is that it does not present the problem of shrinkage that wood does.

Common types of wood girders include *solid,* *built-up,* and *hollow.* Built-up wood girders consist of planks nailed together with two rows of 20d nails. The nails in each row should be spaced about 30″ apart with the end joints over the supports. Fig. 13-5. Glued laminated members are sometimes used. Hollow beams resemble a box made of 2 × 4s, with plywood webs. Fig. 13-6.

The ends of wood girders should bear at least 4″ on masonry walls or pilasters. A ½″ air space should be provided at each end and at each side of a wood girder framed in masonry. (When steel bearing plates are placed at ends of girders between the masonry and the girder, they should be of full bearing size.) Fig. 13-7. Tops of wood girders should be level with tops of sill plates on foundation walls unless there are ledger strips or notched joists. Fig. 13-8.

Size of Girder	Bd. Ft. per Lin. Ft.	Nails per 1000 Bd. Ft.
4 × 6	2.15	53
4 × 8	2.85	40
4 × 10	3.58	32
4 × 12	4.28	26
6 × 6	3.21	43
6 × 8	4.28	32
6 × 10	5.35	26
6 × 12	6.42	22
8 × 8	5.71	30
8 × 10	7.13	24
8 × 12	8.56	20

13-5b. *This table can be used for determining the amounts of lumber and nails necessary for a built-up girder. A 4′ × 6′ girder 20′ long contains 43 board feet of lumber (20 × 2.15 = 43).*

13-6. *Parts of a box girder. Properly built, such girders are much lighter for the load they will carry.*

PLYWOOD WEB

PLYWOOD JOINT

STIFFENER AT PLYWOOD JOINT

LOWER FLANGE

PLYWOOD WEB

UPPER FLANGE

END STIFFENER

Installing a Girder or Beam

If the beam or girder is of metal, it may be installed in one of three locations:

• The top of the beam may be flush with the plate to act as a bearing for the joist. Fig. 13-9.

• A wood plate may be placed on top of the metal beam to carry the joists. Fig. 13-10.

• When the top of the beam is set slightly higher than the wall sill plate, the joist is framed into the beam. Fig. 13-11.

Whether the beam is wood or metal, make sure that it is aligned from end to end and side to side. Also make sure that the length of the bearing post under the girder

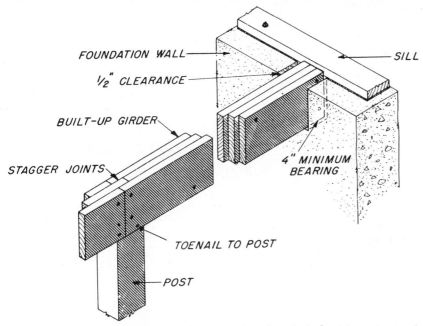

13-7. *A built-up girder set into a masonry pocket. It is best to put a metal bearing plate under the girder.*

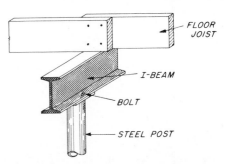

13-9. *This method of running the floor joists over the girder usually requires 2″ × 4″ spacing blocks between the joists to prevent movement on the beam.*

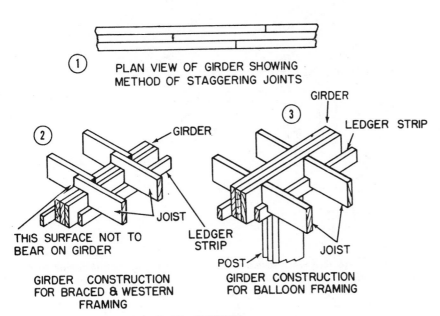

① PLAN VIEW OF GIRDER SHOWING METHOD OF STAGGERING JOINTS

GIRDER CONSTRUCTION FOR BRACED & WESTERN FRAMING

GIRDER CONSTRUCTION FOR BALLOON FRAMING

BUILT UP GIRDERS

13-8. *When the top of the girder is not set even with the top of the sill plate, it is necessary to install a ledger strip and notch the joists at the girder.*

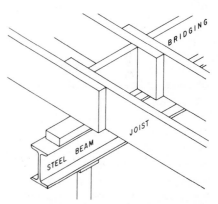

13-10. *This is the method used most frequently to install framing on a steel beam. Fasten a wood sill plate to the beam so that the joists can be nailed to the plate. Note the solid bridging which holds the joist ends vertical.*

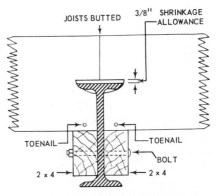

13-11. *A method of notching the joist into a steel girder.*

is correct so that it will properly support the girder.

Installing Joists Over Girders

In the simplest floor and joist framing, the joist rests on top of the girder. In Fig. 13-12, the top of the girder is aligned with the top of the sill plate. This method is used where basement height provides adequate headroom below the girder. The main disadvantage is that shrinkage is greater than

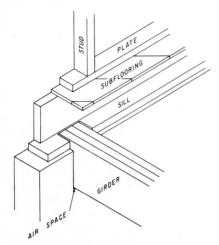

13-12. *First-floor framing at the girder and exterior wall in platform-frame construction. Note that the girder is flush with the top of the sill.*

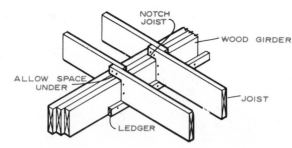

13-13a. *More headroom in the basement is gained by notching the joists and using a ledger strip for support on the girder.*

13-13b. *When using the method shown in Fig. 13-13a, do not notch the joist more than one-third of its depth.*

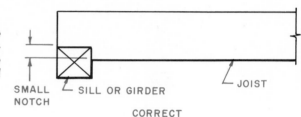

when ledger strips are used. If more clearance is wanted under the girder, ledger strips are securely nailed to each side of the girder to support the joist. The joists are toenailed to the wood girders and nailed to each other where they lap over the girder. Care should be taken to obtain full bearing on the tops of ledger strips. Fig. 13-13.

SPACED GIRDERS

To provide space for heat ducts in a partition supported on the girder, a spaced girder is sometimes installed. Solid blocking is used at intervals between the two members. Fig. 13-14. A single post support for a spaced girder usually requires a brace, preferably of metal, with a span sufficient to support the two members.

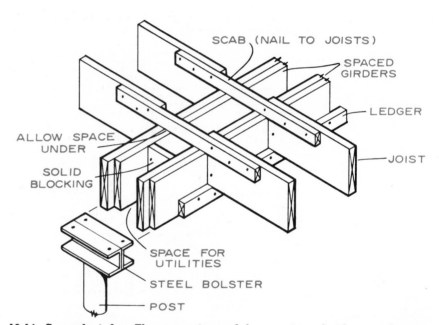

13-14. *Spaced girder. The space is used for running plumbing or heating pipes. A piece of wood called a scab is used to span the opening above the girder.*

WOOD-SILL CONSTRUCTION

There are two types of wood-sill construction over foundation walls—one for platform construction and one for balloon-frame construction.

The *box sill* is usually used in platform construction. It consists of

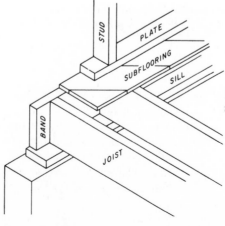

13-15. *First-floor framing at the exterior wall in platform-frame construction.*

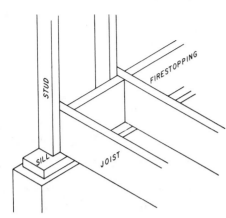

13-16. *First-floor framing at the exterior wall in balloon-frame construction.*

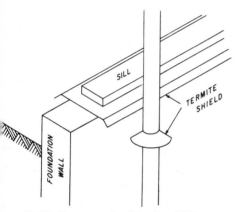

13-17. *Termite shields should be not less than 26-gauge galvanized iron, aluminum, or copper. They should be installed on top of all foundation walls and piers, and around pipes. The outer edges should be bent down slightly.*

a sill or sill plate anchored to the foundation wall for supporting and fastening the joists, with a header (band) at the ends of the joists resting on the foundation wall. Fig. 13-15.

Balloon-frame construction also has a sill plate upon which the joist rests. The studs also bear on this plate and are nailed both to the floor joist and to the plate. Fig. 13-16.

The sill or sill plate is the lowest member of the frame structure that rests on the foundation. Insulation material and a metal termite shield can be placed under the sill, if desired. Fig. 13-17. The sill should consist of one or two thicknesses of 2″ lumber placed on the foundation walls to provide a full and even-bearing surface. Fig. 13-18. Sills should be anchored to the foundation with ½″ bolts spaced approximately 6′ to 8′ apart, with at least two bolts in each pair of sills. Fig. 13-19. The preferred method is to spread a bed of mortar on the foundation and lay the sill upon it at once, tapping gently to secure an even bearing surface along the entire length. The nuts are then put in place over the washers and tightened gently with the fingers. After the mortar has set for a day or two, tighten the nuts. This

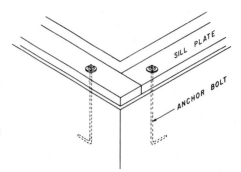

13-19. *The sill plate is anchored to the foundation wall with anchor bolts.*

method provides a good bearing for the sill, and it also prevents leakage of air between the sill and the foundation wall.

Installing Sills or Plates on the Foundation

Establish the building line points at each of the corners of the foundation. Pull a chalk line very tight at these established points and snap a line for the location of the sill. Square up the ends of the sill stock. Then place the sill on edge and mark the location of the anchor bolts. With a square, extend these marks across the width of the sill. The distance "X" in Fig. 13-20a shows how far from the edge of the sill to bore the holes. Locate the midpoints between the lines

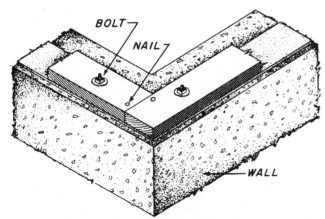

13-18a. *A single sill bolted to the foundation wall. Note the two pieces joined by toenailing at the corner with two 10d nails.*

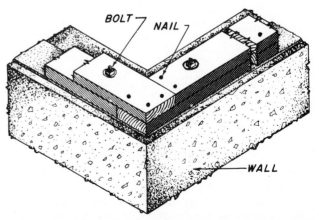

13-18b. *A double sill bolted to the foundation. The top sill is face-nailed to the bottom sill. Note the lap at the corner of the two sills.*

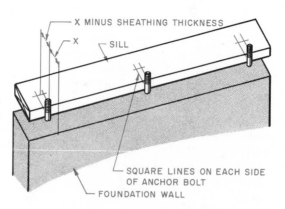

13-20a. *Laying out the location of the bolt holes on the sill.*

13-20b. *Roll out a strip of fiberglass sill sealer on the foundation wall just before laying the sill plate. The insulation will compress under the weight of the building. This fills irregularities and helps keep out dirt. It also keeps out drafts and reduces heat loss.*

representing the bolt locations and bore holes. If the sheathing is to rest on the foundation walls, allow for this by subtracting the sheathing thickness from the distance "X". Locate and bore holes as previously explained. If insulation and termite shields are used, bore holes at the same locations in the insulation and shields. Place the insulation and shields over the anchor bolts. Fig. 13-20b. Then place the sill on top. Start at the high point of the foundation and check the sill for levelness. Shim up as necessary. Add a washer and then tighten the nut to the sill. Apply grout to any openings between sill and foundation.

In hillside construction where there is a step foundation, short sills are placed on each of the steps and then a longer sill is placed in the highest position for the entire length of the building. Cripple studs are then set onto the short sills and cut to a length which will support the longer sill in a level position to carry the floor joists.

Floor Joists for Platform Construction

Check the house plans to determine the size and direction of the joists. Fig. 13-21a & b. On the sill or wall plate lay out the desired joist spacing. Fig. 13-21c. If there is to be a double joist under the outside wall, place a ¾" spacer block between the first and second joists. All other joists should be 16" on center (unless the plan calls for 24" on center).

Lay out a double joist under each cross partition. (A *cross partition* is one which runs parallel to the floor joists.) Fig. 13-22. If the cross partition is to be used for plumbing or heating pipes or ducts, place a solid spacer block between the double joists to allow pipes to run between them. Fig. 13-14. Transfer the joist spacing marks from the sill onto a story pole and set it aside for later use. Place the header joist on edge, making sure it is aligned with the outside edge of the sill. Toenail the header joist in place, nailing as required by code. Now hold the outside joist in a 90-degree vertical position to the header joist along the outside edge of the sill. Spike through the header joist into the end of the outside joist to form the corner. Fig. 13-23.

Any joists having a slight bow edgewise should be so placed that the crown is on top. A crowned joist will tend to straighten out when subfloor and normal floor loads are applied. The largest edge knots should be placed on top since knots on the upper side of a joist

Size Of Floor Joists (Inches)	Spacing Of Floor Joists (Inches)	Maximum Allowable Span (Feet and Inches)							
		Group I		Group II		Group III		Group IV	
		Plastered Ceiling Below	Without Plastered Ceiling Below	Plastered Ceiling Below	Without Plastered Ceiling Below	Plastered Ceiling Below	Without Plastered Ceiling Below	Plastered Ceiling Below	Without Plastered Ceiling Below
2 × 6	12	10-6	11-6	9-0	10-0	7-6	8-0	5-6	6-0
	16	9-6	10-0	8-0	8-6	6-6	7-0	5-0	5-6
	24	7-6	8-0	6-6	7-0	5-6	6-0	4-0	4-6
2 × 8	12	14-0	15-0	12-6	13-6	10-6	11-6	8-0	8-6
	16	12-6	13-6	11-0	11-6	9-0	10-0	7-0	7-6
	24	10-0	11-0	9-0	9-6	7-6	8-0	6-0	6-6
2 × 10	12	17-6	19-0	16-6	17-6	13-6	14-6	10-6	11-6
	16	15-6	16-6	14-6	15-6	12-0	13-0	9-6	10-0
	24	13-0	14-0	12-0	13-0	10-0	10-6	7-6	8-6
2 × 12	12	21-0	23-0	21-0	21-6	17-6	19-0	13-6	14-6
	16	18-0	20-0	18-0	19-6	15-6	16-6	12-0	13-0
	24	15-0	16-6	15-0	16-6	12-6	13-6	10-0	10-6

13-21a. *The group classifications in this table refer to the species and minimum grades of nonstress-graded lumber. See Fig. 13-21b. This table was taken from the Uniform Building Code Manual. If more specific information is needed about design, loading, and deflection, refer to Vol. 1 of the Manual.*

Species	Minimum Grade	Uniform Building Code Standard Number
Group I		
Douglas Fir & Larch[1]	Construction	{25-3 25-4
Group II		
Bald Cypress (Tidewater Red Cypress)	No. 2	25-2
Douglas Fir (South)[1]	Construction	25-4
Fir, White	Construction	{25-3 25-4
Hemlock, Eastern	No. 1	25-5
Hemlock, West Coast & Western[1]	Construction	{25-3 25-4
Pine, Red (Norway Pine)	No. 1	25-5
Redwood, California	Select Heart	25-7
Spruce, Eastern	No. 1	25-8
Spruce, Sitka	Construction	25-3
Spruce, White and Western White	Construction	25-4[2]
Group III		
Cedar, Western	Construction West Coast Studs	25-3
Cedar, Western Red and Incense	Construction	25-4
Douglas Fir & Larch[1]	Standard West Coast Studs	{25-3 25-4
Douglas Fir (South)[1]	Standard	25-4
Fir, Balsam	No. 1	25-8
Fir, White	Standard West Coast Studs	{25-3 25-4

Species	Minimum Grade	Uniform Building Code Standard Number
Group III (Continued)		
Hemlock, Eastern	No. 2	25-5
Hemlock, West Coast & Western[1]	Standard West Coast Studs	{25-3 25-4
Pine, Ponderosa, Lodgepole, Sugar, Idaho White	Construction	25-4
Redwood, California	Construction	25-7
Redwood, California (studs only)	Two Star	25-7
Spruce, Engelmann	Construction	25-4
Spruce, Sitka	Standard West Coast Studs	25-3
Spruce, White and Western White	Standard	25-4[2]
Group IV [See Section 2501 (e) Uniform Bldg. Code]		
Cedar, Western	Utility	25-3
Cedar, Western Red and Incense	Utility	25-4
Douglas Fir & Larch	Utility	{25-3 25-4
Douglas Fir (South)	Utility	25-4
Fir, White	Utility	{25-3 25-4
Hemlock, West Coast & Western	Utility	{25-3 25-4
Pine, Ponderosa, Lodgepole, Sugar, Idaho White	Utility	25-4
Redwood, California	Merchantable	25-7
Redwood, California (studs only)	One Star	25-7
Spruce, Engelmann	Utility	25-4
Spruce, Sitka	Utility	25-3
Spruce, White and Western White	Utility	25-4[2]

[1]Two-inch by 4-inch only.

[2]Spruce (White and Western White) shall be graded under the requirements of Section 25.409 of U.B.C. Standard No. 25-4.

13-21b. *The groups of nonstress graded lumber.*

are on the compression side of the member. Place the joists on the sills and securely nail through the header into the ends of the joists. Now lay the story pole (made earlier) across the tops of the joists, near the center and at right angles to the joists.

After the story pole is laid parallel to the joist header, nail it to each of the joists, using the spacing

13-21c. *Notice that the 16″ spacing is measured from the outside edge of the first joist to the center of the second joist and then to the centers of the other joists.*

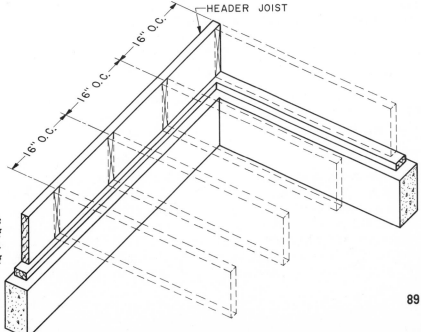

HEADER JOIST

16″ O.C.

89

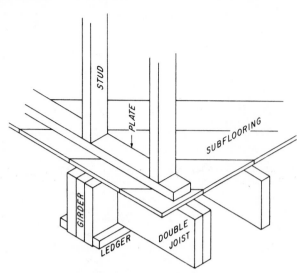

13-22. *Floor joists are doubled under nonbearing partitions.*

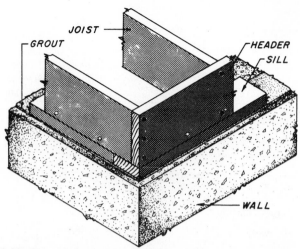

13-23. *Box-sill construction. The header joist is nailed to the other joists with 20d nails. Three nails are driven into the end joist and two into the others. The outside joist and header joist are toenailed to the sill on 16" centers.*

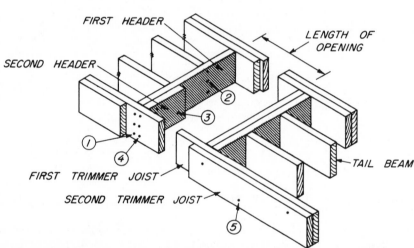

13-24. *Nailing procedure for framing floor openings. 1. First trimmer is nailed to first header with three 20d nails. 2. First header is nailed to tail beams with three 20d nails. 3. Second header is nailed to first header with 16d nails spaced 6" apart. 4. First trimmer is nailed to second header with three 20d nails. 5. Second trimmer is nailed to first trimmer with 16d nails spaced 12" apart.*

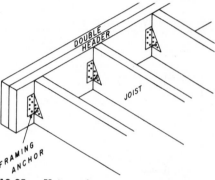

13-25a. *Using framing anchors to secure the tail joists to the header.*

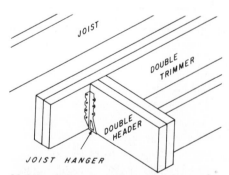

13-25b. *Using a joist hanger to secure a header joist to a trimmer joist. Double trimmers and double headers are used around floor openings.*

marked to obtain the correct spacing of the joists. Leave this strip in place until the subfloor is laid.

After nailing through the header joist (band) into the ends of the joists, pull a line tight along the top edge to check it for alignment. When the band has been properly aligned, nail the other ends of the joists to the girder and to the joists from the other side of the building. Lay out and frame the floor openings and install the bridging as required. The subfloor may then be laid and nailed in place.

Framing Floor Openings

When framing for large openings such as stairwells, fireplaces, and chimneys, the joists and headers framing the opening should be doubled. Fig. 13-24. The proper method of nailing is also shown in Fig. 13-24. Place the first trimmer joist in position. Lay out on this joist the location of the double header. Cut the first header to length and nail it in position. Lay

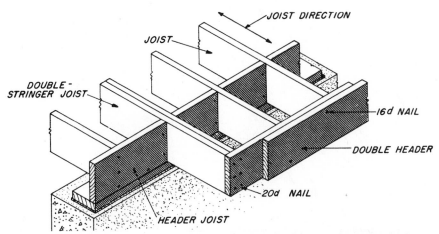

13-26. *Floor framing for a bay window. Nailing procedure: Second header is nailed to first header with 16d nails spaced 6″ apart. First header is end-nailed to each member of double-stringer joist with three 20d nails.*

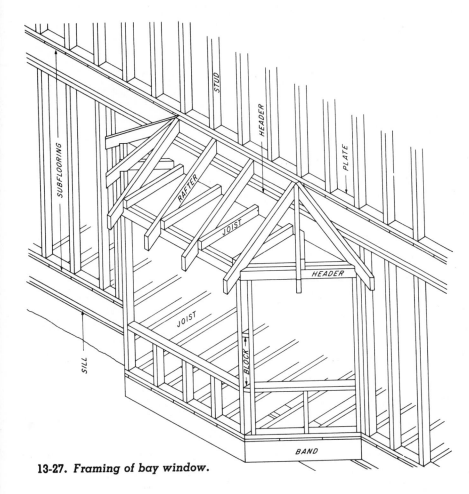

13-27. *Framing of bay window.*

out on this header the position of the tail beams, using regular spacing.

Place the tail beams in position and nail through the header into these joists, using three 20d nails in the end of each joist. Cut and place the second part of the double header in position, nailing it to the first header. Also nail through the first trimmer joist into the end of the second header. Now add the second trimmer joist and nail the two trimmer joists together, keeping the top edges even. Joist hangers and short sections of angle iron are often used as joist supports for the larger openings. Fig. 13-25. For further details of stairwells see "Stairs," Unit 34.

Floor Framing at Bay-Window Projections

The framing for a bay window or similar projection should be arranged so that the floor joists extend beyond the foundation wall. This allows them to carry the necessary loads. Fig. 13-26. This extension should normally not exceed 2′. The joists forming each side of the bay and the header for the bay should be doubled. Nailing, in general, should conform to that for floor openings. The subflooring is extended to the outer framing member and sawed flush with that member. Ceiling joists should be carried by a header framed over the window opening in the projected part of the structure shown in Fig. 13-27.

Bridging

When joists are placed over a long span, they have a tendency to sway from side to side. To help solve this problem, a bracing method called *bridging* is commonly used.

Floor frames are bridged in order to:
● Stiffen the floor frames.
● Prevent unequal deflection (bending) of the joists.
● Enable an overloaded joist to receive some assistance from the joists on either side of it.

Bridging is of two kinds, horizontal (solid) bridging, Fig. 13-28, and cross (diagonal) bridging. Cross bridging is more generally used since it is very effective and requires less material. Lumber 1″

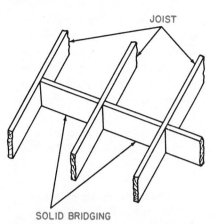

JOIST

SOLID BRIDGING

13-28. *Horizontal or solid bridging. Nailing is easier if the bridging is offset 1⅝″. Nails can then be driven directly through the joist into the end of each piece of bridging.*

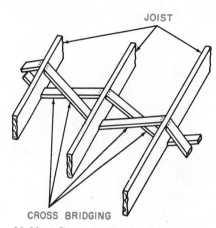

JOIST

CROSS BRIDGING

13-29a. *Cross or diagonal bridging.*

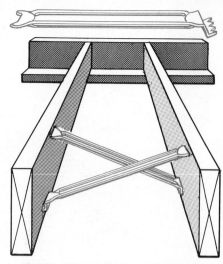

13-29b. *Metal diagonal bridging.*

× 3″ or 2″ × 2″ is usually used for cross bridging. Fig. 13-29a. Rigid metal cross bridging with nailing flanges may also be used. Fig. 13-29b. If the joists are over 8′ long, one row of bridging is installed at the center of the joist span. For joists 16′ and longer, install two rows of bridging equally spaced on the joist span.

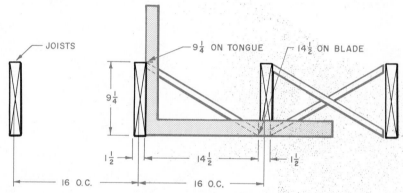

JOISTS

9¼ ON TONGUE   14½ ON BLADE

9¼

1½   14½   1½

16 O.C.   16 O.C.

13-30. *Laying out a piece of bridging. (Shown are 2″ × 10″ joists 16″ on center.)*

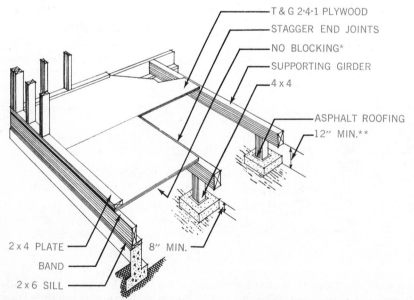

T & G 2·4·1 PLYWOOD
STAGGER END JOINTS
NO BLOCKING*
SUPPORTING GIRDER
4 x 4
ASPHALT ROOFING
12″ MIN.**

2 x 4 PLATE
BAND
2 x 6 SILL
8″ MIN.

13-31. *Girder construction with box-sill framing. The 2·4·1 system is illustrated in Fig. 13-57. *If square-edge 2·4·1 panels are used, blocking is required at unsupported edges. **In areas of termite infestation or under conditions of adverse ground moisture, use 18″ minimum.*

CUTTING AND INSTALLING DIAGONAL BRIDGING

Use the framing square to lay out diagonal bridging. The tongue of the square represents the width of the joists; the blade represents the space between the joists. Place a piece of bridging stock across the square as shown in Fig. 13-30, and mark the angle to be cut along the outside edge of the tongue. Cut the piece off at the marked angle. Place this cut on the body of the square at the 14½″ mark (for 16″ O.C. joist spacing). Lay the other end of the piece of bridging across the 9¼″ mark (for a 10″ joist) and scribe a mark on the outside edge of the tongue. Cut the bridging to the finished length and try it between two joists for fit.

Cut the first piece for a template and use as a pattern for the other pieces. For one row of bridging, locate the center of the span and snap a chalk line across the top of the joists. Before installation, drive two 8d nails at each end of the bridging pieces. Drive the nails until the points just show through.

Start at a wall and nail one piece of bridging in position. Continue by placing one row of bridging on each side of the chalk line on the joist. Complete the nailing at the top of the bridging; however, do not nail the bottom until the subfloor has been laid. This permits the joists to adjust themselves to their final positions. The bottom ends of bridging may then be nailed, forming a continuous truss across the whole length of the floor and preventing any overloaded joists from sagging below the others.

GIRDER FLOOR FRAMING

The girder method of floor framing is widely used in warm climates where homes without basements are built. It is much faster than joist-frame construction, but requires heavier or built-up material for the girders. Fig. 13-31. The correct girder size has to be figured on the basis of the load it is to support, the span, and the type

of material used. This information is found on the building plans or by checking with the local building department for maximum spans of girders. Girder spacing is usually 4' on center, with the maximum spacing of girder posts usually 5' on center. Steel girders are often used for fairly long spans.

Houses of the size and style shown in Fig. 13-32 may have girder floor framing. Such homes are built with 4″ × 6″ girders set on

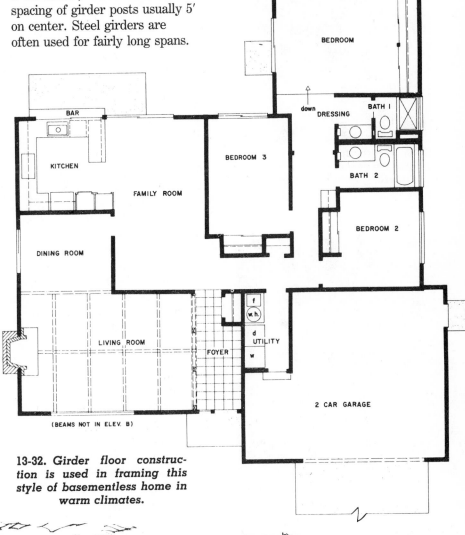

13-32. *Girder floor construction is used in framing this style of basementless home in warm climates.*

13-33. *The footing hole is dug and the precast concrete pier is ready for setting onto the footing.*

13-34. *The piers are set in place on the footings.*

4″ × 6″ posts. First, the building foundation walls and footings are poured. The locations of the piers are established, and the holes for the pier footings are dug. Fig. 13-33. Depending on the soil and local restrictions, these footings may require a built-up form.

The concrete is poured into the cavity provided for the footing, and the piers are set in place. Fig. 13-34. The piers should be in a reasonably straight line, but the height is not critical because the posts will be cut individually to the correct length for supporting the girders. Fig. 13-35. The sill is then cut to size and bolted in place as described on page 87. Fig. 13-36.

The bearing posts must be cut accurately to length to provide a level floor. Pull a line tight from opposite sill plates over the piers. Make certain the line is down tight on the plates. Measure the distance from the line to the top of the pier cap or redwood block on top of the concrete pier. Record this distance, usually on each pier cap or redwood block. Repeat this operation for each line of piers until the height of each bearing post has been determined and recorded. Square one end and cut the bearing posts to length. Care should be taken with material of this size to cut it square with two of its adjacent surfaces.

13-36. *The sill is bolted in place. The girders are laid and ready for installation.*

Treat the end of the post for termites, fungus, and similar problems. Toenail the treated end to the pier cap with two 8d nails on each side (a total of 8 nails). Fig. 13-37. Next square one end of the girder, cut it to length, and toenail it to the posts. Fig. 13-38. If a low house profile is desired or if the finished floor is to have a stepdown area, the tops of the girders in this area are set flush with the top of the sill. When this is done, a special metal hanger must be installed to support the girder end. Fig. 13-39. Two other methods of setting the top of the girder flush with the top of the plate are shown in Fig. 13-40.

13-35. *The foundation wall with the anchor bolts set for the sill. The piers are ready for the posts.*

13-37. *The posts are cut to the correct height and toenailed to the wood pad which is set into the precast pier.*

13-38. *The girders are nailed to the posts. The step in the girders is for a step-down area of the building.*

13-39. *When the top of the girder is to be even with the top of the sill, a metal hanger is used to support the girder.*

When there is a step (two levels) in the floor, the ends of the girders must be headed off to support the subfloor. A 2″ × 6″ is used with 4″ × 6″ girders. Fig. 13-41.

In framing for a fireplace or other openings in the floor, the tail beams and headers are of the same structure material as the girders. Fig. 13-42.

Small openings for heating or air conditioning ducts will require only 2″ × 4″ boards laid flat. Fig. 13-43. Because working space under the girders is small, the plumbing and heating are "roughed in" before the floor is applied. Fig. 13-44. The subfloor is then cut and nailed in place. The surface is now ready for layout and erection of the side-walls. Fig. 13-45.

SUBFLOORING IN JOIST CONSTRUCTION

A subfloor is a wood floor laid over the floor joists, under the finished floor. Ordinarily a subfloor is nailed directly to the floor joists. In conventional joist construction, sound subflooring is virtually a "must". Most modern building codes specify subfloors. Their omission usually is poor economy, even if the finish floor is of strong, durable oak.

13-41. *The girder ends are boxed in for the step in the floor.*

13-42. *The area in which the fire-place is to be located must be "headed off." Note the reinforcing rod which has been set into the fire-place footing.*

13-40a. *The girder may be supported on a post set on the footing.*

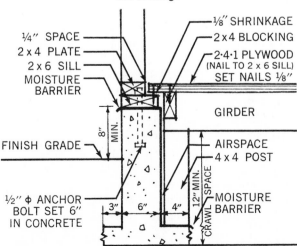

13-40b. *The girder may be supported by a pocket in the foundation.*

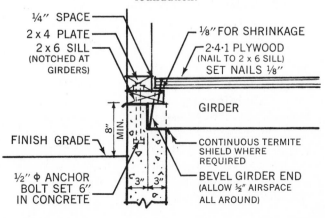

95

13-43. *The heating and air conditioning ducts are installed. They are wrapped with fiberglass insulation, and the joints are taped. Notice that the 2" × 4" boards which support the duct will also support the flooring around the cutout for the register.*

13-44. *Plumbing is roughed in after the girders are in place.*

Purposes Served by Subflooring

Subflooring serves several important purposes. It lends bracing strength to the building. It provides a solid base for the finish floor, making floor sag and squeaks very unlikely. By acting as a bar-

rier to cold and dampness, subflooring helps keep the building warmer and drier in winter. In addition, it provides a safe working surface for building the house.

Selection, Nailing of Subfloor Boards

If strip flooring is to be used for the finish floor, it is generally recommended that the subflooring consist either of softwood plywood or of good quality boards about ¾" thick and not more than 6" wide. Wider boards are likely to expand and contract too much.

Square-edge boards generally are preferred to tongue-and-groove boards. This is because snug joining usually is not desirable in subflooring. This is particularly true of houses in moist climates and summer homes which are not heated in the winter. Square-edge boards are more economical too. Boards should be throughly dry. The use of green subfloor boards frequently causes squeaks and cracks.

Subfloor boards may be applied diagonally or at right angles to the floor joists. When subflooring is at right angles to the joists, the finish floor should then be laid at right angles to the subfloor. However, it is best to lay the subflooring diagonally. This arrangement permits the finish strip flooring to be laid in any direction. For a home of two or more stories, it is best to have the subfloor boards run in opposite diagonal directions on alternate floors. Diagonal subflooring also provides better bracing and stiffness to the building.

Laying a Diagonal Subfloor

When laying diagonal subfloor, lay a relatively long board first. The short cuttings may then be used as you approach a corner and the length decreases. Work can then proceed on both sides of this first diagonal length.

13-45. *Subfloor with roughed-in plumbing ready for wall layout and fabrication. An interior wall will be framed in along the line of the plumbing. All the plumbing pipes shown here will then be totally enclosed within the wall.*

To lay out the first piece, begin at a corner of the building (B in Fig. 13-46) and measure equal distances along the header joist and the first joist (points A and C). Snap a chalk line along the top of the joists between these two points to form a 45° angle. Lay the first board along this line, cut it to fit, and nail it at each joist with two 8d nails, not more than ¾" from the edge.

Often two or more pieces of subfloor board will have to be used to make up one diagonal strip. Ends of these pieces are nailed so they butt together on the top of a joist, forming a joint. Do not allow two of these joints to come side-by-side on the same joist; there should be at least two boards between joints. Generally, the shortest pieces of board should be long enough to span three joists—that is, start from one joist, cross another, and be fastened to a third. Of course, shorter pieces will have

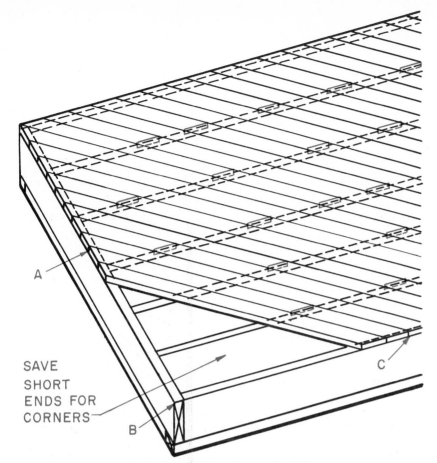

to be used in the corners. Fig. 13-47. Many carpenters will use an 8d nail to space between the edges of the subfloor boards. This allows for drainage and buckling which can be caused by swelling if the floor gets wet during construction. Tongue-and-groove boards should have holes at intervals to allow rainwater to drain off.

Laying a Straight Subfloor

Check the plans to determine the way the finish floor is to be laid in relation to the direction of the floor joists. If the finish floor is to run parallel to the joists, a straight subfloor may be laid. Begin by laying the first piece of subfloor at one end or edge of the building, at right angles to the joists. Make the joints over the centers of the joists. Fig. 13-48. Laying and nailing are then done generally as for a diagonal subfloor.

A

SAVE SHORT ENDS FOR CORNERS

B

C

13-46. *Laying out diagonal subfloor.*

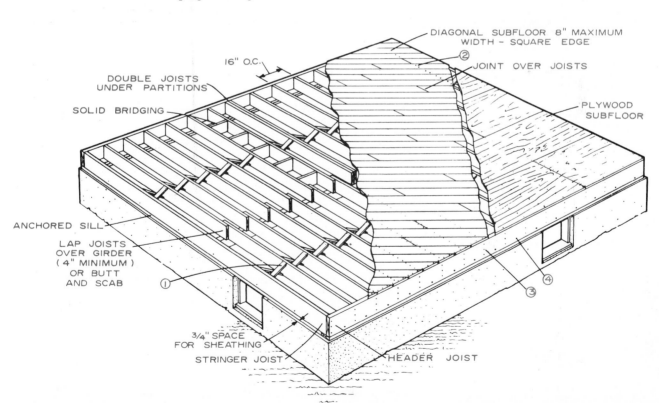

DIAGONAL SUBFLOOR 8" MAXIMUM WIDTH – SQUARE EDGE

② JOINT OVER JOISTS

16" O.C.

DOUBLE JOISTS UNDER PARTITIONS

SOLID BRIDGING

PLYWOOD SUBFLOOR

ANCHORED SILL

LAP JOISTS OVER GIRDER (4" MINIMUM) OR BUTT AND SCAB

①

3/4" SPACE FOR SHEATHING

STRINGER JOIST

HEADER JOIST

④

③

13-47. *Floor framing nailing procedure. 1. Bridging (1″ × 3″) nailed at top and bottom with 8d nails. 2. Subfloor board nailed with two or three 8d nails (plywood subfloor also shown as alternative). 3. Header joist end-nailed to corner joists and intermediate joists with three 20d nails. 4. Header joist toenailed to sill with 10d nails 16″ on center.*

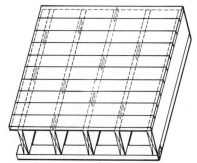

13-48. Straight subfloor. Joints (in color) are over the joists.

Plywood subflooring[1][3]/For direct application of T&G wood strip and block flooring and light weight concrete.[4]

Panel Identification Index[2]	Plywood Thickness (inches)	Maximum Span[5] (inches)	Nail Size & Type	Nail Spacing (inches)	
				Panel Edges	Intermediate
30/12	5/8	12[6]	8d common	6	10
32/16	1/2, 5/8	16[7]	8d common[8]	6	10
36/16	3/4	16[7]	8d common	6	10
42/20	5/8, 3/4, 7/8	20[7]	8d common	6	10
48/24	3/4, 7/8	24	8d common	6	10
1⅛" Groups 1 & 2	1⅛	48	10d common	6	6
1¼" Groups 3 & 4	1¼	48	10d common	6	6

Notes: (1) These values apply for Structural I and II, Standard sheathing and C-C Exterior grades only.

(2) Identification Index appears on all panels except 1⅛" and 1¼" panels.

(3) In some non-residential buildings, special conditions may impose heavy concentrated loads and heavy traffic requiring subfloor constructions in excess of these minimums.

(4) Edges shall be tongue and grooved or supported with blocking for square edge wood flooring, unless separate underlayment layer (¼" minimum thickness) is installed.

(5) Spans limited to values shown because of possible effect of concentrated loads. At indicated maximum spans, floor panels carrying Identification Index numbers will support uniform loads of more than 100 psf.

(6) May be 16" if 25/32" wood strip flooring is installed at right angles to joists.

(7) May be 24" if 25/32" wood strip flooring is installed at right angles to joists.

(8) 6d common nail permitted if plywood is ½".

Use plywood with these kinds of APA grade-trademarks for subfloors.

C-D
32/16 (APA)
INTERIOR
PS 1-74   000
EXTERIOR GLUE

STRUCTURAL I
C-D
20/0 (APA)
INTERIOR
PS 1-74   000
EXTERIOR GLUE

C-C
32/16 (APA)
EXTERIOR
PS 1-74   000

13-51. Recommended plywood subfloor thickness and nailing requirements.

13-49. These girders are set flush with the sill in preparation for the subfloor. Notice the ventilation openings in the foundation wall. These openings are needed in this method of framing.

13-50. Girder construction with 2″ × 6″ tongue-and-groove flooring.

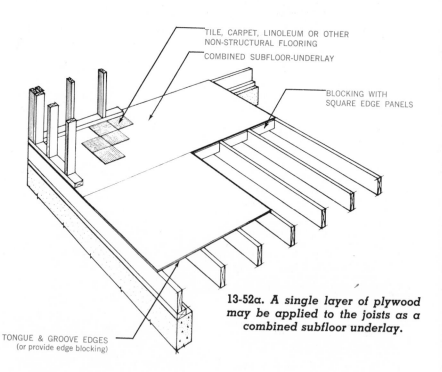

TILE, CARPET, LINOLEUM OR OTHER NON-STRUCTURAL FLOORING

COMBINED SUBFLOOR-UNDERLAY

BLOCKING WITH SQUARE EDGE PANELS

TONGUE & GROOVE EDGES (or provide edge blocking)

13-52a. A single layer of plywood may be applied to the joists as a combined subfloor underlay.

Laying a 2″ Tongue-and-Groove Subfloor

Some local building codes permit the use of 2″ tongue-and-groove (T & G) subfloor over girder floor framing. This eliminates floor joists. When the girder is set into a pocket or hung with a bracket, care should be taken to insure that the top of the girder is even with the top of the sill plate. Fig. 13-49. Sometimes the girders are set on top of the sill, giving the building a high profile; the tongue-and-groove flooring is cut even with the outside of the framing and nailed on top of the girders. Fig. 13-50. Use 16d nails to toenail at the tongue and to face-nail at a joint on all girders. This type of construction requires vents in the foundation.

Laying a Plywood Subfloor

Plywood produces a smooth, solid, stable base for any kind of finish flooring. Big panels cover large areas fast. This reduces the laying time. The correct thickness for the various applications and the nailing requirements can be found in Fig. 13-51.

Plywood can be used as a combined subfloor and underlay for such flooring as tile, carpeting, or linoleum. Either tongue-and-groove or square-edge plywood may be chosen for this combined application. However, if square-edge plywood is used, the joints must be blocked. Fig. 13-52. When a separate underlayment is put over the subfloor, care should be taken to stagger the joints to provide adequate support. This method will not require blocking under the joints. Fig. 13-53. Plywood subflooring under strip flooring does not require blocking and permits the strip flooring to be laid in either direction. Fig. 13-54.

Begin laying the plywood subfloor by placing a full sheet even with one of the outside corners of the joist framing. The grain of the plywood should run at right angles to the joists. Drive just enough nails to hold the panel in place. Place the next full panel in position at the end of the first panel. Be sure the joint is centered over the joist, and leave about 1/32" space between panels.

Plywood Species Group	Subfloor-Underlay		Nail Size (approx.) and Type (set nails 1/16")	Nail Spacing (inches)	
	Min. Plywood Thickness (inches) (1)	Max. Spacing of Supports c. to c. (inches)		Panel edges	Intermediate
Douglas fir or Western larch	1/2	16	6d ring shank or screw type	6	10
	5/8	20	6d ring shank or screw type	6	10
	3/4	24	6d ring shank or screw type	6	10
Groups 1 or 2 WSP	5/8	16	6d ring shank or screw type	6	10
	3/4	24	6d ring shank or screw type	6	10

(1) In some non-residential buildings, special conditions may impose heavy concentrated loads and heavy traffic requiring subfloor-underlay constructions in excess of these minimums.

Note: For certain types of flooring such as wood block or terrazzo, sheathing grades of plywood may be used.

Typical grade-trademarks.

UNDERLAYMENT

GROUP 2 (APA) INTERIOR PS 1-74 000

C-C PLUGGED

GROUP 4 (APA) EXTERIOR PS 1-74 000

13-52b. Specifications for a combined subfloor underlay. (WSP under "Plywood Species Group" stands for Western Soft Pine.)

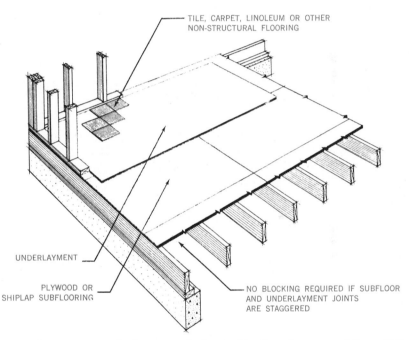

TILE, CARPET, LINOLEUM OR OTHER NON-STRUCTURAL FLOORING

UNDERLAYMENT

PLYWOOD OR SHIPLAP SUBFLOORING

NO BLOCKING REQUIRED IF SUBFLOOR AND UNDERLAYMENT JOINTS ARE STAGGERED

13-53a. A separate underlayment of plywood is placed over a subfloor. If the subfloor is plywood, the joints of the subfloor and the underlayment should be staggered.

Plywood Species Group	Underlayment	Nail Size (approx.) and Type (set nails 1/16")	Nail Spacing (inches)	
	Min. Plywood Thickness (inches)		Panel Edges	Intermediate
Douglas fir or Western larch; Groups 1, 2 or 3 WSP	3/8 (1) Thicker panels to match other floors	3d ring shank (also for 1/2" panels) (4d ring shank for 5/8" and 3/4") 16 gauge staples	6	8 each way
			3	6 each way

(1) FHA accepts 1/4" underlayment.

These are typical grade-trademarks.

UNDERLAYMENT

GROUP 2 (APA) INTERIOR PS 1-74 000

B-C

GROUP 3 (APA) EXTERIOR PS 1-74 000

13-53b. Recommended thickness, nail spacing, and related information for plywood subfloor underlay.

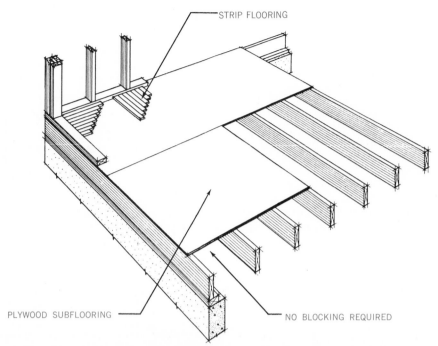

13-54a. *Plywood subflooring under ²⁵/₃₂″ strip flooring allows the strip floor-ing to be laid in either direction.*

13-55. *The subfloor is nailed with a pneumatic nailer.*

Plywood Species Group	Subflooring		Nail Size and Type	Nail Spacing (inches)	
	Min. Plywood Thickness (inches) (3)	Max. Spacing of Supports c. to c. (inches)		Panel Edges	Intermediate
Douglas fir or Western larch; Group 1 WSP Ext. glue, and Ext. C-C only)	½ (1)	16	6d common	6	10
	⅝	20	8d common	6	10
	¾	24	8d common	6	10
Group 2 WSP	⅝ (2)	16	8d common	6	10
	¾	24	8d common	6	10

(1) If ²⁵/₃₂″ wood strips are perpendicular to supports, ½″ can be used on 24″ spans.
(2) If ²⁵/₃₂″ wood strips are perpendicular to supports, ⅝″ can be used on 24″ spans.
(3) In some non-residential buildings, special conditions may impose heavy concentrated loads and heavy traffic requiring subfloor constructions in excess of these minimums.

This is a typical grade-trademark.

C-D PLUGGED

GROUP 2 INTERIOR PS 1-74 000

13-54b. *Specifications for plywood subflooring used under strip flooring.*

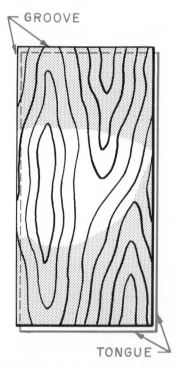

13-56. *A plywood panel 1⅛″ thick with a groove on an end and an edge, and a tongue on the other end and edge. The surface is a full 4′ × 8′ with an additional allowance for the tongue.*

Begin the second row of panels at the end of the building, along-side the first panel laid. Cut a panel in half, lay the end flush with the outside of the building, and nail the half panel to the joists. If the joists are running at right angles to the grain direction of the plywood panel, measure and cut the panel so that the joint will be on the fourth joist. Fig. 13-54. Continue to lay and nail panels in this row. The next (third) row of panels is started with a full panel. This will stagger the joints and provide the strongest floor. Continue to lay,

driving just enough nails in each panel to hold it in position until all panels are laid. Then complete the nailing as required. Fig. 13-55.

TONGUE-AND-GROOVE PLYWOOD FLOORS

As mentioned earlier, some pan-els are made for use as a combined subfloor and underlayment. They are made of interior-type plywood, tongue-and-groove, 1⅛″ thick. These panels are laid over two sup-porting beams placed 48″ on cen-ter. Standard-size T & G panels measure 4′ × 8′ on the face, with

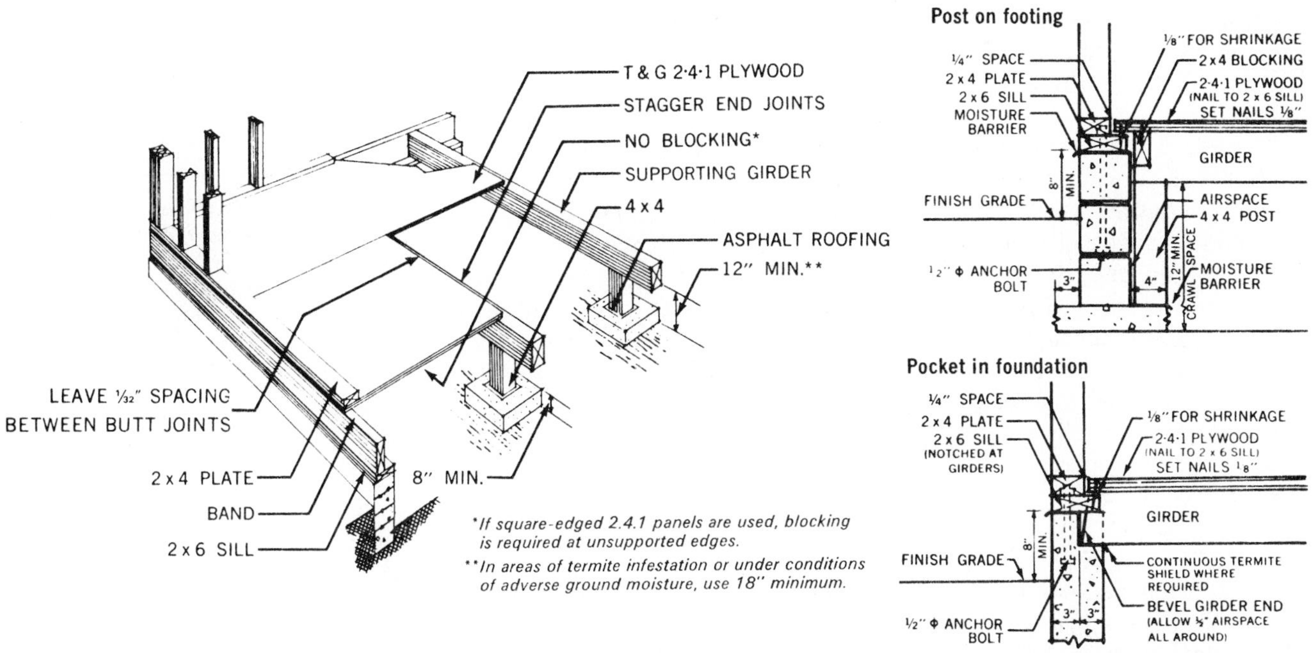
T & G 2·4·1 PLYWOOD
STAGGER END JOINTS
NO BLOCKING*
SUPPORTING GIRDER
4 x 4
ASPHALT ROOFING
12" MIN.**
LEAVE 1/32" SPACING BETWEEN BUTT JOINTS
2 x 4 PLATE
BAND
2 x 6 SILL
8" MIN.
*If square-edged 2.4.1 panels are used, blocking is required at unsupported edges.
**In areas of termite infestation or under conditions of adverse ground moisture, use 18" minimum.
Post on footing
1/4" SPACE
2 x 4 PLATE
2 x 6 SILL
MOISTURE BARRIER
1/8" FOR SHRINKAGE
2 x 4 BLOCKING
2·4·1 PLYWOOD (NAIL TO 2 x 6 SILL) SET NAILS 1/8"
GIRDER
FINISH GRADE
8" MIN.
AIRSPACE
4 x 4 POST
1/2" Φ ANCHOR BOLT
3"
4"
12" MIN.
CRAWL SPACE
MOISTURE BARRIER
Pocket in foundation
1/4" SPACE
2 x 4 PLATE
2 x 6 SILL (NOTCHED AT GIRDERS)
1/8" FOR SHRINKAGE
2·4·1 PLYWOOD (NAIL TO 2 x 6 SILL) SET NAILS 1/8"
GIRDER
FINISH GRADE
8" MIN.
CONTINUOUS TERMITE SHIELD WHERE REQUIRED
BEVEL GIRDER END (ALLOW 1/2" AIRSPACE ALL AROUND)
1/2" Φ ANCHOR BOLT
3"
3"

13-57a. ***Plywood 1⅛" thick is often used with the girder construction system. Girders can be supported in many ways to accommodate the heavy panels.***

2-4-1 Subfloor-underlayment[1]/ For application of tile, carpeting, linoleum or other non-structural flooring; or hardwood flooring
(Live loads up to 65 psf—two span continuous; grain of face plys across supports)

Plywood Species Group	2-4-1 Plywood Thickness (inches)	2-4-1 Maximum Spacing or Supports c. to c. (inches)	Nail Size and Type[2]	Nail Spacing (inches) Panel Edges	Nail Spacing (inches) Inter-mediate
Groups 1, 2, & 3	1⅛ only	48[3]	8d ring shank recommended or 10d common smooth shank (if supports are well seasoned)	6	6

Notes: (1) For additional information, see American Plywood Association standards.
(2) Set nails ⅛" and lightly sand subfloor at joints if resilient flooring is to be applied.
(3) In some non-residential buildings special conditions may impose heavy concentrated loads and heavy traffic, requiring support spacing less than 48".

Typical grade-trademarks.

2·4·1 GROUP 1 INTERIOR PS 1-74 000 APA

13-57b. ***Nailing and related information for 1⅛" subfloor-underlayment panels.***

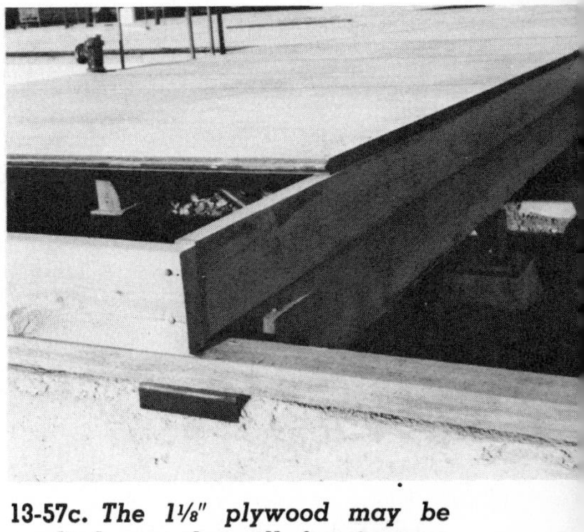

13-57c. ***The 1⅛" plywood may be applied over box-sill framing, as shown. It may also be laid directly on sills and girders which are supported in special brackets.***

additional allowance for the tongue. Fig. 13-56.

Any one of several support systems may be used with this floor system if the beams are spaced up to 48" O.C. Some suggested methods are box beams, 4" lumber, or two 2" joists spiked together. When the beams are set into pockets in the foundation, the plywood can bear directly on the sill. Also conventional box sill construction may be used. Fig. 13-57. Solid lumber beams, particularly when pocketed, should be as dry as possible to minimize shrinkage.

Tongue-and-groove subfloor should be started with the tongue toward the outside of the building. Thus any pounding required to close the joints between the panels can be done on a scrap block against the groove. When the panels are nailed in place, the face grain of the panel should run across the main beams, and wherever possible, cover two openings. End joints should be staggered.

If the panels are square edged, 2" × 4" blocking is required under the edges between beams. If both the sides and ends of the panel are tongue-and-grooved, drive the side joints tight first. Then, drive end joints tight. When the floor covering is of the thin, resilient type, fill any cracks ⅟₁₆″ or wider and sand the joints lightly if they are not absolutely flush.

To achieve the greatest resistance to nail popping, and the maximum in withdrawal strength, use 8d common ring-shanked or helically threaded nails. Space the nails 6″ O.C. at all bearing points. However, 10d common smooth-shanked nails may be substituted if desired. Under resilient tile when the beams are not fully seasoned,

set all nails ⅛″ but do not fill. (*Setting* means driving the heads below the wood surface.) Set the nails just before laying the resilient flooring. This lets you take advantage of the beam seasoning that has taken place. When the panels are tongue-and-grooved at the ends, one line of nails can be used to secure them. Drive the nails through both panels at a point near the middle of the tongue as shown in Fig. 13-58.

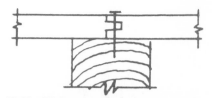

13-58. With tongue-and-groove plywood, a single row of nails at the joint is all that is necessary to nail the two panels together and to the girder.

QUESTIONS

1. List the names of the floor framing parts.
2. What is the purpose of a post?
3. What is considered the best method of framing the joist and girder?
4. How is a sill plate attached to a foundation wall?

5. How is the header joist checked for alignment?
6. What is the purpose of bridging?
7. How does girder floor framing differ from conventional floor framing?

Wall Framing 14

Exterior sidewalls and, in some designs, interior walls normally support the roof load. They will also serve as a framework for attaching interior and exterior coverings. When roof trusses spanning the entire width of the house are used, the exterior sidewalls carry both the roof and ceiling loads. Fig. 14-1. Interior partitions then serve mainly as room dividers. When ceiling joists are used, interior partitions usually sustain some of the ceiling loads. Fig. 14-2.

The exterior walls of a wood-frame house normally consist of *studs, interior and exterior coverings, windows and doors,* and *insulation.* The wall-framing members used in conventional construction are generally 2″ × 4″ studs spaced 16″ on center. Fig. 14-3.

The requirements for wall-framing lumber are *stiffness, nail-holding power, freedom from warpage,* and *ease of working.* Species and grades used for wall framing may, in general, follow

those used for floor-framing materials—for example, Douglas fir, the hemlocks, and southern yellow pine. Also commonly used for studs are spruce, pine, and white fir. No. 1 and No. 2 grades are most often used. Moisture content of framing members usually should not exceed 19% for on-site construction and 12% if shop fabricated.

As with floor construction, there are two general types of wall framing—*platform construction* and *balloon-frame construc-*

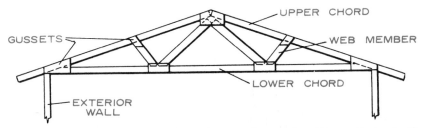

14-1. *Trussed rafters are installed as complete units. They eliminate the need for bearing partitions.*

14-4. *Keep the floor plan readily available for reference when laying out the walls on the subfloor.*

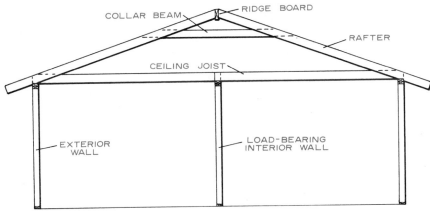

14-2. *Conventional framing requires an interior wall to support the ceiling joists.*

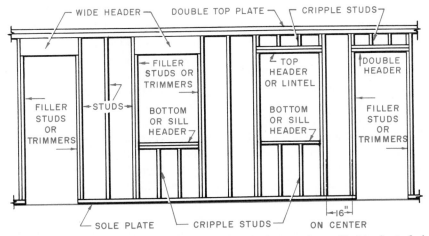

14-3. *Parts of a framed wall section. (Filler studs are also called jack studs.)*

LAYING OUT WALL LOCATIONS

The carpenter responsible for the layout of a building is one of the most important people on the job. Accuracy of layout is most important to the final overall quality of the building. So is knowledge of the structural members of the building. The carpenter must also be aware of special framing requirements for other skilled workers such as plumbers, electricians, and plasterers.

The layout is usually done by two people, a carpenter and an apprentice. The carpenter is responsible for all the measuring and marking. The apprentice observes and learns while holding the end of the measuring tape or chalk line. It is possible for the carpenter to lay out alone, using an awl to hold the end of the tape or chalk line. The area to be laid out must be swept clean and all objects that might be in the way removed. This will help make it possible to make a clear, solid chalk line to which the plates are aligned later.

The carpenter must study the plans for the building and have a thorough understanding of them before layout begins. During the actual layout the carpenter should not have to refer often to the plans, only to check them occasionally to "pick off" dimensions. Fig. 14-4.

A steel tape is used to mark the location of the interior parti-

tion. The platform method is nearly always used by builders throughout the United States and Canada. This is primarily because platform construction is fairly simple and it lends itself to a great many architectural designs.

With the platform method the floor framing should be complete, with subfloor securely fastened in place, before wall framing begins. The first step in wall framing is to lay out the wall locations on the subfloor.

103

14-5. *Laying out on the subfloor. Use a steel tape to measure from the corner to the centerline of all openings and intersecting partitions.*

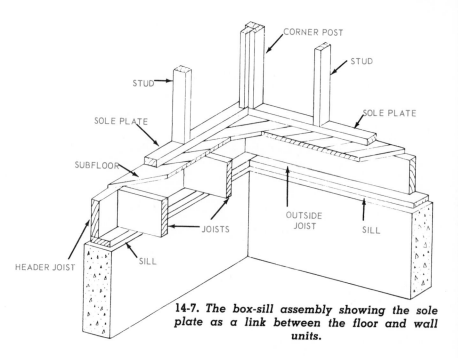

14-7. *The box-sill assembly showing the sole plate as a link between the floor and wall units.*

14-6. *Stretching the chalk line to snap on the subfloor. The chalked line will be used to align the sole plate as it is fastened in place.*

tions and exterior walls. Fig. 14-5. A chalk line is pulled taut on these marks and snapped to indicate the exact location and alignment for the full length of an edge of the sole plate. Fig. 14-6. The carpenter then marks an X on the subfloor to show on which side of the line the plate is located. After the location of the plates is laid out on the subfloor, the various wall members are cut to size in preparation for assembly of the wall sections.

WALL FRAMING

The following is a description of the various parts of the interior and exterior walls in platform construction:

Sole Plate

All partition walls and outside walls are finished with a piece of material corresponding to the thickness of the wall, usually a 2 × 4. Laid horizontally on the subfloor, this member carries the bottom end of the studs. Fig. 14-7. This 2 × 4 is called the "sole" or "sole plate." The sole should be nailed with two 16d nails at each joist that it crosses. If laid lengthwise on top of a girder or joist, there should be two nails every two feet.

For the bottom plate select straight material from the plate stock. Square up the ends of this material, then carefully position the end of the plate on the subfloor at an outside corner. The plate is usually located in from the outside of the building at a distance equal to the thickness of the sheathing. Thus when the sheathing is applied it will rest on the subfloor. Some local codes require that the sheathing cover the joist header and be set on the foundation wall; the sole then is set flush with the outside edge of the subfloor. Fig. 14-8.

Securely nail the plate at the corner. If more than one length of plate stock is necessary, butt the additional lengths up tightly against each other and toenail them together. The joints should be made at the centerline of a stud location. Square and cut off the end of the last length to the desired length and nail the plate securely at this corner.

Continue on around the outside of the building with the sole until complete. Note that the plate is continuous; the door openings are not cut out until the wall is erected and permanently braced.

Before securely nailing between the corners, straighten the sole. Usually the edge of the plate is aligned with the chalk line which was snapped on the subfloor when the walls and partitions were laid out. (See "Laying Out Wall Loca-

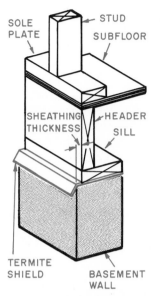

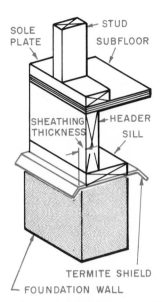

SOLE PLATE — STUD
SUBFLOOR

SHEATHING THICKNESS — HEADER
SILL

TERMITE SHIELD   BASEMENT WALL

SOLE PLATE — STUD
SUBFLOOR

SHEATHING THICKNESS — HEADER
SILL

TERMITE SHIELD
FOUNDATION WALL

14-8. *For the sheathing-application method shown on the left, the sill is set back the thickness of the sheathing. The sheathing rests on the foundation wall and covers the box-sill assembly. The drawing on the right shows platform frame construction on a basementless home. The termite shield extends beyond the interior of the foundation wall.*

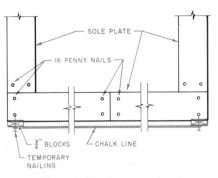

SOLE PLATE
16 PENNY NAILS
¾" BLOCKS   CHALK LINE
TEMPORARY NAILING

14-9. *The chalk line is held out away from the plate with a ¾" block at each end. The plate can then be aligned by measuring ¾" from the plate to the chalk line or by using another ¾" block for a gauge.*

tions.") The sole can also be straightened by driving a nail at each corner and pulling a chalk line along the outside edge of the sole plate. Then place a ¾" spacer block between the chalk line and the plate at each end. Measure in ¾" from the edge of the plate to the chalk line, or use a gauge block between the chalk line and the plate. Fig. 14-9. Nail the plate securely to the subfloor as it is aligned. When more than one length of stock is used, nail at the joints first. Care must be taken to provide a straight plate line and thus a straight wall.

LAYING OUT THE SOLE PLATE

When laying out the sole plate, lay out the door and window openings, the partitions, and then the stud locations, in that order. The studs that are designated where openings occur can then be marked as "cripples." The mark usually used is an *O*. The full-length studs are usually marked with an *X*. Other special studs

which will be discussed later should also be marked. Corner studs are marked with a *C*, partition studs with a *P*, and trimmer or lap studs with an *L* or *S*.

Begin the plate layout by referring to the plans to find the distance from the corner of the building to the center of the first opening. Measure off this distance and square a line across the plate at this point. Mark the line with a centerline symbol (℄) and an identification letter or number for reference when cutting other parts for this opening.

Continue around the outside wall to lay out and mark for identification all of the other openings. Now from the outside corners lay off the centers of all partitions and mark them with a *P*. Fig. 14-10.

Wall Openings

Studs around wall openings require special treatment. Allowances must be made for framing in doors and windows. These al-

lowances, when added to the size of the finished openings, make up what is called a "rough opening." The rough opening is the distance between the trimmer studs. Fig. 14-11. Most window and door schedules will provide the rough opening (R.O.) size for framing. If not, the following allowances can be used for door and window rough opening widths and heights:
• Double-Hung Window (Single Unit):
   Rough opening width = glass width + 6".
   Rough opening height = total glass height + 10".
• Casement Window (Two Sash):
   Rough opening width = total glass width + 11¼".
   Rough opening height = total glass height + 6⅜".
• Doors:
   Rough opening width = width of door + 2½".
   Rough opening height = height of door + 3".

LAYING OUT WALL OPENINGS

After the wall openings have been located on the plate, on each side of the centerline of each opening lay off one-half the rough opening size. Fig. 14-11. For example, for a 2′ 8″ door, the distance between the full-length studs is 37½"—that is, 32" for the door plus 5½" for the side jambs, wedges, and *trimmer studs* (which support the header). Fig. 14-12.

After all openings and partitions have been laid out, lay out all regular and cripple studs.

Top Plate

The top plate (Fig. 14-13) has the following purposes:
- Ties the studding together at the top.
- Forms a finish for the walls.
- Furnishes a support for the lower ends of the rafters.

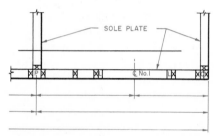

14-10a. Centerlines of wall openings and intersecting walls are located on the plate by measuring from an outside corner. The locations of the wall members are marked for identification so other carpenters can assemble and erect the walls later.

- Serves as a connecting link between wall and roof just as sills and girders are connecting links between floors and the walls.

The plate is made up of one or two pieces of material of the same size as the studs. When placed on top of partition walls, the plate is sometimes called the *cap*. Where the plate is doubled, the first

plate or bottom section is nailed with 16d or 20d nails to the top of the corner posts and to the studs. The connection at the corner is made as in Fig. 14-14a & c. After the single plate is nailed securely and the corner braces are nailed into place, the top part of the plate is nailed to the bottom section with 10d nails, spaced 16"

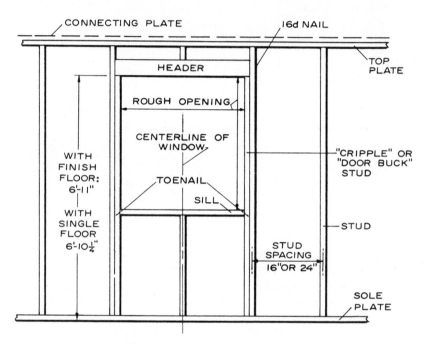

14-11. Rough framing for a window opening.

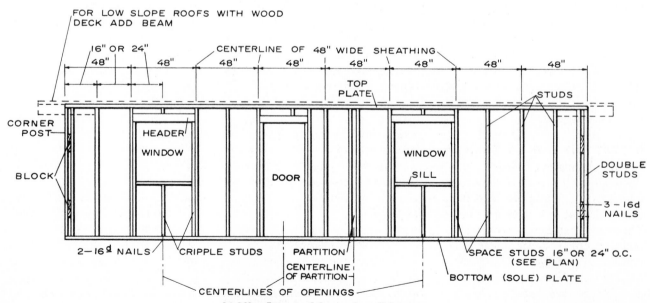

14-10b. A typical framed wall layout.

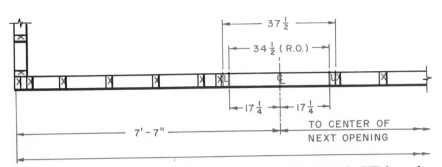

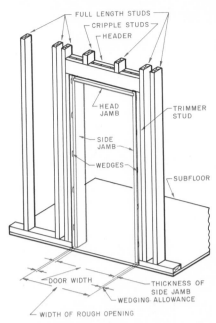

14-12a. *The centerline of a door opening is marked on the sole 7'7" from the outside corner. One-half of the R.O. (17¼") is then laid off on each side of the centerline. The thickness of the trimmer stud is laid off on each side of this. The header will rest on top of the trimmers and between the full studs marked with the X.*

14-12b. *Door framing allowance for rough opening.*

O.C. Fig. 14-15. The edges of the top section should be flush with the bottom section and the corner joints lapped as in Fig. 14-14b.

LAYING OUT TOP WALL PLATES

Select straight material for the top plates. Square off the ends and cut to the same lengths as the bottom plates. Place the top plate next to, or on top of, the bottom plate; nail it just enough to hold it in place. Then transfer all the layout marks from the bottom plate to the top plate. Fig.

14-16. There is another way of laying out these plates. Often it is better to cut top and bottom plates at the same time. The top plate can then be nailed temporarily to the sole plate and all the

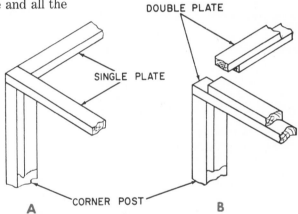

14-14a & b. *Plate construction at a corner.*

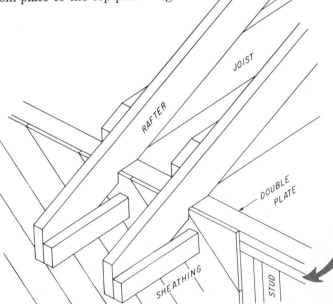

14-13. *The top plate is shown at the arrow. The double or rafter plate sometimes is not fastened until the walls have been erected, plumbed, and straightened.*

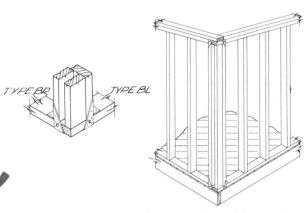

14-14c. *Metal framing anchors may also be used to attach the corner posts to the sole and top plate. Note that these brackets are made in lefts and rights and one of each is necessary.*

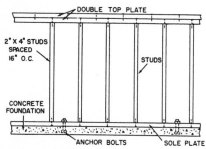

14-15. *A typical wall frame set on a concrete slab.*

14-16. *This carpenter is transferring the layout from the sole plate to the top plate. Note the use of the special template for laying out corners and partitions. Also of special interest are the two pencils in his right hand. The carpenter's pencil is used for location marks and the colored pencil is used for identification marks. The carpenter does not take time to change from one pencil to the other. He writes with either in the position they are held in his hand.*

layout done at one time. Figs. 14-17 & 18.

Studs

Studs are a series of slender wood members placed in a vertical position as supporting parts of a wall or partition. Studs should be 2 × 4s for both one- and two-story buildings. As mentioned earlier, short studs that are some-

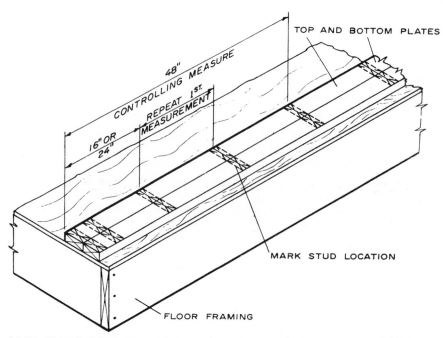

14-17. *This drawing shows measuring to the centers of the stud locations. In actual layout a carpenter will mark only the left edge of the stud and then place an X to the right of this line. The line can then be seen along the edge of the stud for alignment.*

times added above or below the header of doors and windows are called *cripple studs* or *cripples*. Studs which support headers are sometimes called *trimmer studs*.

The number of full studs, trimmers, and cripples can be determined by counting the layout marks on the wall plates. The lengths of these pieces can be found on the story pole or master stud pattern, as will be explained later. Standard studs can also be purchased precut to finished length. Fig. 14-19.

The stud locations are laid out on the sole and top plate, usually 16″ O.C. Begin the layout on the front wall. Measure the first 16″ from the outside-corner edge of the corner post to the center of the first stud; from that point on, measure 16″ O.C. for each of the studs.

When beginning the layout of the first stud on an end wall, the center of the first stud is measured from the outside of the sheathing line on the front wall.

Fig. 14-20. The same is true for any wall parallel to an end wall, as when there is an offset in the building. The layout of the stud locations on the back wall and all parallel walls should begin from the same end of the building as the layout of the front. By using a common measuring point, the studs will bear over the floor joists, while the ceiling joists and rafters bear over the studs. This alignment creates a direct bearing of the rafter right down through to the foundation wall of the building. Fig. 14-21a.

The studs are nailed with either two 16d nails through the plate and into the ends of the stud or they are toenailed into the plate with four 8d nails, two on each side of the stud. Fig. 14-21b.

STORY POLE AND MASTER STUD PATTERN

A *story pole* is a fullsize layout of a cross section of a wall. A *master stud pattern* is the same for a portion of the wall. One or

14-18a. *The outside wall plates have been nailed temporarily to the outside of the joist header (box or band), and the studs are being laid out with a homemade layout template. This template is 4' long, the standard width of wall covering. The template has 4 fingers, 1½" wide and 16" O.C., representing 4 stud markings. The fingers are attached to a piece of angle iron, so that the template works like 4 try squares at once. It is made of aluminum for light weight and protection from moisture.*

14-18b. *Laying out the plates for a partition wall. Note the use, in a different position, of the template described in Fig. 14-18a.*

the other of these layout devices should be developed for any wall-framing job, as a quick reference for use by all the carpenters. The use of such a device speeds construction and helps eliminate costly errors. This is particularly true when there are many carpenters on a single job or when a building contractor is supervising several jobs.

The story pole will show floor level (or levels if the structure is a multi-story or split level), ceiling height, window and door elevations, and thicknesses of many materials used in construction. Usually, however, only a portion of the wall section is laid out full-size. This portion would be from the top of the subfloor to the bottom of the ceiling joist; or in a two-story building it would be the distance between the finished floors. As mentioned, this drawing of a portion of the wall is called the master stud pattern. It includes information about the location and size of the window

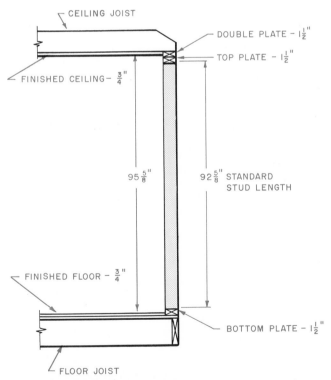

14-19. *A standard stud is 92⅝" long. This stud length will provide a 95⅝" ceiling height with a ¾" finished floor and a ¾" finished ceiling. Some studs are now being cut 93" for a 96" (8') ceiling.*

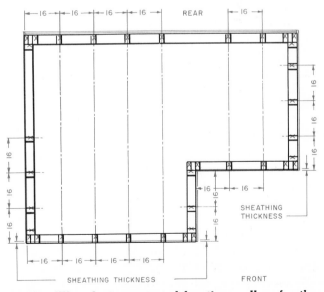

14-20a. *When laying out stud locations, allow for the thickness of the sheathing at the corners. Note that all layout dimensions begin from the left end of the building as one stands in front. This will line up the studs on the front and rear walls.*

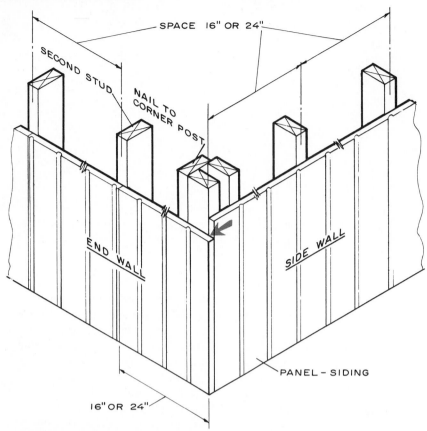

SPACE 16" OR 24"

SECOND STUD

NAIL TO CORNER POST

END WALL

SIDE WALL

PANEL — SIDING

16" OR 24"

14-20b. *The application of panel-siding or sheathing to a stud wall. Notice the center of the first stud on the side wall is measured from the outside of the corner post. However, the first stud on the end wall is measured from the outside surface of the wall covering material to the center of the first stud. This will permit the wall covering material to lap properly at the corner. See arrow.*

headers, sills, door headers, the heights of various openings above the subfloor, and the thicknesses of the ceiling and the finished floor. Fig. 14-22. If there are several different heights to mark off above the subfloor, it may be necessary to avoid confusion by making additional master stud patterns, labeling them for the various rooms or areas.

CORNER POST

A corner post must form an inside corner and an outside corner. The inside corner will provide two good nailing bases for inside wall covering. The outside corner will provide two good nailing bases for outside wall sheathing. Studs at the corners of the frame construction are usually built up from three or more ordinary studs to provide greater strength. Corner posts may be made in several different ways. Two of the more common methods are shown in Figs. 14-23a & b. The number of corner posts required can be determined by counting the number of places on the plan where two walls intersect at right angles. Corner posts are nailed together with 10d and 16d nails. They are distributed on the subfloor where they will be used for assembly of the wall sections. The short pieces of 2 × 4s shown at the base of the corner post in Figs. 14-23a & b are installed after the

wall erection to provide places for nailing the corner ends of the baseboard.

PARTITION CORNER POST

Studs should be arranged at a point where a partition ties into a wall between the corners. Three common types of partition post assemblies are shown in Fig. 14-24a, b, & c. In the type shown in Fig. 14-24a, the regular spacing of the outside wall studs is interrupted by double studs at the point where the partition ties in. The double studs are set 3" apart. This interval allows the partition end stud (which is 3⅝" wide) to lap the others just enough to permit nailing, while leaving most of the inner edges of the others clear to serve as nailing bases for inside wall covering. A variation of the method in Fig. 14-24a is shown in Fig. 14-24b. This method will give more nailing surface for the inside wall covering.

In another type of partition corner assembly, the regular spacing of the outside wall studs is maintained. Fig. 14-24c. The cross blocks are made of stud stock, with the exception of the bottom block, which is made of wider stock to provide a solid nailing base for ends of baseboard.

The number of partition corners can be counted from the plan. The partition corner posts are nailed together with 10d and 16d nails and distributed to their locations on the subfloor for wall assembly.

CRIPPLE STUDS

Cripple studs are those studs which, because of an opening in the wall, do not extend from the sole plate to the top plate. These studs are installed over window, door and fireplace headers and below window sills. Fig. 14-10b. They are located in the same place that a full stud would normally be placed if there were no

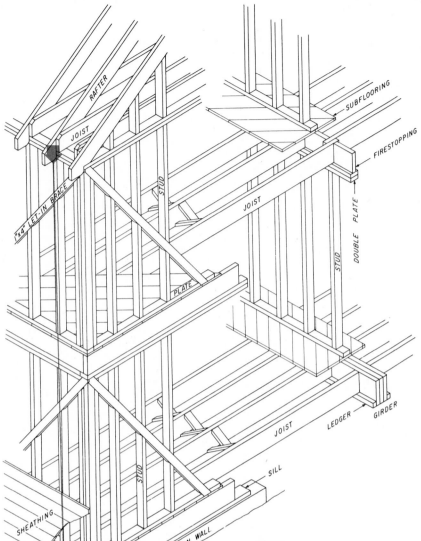

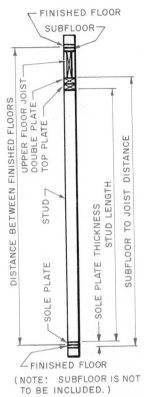

14-22a. A master stud pattern for a two-story dwelling. Additional information may be included on this layout, as is shown in Fig. 14-22b.

14-21a. The layout of the stud locations coincides with the joist layout. Later the rafters can be laid out directly over the studs and joists. This creates a direct bearing from the roof right down to the footings of the building. Although this alignment is not required when a double plate is used, it is recommended.

opening. Cripples are necessary for nailing outside sheathing and inside wall covering. The lengths of the various cripples can be determined by referring to the master stud pattern. These studs are usually precut to length and distributed with the door and win-

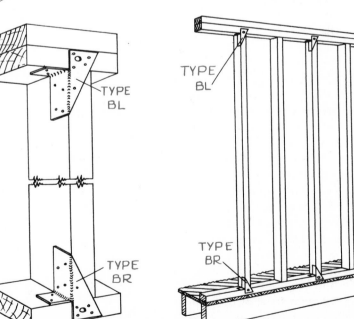

14-21b. Metal framing anchors may be used either inside or outside to attach the studs to the sole and to the top plate. These brackets are normally used only on alternate studs.

111

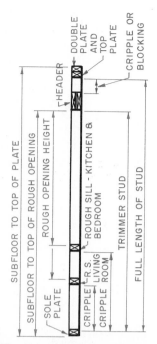

14-22b. *A master stud pattern for a one-story building. Select a clean, straight piece of 1″ × 4″ or 2″ × 4″ stock and lay it out full size. The information for this layout is taken from the building plans.*

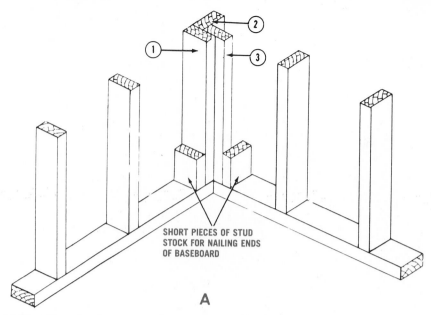

14-23a. *The simplest type of corner post. The pieces numbered 1, 2, and 3 are selected straight standard studs.*

14-23b. *This is the corner post most commonly used. The pieces numbered 1, 2, and 3 are selected straight standard studs. The short blocks are usually 10″ or 12″ long. Studs 1 and 3 are nailed to the blocks with 10d nails. Stud #2 is then nailed to the assembly of 1 and 3. Care should be taken to keep all ends and outside edges flush and even.*

dow headers in readiness for the assembly of the wall sections.

TRIMMER STUDS

As mentioned earlier, trimmers are studs which support the header over an opening. They are shorter than standard studs, so they are sometimes called cripples. Fig. 14-25. Sometimes they are also called *double studs* or *lap studs*. These studs are nailed to a regular stud under the ends of the header. To hold them in place, 10d nails are used, spaced 16″ apart and staggered as shown in Figs. 14-25 & 14-26.

Note that in Fig. 14-26 the trimmers at the window are continuous from the header to the sole plate. The trimmers can also be broken by the bottom or sill header, as shown in Fig. 14-53. There are advantages and disadvantages to each method. The ad-

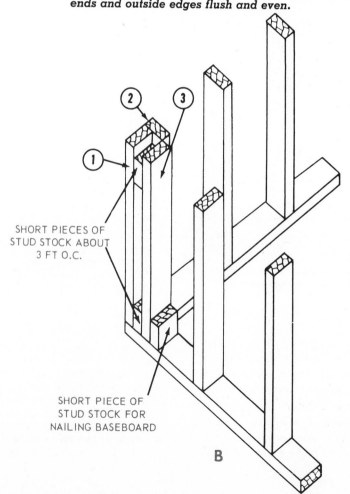

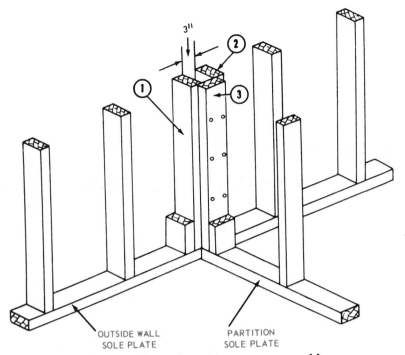

14-24a. *Double-stud partition corner assembly.*

vantage of two-piece trimmers is that less material is needed. Short pieces of stud material from other cuttings can be used. The disadvantage is that the trimmers are broken at the sill. Shrinkage in the sill can allow the header to settle, causing plaster cracks. However, if quality materials are used and if the trimmers and the cripples under the sill are cut to fit tightly, there should be no problem. Using continuous trimmers will reduce shrinkage. It is necessary, however, to install additional cripples under the sill. Check with the local building inspector to make sure the method you select is approved in your area.

The trimmer or double stud for a door may extend from the

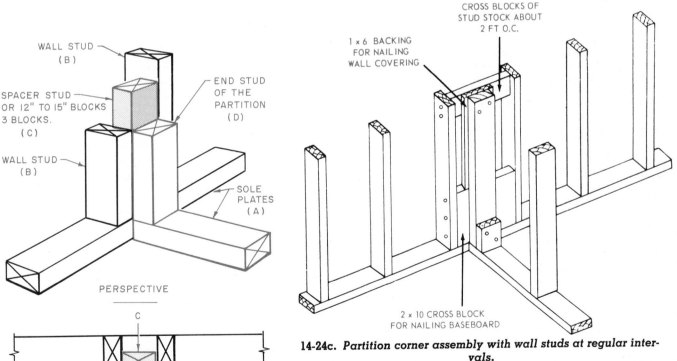

14-24c. *Partition corner assembly with wall studs at regular intervals.*

PERSPECTIVE

PLAN VIEW

14-24b. *This partition corner assembly is the one most commonly used. It will give more nailing on the inside corners than does the one shown in Fig. 14-24a. Nail the two wall studs to the spacer with 10d nails. The end stud of the partition can be nailed to these studs at this time. Or, it can be nailed up as part of the partition wall and then attached to the spacer at the time the partition is erected.*

113

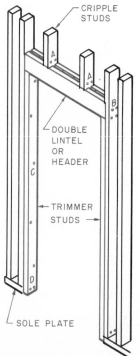

14-25a. *Door opening in a wall or partition. The cripple studs (A) are nailed with four 8d nails, two on each side. The standard studs (B) are nailed to the header with four 16d nails on each side and toe-nailed to the sole plate with two 8d nails. Or the full stud (B) could be nailed from the bottom up through the plate, with two 16d nails if the sole is attached before the wall is erected. The trimmer is nailed with 10d nails at C and staggered 16" O.C. Two 10d nails are driven into the end of the sole at D.*

header to the subfloor instead of resting on the sole plate. Fig. 14-25a.

If this method is used, the sole plate must be cut away between the full studs before the trimmers can be installed. Trimmers should be cut to fit snugly under the header so that they will support it properly. If a header settles, plaster cracks may develop and doors may fit improperly. The double studs in a door opening also form solid supports when the door is slammed (on the latch side) and for the weight of the door (on the hinge side).

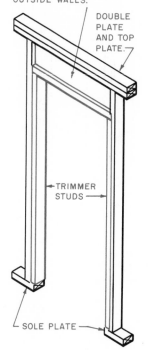

14-25b. *Alternate method of framing a door opening.*

Headers

Where windows or doors occur in outside walls or partitions, parts of some studs must be cut out. It is necessary, therefore, to install some form of header over the doorway to support the lower ends of studs that have been cut. Likewise, at the bottom of a window opening the "rough sill" supports the upper ends of studs that have been cut. Fig. 14-10b. The width of a header is determined by the length of the opening it must span. This information will be available from the building plans or local code requirements. Fig. 14-27.

Headers are sometimes built-up trusses as shown in Fig. 14-28. In some cases 4" stock is used rather than two pieces of 2" material nailed together. This saves work and also allows the thickness of

the header to be exactly the same as the width of a 2" × 4" stud. Fig. 14-29a. (When two 2" members are used for a header, the total thickness is only 3". This requires a ½" spacer to give the header the full 3½" width of the stud.) Fig. 14-29b.

Framing wide openings such as double garage doors which require headers 16' to 18' long can be done with nailed plywood box beams. These headers can be fabricated on or off the building site. The design and construction of these headers is shown in Fig. 14-29c. The ends of the headers should be supported on studs or by framing anchors, depending on the local code requirements. Figs. 14-30 & 14-31. The header lengths are obtained by measuring the layout of the bottom wall plates. The header is measured between the full studs. Fig. 14-11. In the case of the header for the door shown in Fig. 14-12a, the header length would be 37¾".

It is best to number the openings (such as windows, doors, and fireplaces) for identification and then make a cutting schedule for all headers. One person can cut these to length and, if 2" material is used, nail them together. Use 16d nails, two near each end, and stagger the others 16" apart along the length of the header. Don't forget to use ½" spacers between the 2" members where the nailing occurs. Headers are then distributed to their locations on the subfloor in readiness for the assembly of the wall sections.

CONTINUOUS HEADER

A *continuous header*, consisting of two 2" members set on edge, may be used instead of a double top plate. The width of the header will be the same as that required to span the largest opening. Joints in individual members should be staggered at least three

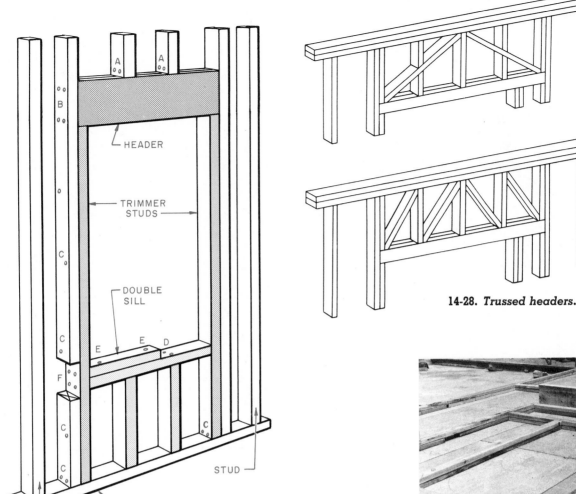

14-28. Trussed headers.

14-26. Window opening. *The cripple studs at A are toenailed with four 8d nails, two on each side. The full stud is nailed to the header at B with four 16d nails, to the trimmer (double stud) at C with 10d nails 16" O.C., and it is toenailed to the sole at the bottom. The lower part of the double sill is nailed with two 10d nails into the ends of the cripples at D. The upper part of the sill is nailed to the lower with 10d nails 8" O.C. and staggered.*

14-29a. *Plates and headers laid out on the subfloor in preparation for the wall assembly. Notice the use of the solid 4" header.*

Nominal depth of lintels made of two thicknesses of nominal two-inch lumber installed on edge	Interior Partitions Or Walls				Exterior Walls		
	Limited attic storage	Full attic storage, or roof load, or limited attic storage plus one floor	Full attic storage plus one floor, or roof load plus one floor, or limited attic storage plus two floors	Full attic storage plus two floors, or roof load plus two floors	Roof, with or without attic storage	Roof, with or without attic storage, plus one floor	Roof, with or without attic storage, plus two floors
4 in.	4 ft.	2 ft.	Not permitted	Not permitted	4 ft.	2 ft.	2 ft.
6 in.	6 ft.	3 ft.	2 ft. 6 in.	2 ft.	6 ft.	5 ft.	4 ft.
8 in.	8 ft.	4 ft.	3 ft.	3 ft.	8 ft.	7 ft.	6 ft.
10 in.	10 ft.	5 ft.	4 ft.	3 ft. 6 in.	10 ft.	8 ft.	7 ft.
12 in.	12 ft. 6 in.	6 ft.	5 ft.	4 ft.	12 ft.	9 ft.	8 ft.

*Supported loads include dead loads and ceiling.

14-27. *These header widths (lintel spans) may be used for most residential buildings. Wider openings often require trussed headers.*

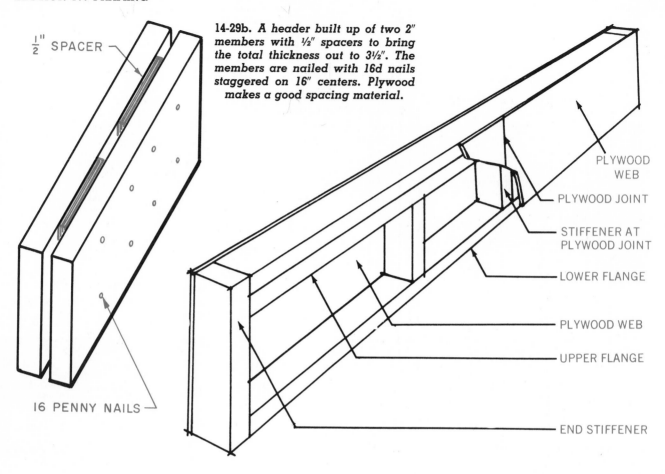

½" SPACER

**14-29b. A header built up of two 2"
members with ½" spacers to bring
the total thickness out to 3½". The
members are nailed with 16d nails
staggered on 16" centers. Plywood
makes a good spacing material.**

PLYWOOD
WEB

PLYWOOD JOINT

STIFFENER AT
PLYWOOD JOINT

LOWER FLANGE

PLYWOOD WEB

UPPER FLANGE

16 PENNY NAILS

END STIFFENER

16' Span Garage Door Header   24-ft-wide building   25 psf L.L.   (420 pounds lineal foot total load)

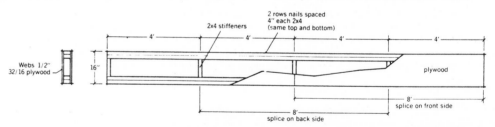

18' Span Garage Door Header   28-ft-wide building   40 psf L.L.   (700 pounds lineal foot total load)

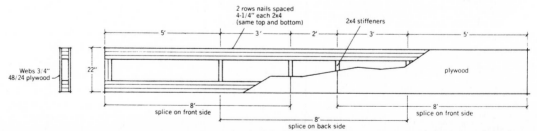

**14-29c. A nailed plywood box beam may be constructed as a header to span a large opening. First, nail the single
piece 2 × 4s together flatwise with 10d nails to form the top and bottom flanges. Then 2 × 4 stiffeners are set in
place at the beam ends and between the flanges behind the plywood web joints. Over this, plywood webs are then
nailed in place with 8d common nails to complete the assembly.**

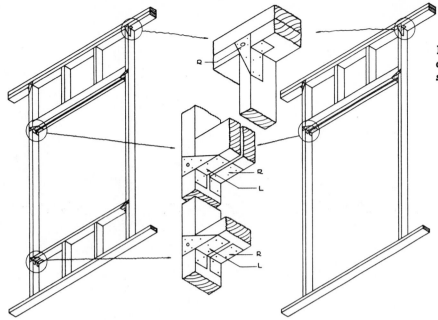

14-30. Window framing with framing anchors used to support the header, sill, and studs. (R and L indicate right and left.)

ever, there are two methods (with slight variation) that are used most often.

In the *first* of the common methods the studs, cripples, trimmers, headers, corners, and partition corners are all precut to length. Then they are distributed to the area on the subfloor where they will be assembled. (Some preassembly of wall members may have been done. Fig. 14-34.)

stud spaces and should not occur over openings. The header is toe-nailed to studs and corners. At intersections with bearing partitions, the members are taped or tied with metal straps. Fig. 14-32.

EXTERIOR WALLS, ASSEMBLY AND ERECTION

Several procedures can be used when assembling wall sections. Whichever method is used, the order in which the exterior walls are assembled and erected must first be determined. Usually the front and rear walls (the longest sections) should be set up first, and the side walls then erected in between. Fig. 14-33. When assembling wall sections, some carpenters first set and plumb the corner posts on the sole plate. Then they raise the studs and toe-nail them in place separately. Next, they nail on the top plate. In a very different procedure, the wall members can be assembled on the subfloor, the wall squared, sheathing applied, windows installed, and even siding applied before the wall is erected. How-

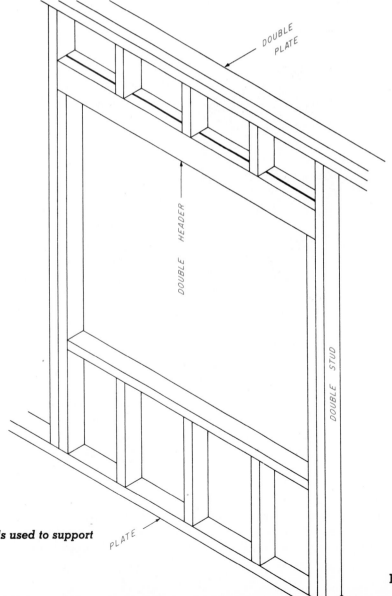

14-31. Window framing with trimmer studs used to support the header.

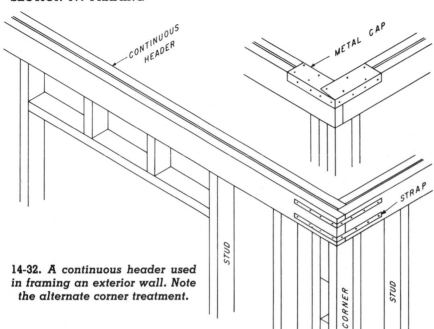

14-32. *A continuous header used in framing an exterior wall. Note the alternate corner treatment.*

14-34. *The rough sill and cripples have been preassembled and laid with the header for wall assembly later.*

14-33. *The front and rear walls have been set up. The materials for the end walls are being distributed in preparation for assembly.*

14-35. *Assembling the wall sections on the subfloor in preparation for erection. This man is nailing the window trimmer in place.*

subfloor about 8′ out from the sole plate. Face the marks on the top plate toward the sole plate. Lay a stud at each mark and place the header so that the rough sill, cripples, and trimmers are in position. Also place the preassembled corners and partition corners at the marked locations. Beginning at one end of the top plate, nail these members at the locations marked. Drive two 16d nails through the top plate into each member. Be careful to keep the edges of the members flush with each other. This is necessary for a smooth application of the interior and exterior wall covering later.

In this method of assembly, the sole plate has already been aligned and nailed securely in place on the subfloor. (See "Sole Plate", page 104.) The wall is now ready to be raised. Since the studs, cripples, trimmers, corners, and partition corners are nailed only to the top plate, the bottom ends will all be loose. Push the loose ends of the assembled wall section up against the

The wall sections are assembled on the subfloor. Fig. 14-35. Begin the assembly by removing the premarked top plate from its temporary nailing on the sole plate. Lay the top plate on edge on the

sole plate. Depending on the size of the wall section, two or more workers are needed to lift the wall to a vertical position. Fig. 14-36.

Nail some braces temporarily so they can be adjusted later when the wall is plumbed and straightened. Fig. 14-37. Each of

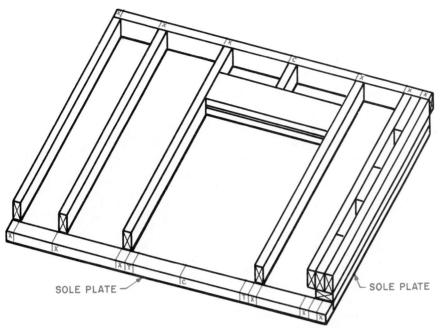

14-36a. *Studs of the assembled wall are pushed up against the sole plate opposite the marks where they are to be nailed.*

14-38. *This carpenter is nailing the sole plate to the ends of the studs before the wall is raised.*

the studs is toenailed down to its mark on the sole plate with 8d nails, two to each side of a stud.

In the *second* common method of assembling wall sections, the sole plate is nailed temporarily to the subfloor for layout and pulled free for assembly. In this method the nails are driven up through the bottom of the sole into the ends of the wall members. Fig. 14-38. The sheathing and siding can be fastened to the studs while the walls are still on the subfloor. If preferred, the wall framing can be set up, squared, and braced, and the sheathing applied later. Figs. 14-39, 14-40, and 14-41. When the sole plate is nailed to the other wall members before erection, a chalk line is snapped

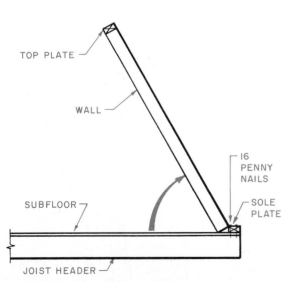

14-36b. *The wall is raised to the vertical position. The ends of the studs are pushed against the sole plate which has been securely nailed in place.*

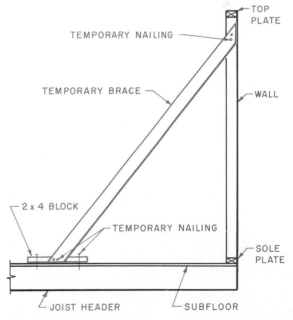

14-37. *The wall is held in a vertical position while a second worker nails the temporary braces.*

14-39. *Checking the assembled wall section for squareness by measuring diagonally across the corners. The two diagonal measurements must be equal.*

on the subfloor to show the exact location of the plate. The sole is straightened as it is nailed in position alongside the chalk line. The bottom plate is fastened to the floor framing with 16d nails

14-41. *These carpenters have completed the assembly of the wall framing, including the double plate, before raising the wall section.*

spaced 16″ apart and staggered when practical. The wall can now be plumbed and temporary bracing added to hold it in place in a true vertical position. Figs. 14-42 & 43.

Temporary Bracing

Temporary bracing may consist of 1″ × 6″ members nailed to one face of a stud and to a 2″ × 4″ block which has been nailed to the subfloor. Braces should be at about a 45° angle. Fig. 14-37. These braces should be nailed to the inside of the walls. This permits attaching the permanent bracing and sheathing without moving any temporary braces. Care should also be taken not to project the ends of the temporary braces above the top plate. Otherwise, the braces could interfere with ceiling and roof framing and would have to be removed. Moving these braces at this time would disturb the plumbed and straightened walls. Use enough nails to hold the wall section securely, but do not drive the nails in all the way. The nailhead should project enough to allow easy withdrawl for removal later. The temporary bracing is left in place until the ceiling and the roof framing are completed and sheathing is applied to the outside walls.

PLUMBING AND STRAIGHTENING THE WALL SECTIONS

Before the walls can be straightened, all exterior and intersecting corners must first be plumbed and temporarily braced. Either a level or a plumb bob may be used to plumb the wall sections.

Using a Plumb Bob

To plumb a corner with a plumb bob, first attach to the bob a string long enough to extend to

14-40. *These carpenters are using another variation of assembly. Both plates have been nailed to the studs while the wall was lying on the subfloor. After the wall is set up, the window framing will be installed either as a preassembled unit or a piece at a time.*

or below the bottom of the corner post. Lay a rule on top of the post so that 2″ of the rule extends over the post on the side to be plumbed; then hang the bob line over the rule so that the line is 2″ from the post and extends to the bottom of it. Fig. 14-44 (1). With

14-42. *Nailing the temporary bracing.*

14-43. *This wall, including the sole plate, was nailed together on the subfloor and raised into position. Notice the pieces of 1″ × 4″ material nailed to the joist header. These pieces keep the assembled wall from sliding off the subfloor as it is being raised.*

another rule, at the bottom of the post measure the distance from the post to the line. If the distance is not 2″, the post is not plumb. Move the post inward or outward until the distance from the post to the line is exactly 2″. Then nail the temporary brace in place. Repeat this procedure from the other outside face of the post. The post is then plumb. This process is carried out for the remaining corner posts and partition posts.

An alternate method of plumbing a post is shown in Fig. 14-44 (2). Attach the plumb bob string securely to the top of the post to be plumbed. Make sure that the string is long enough to allow the plumb bob to hang near the bottom of the post. Use two blocks of wood identical in thickness as gauge blocks. Tack one block near the top of the post between the plumb bob string and the post (gauge block No. 1). Insert the second block between the plumb bob string and the bottom of the

post (gauge block No. 2). If the entire face of the second block makes contact with the string, the post is plumb.

Using a Level

To plumb a corner with a level, do not place the level directly against a stud, because the face or edge of the stud is likely to be irregular in shape. Instead, make a long straightedge. Figs. 14-45 & 46. The straightedge should have a couple of lugs, for placing against the stud, and an edge on which to place the level. Check

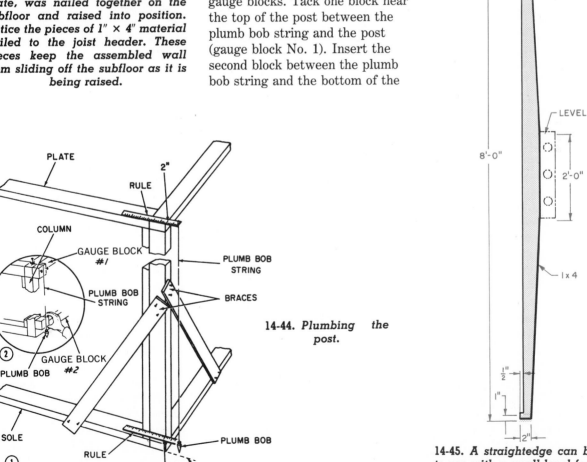

14-44. *Plumbing the post.*

14-45. *A straightedge can be made to use with a small level for plumbing the posts.*

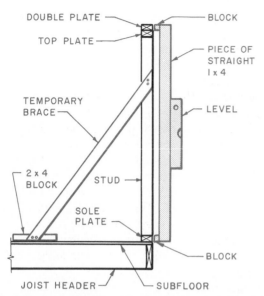

14-46. *The straightedge can simply be a piece of 1" × 4" stock. Make sure the edge on which the level is placed is parallel to a line drawn between the two blocks nailed on the ends.*

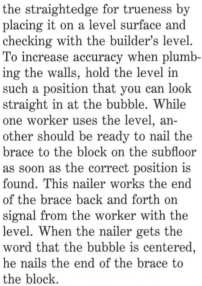

14-47. *Using a chalk line as a reference to straighten the walls.*

the straightedge for trueness by placing it on a level surface and checking with the builder's level. To increase accuracy when plumbing the walls, hold the level in such a position that you can look straight in at the bubble. While one worker uses the level, another should be ready to nail the brace to the block on the subfloor as soon as the correct position is found. This nailer works the end of the brace back and forth on signal from the worker with the level. When the nailer gets the word that the bubble is centered, he nails the end of the brace to the block.

Outside corners must be plumbed and braced both in and out and side to side. After all exterior corners have been plumbed and braced, the intersecting interior partition posts are plumbed and braced. The partition posts need to be plumbed in one direction only (in and out). This will plumb the exterior wall at the point of intersection. If the top and bottom plates have been cor-

rectly cut and laid out, the partition corners will be plumb from side to side after the exterior corner posts are plumbed. After all exterior and partition corner posts have been plumbed and nailed in place with temporary bracing, the wall sections between the posts should be straightened. If necessary, they should be held with additional temporary bracing.

Straightening Walls

To straighten walls, fasten a chalk line to the outside of one of the corner posts at the top. Stretch the line to the corner post at the opposite end of the building, and fasten the line to this post in the same manner as for the first post. Place a small wooden block, ¾″ thick, under each end of the line to give clear-

ance. Fig. 14-47. Place additional temporary braces at intervals close enough to hold the wall straight. When the wall is far enough away from the line to permit a ¾″ block to slide between the line and the plate, the brace is nailed. This procedure is carried out for the entire perimeter of the building. Inside partition walls should be straightened later in the same manner.

PARTITIONS—ASSEMBLING AND ERECTING

Partition walls divide the inside space of a building. These walls in most cases are framed as part of the building. Partition walls are of two types—*bearing* and *nonbearing*. A bearing wall supports ceiling joists, while a nonbearing wall supports only itself. Partition walls are framed in the same

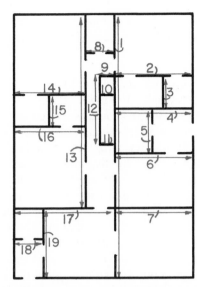

14-48. *Sequence of assembling and erecting interior partitions. After reading the text you will understand why the sequence was suggested. There are other possible sequences that would be worth considering. Can you suggest one that might have its own advantages?*

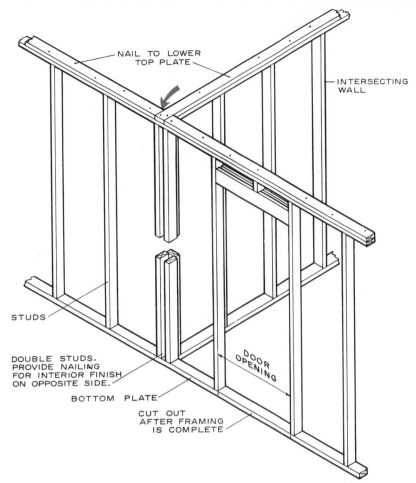

14-49. *The top plate is usually fastened in place after the walls have been plumbed and straightened. Ten-penny nails are used in the nailing pattern to attach the double plate.*

manner as outside walls. After all the exterior walls are set up, plumbed, braced and the sole plate securely nailed, the interior walls (partitions) are assembled and erected. The top and bottom plates for the partitions are cut and laid out in the same way as described for exterior walls (pages 105 and 107). The sizes of the various partition parts (such as headers, trimmers, and cripples) can be learned from the master stud pattern and the building plans. These parts are cut to size, marked for identification, and then distributed to the areas on the subfloor where they are to be assembled. Assembly of the partitions is also the same as described for outside walls. Careful planning of the order in which the partitions are assembled and erected is very important. A floor plan with a suggested sequence of assembly and erection is shown in

Fig. 14-48. The next two paragraphs will explain some of the reasons for the order in which the steps are carried out.

The first operation is to raise, fasten, and temporarily brace the longest center partition. Work then proceeds from one end of the building to the other and from the center wall out to the exterior walls. To save time and effort, operations in one area are completed before moving to the next areas.

Still referring to Fig. 14-48, notice that partition 2 helps support the center partition and connects it to the previously plumbed exterior wall. The third partition is at right angles to the second, and the fourth is at right angles to the

third. This pattern continues to the rear of the building. This is better than erecting two parallel partitions (such as 2 and 4) and then having to work in a confined area to erect the connecting partition (3).

The Double Plate or Rafter Plate

The double plate or rafter plate is applied over the top plate. Fig. 14-49. The same kind of material is used as for the top plate. The pieces must be cut accurately to length. This can be done either at the same time when the sole and top plates are cut or after the walls are erected. Note, however, that the rafter plates are not cut exactly the same length as the

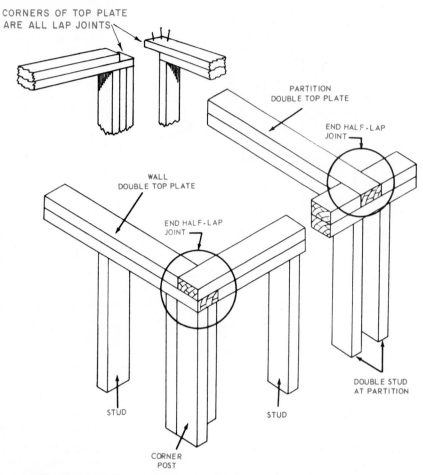

CORNERS OF TOP PLATE
ARE ALL LAP JOINTS

PARTITION
DOUBLE TOP PLATE

END HALF-LAP
JOINT

WALL
DOUBLE TOP PLATE

END HALF-LAP
JOINT

DOUBLE STUD
AT PARTITION

STUD

STUD

CORNER
POST

14-50. *The double top plates are joined together with half-lap joints.*

other two plates. Since one of the primary purposes of this plate is to tie the walls together at the top, the rafter plate laps over the joint formed at a corner of intersecting wall. Fig. 14-50. On a long wall, if joints are necessary in the rafter or double plate, they should be at least 4' from any joint in the top plate. Fasten the double plate with 10d nails space 16″ O.C. End laps between adjoining plates are nailed with two 16d nails on each lap. (See arrow, Fig. 14-49.)

BRACING

Bracing stiffens framed construction and makes it rigid. Good bracing keeps corners square and plumb. It also prevents warping, sagging, and shifting. Without

bracing, the shape changes in the frame would cause badly fitting doors and windows and cracked plaster.

It is important to have the frame properly nailed together. The bracing is then securely nailed to hold the framing rigid. Figure 14-51 provides a review of correct nailing procedure and shows the bracing in place: (1) The sole plate is nailed to a joist or header joist with 16d nails spaced 16″ O.C. (2) The top plate is end-nailed to the studs with two 16d nails. (3) Studs are toe-nailed to the sole plate with two 8d nails on each side. (4) Doubled studs are nailed together with 10d nails 16″ O.C. (5) Top plates are spiked together with 10d nails 16″ O.C. (6) The top plates, laps, and

intersections are nailed together with two 16d nails. (7) A 1″ brace is nailed to each stud and plate with two 8d nails. (8) Corner studs and multiple studs are nailed with 10d nails 12″ O.C. Other joints should be nailed to provide proportional strength.

Frame walls may be braced at the corners by diagonal members (usually 1″ × 4″) set in gains cut into the plates, studs, and corner posts. Fig. 14-51. To lay out these gains, place the bracing members in position against the framing members. Then score (mark) the outline on each stud or plate with a scratch awl. Diagonal braces must be properly nailed at each crossing stud with not less than two 8d nails and at the ends with three 8d nails. Fig. 14-52. A single diagonal extending from floor line to ceiling line should cross at least three stud spaces. When a wall opening makes this impossible, a K-brace is installed. Fig. 14-53.

Sometimes diagonal bracing gains are laid out and cut into the framing before the walls are erected. Figs. 14-54 & 55. Then the braces are also nailed to the sole plate before wall erection. Fig. 14-56. When the wall sheathing provides adequate corner bracing, diagonal bracing may be omitted. Diagonal-laid wood sheathing or plywood is usually considered adequate bracing. Figs. 14-57 & 58. When other sheathing materials are used, a sheet of plywood is sometimes fastened to each side of each corner instead of diagonal bracing. Also if there is not enough room to brace adequately because of an opening in the exterior wall, a sheet of plywood will provide the necessary strength for the corner. Fig. 14-59.

SPECIAL FRAMING

Special framing includes such

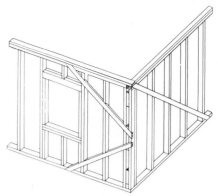

14-53. *A diagonal brace should reach from the double plate to the sole plate and cross 3 stud spaces. If this is not possible because of an opening, a K-brace should be used.*

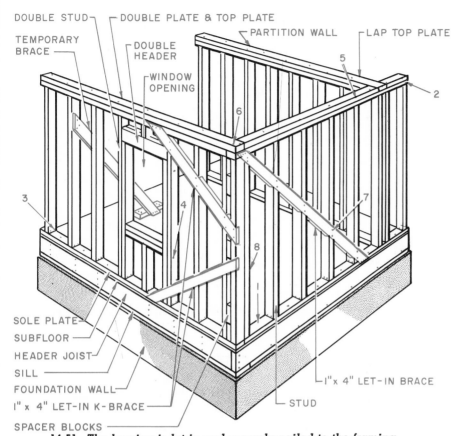

DOUBLE STUD
DOUBLE PLATE & TOP PLATE
TEMPORARY BRACE
DOUBLE HEADER
PARTITION WALL
LAP TOP PLATE
WINDOW OPENING
5
6
2
3
4
7
8
SOLE PLATE
SUBFLOOR
HEADER JOIST
SILL
FOUNDATION WALL
1" x 4" LET-IN K-BRACE
SPACER BLOCKS
1" x 4" LET-IN BRACE
STUD

14-51. *The bracing is let-in and securely nailed to the framing.*

14-52. *Diagonal bracing at the corner. Note the nailing pattern.*

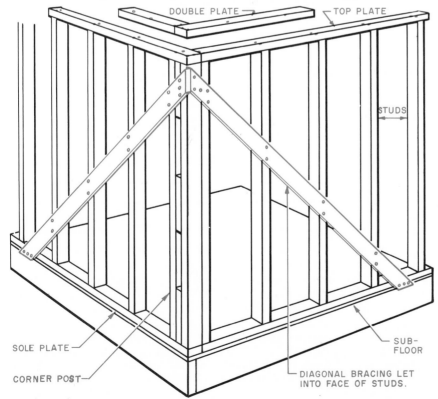

DOUBLE PLATE
TOP PLATE
STUDS
SOLE PLATE
CORNER POST
DIAGONAL BRACING LET INTO FACE OF STUDS.
SUB-FLOOR

14-54. *This carpenter has the power saw set to make the first cut into the studs. These cuts are made to a depth equal to the thickness of the brace stock. The saw blade is placed against the brace, which serves as a guide for cutting the gain.*

14-55. *The second cut is made with the saw held on its side to complete the gain for the brace.*

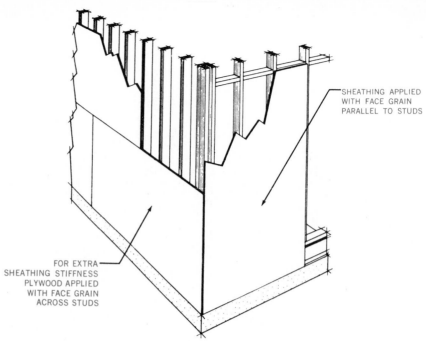

SHEATHING APPLIED WITH FACE GRAIN PARALLEL TO STUDS

FOR EXTRA SHEATHING STIFFNESS PLYWOOD APPLIED WITH FACE GRAIN ACROSS STUDS

14-56. *The brace is nailed securely at the bottom only (near top of photo). All other nails are just started. After the wall is set up and plumbed all the nails are driven home to hold the wall plumb.*

14-57. *Plywood used as sheathing eliminates the need for diagonal bracing.*

Wall Sheathing[1]

Panel Identification Index	Minimum Thickness (inch)	Maximum Stud Spacing (inches) Exterior Covering Nailed to:		Nail Size [2]	Nail Spacing (inches)	
		Stud	Sheathing		Panel Edges (when over framing)	Intermediate (each stud)
12/0, 16/0, 20/0	5/16	16	16[3]	6d	6	12
16/0, 20/0, 24/0	3/8	24	16 24[3]	6d	6	12
24/0, 32/16	1/2	24	24	6d	6	12

Notes: (1) When plywood sheathing is used, building paper and diagonal wall bracing can be omitted

(2) Common smooth, annular, spiral thread, galvanized box or T-nails of the same diameter as common nails (0.113" dia. for 6d) may be used. Staples also permitted at reduced spacing.

(3) When sidings such as shingles are nailed only to the plywood sheathing, apply plywood with face grain across studs.

Recommended minimum stapling schedule for plywood/All values are for 16-gauge galvanized wire staples having a minimum crown width of 3/8"

Plywood wall sheathing/Without diagonal bracing

Plywood Thickness	Staple Leg Length	Spacing Around Entire Perimeter of Sheet	Spacing at Intermediate Members
5/16"	1¼"	4"	8"
3/8"	1⅜"	4"	8"
½"	1½"	4"	8"

Typical grade-trademarks.

STRUCTURAL I
C-D
20/0 APA
INTERIOR
PS 1-74 000
EXTERIOR GLUE

C-D
32/16 APA
INTERIOR
PS 1-74 000
EXTERIOR GLUE

C-C
32/16 APA
EXTERIOR
PS 1-74 000

14-58. *Plywood paneling will have one of the four grade stamps shown here (arrow #1). After finding the grade stamp, use the top chart for information on nailing. For example, for a 12/0 panel, 5/16" thick (arrow #2) the maximum stud spacing is 16". The panel should be attached with 6d nails, with 6" spacing around the edges of the panel and 12" spacing on the studs in between. If the panels are to be stapled, refer to the bottom chart (arrow #3).*

14-59. *Using plywood for corner bracing.*

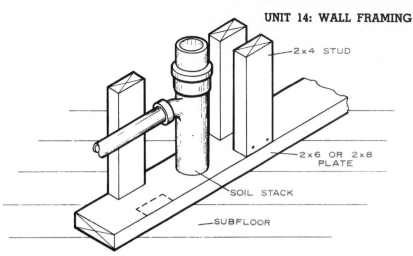

14-60. *A piece of material somewhat larger than 2″ × 4″ is used for a sole plate to provide room for a 4″ soil pipe. The 2″ × 4″ studs are then installed flatwise.*

operations as providing openings for heating vents and plumbing pipes, adding support for heavy items (such as bathtubs), and providing bases for nailing paneling or other covering materials. Such framing makes a carpenter's job easier and faster in the finishing stages and adds strength and quality to the construction.

Plumbing Stack Vents

Plumbing stack vents are usually installed by using a nominal 2″ × 6″ plate and placing the studs flatwise at each side. Fig. 14-60. This provides the needed wall thickness for the bell (large end) of a 4″ cast-iron soil pipe, which is larger than the thickness of a 2″ × 4″ stud wall. It is also possible to fur out several studs to a 6″ width at the soil stack, rather than thickening the entire wall. In areas where building regulations permit the use of 3″ vent pipe, a 2″ × 4″ stud wall may be used, but it requires reinforcing scabs at the top plate. Fig. 14-61. Use 12d nails to fasten the scabs.

Bathtub Framing

The floor joists in the bathroom which support the tub or shower should be arranged so that no cutting is necessary in connecting the drainpipe. This may require only a small adjustment in spacing the joists. Fig. 14-62. When joists are parallel to the length of

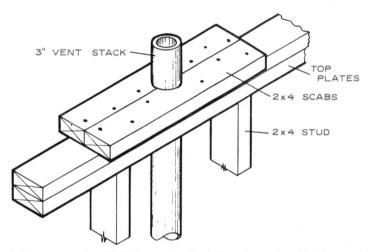

14-61. *A 3″ vent stack will fit in a standard 2″ × 4″ stud wall. The top plate, however, will have to be reinforced as shown here.*

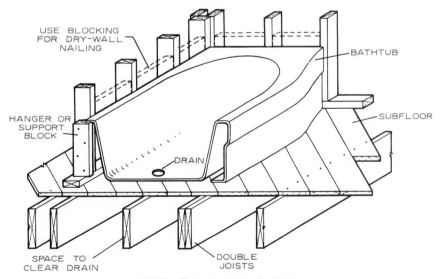

14-62. *Framing for a bathtub.*

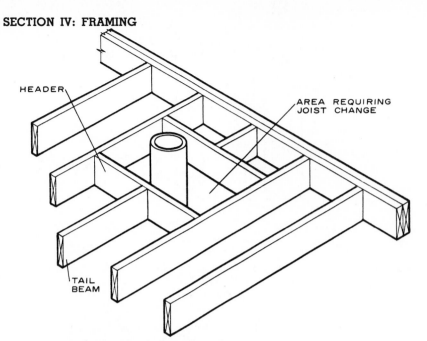

14-63. Headers are used to support joists which must be cut.

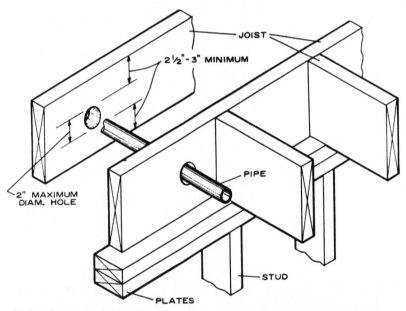

14-64. Do not weaken the joists by cutting or boring too large a hole.

Notching the top or bottom of the joist should be done only in the end quarter of the span. The notch should not be more than one-sixth the depth of the joist. Thus, for a nominal 2″ × 8″ joist, 12′ long, the notch should be not more than 3′ from the end support and about 1⅔″ deep. When a joist would require greater alteration, headers and tail beams can be used instead. Fig. 14-63. Proper planning will reduce the need for altering joists.

When necessary, holes may be bored in joists if the diameters are no greater than 2″ and the edges of the holes are not less than 2½″ from the top or bottom edges. Fig. 14-64. This usually limits a 2″-diameter hole to joists of nominal 2″ × 8″ size and larger.

Plumbing Fixtures

The weight and eventual use of plumbing fixtures requires a secure anchorage. This backing is provided at the necessary locations as shown in Fig. 14-65.

Cabinets and Utility Boxes

Special support and blocking must also be provided for most cabinets. Some cabinets are designed to fit between studs 16″ O.C. and flush with the wall covering. These are usually designed to be fastened from the inside of the cabinet, through the sides, directly into the studs. Backing must be provided at the top and bottom of the cabinet for nailing the wall covering.

Bathroom vanities, kitchen cabinets, and other projecting cabinets must be securely fastened. Thus they require special blocking. The location of this blocking can be determined by studying the building plans. One method of blocking for cabinets is shown in Fig. 14-66. In some cases it is advisable to install an extra full-

the tub, they are usually doubled under the outer edge. Tubs are supported at the enclosing walls by hangers or woodblocks. The wall in which the fixture plumbing is located should also be framed to allow for a small access panel.

CUTTING FLOOR JOISTS

Floor joists should be cut,

notched, or drilled only where they will not be greatly weakened. While it is best to avoid cutting, alterations are sometimes required. Joists should then be reinforced by nailing a 2″ × 6″ scab to each side of the altered member, using 12d nails. An additional joist adjacent to the cut joist can also be used.

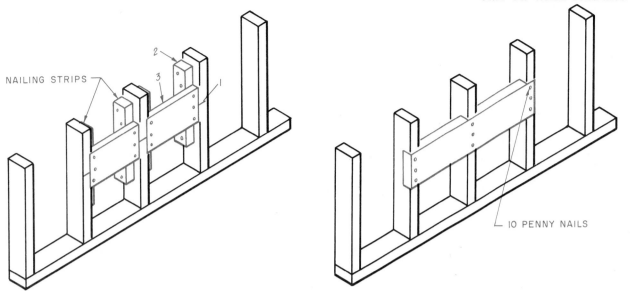

NAILING STRIPS

10 PENNY NAILS

14-65. *Blocking for hanging plumbing fixtures: 1. Determine the height of the fixture and mark the location. 2. Nail a block on the side of the stud, back from the edge a distance equal to the thickness of the backing material. 3. Cut the backing to fit between the studs and nail it in place. Note: The backing material could be gained into the studs (see "Corner Bracing") at the correct height and face-nailed with 10d nails.*

length stud or even a 2″ × 6″ flatwise in the wall. This is especially true in kitchens in which upper and lower cabinets are to be installed.

Trim Backing

Baseboard, chair rail, ceiling, and other moldings require additional backing for nailing at the ends. Without additional backing the nails must be driven very near the ends of the molding and usually at a slight angle to anchor into the corner posts. This often results in splitting the ends of the molding. Backing such as shown in Figs. 14-23 & 24 will permit correct nailing procedures.

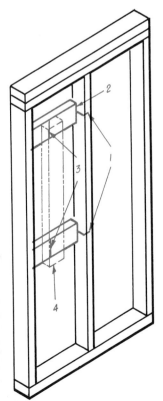

14-66. *Blocking for hanging cabinets: 1. Mark the top and bottom of the cabinets on the studs. 2. Fasten blocks between the studs for attaching the cabinet backing. These blocks must be back from the edge of the studs a distance equal to the thickness of the cabinet backing. 3. Mark the position of the cabinet backing on the blocks. 4. Fasten the cabinet backing to the blocks on the location marks.*

Backing for Accessories

Towel bars, shower curtain rods, soap dishes, tissue-roll holders and similar items should be supported by backing. These

items do not require backing as heavy as that for cabinets or plumbing fixtures. Usually a piece of 1″ × 4″ material located and fastened as in Fig. 14-65 will provide much better support than the wall covering alone.

Heating Ducts

Heating ducts require openings in the ceiling, floor, or wall. Backing must therefore be provided for fastening the covering material. Fig. 14-67. An opening in the wall larger than the distance between the studs will require cutting off one or more studs. It also requires a header to support the shortened stud, which serves as a nailing surface for the wall covering. Fig. 14-68. An opening in the floor larger than the distance between the joists is formed as shown in Fig. 14-63.

Cabinet Soffits

The cabinet soffit is sometimes called a *bulkhead*. There are two common types. The first allows the cabinet to be constructed flush with the front of the soffit.

129

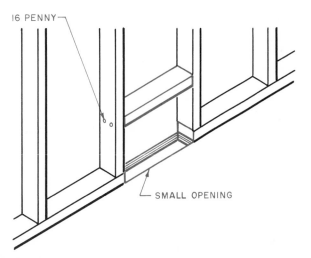

16 PENNY

SMALL OPENING

14-67. *Recommended framing for a small heating duct opening.*

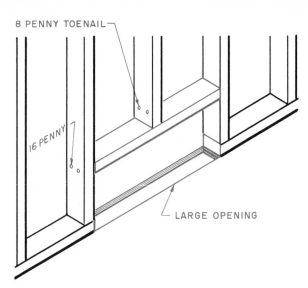

8 PENNY TOENAIL

16 PENNY

LARGE OPENING

14-68. *Recommended framing for a large heating duct opening.*

Fig. 14-69. This type is used mainly when the cabinets are built on the job. In the second type, the cabinets are set back about 2″. This method which is more common, is used with prefabricated cabinets. Such cabinets are usually not installed until after the interior painting is completed. Details can be seen in Fig. 14-70.

CONSTRUCTING SOFFITS FOR FLUSH-MOUNTED CABINETS

Snap a chalk line along the wall studs to represent the bottom edge of the soffit (arrow 1, Fig. 14-69). This line will usually be 84″ plus the thickness of the lath and plaster or dry wall above the finished floor. With 10d nails, nail a 2″ × 2″ along the top edge of this line into each wall stud.

Assume the shelves are to be 11¼″ deep. Snap a chalk line along the underside of the ceiling joists 11¼″ out from the double

14-69. *Soffit construction details for flush-mounted cabinets.*

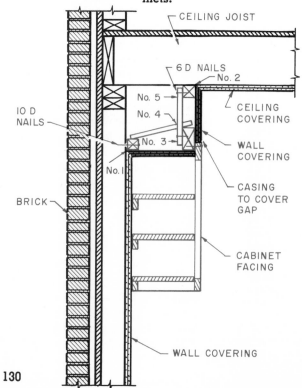

CEILING JOIST

6 D NAILS
No. 2
No. 5
No. 4
No. 3
No. 1

CEILING COVERING

WALL COVERING

CASING TO COVER GAP

CABINET FACING

10 D NAILS

BRICK

WALL COVERING

14-70. *Construction details for a conventional soffit.*

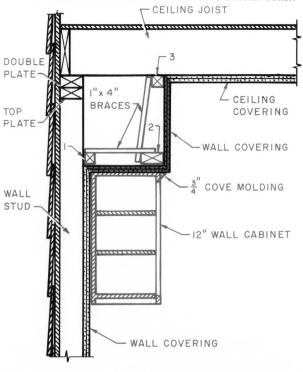

CEILING JOIST

DOUBLE PLATE

TOP PLATE

1″ x 4″ BRACES

3

2

CEILING COVERING

WALL COVERING

¾″ COVE MOLDING

12″ WALL CABINET

WALL STUD

WALL COVERING

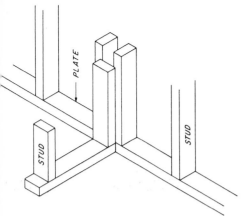

14-71. *This assembly of studs where a partition meets a wall will provide adequate nailing for the wall finish material.*

plate (arrow 2). This allows ¾″ for the lath and plaster to be applied later. With 10d nails fasten a 2″ × 2″ along this chalk line into the bottom edge of each ceiling joist.

(If the ceiling joists run the same direction as the soffit, a ladder-type construction, similar to that used when installing non-bearing partitions, will have to be used. The upper 2″ × 2″ for the bulkhead is nailed to the underside of the 2″ × 4″ ladder construction.)

Cut the 2″ × 4″ for the front bottom edge of the soffit to finished length. Nail 1″ × 4″ × 12″ pieces to the back (arrow 3) and top edge (arrow 4) of the 2″ × 4″. Temporarily tack the 2″ × 4″ in position by nailing it with 6d nails to the top 2″ × 2″ (arrow 5). With a level make sure that the 2″ × 4″ is level along its length and that its bottom edge is level with the bottom edge of the 2″ × 2″ nailed to the wall.

When the 2″ × 4″ is in position, use 6d nails to fasten the 1″ × 4″ pieces securely to the two 2″ × 2″. Additional 2″ × 4″ blocking may be required along the face or bottom edge of the soffit in order to join cabinets or install electrical outlet boxes.

CONSTRUCTING THE CONVENTIONAL CABINET SOFFIT

This type of soffit framing is used when the cabinets are prefabricated and installed later. Two 2″ × 2″ boards are installed, one on the wall (Fig. 14-70), (arrow 1) and the second on the ceiling (arrow 3). These are positioned and nailed in place as in the previous method. The differences are that the 2″ × 2″ along the ceiling is positioned 14″ from the face of the wall studs to the outside edge of this 2″ × 2″, and the 2″ × 4″ at the bottom corner is turned flat rather than on edge (arrow 2). After the wall covering has been applied and painted, the cabinets are attached to the wall and to the bottom of the soffit. The top front edge of the cabinet is securely attached to the 2″ × 4″ laid flat in the soffit. Since prefabricated cabinets are usually 12″ deep, this allows a 2″ overhang of the soffit at the front edge. A piece of cove or quarter-round molding may be used to close the joint between the cabinet and the bulkhead.

Wall and Ceiling Finish Nailers

Wall and ceiling finish nailers are horizontal or vertical members to which paneling, dry wall, rock lath, wall board, or other covering materials are nailed. These pieces are installed at the interior corners of walls and at the junction of the wall and ceiling. The vertical nailers located at the interior corners of the walls may be made up of studs placed where they will make a good base for nailing. Fig. 14-71. This method of construction also provides a good tie between walls.

A second method of installing a vertical nailer at the interior wall corners is provided by a 2″ × 6″ nailed to the stud of the intersect-

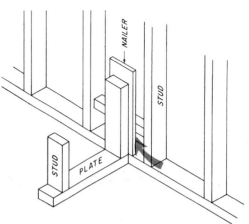

14-72. *Attach a 6″ board to the back of the last 2″ × 4″ and the wall partition to provide adequate nailing for the wall finish material. Note that a header is nailed between the wall studs so that the partition wall may be toenailed through the nailer into the header to secure the partition wall to the main wall.*

ing wall. In this method, a header nailed between the studs of the main wall is necessary to back up the nailer. Fig. 14-72.

At the junction of a wall and ceiling, doubling of the ceiling joists over the wall plates provides a nailing surface for the interior wall finish. Fig. 14-73. The walls are tied to the ceiling framing by toenailing through the ceiling joists into the wall plates.

Another method of providing nailing at the ceiling line is similar to that used on the walls. A 2″ × 6″ nailer is nailed to the top of the wall plate. Fig. 14-74. A header is nailed between the ceiling joists and then toenailed to the wall plate through the nailer.

Nailers need not be continuous. Several short pieces of 2″ × 6″ can be used rather than new material. Nailers should be firmly secured with 16d common nails so that they will not be hammered out of position when the wall finish is nailed in place. If nailers are out of position, the wall finish material will be out of square at the corner.

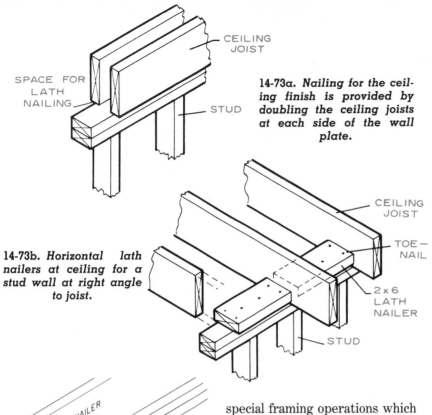

14-73a. Nailing for the ceiling finish is provided by doubling the ceiling joists at each side of the wall plate.

14-73b. Horizontal lath nailers at ceiling for a stud wall at right angle to joist.

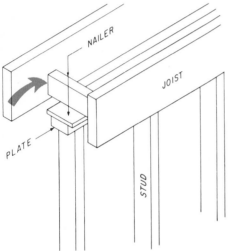

14-74. A nailer attached to the top of a wall plate. The header nailed between the joists (arrow) can then be toenailed into the top of the nailer to secure the partition to the ceiling.

Schedule for Special Framing

Usually there is a need to enclose a building quickly, to protect it from the weather. However, special framing is time consuming. Therefore during the regular framing stage a builder will take time to do only those special framing operations which must be done at that time. The rest of the special framing often is done as fill-in work during slack periods in later construction stages. The climate, other projects, and available manpower affect the scheduling of special framing.

MULTILEVEL FRAMING

The use of platform-frame construction simplifies the framing of multilevel structures. The first level of the two-story house is constructed in the same manner as for a one-story home. However, the ceiling joists of the first story become the floor joists of the second story. Therefore these joists must be wider to support the additional load.

A subfloor is applied to the top of the joists, the second-story floor plan is laid out, and the walls are constructed and raised in the same way as for the first floor. Fig. 14-75. Some architectural designs require an overhang of the second floor joists beyond the first story walls. The framing of the floor joists can be seen in Fig. 14-76.

Split levels and other variations of architecture require a slightly different treatment of the wall framing. After consulting the sectional elevations in the building plans, make a master stud pattern for each level. This pattern will show the true lengths of the studs for the stub walls found in split-level homes. Keep in mind that each floor platform is supported by a wall section. Thus, a second-level floor platform is supported by the walls which enclose the rooms of the lower level. Fig. 14-77.

Bay-window framing is sometimes associated with multilevel homes. Fig. 14-78 (page 136). A bay-window projection requires that the floor also be framed out beyond the foundation wall, as for a second-floor overhang. Fig. 14-76.

FRAMING FOR A FIREPLACE

Fireplace construction is basically the same regardless of design. The construction of a typical fireplace and recommended dimensions for essential parts or areas of fireplaces of various sizes are shown in Fig. 14-79 shown on page 137.

Footings

Foundation and footing construction for chimneys with fireplaces is similar to that for chimneys without fireplaces. The footings should extend at least 6″ beyond the chimney on all sides and should be 8″ thick for one-story houses and 12″ thick for two-story houses with basements. Be sure the footings rest on good firm soil below the frostline.

Hearth

The hearth consists of two parts: the front, or finish, hearth

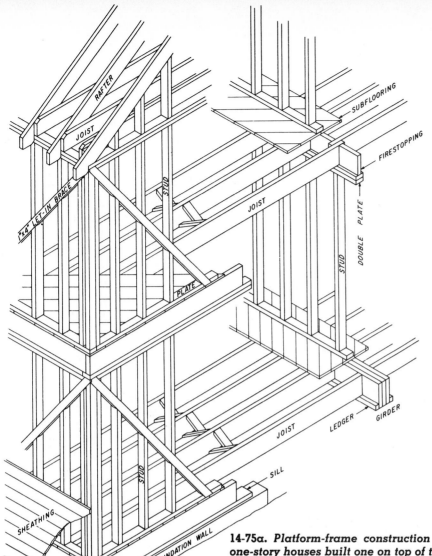

and the back hearth, under the fire. Because the back hearth must withstand intense heat, it is built of heat-resistant materials, such as firebrick. The front hearth is simply a precaution against flying sparks. While it must be noncombustible, it need not resist intense prolonged heat. The hearth should project at least 16″ from the front of the fireplace and should be 24″ wider than the fireplace opening. (That is, the hearth should be 12″ wider than the fireplace on each side.)

The hearth can be flush with the floor so that sweepings can be brushed into the fireplace, or it can be raised. Raising the hearth to various levels and extending its length as desired is presently common practice, especially in contemporary design. If there is a basement, a convenient ash dump can be built under the back of the hearth. Fig. 14-80 (page 138).

14-75a. Platform-frame construction in a two-story house. This is just two one-story houses built one on top of the other under a single roof. Remember, however—because of the additional weight of the second level—larger support members are needed to carry the load.

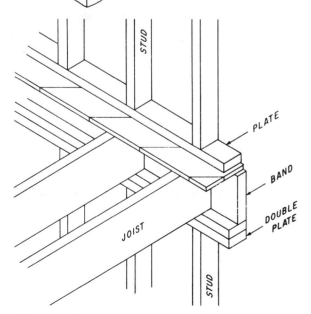

14-75b. Detail of the second-floor framing at the exterior wall.

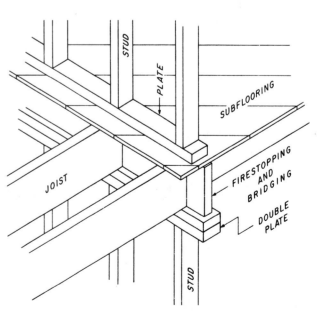

14-75c. Detail of the second-floor framing over a bearing partition.

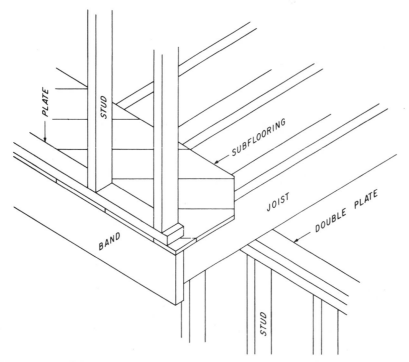

14-76a. *Second-floor overhang on an exterior wall. Since these joists run at right angles to the wall below, they can just be extended.*

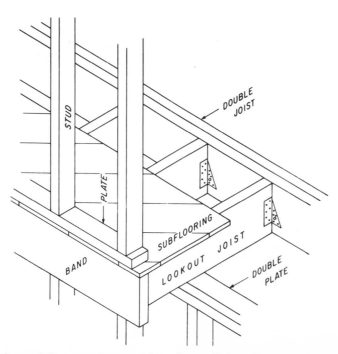

14-76b. *Second-floor overhang with wall parallel to the joists. Lookout joists are used and must be attached to a double joist. The double joist should be located a distance of twice the overhang back from the lower wall.*

In buildings with wooden floors, the hearth in front of the fireplace should be supported by masonry trimmer arches or other fire-resistant construction. Fig. 14-81 (page 138). Wood centering under the arches used during construction of the hearth and hearth extension should be removed when construction is completed. Figure 14-82 on page 138 shows the recommended method of installing floor framing around the fireplace hearth.

Walls

Building codes generally require that the back and sides of fireplaces be constructed of solid masonry or reinforced concrete at least 8″ thick and that they be lined with firebrick or other approved noncombustible material not less than 2″ thick or with steel lining not less than ¼″ thick. Such lining may be omitted when the walls are of solid masonry or reinforced concrete at least 12″ thick.

Jambs

The jambs of the fireplace should be wide enough to provide stability and to present a pleasing appearance. For a fireplace opening 3′ wide or less, the jambs can be 12″ wide if a wood mantel will be used or 16″ wide if they will be of exposed masonry. For wider fireplace openings, or if the fireplace is in a large room, the jambs should be proportionately wider. Fireplace jambs are frequently faced with ornamental brick or tile. Fig. 14-79.

No woodwork should be placed within 6″ of the fireplace opening. Woodwork above and projecting more than 1½″ from the fireplace opening should be placed at least 12″ above the top of the fireplace opening. The mantel height above the opening can also be figured by adding 6″ to the width of the mantel. Fig. 14-83 (page 139).

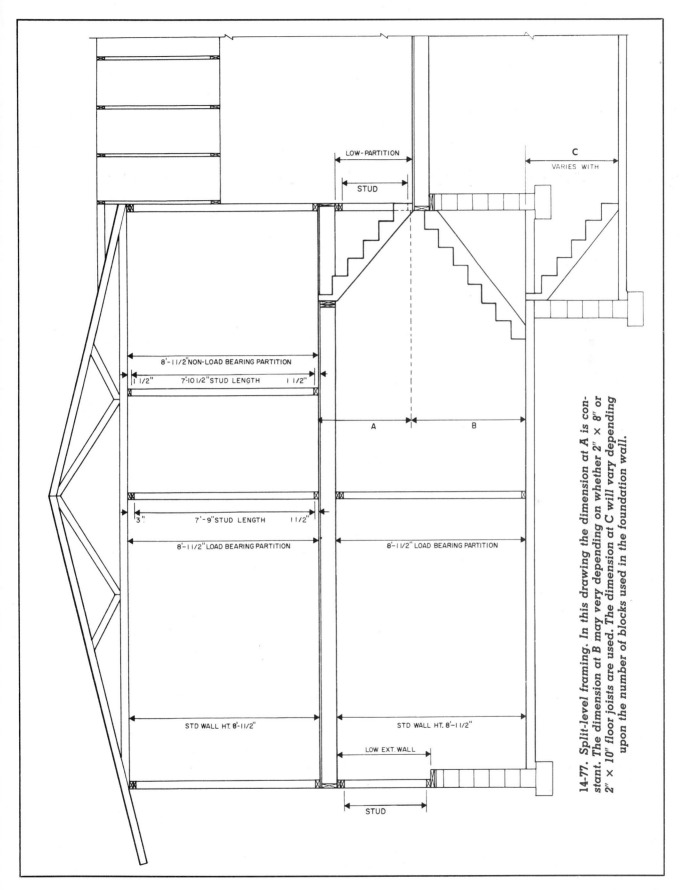

14-77. *Split-level framing. In this drawing the dimension at A is constant. The dimension at B may very depending on whether 2" × 8" or 2" × 10" floor joists are used. The dimension at C will vary depending upon the number of blocks used in the foundation wall.*

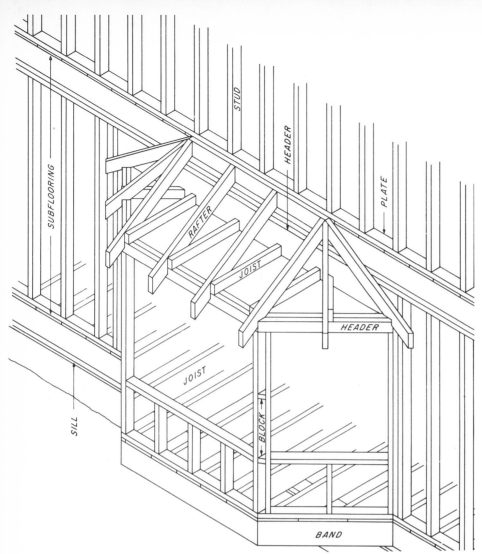

Labels on diagram: SUBFLOORING, STUD, HEADER, RAFTER, JOIST, PLATE, HEADER, JOIST, BLOCK, SILL, BAND

14-78. *Bay window framing. The ceiling joists in the framing of this bay are set on top of the window headers. The top of a bay window should be kept in line with the other windows and doors in the room. Therefore the wall header will not be a standard header height. It will have to be raised so that its bottom is in line with the bottom of the bay ceiling joists.*

Lintels

A lintel must be installed across the top of the fireplace opening to support the masonry. For fireplace openings 4′ wide or less, ½″ × 3″ flat steel bars, 3½″ × 3½″ × ¼″ angle irons, or specially designed damper frames may be used. Wider openings will require heavier lintels.

If a masonry arch is used over the opening, the fireplace jambs must be heavy enough to resist the thrust of the arch.

Throat

Proper construction of the throat area is essential for a satisfactory fireplace. Fig. 14-79, f-f. The sides of the fireplace must be vertical up to the throat, which should be 6″ to 8″ or more above the bottom of the lintel. The area of the throat must be not less than that of the flue. The length must be equal to the width of the fireplace opening, and the width of the throat will depend on the width of the damper frame (if a damper is installed). Five inches above the throat (at e-e in Fig. 14-79) the sidewalls should start sloping inward to meet the flue (at t-t in Fig. 14-79).

Recommended Dimensions For Fireplaces And Size of Flue Lining Required

Size of fireplace opening		Depth	Minimum width of back wall	Height of vertical back wall	Height of inclined back wall	Size of flue lining required	
Width w Inches	Height h Inches	d Inches	c Inches	a Inches	b Inches	Standard rectangular (outside dimensions) Inches	Standard round (inside diameter) Inches
24	24	16–18	14	14	16	8½ x 8½	10
28	24	16–18	14	14	16	8½ x 8½	10
30	28–30	16–18	16	14	18	8½ x 13	10
36	28–30	16–18	22	14	18	8½ x 13	12
42	28–32	16–18	28	14	18	13 x 13	12
48	32	18–20	32	14	24	13 x 13	15
54	36	18–20	36	14	28	13 x 18	15
60	36	18–20	44	14	28	13 x 18	15
54	40	20–22	36	17	29	13 x 18	15
60	40	20–22	42	17	30	18 x 18	18
66	40	20–22	44	17	30	18 x 18	18
72	40	22–28	51	17	30	18 x 18	18

Damper

A damper consists of a cast iron frame with a hinged lid. This hinged lid opens or closes to vary the throat opening. Dampers are not always installed, but they are recommended, especially in cold climates.

With a well-designed, properly installed damper, you can:
• Regulate the draft.
• Close the fireplace flue to prevent loss of heat from the room when there is no fire in the fireplace. In the summer, the flue should be closed to prevent loss of cool air from the air-conditioning system.
• Adjust the throat opening according to the type of fire and thus reduce heat loss. For example, a roaring pine fire may require a full throat opening, but a slow-burning hardwood log fire may require an opening of only 1″ or 2″. Closing the damper to the opening will reduce loss of heat up the chimney.
• Close or partially close the flue to prevent loss of heat from the main heating system. When air heated by a furnace goes up a chimney, an excessive amount of fuel may be wasted.
• Close the flue in the summer to prevent insects from entering the house through the chimney.

Dampers of various designs are on the market. Some support the

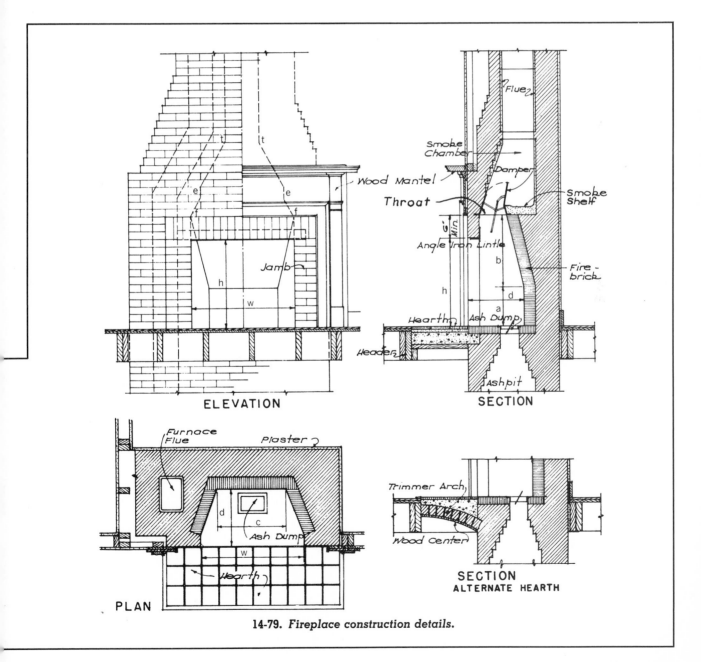

14-79. *Fireplace construction details.*

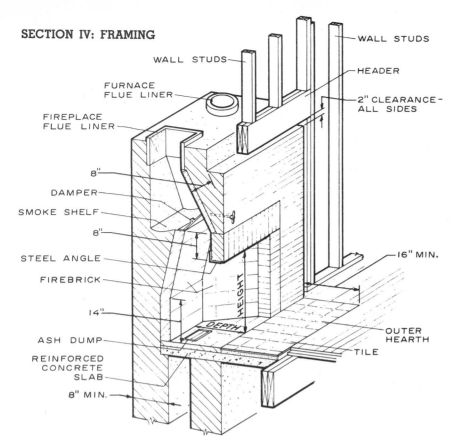

WALL STUDS

FURNACE FLUE LINER

FIREPLACE FLUE LINER

WALL STUDS

HEADER

2" CLEARANCE- ALL SIDES

8"

DAMPER

SMOKE SHELF

8"

STEEL ANGLE

FIREBRICK

14"

ASH DUMP

REINFORCED CONCRETE SLAB

8" MIN.

HEIGHT

DEPTH

16" MIN.

OUTER HEARTH

TILE

14-80. *Fireplace construction and framing details.*

14-82b. *A fireplace opening framed in a house without basement. Note the reinforcing rod which will be used to tie the exterior fireplace masonry to the building.*

JOIST

DOUBLE HEADER

CONCRETE SLAB

CENTERING

14-81. *Hearth centering detail.*

masonry over fireplace openings, thus replacing ordinary lintels. It is important that the full damper opening equal the area of the flue.

Smoke Shelf and Chamber

A smoke shelf prevents downdraft. Fig. 14-79. It is made by setting the brickwork at the top of the throat back to the line of the flue wall for the full length of the throat. Depth of the shelf may be 6″ to 12″ or more, depend-

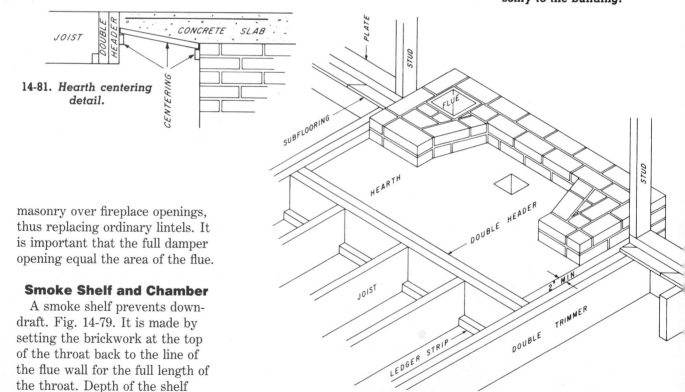

PLATE

STUD

FLUE

SUBFLOORING

HEARTH

DOUBLE HEADER

STUD

JOIST

2" MIN

LEDGER STRIP

DOUBLE TRIMMER

14-82a. *Floor framing details around a fireplace.*

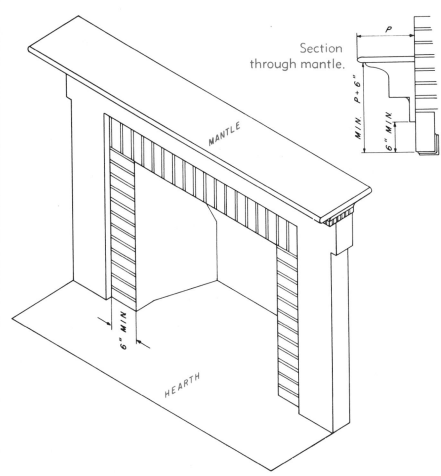

14-83. *Trim clearance around a fireplace opening. The mantel should be 6″ plus the width of the mantel above the top of the fireplace opening.*

ing on the depth of the fireplace. The smoke shelf is concave to retain any slight amount of rain that may enter.

The smoke chamber is the area from the top of the throat (e-e in Fig. 14-79) to the bottom of the flue (t-t in Fig. 14-79). As stated under "Throat" the sidewalls should slope inward to meet the flue. The smoke shelf and the smoke chamber walls should be plastered with cement mortar at least ½″ thick.

Fireplace Flue

Proper proportion between the area of the fireplace opening, area of the flue, and height of the flue is essential for satisfactory operation of the fireplace. The area of a lined flue 22′ high should be at least one-twelfth of the area of the fireplace opening. The area of an unlined flue or a flue less than 22′ high should be one-tenth of the area of the fireplace opening. The table in Fig. 14-79 lists dimensions of fireplace openings and indicates the size of flue lining required.

QUESTIONS

1. What is the purpose of the exterior side wall in a structure?

2. What are the requirements for wall framing lumber?

3. Which of the two general types of wall framing is used most often in the United States?

4. Describe briefly the difference between balloon and platform framing.

5. List the names of several parts used in wall framing.

6. What is meant by a rough opening?

7. The width of a rough opening is measured between what two wall members?

8. What is the purpose of the top plate?

9. When laying out stud locations, why is it important for the carpenter to begin from the same side on the back and front of the house?

10. What is a story pole?

11. Describe a header and its purpose.

12. Which method of wall assembly would you prefer and why?

13. Describe how the carpenter can determine when the exterior wall is straight.

14. What is the difference between a bearing and a nonbearing wall?

15. List several examples of special framing.

16. Describe briefly the difference between the two types of cabinet soffits.

17. Why is platform framing easier than balloon framing for multilevel structures?

Wall Sheathing

15

Exterior coverings over wall framing commonly consist of a sheathing material to which is added finish siding, brick, stone, or other exterior wall material. The inner layer of the outside wall covering on a frame structure is called the *sheathing* (usually pronounced "sheeting"). The outer layer is called the *siding*. Siding, because it is not a structural element, is considered a part of the exterior finish. The sheathing, because it strengthens and braces the wall framing, is considered a structural element. It is therefore a part of the framing. Sheathing forms a flat base upon which the siding is applied and adds not only strength but also insulation to the house. If the type of sheathing used does not provide stiffness and rigidity, diagonal corner bracing should be used. Sheathing is sometimes eliminated from houses built in mild climates.

TYPES OF SHEATHING

The four most common types of sheathing used on modern structures are wood, plywood, fiberboard, and gypsum. Fig. 15-1.

Wood

Wood sheathing consists usually of $1'' \times 6''$ or $1'' \times 8''$ boards, but thicker and/or wider stock is sometimes used. Boards may be square-edged for ordinary edge-butt joining, or they may be shiplapped or dressed-and-matched.

Fig. 15-2. *Dressed-and-matched* is simply a term which is used instead of *tongue-and-groove* with reference to sheathing, siding, or flooring.

Plywood

Plywood wall sheathing covers large areas fast and adds great strength and rigidity to the structure of the house. A plywood-sheathed wall is twice as strong and rigid as a wall sheathed with diagonal boards. Therefore let-in corner bracing is not needed with plywood wall sheathing. Plywood also holds nails well and makes a solid nailing base for the finished siding. Plywood sheathing comes in sheets 4' wide and 8' or more long and is squared-edged. It may be applied either vertically or horizontally. Fig. 15-3. Plywood and hardboard are also used as exterior coverings without sheathing, but grades, thicknesses, and types vary from normal sheathing requirements. This phase of wall

construction will be discussed in Unit 28, "Exterior Wall Coverings."

Fiberboard

Fiberboard (sometimes called insulation board) is a synthetic

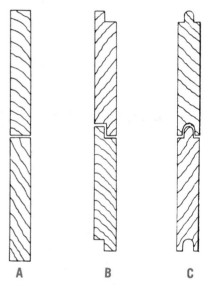

15-2. *Types of edges on wood sheathing: A. Square. B. Shiplap. C. Dressed-and-matched.*

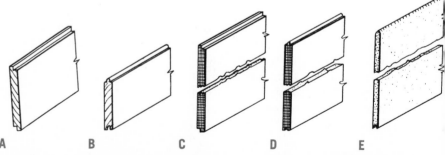

15-1. *Some common types of sheathing: A. Wood shiplap. B. Wood tongue-and-groove. C. Fiberboard shiplap. D. Fiberboard tongue-and-groove. E. Gypsum.*

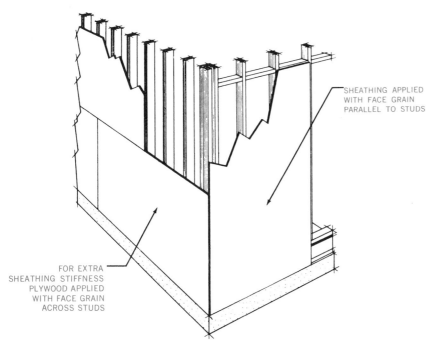

15-3. *Plywood sheathing may be applied either horizontally or vertically. When applied horizontally, additional blocking should be included at the horizontal joint between the studs as a base for nailing.*

SHEATHING APPLIED WITH FACE GRAIN PARALLEL TO STUDS

FOR EXTRA SHEATHING STIFFNESS PLYWOOD APPLIED WITH FACE GRAIN ACROSS STUDS

15-4. *Installing asphalt-impregnated insulating sheathing vertically.*

one of several stages during construction:

● When the wall is lying on the subfloor, completely framed and squared. The advantage in applying the sheathing at this time is that it can be nailed in place while the wall sections are lying flat, thus eliminating ladders or scaffolding. The disadvantage is the added weight that must be lifted when erecting the walls.

● When the wall frames have been erected, plumbed, and braced.

● When the ceiling joists have been installed and the wall frames plumbed and braced.

● When the roof has been framed or after the roof frame has been covered.

Most carpenters apply the sheathing as soon as possible because it adds strength and rigidity to the structure. Walls that have been covered with sheathing provide a more solid structure for the ceiling and roof members. Scaffolding is usually erected as soon as the sheathing is applied. Most workers apply the sheathing before the ceiling joists or roof framing so that they will have the scaffolding to stand on when framing the roof.

Wood

Wood sheathing may be applied either horizontally or diagonally. Fig. 15-5. Horizontal sheathing is more often used because it is easy to apply and there is less lumber waste than with diagonal sheathing. However, horizontal sheathing requires diagonal corner bracing in the wall framework. Diagonal sheathing is applied at a 45° angle. This method of sheathing adds greatly to the rigidity of the wall and eliminates the need for corner bracing. There is more lumber waste than with horizontal sheathing because of angle cuts. As stated, the application is

material usually coated or impregnated with asphalt to increase water resistance. Fiberboard sheathing is commonly used in 2′ × 8′ sheets, which are usually applied horizontally, and in 4′ × 8′ sheets, which are applied vertically. Fig. 15-4. Edges are usually shiplapped or dressed-and-matched for joining. Thickness is normally 25/32″.

Gypsum

Gypsum sheathing consists of a treated gypsum filler faced on both sides with lightweight paper. Gypsum sheathing comes in 2′ × 8′ sheets which are applied horizontally. Sheets are usually dressed-and-matched, with V-shaped grooves and tongues. This makes application easier and adds a small amount of tie between sheets. E, Fig. 15-1.

APPLICATION OF SHEATHING

Sheathing may be applied at any

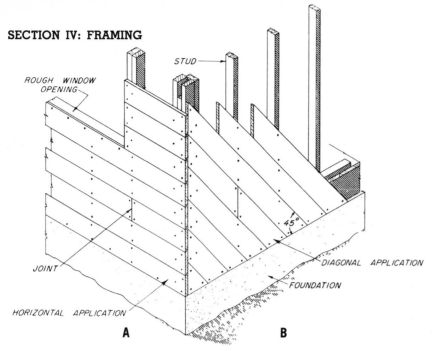

15-5. Wood sheathing application: A. Horizontal. B. Diagonal.

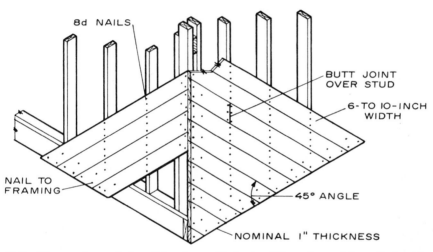

15-6. Diagonal wood sheathing is applied at a 45° angle and is nailed with 8d nails to the framing. Note that the pieces are joined at the center of a stud.

meaning that no two end joints may lie next to each other on the same stud. Fig. 15-6. If end-matched boards are used, end joints may lie in the spaces between studs, but end joints in adjacent courses (each strip of sheathing is called a course) must not lie in the same stud space.

Before nailing, each board should be driven tightly against the board that is already in place. Boards at openings should be laid or cut to bring the boards exactly flush with trimmers, headers, or subsills. Sheathing should normally be carried down over the outside floor framing members. This provides an excellent tie between wall and floor framing. Fig. 15-7.

Fiberboard

The most popular size of fiberboard and gypsum sheathing is 2′ × 8′. This size is applied horizontally with the vertical joints staggered. Fig. 15-8. Fiberboard sheathing should be nailed at all studs with 2″ galvanized roofing nails or other types of noncorrosive nails. Space nails 4½″ on center or 6 nails for every 2′ of height. Nails should be kept at least ⅜″ from the edge of the sheet. Fig. 15-8.

Fiberboard in 4′ × 8′ sheets is usually applied vertically because

somewhat more difficult. The building specifications will state whether horizontal or diagonal application is required.

With either method of application, 6″ or 8″ boards should be nailed with two 8d nails at each stud crossing. Wider boards should be nailed with three 8d nails. Unless boards are end-matched (shaped on the ends for tongue-and-groove joining), end joints must lie on the centers of studs. End joints must be broken,

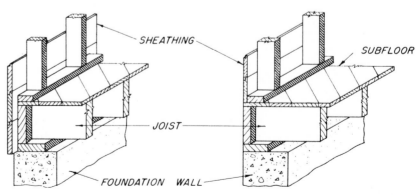

15-7. Location of sheathing: A. Sheathing may be set on the foundation wall. B. Sheathing may be set on the subfloor.

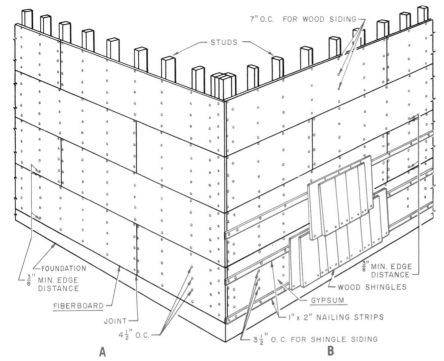

15-8. *Horizontal application of 2' × 8' sheathing: A. Fiberboard. B. Gypsum.*

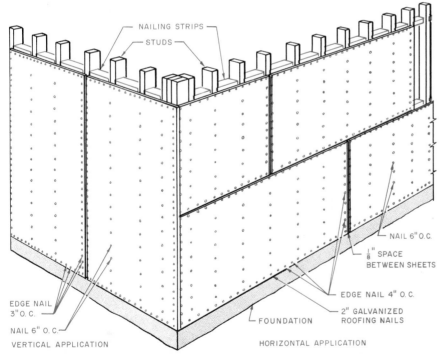

15-9. *Fiberboard in 4' × 8' sheets may be applied either vertically or horizontally. When the application is horizontal, the edge nailing is 4" on center. When the application is vertical, the edge nailing is 3" on center.*

nails spaced 3" on center at the edges and 6" on center at intermediate framing members. The minimum edge distance is ⅜". It is also recommended that the sheets be spaced ⅛" apart to allow for expansion and avoid buckling. Joints should come on the centerline of framing members. Fig. 15-9.

Gypsum

Gypsum sheathing is nailed with 1¾" or 2" galvanized roofing nails. When the exterior finish to be used is one in which siding nails carry through into the studs, the roofing nails for the sheathing may be spaced either 7" on center or 4 nails for every 2' of height. Otherwise, nails should be spaced 3½" on center (7 nails in 2' height). Fig. 15-8.

Plywood

Plywood used for sheathing is usually in 4' × 8' sheets and should be a minimum of 5⁄16" thick for studs spaced 16". Six-penny nails should be used, spaced not more than 6" apart on edge members and 12" on intermediate members. Fig. 15-10.

Plywood is usually applied vertically, using perimeter nailing with no additional blocking. Fig. 15-11. When the sheathing is started at the foundation wall rather than the subfloor, use a 2" × 4" nailing strip between the studs for backing. Fig. 15-12.

Plywood is applied horizontally in the same way as 4' × 8' fiberboard sheathing. Fig. 15-9. However, the type of nail and nail spacing is different for plywood than for fiberboard. Fig. 15-10a & b. Blocking is desirable at the horizontal joint between studs as a base for nailing.

When finish siding requires nailing between studs (as with wood shingles), the plywood should be ⅜" thick. If 5⁄16" ply-

perimeter nailing (nailing around outside edges) is possible. When the sheathing is started at the foundation wall rather than the

subfloor, use a 2" × 4" nailing strip between the studs for backing. The sheathing should be nailed with 2" galvanized roofing

Wall sheathing[1]

Panel Identification Index	Minimum Thickness (inch)	Maximum Stud Spacing (inches) Exterior Covering Nailed to:		Nail Size [2]	Nail Spacing (inches)	
		Stud	Sheathing		Panel Edges (when over framing)	Intermediate (each stud)
12/0, 16/0, 20/0,	5/16	16	16[3]	6d	6	12
16/0, 20/0, 24/0	3/8	24	16 24[3]	6d	6	12
24/0, 32/16,	1/2	24	24	6d	6	12

Notes: (1) When plywood sheathing is used, building paper and diagonal wall bracing can be omitted

(2) Common smooth, annular, spiral thread, galvanized box or T-nails of the same diameter as common nails (0.113″ dia. for 6d) may be used. Staples also permitted at reduced spacing.

(3) When sidings such as shingles are nailed only to the plywood sheathing, apply plywood with face grain across studs.

Look for these APA grade-trademarks on wall sheathing.

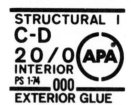

15-10a. Plywood wall sheathing application details.

Recommended minimum stapling schedule for plywood / All values are for 16-gauge galvanized wire staples having a minimum crown width of 3/8″

Plywood wall sheathing / Without diagonal bracing

Plywood Thickness	Staple Leg Length	Spacing Around Entire Perimeter of Sheet	Spacing at Intermediate Members
5/16″	1 1/4″	4″	8″
3/8″	1 3/8″	4″	8″
1/2″	1 1/2″	4″	8″

15-10b. Stapling schedule.

15-11. Installing plywood sheathing vertically.

wood is used for sheathing, the wood shingles must either be nailed to stripping or attached with barbed nails.

BUILDING PAPER

Building paper is applied between the sheathing and the siding. It prevents the passage of air through the walls, but it is of relatively little value as a heat insulator because of its thinness.

Building paper should be used behind exterior stucco finish and also over wood sheathing. It should be provided whether or not the sheathing is tongued and grooved. For other than stucco finish it may be omitted when any of the following sheathings are used:
- Plywood.
- Fiberboard that has been treated at the factory to render it water-resistant.
- Core-treated water-repellent gypsum.

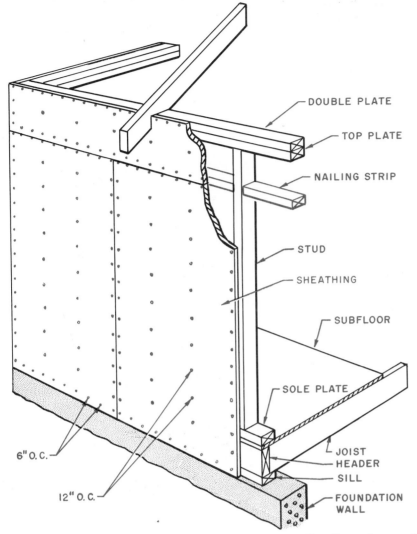

DOUBLE PLATE

TOP PLATE

NAILING STRIP

STUD

SHEATHING

SUBFLOOR

SOLE PLATE

6" O. C.

12" O. C.

JOINT
HEADER
SILL
FOUNDATION
WALL

15-12. *Plywood sheathing started at the foundation wall will require a nailing strip along the top joint.*

- Non-core-treated gypsum that has been treated at the factory to make it moisture-resistant.

In general, the soft, porous, relatively thick machine-finished building papers should be avoided in favor of asphalt-saturated paper or paraffined paper. These papers usually have sufficient water and air tightness and can usually stand handling in all kinds of weather. Building paper should be water-resistant but not vapor-resistant. Vapor-resistant paper might trap moisture between the walls.

The paper should be applied smoothly, and the joints should be lapped and nailed without bulges or cracks through which the air could find its way. Care should be taken around window and door openings to close all cracks.

Building paper should be applied horizontally, starting at the bottom of the wall. Succeeding layers should overlap about 4″ and should cover strips previously applied around openings. Strips about 6″ wide should be installed behind all exterior trim of exterior openings.

QUESTIONS

1. Why is wall sheathing considered a structural element and a part of the framing?

2. List four common types of sheathing.

3. Why isn't let-in corner bracing required when plywood wall sheathing is used?

4. Wall sheathing is usually applied at one of four stages of construction. At which stage would you apply wall sheathing? List reasons.

5. Why is building paper applied between sheathing and siding?

6. Which of the various kinds of building papers available is recommended for application between the sheathing and the siding? Why?

Ceiling Framing

16

When the wall framing has been completed, it is ready to be tied together with the roof framing. There are two basic methods of roof framing for platform-frame construction: trussed roof construction and conventional roof construction.

Trusses are prefabricated assemblies placed on the building and attached as a unit. Fig. 16-1. The lower chords in the trusses form the ceiling of the room and support the ceiling finish. Since trusses are the completed roof frame, they will be discussed in a separate chapter on roof framing (Unit 22).

In conventional roof construction, the *ceiling joists* and *rafters* are laid out, cut, and fastened one piece at a time to the building. Fig. 16-2. *Ceiling joists* are the parallel beams which support ceiling loads. They are supported in turn by larger beams, girders, or bearing walls. *Rafters* are the inclined members of a roof framework. They support the roof loads. Conventional roof framing requires ceiling joists which serve as a tie between the exterior walls and the interior partitions. The ceiling joists also serve as floor joists for an attic or a second story.

CEILING JOISTS

Size

The size of the ceiling joists is determined by the distance they must span and the load they must carry. The species and grade of wood are also factors to be considered. The correct size for the joists will be found on the building plans as recommended by the building code. As a general reference, the table in Fig. 16-3 shows the joist sizes and the spacing and span limitations, but be sure to confirm these with the local building code.

Layout

Ceiling joists are usually located across the width of the building and parallel to the rafters. The ends of the joists which rest on the exterior wall plates next to the rafters will usually project above the top edge of the rafter. These ends must be cut off on a slope that is equal to the roof pitch. Fig. 16-4.

The spacing of the joists is 16″ or 24″ on center, depending on

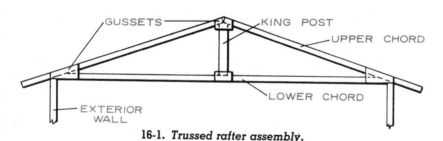

16-1. *Trussed rafter assembly.*

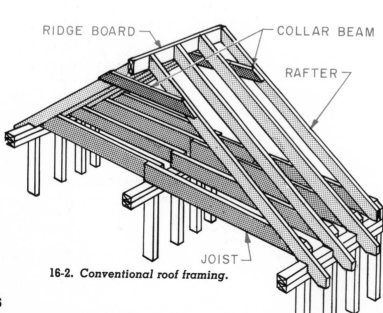

16-2. *Conventional roof framing.*

Size of Ceiling Joists (Inches)	Spacing of Ceiling Joists (Inches)	Maximum Allowable Span (Feet and Inches)			
		Group I	Group II	Group III	Group IV
2 × 4	12	11-6	11-0	9-6	5-6
	16	10-6	10-0	8-6	5-0
2 × 6	12	18-0	16-6	15-6	12-6
	16	16-0	15-0	14-6	11-0
2 × 8	12	24-0	22-6	21-0	19-0
	16	21-6	20-6	19-0	16-6

16-3. *Allowable spans for ceiling joists using non-stress-graded lumber.*

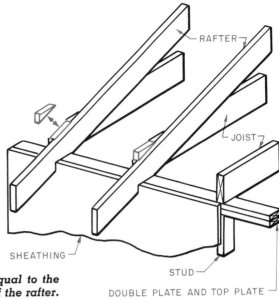

16-4. *The ends of the ceiling joists must be cut off on a slope equal to the roof pitch. It is best to cut them off about ⅛″ below the top edge of the rafter.*

the building specifications. Installation is begun at one end of the building and continued across the structure. Extra joists, if needed, are placed without altering the spacing of the prime joists. The first joist is located at the inside edge of the plate on an end wall. This provides edge nailing for the ceiling finish. Fig. 16-5. The second joist is usually located over the stud in the side wall. The distance between the first two joists will thus be less than 16″ or 24″, depending on the center spacing used. Fig. 16-6. Each succeed-

ing joist is spaced 16″ or 24″ on center.

If the layout of the ceiling joists places a joist over a stud and the studs for the two sidewalls have been laid out from the same end of the building, the ends of the joists will butt against each other over the bearing wall. Fig. 16-6. The total length of the two joists will equal the width of the building, and the ends that butt will have to be squared and cut off to length. Since each of the joist

ends will be resting on just half of the partition wall plate, a plywood joist splice should be nailed securely to both sides of the joists. Fig. 16-7a. Metal connectors are also available for this purpose. Fig. 16-7b.

An alternate method of joist layout is to offset the joists 1½″ on the two outside walls so that they lap each other when they join over the bearing partition. Fig. 16-8. This lap is faced-nailed with three 16d nails, and the joists are toenailed to the bearing partition wall plate with two 10d

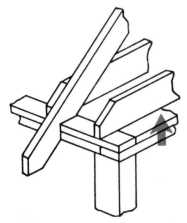

16-5. *The first ceiling joist is set on the inside edge of the end wall to permit nailing of the material for the ceiling surface. (See arrow.)*

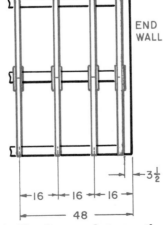

16-6. *The distance between the first two joists is less than 16″. The joists shown here are butted end to end on the bearing wall. The butt joint must be reinforced.*

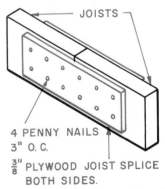

16-7a. *Ceiling joists butted end to end must be spliced together for strength. When lumber is used instead of plywood for a splice, it must be ¾″ thick and at least 24″ long.*

147

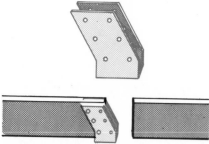

16-7b. *A metal connector may be used to reinforce the butt end joint of the ceiling joists.*

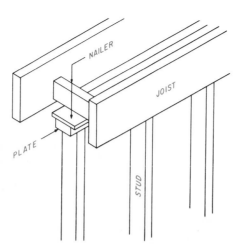

16-9. *A nonbearing partition wall is fastened to a block which has been nailed between the joists. Notice the backing which has been attached to the top of the partition for nailing the ceiling material.*

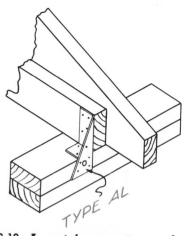

16-10. *A metal connector used to fasten the ceiling joist to the double plate.*

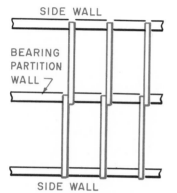

16-8. *Ceiling joists lapped on the bearing wall.*

nails. Nonbearing partitions which run parallel to the ceiling joists are nailed to blocks installed between the joists. Fig. 16-9.

Installation

Sight down the edge of the joist to determine the crown and place the crown, or camber, up. When attaching the ceiling joist to the exterior wall plate, keep the end of the joist even with the outside edge of the plate. Fasten this end of the joist first. Toenail three 10d nails through the joist into the plate or use a metal bracket and the special nails furnished with the bracket. Fig. 16-10. Make sure the walls are straight along the top edge and plumb. Fasten the joists together where they join and then to any other

double plates they may cross over. This will tie the building together at the top and make it ready for the roof framing or the sheathing.

SPECIAL FRAMING

When framing a low-pitched hip roof, the first ceiling joist will in-

terfere with the bottom edge of the rafters. Stub joists installed at right angles to the regular joists will correct this situation. Fig. 16-11. Space the stub 16" on center for attaching the finished ceiling. Locate them so that the rafters, when installed, may be

16-11a. *Stub joists are securely anchored to the regular joist with metal straps.*

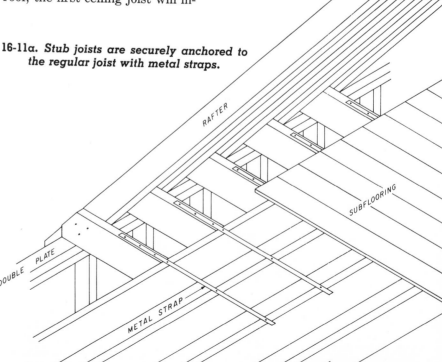

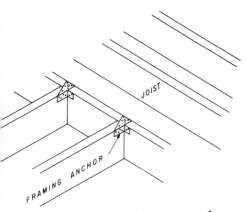

16-11b. *A metal framing anchor may be used in place of the metal straps to secure the stub joists to the regular joist.*

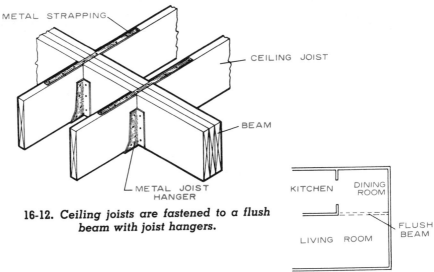

16-12. *Ceiling joists are fastened to a flush beam with joist hangers.*

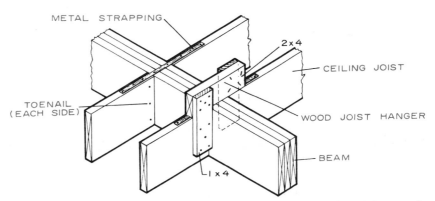

16-13. *Wood brackets may also be used to attach the ceiling joists to the beam.*

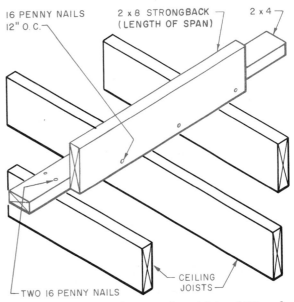

16-14. *A strongback is used to give long joists additional support.*

nailed directly to the side of the stubs.

Openings in the ceiling that are larger than the spacing between the joists are often necessary. They may be needed for a chimney or for access to the attic area. An enlarged opening will require the cutting of one or more joists. Such joists will need to be supported and framed as described earlier in the section "Framing Floor Openings" in Unit 13.

FRAMING FOR FLUSH CEILING AREA

A living-dining-kitchen group or kitchen-dining-family room group is often designed as one open area with a flush ceiling throughout. This makes the rooms appear much larger than they actually are. When trusses are used, there is no problem because they span from one exterior wall to the other. However, if ceiling joists and rafters are used, some type of beam is needed to support the interior ends of the ceiling joists.

This support can be provided by a flush beam, which spans from an interior cross wall to an exterior end wall. Joists are fastened to the beam with joist hangers. Fig. 16-12. These hangers are nailed to the beam with 8d nails and to the joist with nails furnished with the hangers or 1½" roofing nails. Hangers are perhaps most easily fastened by nailing them to the end of the joist before the joist is raised into place.

An alternate method of framing utilizes a wood bracket at each pair of ceiling joists, tying them to a beam which spans the open area. Fig. 16-13. This beam is blocked up and fastened at each end to the double plate, at a height equal to the depth of the ceiling joists.

On a long span with a continuous joist, the joists sometimes need additional support. In this case, a ceiling joist strongback is constructed. Fig. 16-14. The joists should be aligned and properly spaced where they pass under the strongback before they are blocked and nailed.

QUESTIONS

1. What are the functions of the ceiling joists?
2. When laying out the ceiling joists, why are the first two ceiling joists less than 16″ on center?
3. What are two methods used to join the ceiling joists on the partition plate?

4. What must be done to the building before the joists are nailed together where they join to the partition plates?
5. What is a ceiling joist strongback?

Roof Framing 17

The primary function of a roof is to protect the house in all types of weather with a minimum of maintenance. Roof construction should be strong in order to withstand snow and wind loads. Roofing members should be securely fastened to each other to provide continuity across the building, and they should be anchored to exterior walls.

A second consideration is appearance. Besides being practical, a roof should add to the attractiveness of the home. Various roof styles are used to create different architectural effects. A carpenter must understand and be able to frame these various styles.

ROOF STYLES

The basic roof styles used for homes and small buildings are as follows:
- Flat.
- Shed.
- Gable.
- Gable with dormer.
- Gable and valley.
- Hip.
- Hip and valley.

Variations of these roofs are associated with architectural styles of different countries or geographic regions. Some of these variations include the following. Fig. 17-1.
- Gambrel.
- Mansard.
- Butterfly.
- Dutch hip.

Flat Roof

Roof joists for flat roofs are laid level or at a slight slope for drainage. Sheathing and roofing are applied to the top of the joists, which in this case serve as rafters. The ceiling material is applied to the underside. Fig. 17-1.

Shed Roof

Sometimes called a lean-to, this roof is nearly flat, and its slope is in one direction only. The shed roof is used for contemporary homes and for additions to existing structures. When it is used as an addition, the roof may be attached to the side of the existing structure or to the existing roof. Figs. 17-1 & 17-2.

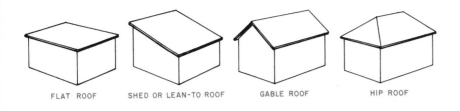

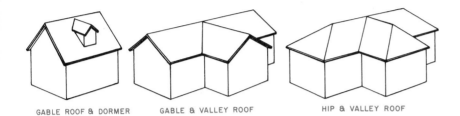

17-2. *A shed roof on an addition tied into the main roof of the existing structure.*

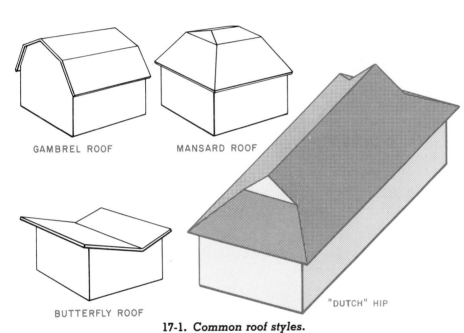

17-1. *Common roof styles.*

continues toward the center of the building where it meets with the slopes from the other sides. The mansard roof style was brought to this continent by the French when they settled in Quebec. Fig. 17-1.

Gambrel Roof

The gambrel roof is a variation of the gable roof. It has steep slopes on two sides. Partway up a second slope is developed which continues toward the center of the building, where it meets with the roof from the other side. This roof style was first used in the United States by German settlers in New York and Pennsylvania. Fig. 17-1.

Butterfly and Dutch Hip Roofs

A butterfly roof is an inverted gable roof. The Dutch hip roof is a hip roof with a small gable at the ridge. Construction details are provided in the architectural plans.

ROOF FRAMING TERMS

Span. The distance between the outside edge of the double plates. It is measured at right angles to the ridge board. Fig. 17-3.

Run. One-half the span distance (except when the pitch of the roof is irregular). Fig. 17-4.

Gable Roof

The gable roof is the most common. It has two roof slopes which meet at the top, or ridge, to form a gable at each end. For variation, the gable may include dormers which add light and ventilation to second-floor rooms. Fig. 17-1.

Hip Roof

The hip roof slopes at the ends of the building as well as at the two sides. This slope to all sides makes possible an even overhang all around the building. The low appearance of this roofline and the fact that it minimizes maintenance (there is no siding above the eaves) make it a popular choice. Fig. 17-1.

Mansard Roof

The mansard is a variation of the hip roof. It has steep slopes on all four sides but these do not meet at the center as in a hip roof. Partway up on each side, a second slope is developed. The second slope is almost flat and

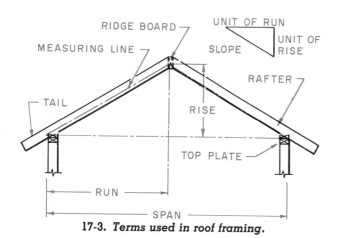

17-3. Terms used in roof framing.

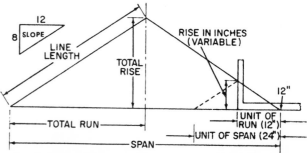

17-4. A comparison of total and unit terms. The unit of run is always 12". Therefore the unit of span is always 24". The rise in inches is variable, depending on the pitch assigned to the roof. In the example shown here, there are 8" of rise per (foot) unit of run.

Unit of Run. One unit of run is equal to 1′, or 12″, of run. Fig. 17-4.

Measuring line. An imaginary line running lengthwise from the outside wall to the ridge. Fig. 17-4.

Rise. The vertical distance from the top of the double plate to the upper end of the measuring line. Fig. 17-4.

Unit rise. The number of inches that a roof rises for every foot of run (unit of run). Fig. 17-5.

Pitch. The angle which the roof surface makes with a horizontal plane. It is the ratio of the rise to the span. Fig. 17-5.

For a example, a roof may have a rise of 6′ and a span of 24′. Fig. 17-6. Such a roof has ¼ pitch:

$$\frac{6' \text{ (rise of rafter)}}{24' \text{ (span of building)}} = \text{¼ pitch}$$

A ½ pitch roof rises one-half the distance of the span. For a 24′ span, the rise of a ½ pitch roof would be 12′. Fig. 17-7. The common roof pitches are ¼, ½, and ⅓. Figs. 17-6, 17-7 & 17-8.

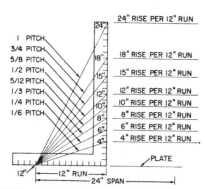

17-5. A comparison of the two methods of expressing the pitch.

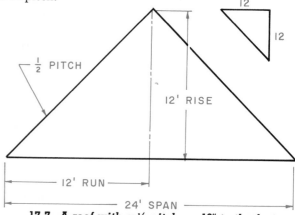

17-7. A roof with a ½ pitch, or 12″ to the foot.

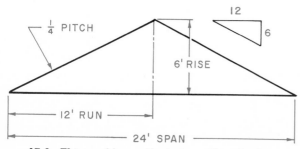

17-6. This roof has a ¼ pitch, or 6″ to the foot.

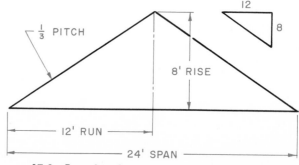

17-8. A roof with a ⅓ pitch, or 8″ to the foot.

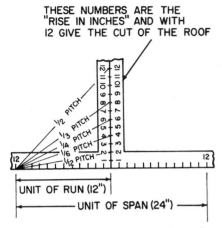

17-9. Pitch can be expressed as the rise in inches per unit of run.

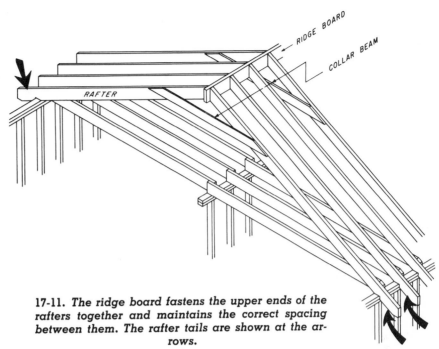

17-11. The ridge board fastens the upper ends of the rafters together and maintains the correct spacing between them. The rafter tails are shown at the arrows.

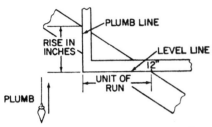

17-10. The framing square is used to lay out the plumb and level lines on a rafter. The plumb line is drawn along the tongue of the square and the level line along the body (sometimes called the blade).

Cut of a roof. The rise in inches and the unit of run. It is used when referring to the roof pitch. Fig. 17-9. For example, the cut may be 6-12 (6″ of rise per foot of run), 8-12 (8″ of rise per foot of run), or 12-12 (12″ of rise per foot of run). Fig. 17-6, 17-7, & 17-8.

Slope. The incline of a roof. Slope is the inches of vertical rise in twelve inches of horizontal run. It is expressed sometimes as a fraction (×/12), but typically as "× in 12." For example, a roof that rises at the rate of 4″ for each foot (12″) of run is designated as having a 4-in-12 slope (or 4/12). The triangular symbol above the roof in Fig. 17-6 conveys this information.

Plumb and level lines. These terms refer to the direction of a line on a rafter, not to any particular rafter cut. Any line that is vertical when the rafter is in its proper position is called a plumb line. Any line that is level when the rafter is in its proper position is called a level line. Fig. 17-10.

Ridge board. The horizontal piece that connects the upper ends of the rafters. Fig. 17-11.

Tail. The portion of the rafter which extends beyond the wall of the building to form the overhang or eave. Fig. 17-11, arrows.

Rafters. The inclined members of the roof framework. They serve the same purpose in the roof as joists in the floor or studs in the wall and are usually spaced 16″ or 24″ apart. They vary in width depending on their length, the distance they are spaced apart, their slope, and the kind of roof covering to be used. Rafters sometimes extend beyond the wall of the building to form eaves and protect the sides of the building.

Wood members used for roof framing should normally not exceed 19% moisture content. There are several kinds of rafters necessary for framing the many different roof styles.

Common rafters extend from the plate to the ridge board at 90° to both. Fig. 17-12.

Hip rafters extend diagonally from the corners formed by the plate to the ridge board. Fig. 17-12.

Valley rafters extend diagonally from the plates to the ridge board along the lines where two roofs intersect. Fig. 17-12.

Jack rafters never extend the full distance from the plate to the ridge board. There are three kinds of jack rafters: hip jacks extend from the plate to a hip rafter, valley jacks extend from the ridge to a valley rafter, and cripple jacks extend between a hip rafter and a valley rafter or between two valley rafters. Fig. 17-12.

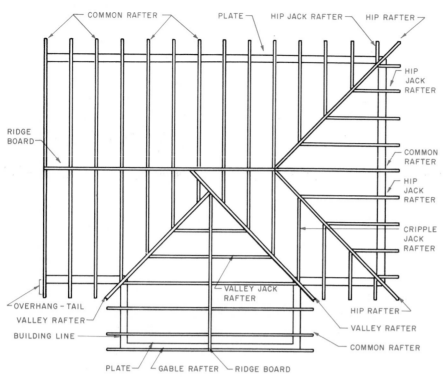

17-12. *The parts of a roof frame.*

which will usually be of concern to the carpenter:

- Gable.
- Hip.
- Gable and valley.
- Hip and valley.

The lean-to or shed roof is one-half of a gable roof. It extends from ridge board to plate on one side only. The gambrel roof plan is the same as for a gable roof. The mansard roof is a combination of a hip and a flat roof or two hip roofs. The first hip off the plate has a steep slope and the second is either flat or has a very low slope.

Always make a roof frame plan to determine the kinds of rafters that will be needed for framing. If the plan is drawn to scale, the exact number of each kind of rafter can also be determined. However, the actual length of each rafter should be figured from the dimensions taken directly off the building. The roof frame plan for each of the more common roof styles can be made as follows:

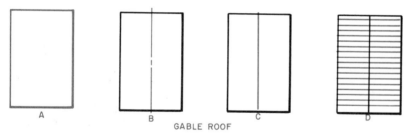

GABLE ROOF

17-13. *Frame plan for gable roof. The frame plan for a shed roof would be one-half of this.*

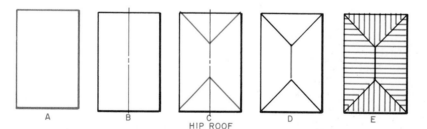

HIP ROOF

17-14. *Frame plan for hip roof.*

LAYING OUT A ROOF FRAME PLAN

Before cutting rafters, the carpenter must determine what kinds are necessary to frame the roof. A roof framing diagram may be included among the working drawings; if not, you should lay one out for yourself.

There are four types of roofs

Gable Roof

1. Lay out the outline of the building. A, Fig. 17-13.

2. Determine the direction in which the rafters will run and draw the center line at right angles to this direction. B, Fig. 17-13.

3. The center line determines the location of the ridge line. C, Fig. 17-13.

4. Determine the distance between the rafters and lay out the roof frame plan. D, Fig. 17-13.

Hip Roof

1. Lay out the outline of the building. A, Fig. 17-14.

2. Locate and draw a center line. B, Fig. 17-14.

3. At each corner, draw a 45° line from corner to center line. This establishes location of hip rafters. C, Fig. 17-14.

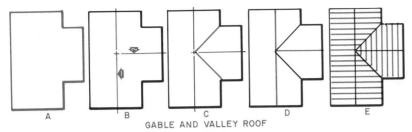

17-15. *Gable and valley roof frame plan.*

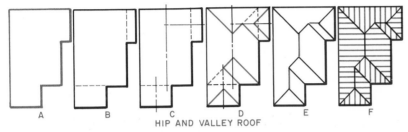

17-16. *Hip and valley roof frame plan.*

4. Draw the ridge line between the intersecting points of the hip rafters. D, Fig. 17-14.

5. Determine the distance between the rafters and lay out the roof frame plan. E, Fig. 17-14.

Gable and Valley Roof

1. Lay out the outline of the building. A, Fig. 17-15.

2. Draw the center line of the larger rectangle. B, Fig. 17-15 (arrow 1).

3. Draw the center line of the smaller rectangle. B, Fig. 17-15 (arrow 2).

4. Draw a line at 45° between the interior corners of the building outline and the ridge line. C, Fig. 17-15.

5. Draw in the ridge lines. D, Fig. 17-15.

6. Determine the distance between the rafters and lay out the roof frame plan. E, Fig. 17-15.

Hip and Valley Roof

1. Lay out the outline of the building. A, Fig. 17-16.

2. Outline the largest possible rectangle inside the building outline. B, Fig. 17-16.

3. Draw center lines for each rectangle formed inside the building outline. C, Fig. 17-16.

4. Draw a line at 45° from each corner, both inside and outside, and extend these lines to intersect with the center lines drawn in C. The solid lines indicated the location of the hip rafters on outside corners and valley rafters on inside corners. D, Fig. 17-16.

5. The center lines drawn in C connect the hip and valley rafters. Draw these in as solid lines to indicate the location of the ridges. E, Fig. 17-16.

6. Figure distance between rafters and lay out roof frame plan. F, Fig. 17-16.

QUESTIONS

1. What is the main purpose of a roof?
2. List several roof styles.
3. What is the ridge board's function?
4. What pieces in the roof serve the same purpose as joists in the floor?

5. What is the difference between a hip rafter and a valley rafter?
6. What is the purpose of a roof frame plan?

Conventional Roof Framing With Common Rafters

18

There are two methods of roof framing for the pitched roof styles discussed in the previous unit: conventional and trussed roof construction. In conventional roof construction, the carpenter builds the roof with ceiling joists and rafters, a piece at a time, on the building's walls. Fig. 18-1. In trussed roof construction the trusses are usually prefabricated and are attached to the building as units. Fig. 18-2. These two framing methods are used most often for roof slopes of 4 in 12 (⅙ pitch) and greater. The framing of a low-pitched or flat roof will be discussed later.

In this unit the layout and cutting procedures for framing a conventional roof will be discussed. The procedure for constructing a pitched roof using trusses will be discussed in Unit 22.

The joist and rafter method is known by most carpenters and therefore is used frequently. Common types of sheathing and finish materials are used, insulation is easily installed between the joists, and the roof load is carried on the walls without causing the ceiling to deflect.

There are some disadvantages to this type of construction. It takes longer, and the building is therefore exposed to the weather longer. Also, the carpenters building a roof with joists and rafters must stand on scaffolding and ceiling joists.

In conventional roof construction, the rafters should not be erected until the ceiling joists have been fastened in place. The ceiling joists act as a tie and prevent the rafters from spreading and pushing out on the exterior walls.

LAYING OUT COMMON RAFTERS

The rafters are the skeleton of the roof and must be carefully made and fitted if they are to support the roof weight. The top of the rafter rests against the ridge board and is called a top or plumb cut. The bottom of the rafter rests on the plate; this is a level or seat cut. Fig. 18-3. These cuts must be made accurately if the rafter is to fit properly.

A plumb cut line is drawn with the framing square as a guide. The unit run (12″ mark) on the body of the square is aligned with the edge of the rafter. The unit

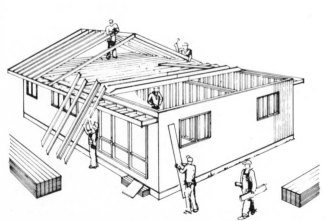

18-1. Conventional roof framing.

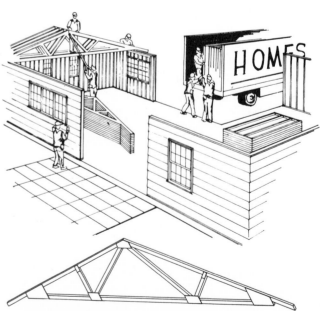

18-2. Framing the roof with trussed rafters.

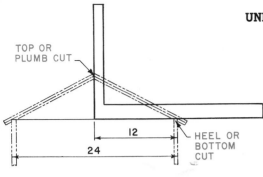

18-3. *This framing square was enlarged to show its relationship to the roof and to the top and bottom cuts.*

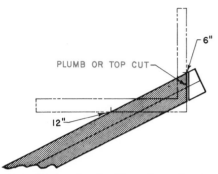

18-4. *This plumb line has been drawn for the top cut on a roof with a 6" unit rise (¼ pitch roof).*

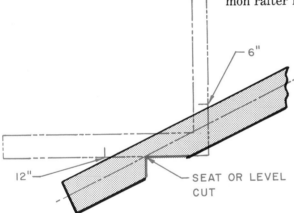

18-5. *A level line drawn for the seat cut of a bird's-mouth. This cut is made for a roof with a 6" unit rise.*

rise on the tongue of the square (number used will correspond to the slope of the roof) is aligned on the same edge of the rafter. The plumb line is then drawn along the edge of the tongue. Fig. 18-4. A level line is drawn for the same roof pitch with the square in the same position on the rafter except that the line is drawn along the body of the framing square. Fig. 18-5.

The theoretical length of a common rafter is the shortest dis-

tance between the outer edge of the plate (A) and a point where the measuring line of the rafter comes in contact with the ridge line (B). Figs. 18-6 & 18-7. This length is found along the measuring line and may be calculated in several ways:

• By using the Pythagorean theorem.
• By applying the unit length obtained from the rafter table on the framing square.
• By stepping off the length with the framing square.

Pythagorean Theorem Method

The Pythagorean theorem states that the square of the hypotenuse of a right triangle is equal to the sum of the squares of the other two sides ($A^2 = B^2 + C^2$). The rise, the run, and the rafter of a roof form a right triangle, with the rafter as the hypotenuse. Fig. 18-8. The length of the rafter (A) can thus be calculated from the rise (B) and the run (C). Fig. 18-9.

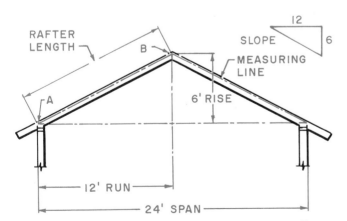

18-6. *The theoretical rafter length is from point A to point B.*

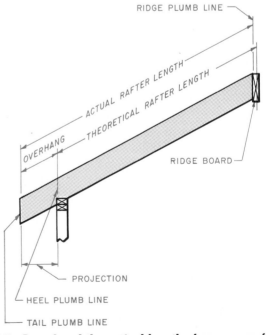

18-7. *Actual and theoretical length of common rafter.*

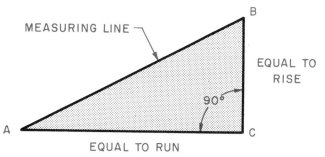

18-8. *The measuring line is the hypotenuse of the right triangle and represents the length of the rafters.*

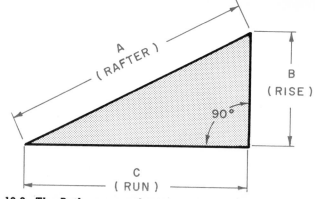

18-9. *The Pythagorean theorem states that the length of the hypotenuse (A) will be the square root of the sum of the squares of the other two sides ($\sqrt{B^2 + C^2}$).*

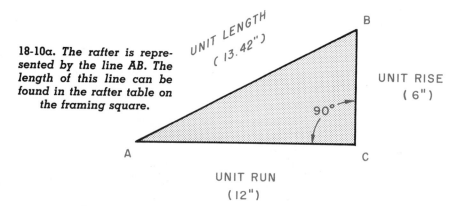

18-10a. *The rafter is represented by the line AB. The length of this line can be found in the rafter table on the framing square.*

example is 3′. Therefore the total length of the common rafters is 39″. Fig. 18-11.

$$13 \text{ (inches per unit of run)} \times$$
$$3 \text{ (units of run)} = 39″$$

Step-Off Method

A third method for finding the theoretical rafter length is by using the framing square to "step off" the length. Fig. 18-12. Place the square on the rafter with the tongue on the plumb cut. Step off the cut of the roof (for example in Fig. 18-12, 6″ on the tongue and 12″ on the blade) on the rafter stock as many times as there are feet in the total run. In this case, it would be three times.

Often the run of a building will not come out in even feet. For example, the run might be 3′4″. The extra 4″ is taken care of in the same manner as the full foot run. With the square at the first step position, draw a line along the edge of the tongue to represent the center line of the ridge board. At the 4″ mark of the blade, make a mark on the rafter along the level line—not along the edge of

Because of the ease and convenience of other methods, the Pythagorean theorem method is seldom used. However, the other methods commonly used on the job are based on this mathematical formula.

Unit Length Method

The unit length is the hypotenuse of a right triangle with the unit of run (12″) as the base and the unit rise (rise in inches per foot of run) as the altitude. Fig. 18-10a. The unit length is found on the rafter table of the framing square. Fig. 18-10b. The inch markings along the top of the table represent unit rise. The top

line of the table reads: "Length Common Rafters per Foot Run." If you follow across the top line to the figure under 6 (for a unit rise of 6″), you will find the figure 13.42. This is the unit length for a roof triangle with a unit run of 12″ and a unit rise of 6″.

Let's figure the total length of a rafter for a small building with a unit rise of 5″, a span of 6′, and a run of 3′. Look at the rafter table to obtain the unit length. Fig. 18-10b.

For a unit rise of 5″, the unit length is 13″ per unit of run (one foot). The total length is the unit length times the total run. The total run of the building in this

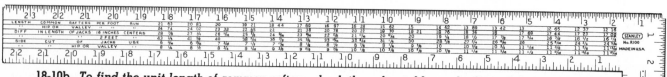

18-10b. *To find the unit length of common rafters, check the rafter table on the face of the steel square.*

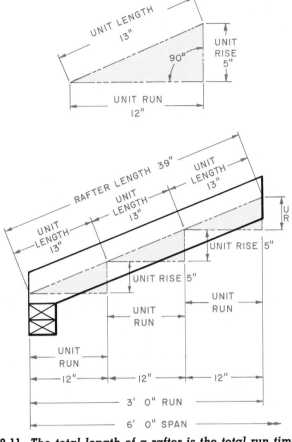

18-11. *The total length of a rafter is the total run times the unit length. In this example the total run is 3' and the unit length is 13". Therefore the theoretical length of the rafter is 39".*

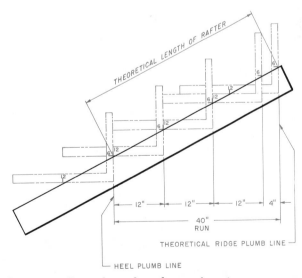

18-12. *Stepping off the length of a common rafter.*

the rafter. Fig. 18-13. Then, as in Fig. 18-12, step off the cut three more times, for a total run of 3'4". This is the theoretical length of the rafter. The ridge board thickness and overhang can now be figured and laid out.

CUTTING COMMON RAFTERS

Common Rafter Ridge Allowance

The theoretical length does not take into account the thickness of the ridge board or the length of the overhang, if there is one. To cut a rafter without an overhang to its actual length, you must deduct one-half the thickness of the ridge board from the ridge end. Fig. 18-14. For example, if 2" ma-

18-13. *Stepping off a rafter when the run is not an even number of feet.*

terial is used for the ridge board, the actual thickness is 1½". One-half of this is ¾". The ¾" is laid

off along the level line, and the line for the actual ridge plumb cut is made. Fig. 18-15.

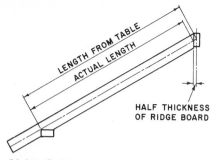

18-14. *Subtract one-half the actual thickness of the ridge board from the theoretical length of the rafter to obtain the rafter's actual length. If there is to be an overhang, this will be added later.*

18-15. *Lay off one-half the thickness of the ridge board at right angles to the tongue of the square (along the level line). Do not lay it off along the edge of the rafter.*

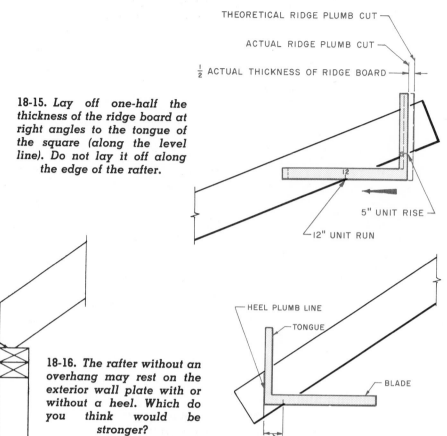

18-16. *The rafter without an overhang may rest on the exterior wall plate with or without a heel. Which do you think would be stronger?*

18-17. *Laying out the rafter seat.*

Common Rafter Overhang

A roof may or may not have an overhang. If not, the rafter must be cut so that its lower end is even with the outside of the exterior wall. Fig. 18-16. The portion of the rafter which rests on the plate is called the seat. To lay out the seat, place the tongue of the framing square on the heel plumb line with the rafter edge intersecting the correct seat width on the blade. Fig. 18-17. Draw a line from the heel plumb line along the blade.

A roof with a wide overhang at the cornice and the gable ends not only enhances appearance but also provides protection to side and end walls. Thus even in lower-cost houses, when style and design permit, wide overhangs are desirable. Though it adds slightly to the initial cost, future savings on maintenance usually merit this type of roof extension.

If the roof does have an overhang, or eave, the overhanging part of the rafter is called the tail and must be added to the length of the rafter. The length of the tail may be calculated as if it were a separate little rafter. Any of the methods used for finding rafter length may be used to find the length of the tail. Suppose the run of the overhang (sometimes called the projection) is 2′ and the unit rise of the roof is 8″. Fig. 18-18. Look at the rafter table and find the unit length for a common rafter with a unit rise of 8″. Fig. 18-10b. The unit length of the rafter is 14.42″. Since the total run of the overhang is 2′, the tail (length of overhang) is 28.84″, or $28\frac{27}{32}''$:

14.42 (inches per unit of run) ×
2 (units of run) = 28.84″

Another way to lay off the overhang is with the framing square. Suppose the run of the overhang is 10″. Fig. 18-19. Start the layout by placing the tongue of the square along the heel plumb line and setting the square to the cut of the roof. In Fig. 18-19, the square is set to a unit rise of 8″ and a unit run of 12″. Move the square in the direction of the arrow in Fig. 18-19 until the 10″ mark of the blade is on the heel plumb line. Draw a line along the tongue. This will be the tail cut.

Many carpenters do not cut the

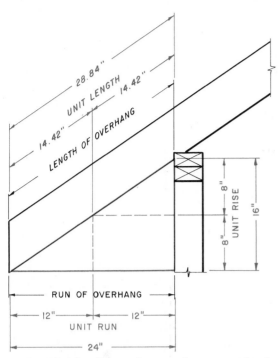

18-18. *The length of the rafter overhang may be found by using the rafter table on the framing square.*

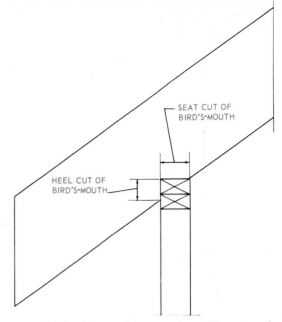

18-20. *The bird's-mouth on a rafter with an overhang.*

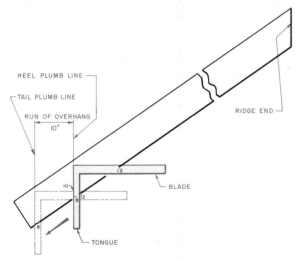

18-19. *Laying out the run of the overhang directly on the rafter with the framing square.*

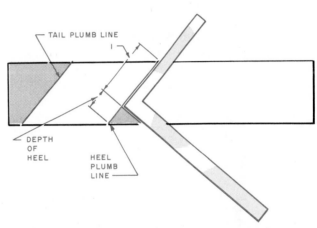

18-21. *Using the square to lay out a bird's-mouth. For a common rafter on a gable roof, the depth of the heel cut is laid out along the heel plumb line. This completes the layout of the bird's-mouth. The length of the line at arrow #1 will be the important dimension when laying out the bird's-mouth for a hip and valley rafter, as you will learn later.*

tail to the finished length until after the rafters have been fastened in place. The length of the tail is calculated, and a sufficient amount of material is left beyond the bird's-mouth for the overhang. Fig. 18-20. All other cuts except the tail plumb cut are made. After the rafters are fastened in place, the exact length of the tail is marked on the end rafters. A chalk line is snapped on the top edge of all the rafters. A tail plumb line is then drawn down from this chalk line on each rafter and the tail is cut along the line.

Bird's-mouth

A rafter with an overhang has a notch in it called a bird's-mouth. Fig. 18-20. The plumb cut of the bird's-mouth, which bears against the side of the rafter plate, is called the heel cut. The level cut, which bears on the top of the rafter plate, is called the seat cut.

The size of the bird's-mouth for a common rafter is usually stated in terms of the depth of the heel cut rather than the width of the seat cut. The bird's-mouth is laid out much the same way as the seat on a rafter without an overhang. Measure off the depth of the heel on the heel plumb line, set the square, and draw the seat line along the blade. Fig. 18-21.

Common Rafter Pattern

Calculate the actual length of a common rafter and lay one out on a piece of stock. When laying out rafters, remember to use the crown of the rafter member for the top edge. Carefully cut out the rafter. Use this rafter as a pattern for cutting a second rafter.

Try the two rafters on the building with the ridge board or a scrap piece of the same size material as the ridge board between to see how the heel cut and the top cut fit. If they are all right, use one of these rafters as a pattern to cut all others needed. Distribute the rafters to their locations on the building. The rafters are usually leaned against the building with the ridge cut up. The workers on the building can then pull them up as needed and fasten them in position. Fig. 18-22.

SADDLE BRACE ROOF FRAMING

A metal bracket used for roof framing permits the use of square-end lumber for rafters,

eliminating both the plumb cut at the ridge and the bird's-mouth at the plate. This bracket will adjust to any pitch. Fig. 18-23. Hip and jack rafter brackets are also available. Fig. 18-24. This "saddle brace" produces a strong roof which exceeds regulations of federal, state, and local building codes and also meets FHA and CMHC (Canada) requirements. (CMHC is the Central Mortgage and Housing Corporation. It can be compared to the FHA in the United States.) To use the metal brackets on a gable roof follow this procedure:

1. Lay out the rafter spacing on the top plates and nail the anchor brackets to the top plates with two 1½" roofing nails.

2. Install a ceiling joist alongside each anchor bracket. Fig. 18-25.

3. Set the ridge in the center of the building at the right height for the required pitch. Fig. 18-26. (Installation of the ridge board is discussed in Unit 21.)

4. Lay out the rafter spacing on the ridge board and install the brackets over the ridge. Fig. 18-26. Nail each bracket in place with three 1½" roofing nails.

5. Insert a square-end lumber rafter into the saddle brace plate anchor and ridge bracket. Fig. 18-27. Make sure the rafter is pushed firmly against the ridge.

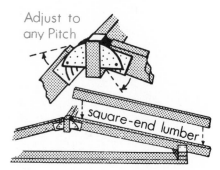

18-23. Square-end lumber is used with the saddle brace, which will adjust to any pitch.

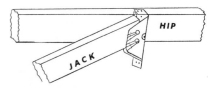

18-24. Saddle braces may also be used for framing hip and jack rafters.

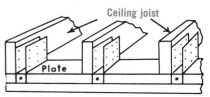

18-25. Anchor brackets are used to attach the rafters to the ceiling joists and the double plate.

18-22. Framing a gable roof. The rafters are leaned against the building and pulled up as needed. It is best to have three workers when framing a roof: one at the ridge and one at each plate where the rafters are to be fastened. The rafters are erected alternately: one from the front, then one from the back.

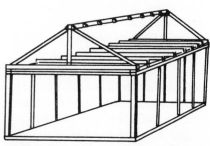

18-26. The ridge is set in place ready to receive the rafters. Brackets are installed over the ridge.

162

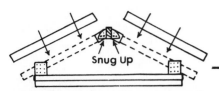

18-27. *Place the rafter in the saddle brace at the ridge and the anchor bracket at the plate. Push it up firmly against the ridge.*

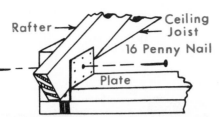

18-28. *Nail through the anchor plate into the rafter from one side. From the other side, nail through the ceiling joist into the rafter.*

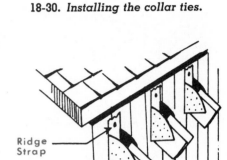

18-30. *Installing the collar ties.*

6. With the rafter and ceiling joist in place at the plate, drive a 16d nail from the ceiling joist side through to the rafter. Drive another 16d nail from the rafter side through to the ceiling joist. Fig. 18-28.

7. Make sure the bottom of the saddle brace at the ridge is snug against the bottom of the rafter. Nail each rafter face through the ridge bracket with two 1½" roofing nails. Fig. 18-29.

8. After installing all rafters, attach a collar tie to every fourth rafter if the spacing is 16" on center and every third rafter if the

18-29. *Nail the rafter to the ridge bracket with two nails on each side.*

spacing is 24" on center. This is a minimum standard. Local building codes may vary. Fig. 18-30.

Attaching a Shed Roof

A shed roof may be attached to

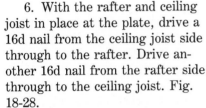

18-31. *Using the saddle brace to attach a shed roof to an existing building.*

an existing building by using these metal brackets. Cut the saddle brace in half at the ridge strap and bend the strap up. Nail it against the existing wall. Fig. 18-31.

QUESTIONS

1. What are some of the disadvantages of conventional roof framing?

2. What prevents the rafters from spreading and pushing out on the exterior walls?

3. What is the cut at the top of the rafter called?

4. What is the name of the cut that rests on the plate?

5. When laying out the rafter, what line is drawn along the edge of the tongue on the framing square?

6. What are three ways that the length of a common rafter can be calculated?

7. A building is 24' wide and has a ⅓ pitch. What is the theoretical length of a common rafter?

8. Describe the step-off method for finding the length of a rafter.

9. What is meant by ridge allowance?

10. What is the bird's-mouth?

11. What is a saddle brace? If saddle braces are used, will the roof meet the FHA requirements?

Hip and Valley Rafters

19

The hip rafter is a roof member that forms a raised area or "hip" in the roof, usually extending from the corner of the building diagonally to the ridge. Fig. 19-1a and b. The valley rafter is similar, but it forms a depression in the roof instead of a hip. Fig. 19-1 a and c. Like the hip rafter, it extends diagonally from plate to ridge.

The total rise of hip and valley rafters is the same as that of common rafters. Fig. 19-1a. Hip and valley rafters may be the same thickness as common rafters, but they should be 2" deeper to permit full bearing with the beveled end of the jack rafter. Fig. 19-2.

HIP RAFTER LAYOUT

The length of a hip rafter, like the length of a common rafter, is calculated on the basis of the unit run and unit rise and/or the total run and total rise. Any of the methods previously described for determining the length of a common rafter may be used. However, some of the basic data for hip and valley rafters is different.

Figure 19-3 shows part of a roof framing plan for a hip roof. On a hip roof framing plan, the lines which indicate the hip rafters (EC, AC, KG, and IG in Fig. 19-3) form 45° angles with the building lines. A line which indicates a rafter in the roof framing diagram corresponds to the total run (not length) of the rafter it represents. You can see from the

diagram that the total run of a hip rafter is the hypotenuse of a right triangle, with the shorter sides each equal to the total run of a common rafter. Fig. 19-3. In Fig. 19-4 one corner of the roof framing plan (ABCF in Fig. 19-3) has been drawn in perspective to show the relative position of the hip rafter to the common rafter.

The unit run of a hip rafter is the hypotenuse of a right triangle with the shorter sides each equal to the unit run of a common rafter. Fig. 19-5. The unit run of a common rafter is 12". By the Pythagorean theorem, the unit run of a hip rafter is the square root of $12^2 + 12^2$, which is 16.97, or 17. Fig. 19-6. The unit run of a valley rafter is also 17".

Like the unit length of a common rafter, the unit length of a hip rafter may be obtained from the rafter table on the framing square. In Fig. 18-10b, the second line in the table is headed "Length Hip or Valley per Foot Run." This means "per foot run of a common rafter in the same roof." Another way to state this would be "per 16.97" run of hip or valley rafter." For example, the unit length for a unit rise of 8" is 18.76". To calculate the length of a hip rafter, multiply the unit length by the number of feet in the total run of a common rafter.

In Fig. 19-5, the corner of the building from Fig. 19-3 is shown. In this example the run of a common rafter is 5′. The unit rise is

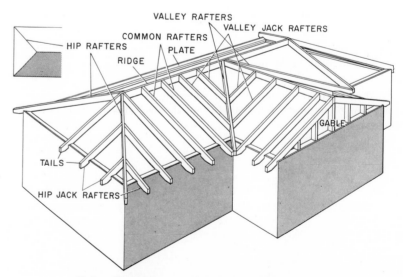

19-1a. *Roof frame with hip and valley rafters.*

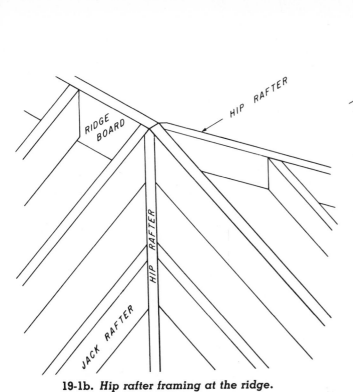

19-1b. *Hip rafter framing at the ridge.*

19-1c. *Valley rafter framing at the junction of two ridges.*

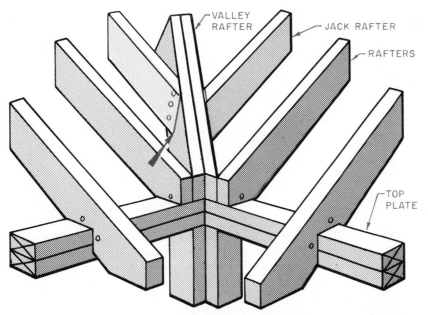

19-2. *Because the beveled cut on the ends of jack rafters creates a longer cut (see arrow), the hip and valley rafters must be 2″ deeper than common rafters to permit a full bearing surface. The valley rafter in this drawing has been cut off at the plate. Normally it is extended to become part of the overhang.*

Plumb and Level Lines

The plumb and level lines on a hip or valley rafter are also referred to as the top and bottom cuts. The top cut is the plumb line and the bottom cut is the level line. To obtain the top and bottom cuts of the hip or valley rafters, set off 17″ on the body of the square. On the tongue set off the rise per foot of common rafter run. A line drawn along the body will be the level or seat cut, and a line drawn along the tongue will be the plumb or top cut. Fig. 19-7.

Ridge Allowance

As is the case with a common rafter, the theoretical length of a hip rafter does not take into account the thickness of the ridge board. The ridge-end shortening allowance for a hip rafter depends on the manner in which the ridge end of the hip rafter is joined to the other structural members. The ridge end of the hip rafter may be framed against the ridge board or against the ridge end

8″ and the unit length of the hip rafter for this unit rise is 18.76″. The unit length multiplied by the total run in feet is the length of the hip rafter in inches (18.76″ × 5 = 93.8″, or 7′9¹³⁄₁₆″). As in

the case of common rafters, this is the theoretical length. To obtain the actual length, the ridge board shortening allowance and the rafter tail will have to be calculated and laid out.

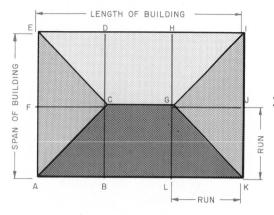

19-3. *Hip roof framing diagram.*

of common rafters. Figs. 19-8 & 19-9.

If the hip rafter is framed against the ridge board, the shortening allowance is one-half the 45° thickness of the ridge piece. The 45° thickness of a piece of stock is the length of a line laid at 45° across the thickness of the stock. If the hip rafter is framed against the common rafters, the shortening allowance is one-half the 45° thickness of a common rafter.

To lay off the shortening allow-

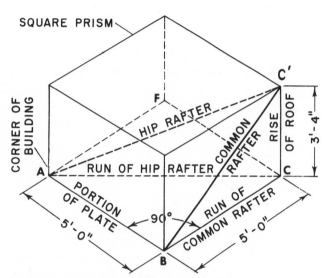

19-4. *The relative position of a hip rafter to a common rafter is shown in this perspective drawing of a corner from the roof framing diagram in Fig. 19-3.*

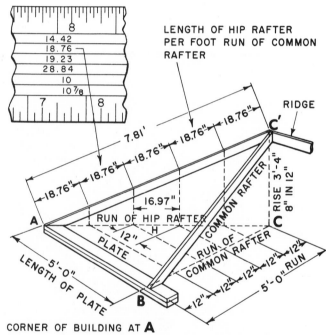

19-5. *The relationship between the unit run of a hip rafter and the unit run of a common rafter.*

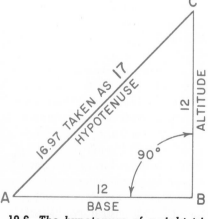

19-6. *The hypotenuse of a right triangle whose shorter sides each equal 12″ is 16.97″. This can be rounded off to 17″.*

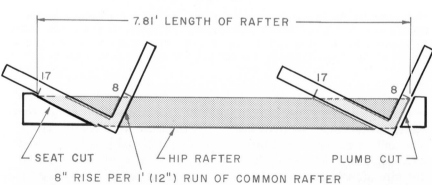

19-7. *Marking the top (plumb) cut and the seat (level) cut of a hip rafter.*

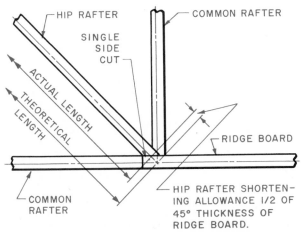

19-8. *A hip rafter framed against the ridge board re-quires a single side cut. However, the end common raf-ter must have a 45° angle cut for framing against the side of the hip rafter.*

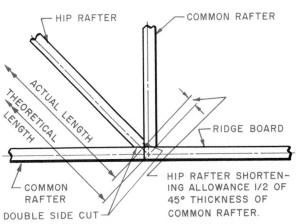

19-9. *A hip rafter framed against the ridge-end common rafters requires a double side cut.*

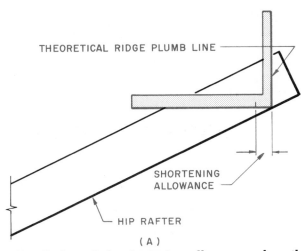

19-10a. *To lay off the shortening allowance, place the tongue of the square along the theoretical ridge plumb cut line and measure off the shortening allowance along the blade of the square (level line).*

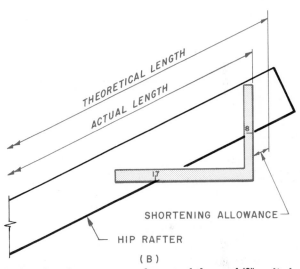

19-10b. *Set the square to the cut of the roof (8″ unit rise for this example) with the tongue on the shortening al-lowance mark. Draw the actual ridge plumb line along the edge of the tongue.*

ance, set the tongue of the framing square to the theoretical ridge plumb cut line. Measure off the shortening allowance along the blade. Fig. 19-10a. Set the square at the mark to the cut of the rafter (unit rise and unit run) and draw the actual ridge plumb cut line. Fig. 19-10b. Remember that the cut of the common rafter is based on a 12″ unit run whereas the unit run of the hip or valley rafter is 17″. Therefore, to set the

square at the cut of the hip rafter, the tongue is set at the unit rise and the blade is set at the 17″ mark.

Side Cuts

Since a common rafter runs at 90° to the ridge board, the ridge end of a common rafter is cut square, or at 90° to the length-wise line of the rafter. A hip rafter, however, joins the ridge piece or the ridge ends of the

common rafters at an angle. The ridge end of a hip rafter must therefore be cut to a correspond-ing angle. This cut is called a *side cut*. Figs. 19-8 & 19-9. The side cut may be laid out in one of two ways.

One method is illustrated in Figs. 19-11 & 19-12. Place the tongue of the framing square along the actual ridge plumb cut line and measure off one-half the thickness of the hip rafter along the blade (level line). Shift the tongue to the mark, set the

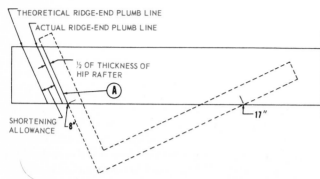

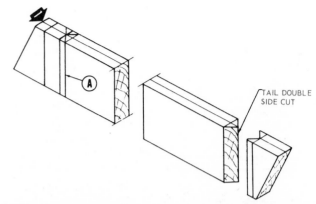

19-11. *To lay out the side cut, at a right angle to the ridge plumb cut line measure off one-half the thickness of the hip rafter from the actual ridge plumb cut line.*

19-12. *Draw a centerline on the edge of the rafter (arrow #1). Extend the plumb lines from the face of the rafter to intersect the centerline at 90°. The side cut line is drawn from line A through the intersection of the centerline and the actual ridge-end plumb line.*

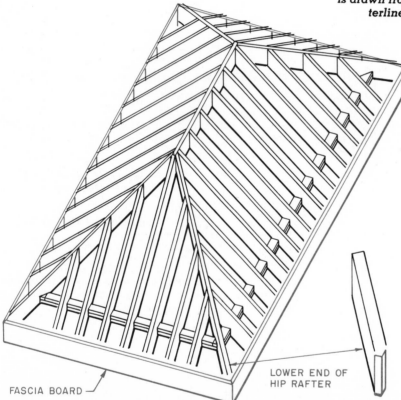

19-13. *Hip roof framing. The end of the hip rafter has a double side cut. The fascia boards from the side and end will be fastened along the ends of the rafters and mitered to form an outside corner at the hip rafter. See Fig. 19-12 for detail of hip rafter tail double side cut.*

framed against the ridge piece, there will be only a single side cut. Fig. 19-8. For a hip rafter which is to be framed against the ridge ends of the common rafters, there will be a double side cut. Fig. 19-9. In either case, the tail of the rafter must have a double side cut at the same angle, but in the reverse direction, to allow attachment of the fascia board. Fig. 19-13.

A second method of laying out the angle of the side cut on a hip rafter is by referring to the rafter table on the framing square. In Fig. 18-10b, the bottom line of the table is headed "Side Cut Hip or Valley Use." Follow this line

square to the cut of the rafter (17″ and 8″ in this example), and draw the plumb line. A, Fig. 19-11. Turn the rafter edge up, draw an edge center line, and draw in the angle of the side cut. Fig. 19-12. For a hip rafter which is to be

19-14. *Framing square in position on the back edge of the hip rafter for a unit rise of 8″. A single side cut will be made for framing against the ridge board.*

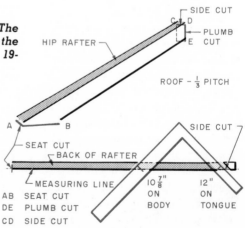

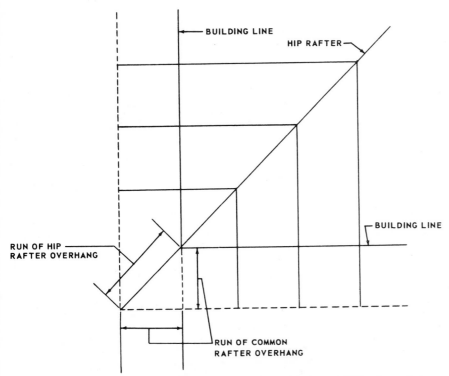

BUILDING LINE

HIP RAFTER

BUILDING LINE

RUN OF HIP RAFTER OVERHANG

RUN OF COMMON RAFTER OVERHANG

19-15. *Run of hip rafter overhang. For each unit of run (12″) of a common rafter, the unit of run for the hip rafter is 17″. Therefore if the run of the common rafter overhang in this drawing is 2′ (24″), the run of the hip rafter overhang will be 34″ (2 × 17).*

over to the column headed by the figure 8 (for a unit rise of 8″). The number shown is 10⅞. Place the framing square face up on the rafter edge, with the tongue on the ridge-end plumb cut line. (This is line A in Fig. 19-12.) Set the square to a cut of 10⅞″ on the blade and 12″ on the tongue. Draw the side cut angle along the tongue. Fig. 19-14.

Overhang

A hip or valley rafter overhang, like a common rafter overhang, is figured as a separate rafter. The run of the overhang, however, is not the same as the run of a common rafter overhang in the same roof. The run of the hip or valley rafter overhang is the hypotenuse of a right triangle whose shorter sides are each equal to the run of a common rafter overhang. Fig. 19-15. If the run of the common rafter overhang is 2′ for a roof

with an 8″ unit rise, the length of the hip or valley rafter tail is figured as follows:

1. Find the unit length of the hip or valley rafter on the framing square. Fig. 18-10b. For this roof it is 18.76″.

2. Multiply the unit length of the hip or valley rafter by the run of the common rafter overhang:

18.76″ (unit length of hip or valley rafter) × 2 (feet of run in common rafter overhang) = 37.52″, or 37½″

3. Add this product to the theoretical rafter length.

The overhang may also be stepped off as described in Unit 18 for a common rafter. When stepping off the length of the overhang, set the 17″ mark on the blade of the square even with the edge of the rafter. Set the unit rise, whatever it might be, on the

tongue even with the same rafter edge.

Bird's-mouth

Laying out the bird's-mouth for a hip rafter is much the same as for a common rafter. However, there are a couple of things to remember. When the plumb (heel cut) and level (seat cut) lines are laid out for a bird's-mouth on a hip rafter, set the body of the square at 17″ and the tongue to the unit rise (depending on the roof pitch). Fig. 19-7. When laying out the depth of the heel for the bird's-mouth, measure along the heel plumb line down from the top edge of the rafter a distance equal to the same dimension on the common rafter. Fig. 19-16. This must be done so that the hip rafters, which are usually wider than the common rafters, will be level with the common rafters.

Backing or Dropping a Hip Rafter

If the dimension above the bird's-mouth is exactly the same on a hip rafter as on a common rafter, the edges of the hip rafter will extend above the upper ends of the jack rafters and interfere with the application of the sheathing. A, Fig. 19-17. This can be corrected by either backing or dropping the hip rafter. *Backing* means to bevel the upper edge of the hip rafter. B, Fig. 19-17. *Dropping* means to deepen the bird's-mouth so as to bring the top edge of the hip rafter down to the upper ends of the jacks. C, Fig. 19-17.

The amount of backing or drop required is calculated as shown in A of Fig. 19-18. Set the framing square to the cut of the rafter (8″ and 17″ in this example) on the upper edge, and measure off one-half the thickness of the rafter from the edge along the blade (arrow 1). A backing line drawn

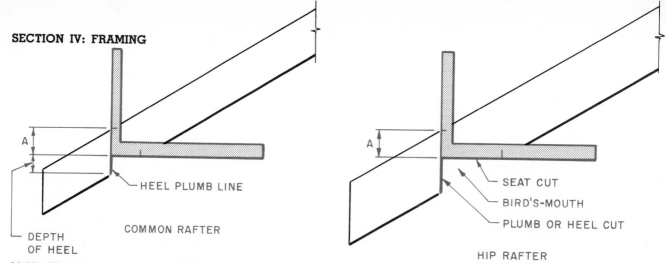

19-16. *When laying out the bird's-mouth on a hip rafter, measure down from the top edge. Dimension A in the drawing must be the same for common and hip rafters so that the tops of all the rafters will be in line for the application of sheathing.*

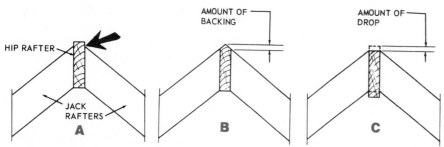

19-17. *A. The edge of a hip rafter may extend above the upper ends of the jack rafters. B. Backing a hip rafter. C. Dropping a hip rafter.*

through this mark, parallel to the edge, will indicate the bevel angle, if the rafter is to be backed. B, Fig. 19-18. The perpendicular distance between the backing line and the edge of the rafter will be the amount of drop (arrow 2). A, Fig. 19-18. This is the amount by which the depth of the hip rafter bird's-mouth should exceed the depth of the common rafter bird's-mouth. C, Fig. 19-18.

VALLEY RAFTER LAYOUT

A valley rafter follows the line of intersection between a main roof surface and a gable roof addition or gable roof dormer surface. Most roofs which contain valley rafters are equal-pitch roofs; that is, the pitch of the addition or dormer roof is the same as the

pitch of the main roof. In an equal-pitch roof the valley rafters always run at 45° to the building line and the ridge boards.

Framing an Equal-Span Roof Addition

In equal-span framing, the span of the addition is the same as the span of the main roof. Fig. 19-19. When the pitch of the addition roof is the same as the pitch of the main roof, equal spans bring the ridge pieces to equal heights.

Look at the roof framing diagram in Fig. 19-19. The total run of a valley rafter (indicated by AB and AD in the diagram) is the hypotenuse of a right triangle whose shorter sides are each equal to the total run of a common rafter in the main roof. The

unit run of a valley rafter is therefore 16.97", the same as the unit run for a hip rafter. Figuring the length of an equal-span addition valley rafter is thus the same as figuring the length of a hip rafter.

The ridge-end shortening allowance for an equal-span addition valley rafter is one-half the 45° thickness of the ridge board. Fig. 19-20. Side cuts are laid out as they are for a hip rafter. The valley rafter tail has a double side cut, like the hip rafter tail, but in the reverse direction, since the tail cut on a valley rafter must form an inside rather than an outside corner. Fig. 19-21. The overhang, if any, and the bird's-mouth are figured just as they are for a hip rafter. A valley rafter, however, does not require backing or dropping.

Framing an Unequal-Span Roof Addition

Sometimes the span of the roof addition is shorter than the span of the main roof. Fig. 19-22. When the pitch of the addition roof is the same as the pitch of the main roof, the shorter span of the addition brings the addition ridge board to a lower level than the main roof ridge board.

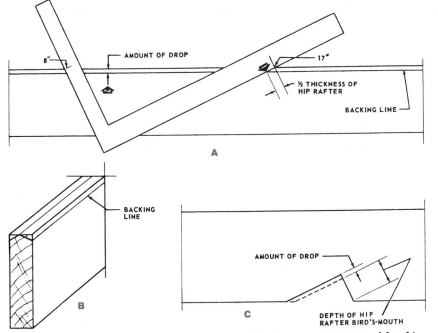

19-18. *Backing or dropping a hip rafter: A. Determining amount of backing or drop. B. Bevel angle for backing rafter. C. Deepening the bird's-mouth for dropping the rafter.*

potenuse of a right triangle with shorter sides each equal to the total run of a common rafter *in the addition.* The total run of a common rafter in the main roof is equal to one-half the span of the main roof. The total run of a common rafter in the addition is equal to one-half the span of the addition.

DETERMINING THE LENGTH OF A VALLEY RAFTER

When the total run of any rafter is known, the theoretical length can be found by multiplying the unit length by the total run. Suppose, for example, that the addition in Fig. 19-22 has a

There are two ways of framing an addition of this type. In one method, a full-length valley rafter (AD in Fig. 19-22) is framed between the rafter plate and the ridge board, and a shorter valley rafter (CB in the figure) is then framed to the longer one. The total run of the longer valley rafter is the hypotenuse of a right triangle whose shorter sides are each equal to the total run of a common rafter *in the main roof.* The total run of the shorter valley rafter, on the other hand, is the hy-

19-20. *Ridge-end shortening allowance for an equal-span addition valley rafter.*

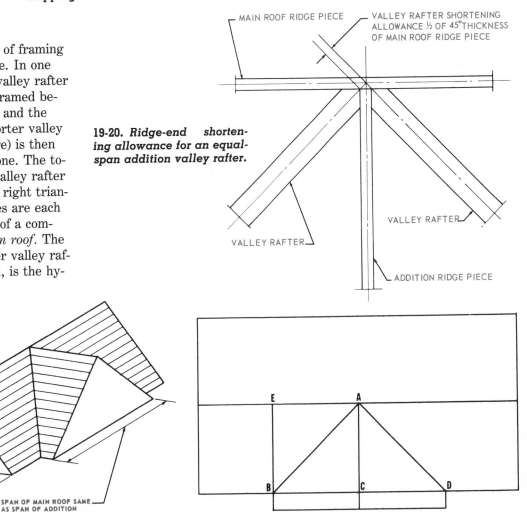

19-19. *Roof with an equal-span addition.*

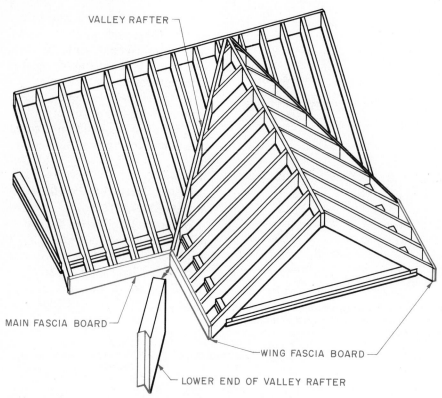

19-21. *Valley rafter framing. Notice the inside corner formed by the fascia boards.*

span of 30′ and that the unit rise of a common rafter in the addition is 9″. The rafter table in Fig. 18-10b shows that the unit length for a valley rafter in a roof with a common rafter unit rise of 9″ is 19.21″. To find the theoretical length of the valley rafter, multiply its unit length by the number of feet in a common rafter of the roof to which it belongs. The total run of a common rafter is equal to one-half the span. Therefore the length of the longer valley rafter in Fig. 19-22 would be 19.21″ times one-half the span of the main roof. The length of the shorter valley rafter would be 19.21″ times one-half the span of

the addition. Since one-half the span of the addition is 15′, the length of the shorter valley rafter is 19.21″ × 15, or 288.15″. Converted to feet, this is 24.01′.

The shortening allowances for the long and short valley rafters are shown in Fig. 19-23. Note that the long valley rafter has a single side cut for framing to the main roof ridge piece, while the short valley rafter is cut square for framing to the addition ridge piece.

A Second Method of Framing an Unequal-Span Addition

Another method of framing an equal-pitch, unequal-span addition is to nail the inboard end of the addition ridge piece to a piece of stock which hangs from the main roof ridge board. Fig. 19-24. This method calls for two short valley rafters, each of which extends from the rafter plate to the addition ridge piece. The total run of each of these valley rafters is the hypotenuse of a right triangle whose shorter sides are each equal to the total run of a common rafter in the addition.

The shortening allowance for each short valley rafter is one-half the 45° thickness of the addition ridge piece. Fig. 19-25. Each raf-

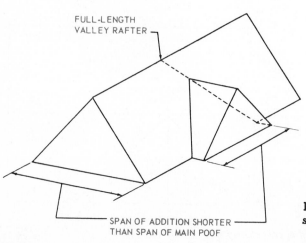

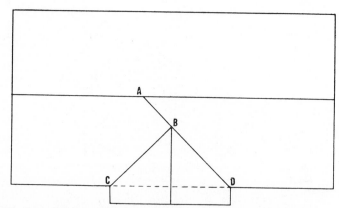

19-22. *An addition with a span less than the main roof span. This addition is formed with a long and a short valley rafter.*

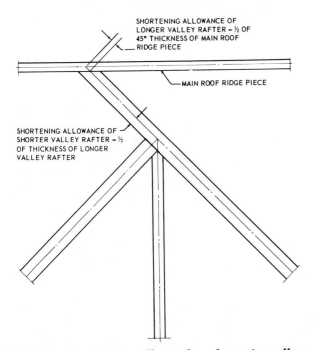

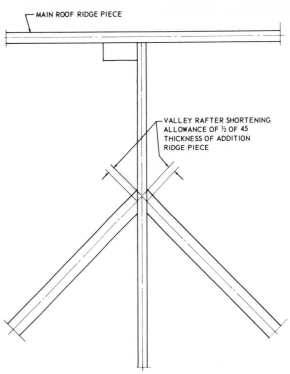

19-23. *Long and short valley rafter shortening allowances.*

SHORTENING ALLOWANCE OF LONGER VALLEY RAFTER = ½ OF 45° THICKNESS OF MAIN ROOF RIDGE PIECE

MAIN ROOF RIDGE PIECE

SHORTENING ALLOWANCE OF SHORTER VALLEY RAFTER = ½ OF THICKNESS OF LONGER VALLEY RAFTER

MAIN ROOF RIDGE PIECE

VALLEY RAFTER SHORTENING ALLOWANCE OF ½ OF 45 THICKNESS OF ADDITION RIDGE PIECE

19-25. *Shortening allowance of valley rafters in suspended-ridge method of addition roof framing.*

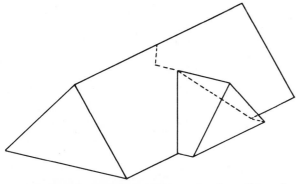

19-24. *Another addition with a span less than the main roof span. This addition is framed with the addition ridge board suspended from the main roof ridge board. The two valley rafters (AB and AC) are the same length.*

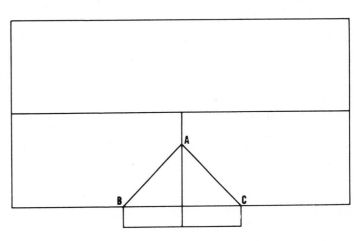

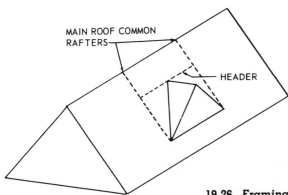

MAIN ROOF COMMON RAFTERS

HEADER

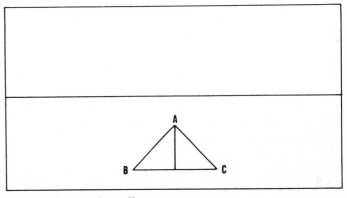

19-26. *Framing a dormer without side walls.*

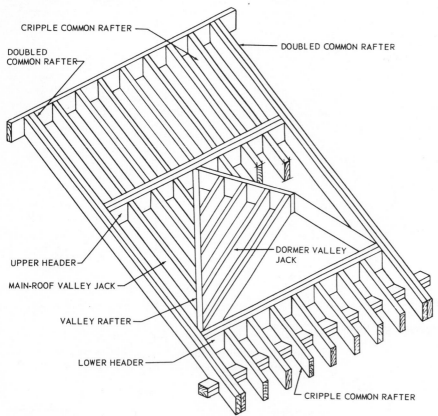

CRIPPLE COMMON RAFTER

DOUBLED COMMON RAFTER

DOUBLED COMMON RAFTER

UPPER HEADER

MAIN-ROOF VALLEY JACK

VALLEY RAFTER

LOWER HEADER

DORMER VALLEY JACK

CRIPPLE COMMON RAFTER

19-27. *Arrangement and names of framing members for a dormer without side walls.*

ter is framed to the addition ridge piece with a single side cut.

Framing a Gable Dormer without Side Walls

When a gable dormer without side walls is framed, the dormer ridge piece is fastened to a header set between a couple of doubled main roof common rafters. Fig 19-26. The valley rafters are framed between this header and a lower header. The total run of a valley rafter is the hypotenuse of a right triangle whose shorter sides are each equal to the total run of a common rafter in the dormer.

The arrangement and names of framing members in this type of dormer framing are shown in Fig. 19-27. Note that the upper edges of the headers must be beveled to the cut of the main roof.

In this framing method, the

shortening allowance for the upper end of a valley rafter is one-half the 45° thickness of the inside member in the upper doubled header. Fig. 19-28. The shortening allowance for the lower end is one-half the 45° thickness of the inside member in the doubled common rafter. Each valley rafter has a double side cut at the upper and the lower end.

Framing a Gable Dormer with Side Walls

A method of framing a gable dormer with side walls is illustrated in Fig. 19-29. As indicated in the framing diagram, the total run of a valley rafter is again the hypotenuse of a right triangle whose shorter sides are each equal to the run of a common rafter in the dormer. Figure the lengths of the dormer corner posts and side studs just as you do the lengths of gable-end studs (see page 188). Lay off the lower-end cutoff angle by setting the square to the cut of the main roof. The valley rafter shortening allowances for this method of framing are shown in Fig. 19-30.

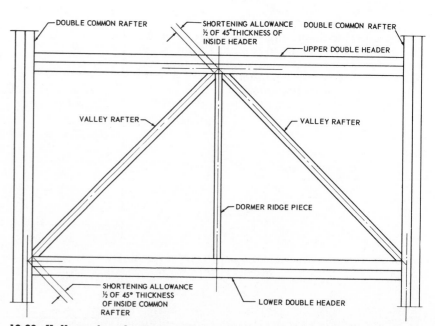

DOUBLE COMMON RAFTER

SHORTENING ALLOWANCE ½ OF 45° THICKNESS OF INSIDE HEADER

DOUBLE COMMON RAFTER

UPPER DOUBLE HEADER

VALLEY RAFTER

VALLEY RAFTER

DORMER RIDGE PIECE

SHORTENING ALLOWANCE ½ OF 45° THICKNESS OF INSIDE COMMON RAFTER

LOWER DOUBLE HEADER

19-28. *Valley rafter shortening allowances for a dormer without side walls.*

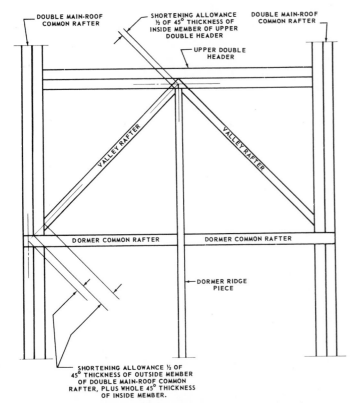

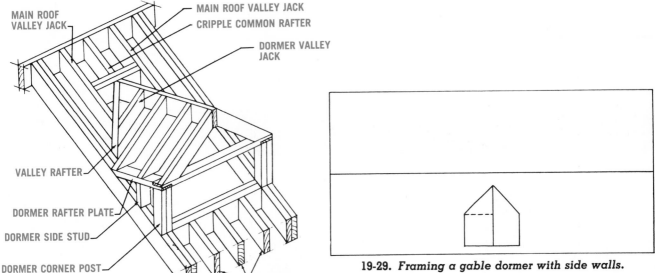

19-29. Framing a gable dormer with side walls.

19-30. Valley rafter shortening allowances for a dormer with side walls.

QUESTIONS

1. What is a hip rafter?

2. What is a valley rafter?

3. Explain why the unit run of a hip rafter is 16.97 when the unit run of a common rafter is 12.

4. What number is used on the body of the square when making a cut for a hip or valley rafter?

5. What is the shortening allowance at the ridge for a hip rafter when the ridge end is framed against the ridge board?

6. Why is the run of a hip rafter overhang greater than the run of a common rafter overhang in the same roof?

7. When laying out the depth of the heel for the bird's-mouth on a hip rafter, why must you measure down from the top edge of the rafter a distance equal to the same dimension used on the common rafter?

8. Why must the upper edge of the hip rafter be beveled or the bird's-mouth cut deeper?

9. Describe two methods of framing an unequal-span roof addition.

10. In a right triangle, what is the side opposite the right angle called?

Jack Rafters

20

A jack rafter is a shortened common rafter that may be framed to a hip rafter, a valley rafter, or both. This means that in an equal-pitch framing situation, the unit rise of a jack rafter is always the same as the unit rise of a common rafter.

TYPES OF JACK RAFTERS

A *hip jack* rafter extends from a hip rafter to a rafter plate. Fig. 20-1.

A *valley jack* rafter extends from a valley rafter to a ridge board. Fig. 20-1.

A *cripple jack* rafter does not contact either a plate or a ridge piece. There are two kinds of cripple jack rafters: (1) the *valley cripple jack* extends between two valley rafters in the long-and-short-valley-rafter method of addition framing and (2) the *hip-valley cripple jack* extends from a hip rafter to a valley rafter. Fig. 20-2.

LENGTHS OF HIP JACK RAFTERS

A roof framing diagram for a series of hip jack rafters is shown in Fig. 20-3. The jacks are always on the same spacing as the common rafters. The spacing in this instance is 16″ on center. You can see from the arrow in the diagram that the total run of the shortest jack is also 16″.

Suppose the unit rise of a common rafter in this roof is 8″ per 12″ of run. The jacks have the

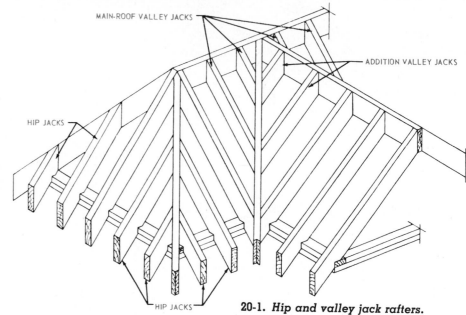

20-1. *Hip and valley jack rafters.*

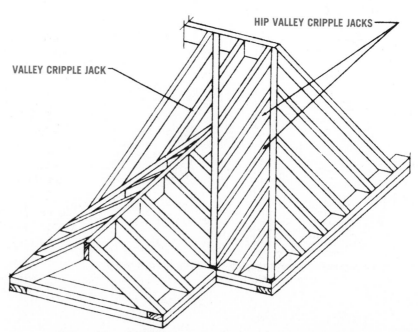

20-2. *Valley cripple jack and hip-valley cripple jacks.*

same unit rise as a common rafter. The unit length of a rafter is the hypotenuse of a right triangle with the unit run as base and the unit rise as height. The unit length of a jack rafter in the example is therefore the square root of $(12^2 + 8^2)$, or 14.42. This means that a jack is 14.42″ long for every 12″ of run.

The theoretical total length of the shortest jack rafter can now be calculated:

$$\frac{12'' \text{ (unit run)}}{14.42'' \text{ (unit length)}}$$

$$= \frac{16'' \text{ (total run)}}{X \text{ (total length)}} \quad X = 19.23''$$

This is the length of the shortest hip jack when the jacks are spaced 16″ on center and the unit rise is 8″. It is also the *common difference* of these jacks. This means that the next hip jack will be 2 × 19.23″ long, the one after that 3 × 19.23″ long, and so on.

The common difference for hip jacks spaced 16″ on center and for hip jacks spaced 24″ on center can be found in the rafter table on the framing square. Fig. 18-10b. The third line of the table reads "Difference in Length of Jacks 16 Inches Centers." Follow this line to the column headed 8 (for a unit rise of 8″) to find the length of the first jack rafter and the common difference, 19¼″.

LENGTHS OF VALLEY AND CRIPPLE JACKS

The best way to figure the total lengths of valley jacks and cripple jacks is to lay out a framing diagram. Figure 20-4 shows part of a framing diagram for a main hip roof with a long-and-short-valley-rafter gable addition. By studying the diagram you can figure the total lengths of the valley jacks and cripple jacks as follows:

The run of valley jack No. 1 is

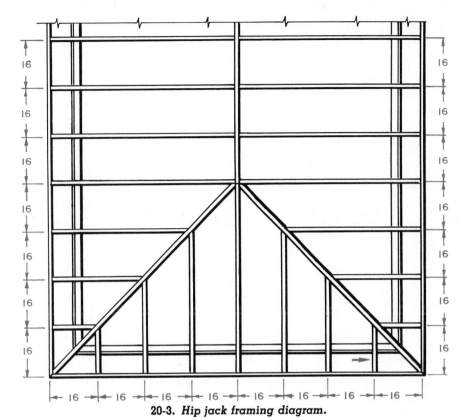

20-3. *Hip jack framing diagram.*

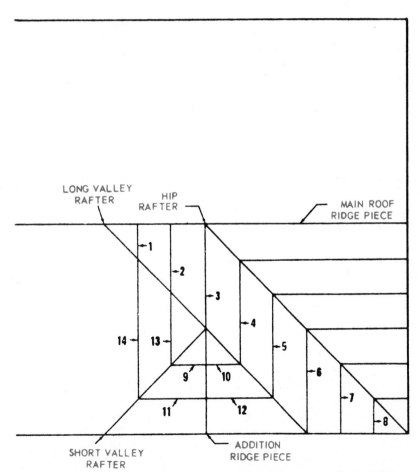

20-4. *Jack rafter framing diagram for a hip roof with a gable addition.*

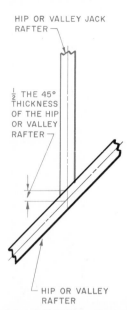

20-5. *The shortening allowance for the upper end of a hip jack or the lower end of a valley jack rafter is one-half the 45° thickness of the hip or valley rafter, whichever the jack rafter intersects.*

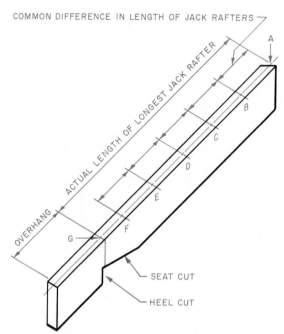

20-6. *Use the longest jack rafter as a pattern. The second jack rafter if BG, the third jack rafter is CG, and so on.*

the same as the run of hip jack No. 8, which is the shortest hip jack. The length of valley jack No. 1 is therefore equal to the common difference of jacks.

The run of valley jack No. 2 is the same as the run of hip jack No. 7, and the length is therefore twice the common difference of jacks.

The run of valley jack No. 3 is the same as the run of hip jack No. 6. The length is therefore three times common difference of jacks. The run of hip-valley cripple No. 4, and also of hip-valley cripple No. 5, is the same as the run of valley jack No. 3. The length of these rafters is thus the same as the length of No. 3.

The run of valley jack No. 9, and also of valley jack No. 10, is equal to the spacing of jacks on center. Therefore the length of each of these jacks is equal to the common difference of jacks. The run of valley jacks Nos. 11 and 12 is twice the run of valley jacks

Nos. 9 and 10. The length of each of these jacks is therefore twice the common difference of jacks.

The run of valley cripple No. 13 is twice the spacing of jacks on center, and the length is therefore twice the common difference of jacks. The run of valley cripple No. 14 is twice the run of valley cripple No. 13, and the length is therefore twice the common difference of jacks.

JACK RAFTER SHORTENING ALLOWANCES

A hip jack rafter has a shortening allowance at the upper end consisting of one-half the 45° thickness of the hip rafter. Fig. 20-5. A valley jack rafter has a shortening allowance at the upper end, consisting of one-half the thickness of the ridge board (Figs. 19-11 & 19-12) and another at the lower end, consisting of one-half the 45° thickness of the valley rafter. Fig. 20-5. A hip-valley cripple has a shortening allowance at the upper end, consisting

of one-half the 45° thickness of the hip rafter, and another at the lower end, consisting of one-half the 45° thickness of the valley rafter. A valley cripple has a shortening allowance at the upper end, consisting of one-half the 45° thickness of the long valley rafter, and another at the lower end, consisting of one-half the 45° thickness of the short valley rafter.

JACK RAFTER SIDE CUTS

The side cut on a jack rafter can be laid out by the method illustrated in Figs. 19-11 & 19-12 for laying out the side cut on a hip rafter. Another method is to use the rafter table on the framing square. Fig. 18-10b. Find the line headed "Side Cut of Jacks Use" and read across to the figure under the unit rise. For a unit rise of 8, the figure given is 10. To lay out the side cut on a jack with this unit rise, set the square face-up on the edge of the rafter to 12″ on the tongue and 10″ on the blade. Draw the side cut line along the tongue, as was described earlier for side cuts on hip rafters. Fig. 19-14.

JACK RAFTER BIRD'S-MOUTH AND OVERHANG

A jack rafter is a shortened common rafter. Consequently the bird's-mouth and overhang on a jack rafter are laid out just as they are on a common rafter.

JACK RAFTER PATTERN

Lay out and cut the longest jack rafter first. Be careful to calculate and make all necessary allowances to determine the actual length. Set the rafter in place on the building and check the fit of all the cuts. See that the spacing between the centers of the rafters is correct. When everything is correct, use this rafter as a pattern. On the top edge of the rafter, measure down the center line from the ridge end a distance equal to the common difference measurement (found on the framing square rafter table). This is the length of the second longest jack rafter. Continue to mark the common difference measurements along the top edge until the lengths of all the jacks have been laid out. Fig. 20-6. Using this pattern, mark off all the jack rafters. When all the rafters have been cut, the pattern is used as a part of the roof frame.

QUESTIONS

1. What is a jack rafter?
2. What is a hip jack rafter?
3. What is a valley jack rafter?
4. What is a cripple jack rafter?
5. If the shortest hip jack rafter is 19¼″ long, how long will the third hip jack rafter be?

Layout and Erection of the Roof Frame 21

When the building has been framed, plumbed, and squared and the ceiling joists are in place, the structure is ready for roof framing. Lay out and cut two common rafters for trial purposes as discussed in Unit 18. With a scrap piece of material the same thickness as the ridge board, set the two common rafters on the building and check all the cuts to make certain the rafters fit properly. If necessary, make corrections on the trial rafters. When they fit properly, use one as a pattern and cut the required number of common rafters for the roof frame. Lean the rafters in position against the building with the ridge end up. Calculate and lay out the actual length of the ridge board. Then lay out the rafter locations on the double plates and ridge board in preparation for the roof frame erection.

LAYING OUT THE RIDGE BOARD

Gable Roof

Laying out the ridge piece for a gable main roof presents no particular problem, since the theoretical length of the ridge piece is equal to the length of the building. The actual length would include any overhang. Fig. 21-1.

Hip Roof

For a hip main roof, the ridge piece layout requires a certain amount of calculation. In an equal-pitch hip roof the theoretical length of the ridge piece amounts to the length of the building minus twice the total run of a main roof common rafter. The actual length, however, depends upon the way in which the hip rafters are framed to the ridge.

21-1. *Framing a gable roof with an unequal-span addition and a Dutch hip on the front of the addition. The workers are framing the overhang on the gable end.*

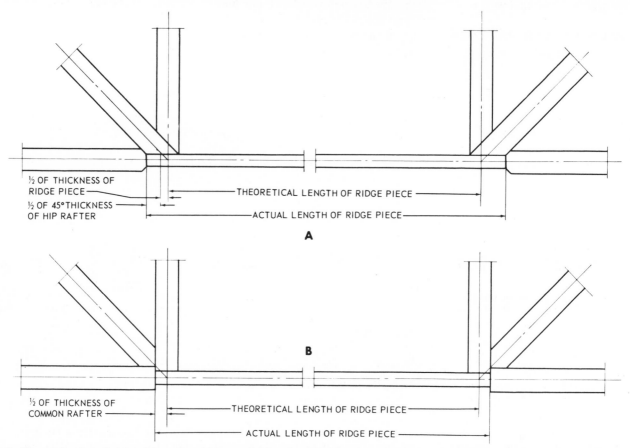

21-2. *Theoretical and actual lengths of hip roof ridge boards. A. Hip rafter framed against ridge board. B. Hip rafter framed between common rafters. In this drawing the ridge board is 1″ material. Usually it is 2″, the same as the ridge-end common rafters, and the side cuts on these rafters (A) are not needed.*

The theoretical ends of the ridge board are at the points where the ridge center line and the hip rafter center lines cross. Fig. 21-2. If the hip rafter is framed against the ridge board, the actual length of the ridge board exceeds the theoretical length, at each end, by one-half the thickness of the ridge board plus one-half the 45° thickness of the hip rafter. A, Fig. 21-2. If the hip rafter is framed between the common rafters, the actual length of the ridge board exceeds the theoretical length, at each end, by one-half the thickness of a common rafter. B, Fig. 21-2.

Equal-Span Addition

For an equal-span addition, the length of the ridge board is equal to the length the addition projects beyond the building, plus one-half the span of the building, minus the shortening allowance at the main roof ridge. The shortening allowance amounts to one-half the thickness of the main roof ridge piece. Fig. 21-3.

Unequal-Span Addition

The length of the ridge board

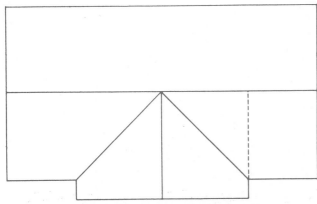

21-3. *Determining the length of the ridge board for an equal-span addition.*

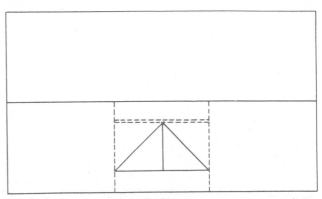

21-5. *Determining length of ridge board on dormer without side walls.*

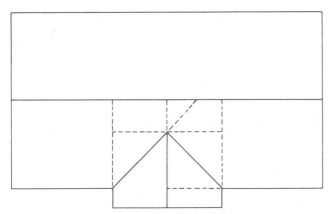

21-4. *Determining the length of the ridge board for an unequal-span addition.*

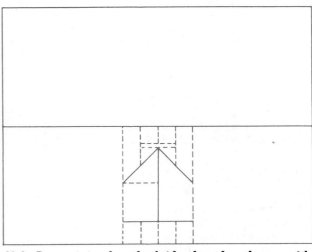

21-6. *Determining length of ridge board on dormer with side walls.*

for an unequal-span addition varies with the method of framing the ridge piece. Fig. 21-4. If the addition ridge board is suspended from the main roof ridge board, the length is equal to the length the addition projects beyond the building, plus one-half the span of the main roof.

If the addition ridge board is framed by the long-and-short-valley-rafter method, the length is equal to the length the addition projects beyond the building, plus one-half the span of the addition, minus a shortening allowance consisting of one-half the 45° thickness of the long valley rafter.

If the addition ridge piece is framed to a double header set between a couple of double main roof common rafters, the length of the ridge piece is equal to the

length the addition projects beyond the building, plus one-half the span of the addition, minus a shortening allowance consisting of one-half the thickness of the inside member of the double header.

Dormer without Side Walls

The length of the ridge piece on a dormer without side walls is equal to one-half the span of the dormer, less a shortening allowance consisting of one-half the thickness of the inside member of the upper double header. Fig. 21-5.

Dormer with Side Walls

The length of the ridge board on a dormer with side walls

amounts to the length of the dormer side-wall top plate, plus one-half the span of the dormer, minus a shortening allowance consisting of one-half the thickness of the inside member of the upper double header. Fig. 21-6.

LAYING OUT THE RAFTER LOCATIONS

The layout of the rafter spacing on the wall plates and ridge board is determined by checking either the building plans or the roof frame plan. Rafter locations are laid out on plates, ridge board, and other rafters with the same lines and X's used to lay out stud and joist locations. (See Units 14 and 16). In most cases the rafters are located next to the ceiling

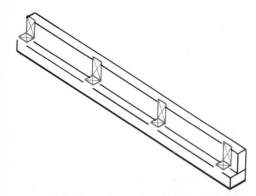

21-7. *Lay the ridge board on edge on the top plate and extend the layout lines from the plate onto the ridge board.*

joists. The rafters can then be fastened to the side of the joists to tie the building together.

Gable Roof

For a gable roof the rafter locations are laid out on the top plates first. The locations are then transferred to the ridge piece by matching the ridge board against a top plate. Fig. 21-7.

On a gable roof the first rafters on each end are usually set even with the outside wall to permit a smooth unbroken surface for the application of the sheathing. Since the first ceiling joist was set on the inside edge of the wall, it will be necessary to place a spacer block between the first rafter and the first ceiling joist. Fig. 21-8. The other rafters are fastened to the side of the joists along the length of the building.

If the rafters are on 24″ centers and the ceiling joists are on 16″ centers, the first rafter will be placed as shown in Fig. 21-8. The second rafter will rest on the plate between the second and third joists. The third rafter will fasten to the side of the fourth joist. The rafters will continue to alternately rest on the plate between two joists and fasten to the side of a joist along the remaining length of the building. Fig. 21-9.

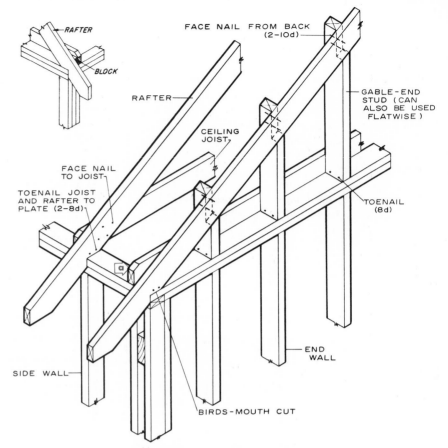

21-8. *Gable roof rafter locations.*

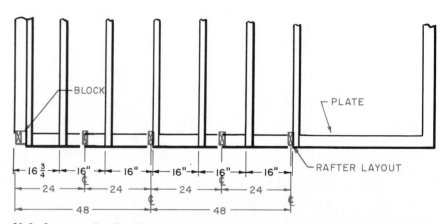

21-9. *Layout of a building with the rafters on 24″ centers and the ceiling joists on 16″ centers.*

Always begin the rafter layout on the plates from the same end of the building for the two opposite walls, and continue along the length of the building. Fig. 21-10. This will insure direct bearing down through the walls to the foundation wall. It will also make the rafters butt directly opposite each other on the ridge board. Fig. 21-11.

Hip Roof

The top plate locations of the

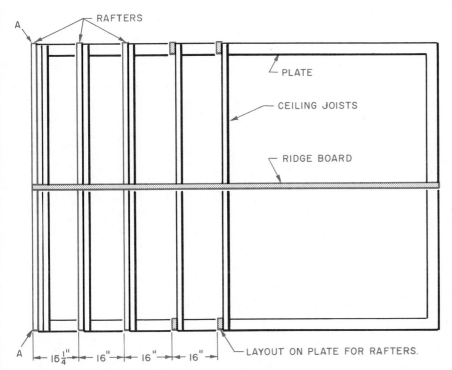

ridge-end common rafters in an equal-pitch hip roof measure one-half of the span (or the run of a main roof common rafter) away from the building corners. These locations, plus the top plate locations of the rafters lying between the ridge-end common rafters, can be transferred to the ridge board by matching the ridge board against the top plates. Fig. 21-12.

Addition Roofs

In an equal-span addition the valley rafter locations on the main roof ridge board lie alongside the addition ridge board location. In Fig. 21-13 the distance between the end of the main roof ridge board and the addition ridge piece location is equal to distance A plus distance B, distance B being

21-10. *Begin the layout of the rafters from the same end of the building as the layout of the floor joists, wall studs, and ceiling joists. In this drawing the layout for each phase began at arrow A on the two side walls.*

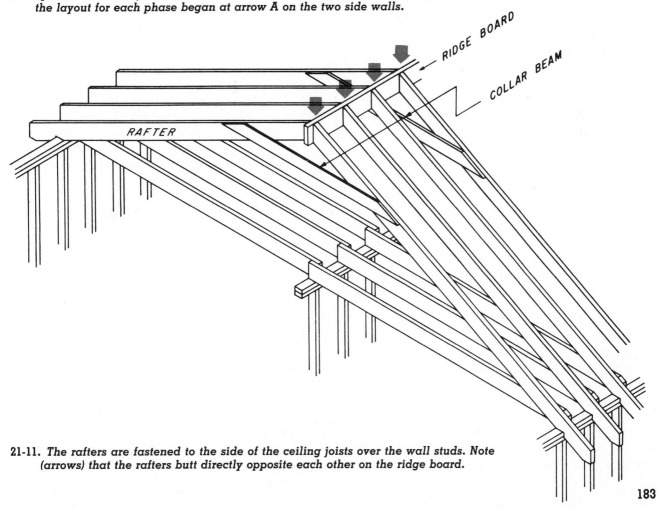

21-11. *The rafters are fastened to the side of the ceiling joists over the wall studs. Note (arrows) that the rafters butt directly opposite each other on the ridge board.*

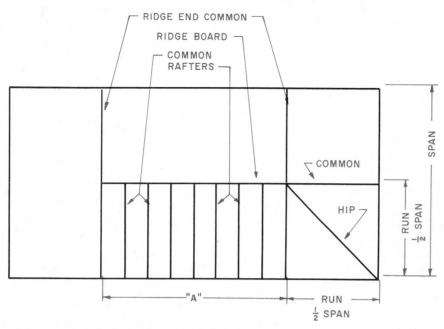

21-12. *The locations of the rafters in the area "A" are transferred to the ridge board from the top plate.*

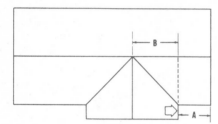

21-13. *Ridge board location for equal-span addition on a gable roof.*

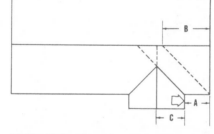

21-15. *Ridge board and valley rafter locations for unequal-span addition.*

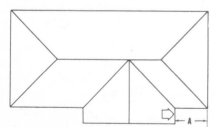

21-14. *Ridge board location for equal-span addition on a hip roof.*

one-half the span of the addition. In Fig. 21-14 the distance between the *theoretical* end of the main roof ridge board and the addition ridge board location is the same as distance A.

In an unequal-span addition, if framing is by the long-and-short-valley-rafter method, the distance from the end of the main roof ridge board to the upper end of the longer valley rafter is equal to distance A plus distance B, distance B being one-half the span of the main roof. Fig. 21-15. The location of the inboard end of the shorter valley rafter on the longer valley rafter can be determined as follows:

• Obtain the unit length of the longer valley rafter from the rafter table. Fig. 18-10b. Suppose that the common rafter unit rise is 8″. In that case the unit length of a valley rafter is 18.76″.

• Between the point where the shorter rafter ties in and the top plate, the total run of the longer valley rafter is the hypotenuse of a right triangle whose shorter sides are each equal to the total run of a common rafter in the addition. The total run of a common rafter in the addition is one-half the span. Suppose the addition is 20′ wide; the run of a common rafter would be 10′. C, Fig. 21-15.

• You know that the valley rafter is 18.76″ long for every foot of common rafter run. The location mark for the inboard end of the shorter valley rafter on the longer valley rafter can thus be calculated:

18.76 (in. per ft. of run) × 10 (ft. of run) = 187.6″; 187.6″ = 15.63′, or 15′7⁹/₁₆″

This is the distance from the heel plumb cut line of the longer valley rafter to the location mark.

If framing is by the suspended-ridge method, the distance between the suspension point on the main roof ridge board and the end of the main roof ridge piece is equal to distance A plus distance C. Fig. 21-15. Distance C is one-half the span of the addition. The distance between the outboard end of the addition ridge board and where the valley rafters (both short in this method of framing) tie into the addition ridge board is equal to one-half the span of the addition plus the length of the addition side-wall top plate.

ERECTING THE RIDGE BOARD

Many carpenters raise the ridge board and the gable-end rafters all at one time. Each member supports the other. However, with the gable-end rafters nailed in place, it is difficult to make adjustments.

It is possible to put the ridge board in place before raising any rafters. Nail uprights on the walls and cross partitions below the center line of the ridge board for support. Erect the ridge board,

184

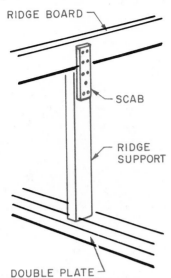

21-16a. *An upright (leg) supports the ridge board in position for the rafter erection.*

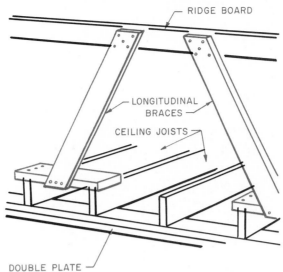

21-16b. *Longitudinal bracing for the ridge board.*

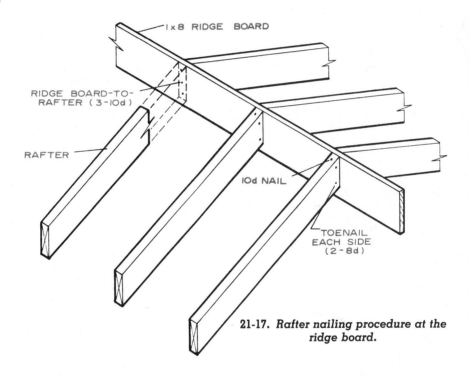

21-17. *Rafter nailing procedure at the ridge board.*

level and align it, and nail it in place. Fig. 21-16a. The ridge board should also be braced longitudinally to prevent the roof from swaying. This is particularly important on a gable roof. Fig. 21-16b. After the ridge board is nailed in position, begin the erection of the rafters.

ERECTING THE RAFTERS

Roof framing should be done from a scaffold with planking not less than 4′ below the level of the main roof ridge board. The usual type of roof scaffold consists of diagonally braced, two-legged horses, spaced about 10′ apart and extending the full length of the ridge piece.

If the building has an addition, as much as possible of the main roof is framed before the addition framing is started. All types of jack rafters are usually left out until after the headers, hip rafters, valley rafters, and ridges to which they will be framed have been installed.

Gable Roof

For a gable roof the two pairs of gable-end rafters and the ridge piece are usually erected first. Two people, one at either end of the scaffold, hold the ridge board in position, while a third person sets the gable-end rafters in place and toenails them at the top plate with 8d nails, two on one side and one on the other side. Make certain the person standing at the ridge pulls the rafter up so that the (plumb) heel cut of the bird's-mouth is tight against the side of

the building when the rafter is nailed at the plate. Each worker on the scaffold then end-nails the ridge piece to one of the rafters with three 10d nails driven through the ridge piece into the end of the rafter. The other rafter is toenailed to the ridge piece and to the first rafter with four 8d nails, two on each side of the rafter.

If the ridge board has not been

21-18. *Toenailing a rafter to the ridge board and to the rafter opposite it.*

21-19. *Face-nailing a jack rafter through the ridge board.*

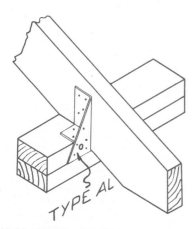

21-20. *Metal brackets are sometimes used to fasten the rafter to the plate. These brackets are fastened with special nails (11 gauge—1¼" long) which are furnished with the bracket.*

previously erected and braced, temporary braces like those for a wall should be installed at the ridge ends to hold the rafters approximately plumb, after which the rafters between the end rafters should be erected. Figs. 21-17, 18, & 19. The braces should then be released, and the pair of rafters at one end should be plumbed. The braces are then reset and left in place until enough sheathing has been installed to hold the rafters plumb.

Ceiling-joist ends are nailed to adjacent rafters with four 10d nails, two to each side. Metal brackets may also be used to attach the rafters to the plate. Fig. 21-20.

Hip Roof

On a hip roof the ridge board and the common rafters extending from the ridge ends to the side walls are erected first, in about the same manner as for a gable roof. The intermediate common rafters are then filled in. After that, the ridge-end common rafters extending from the ridge ends to the mid-points on the end walls are erected. The hip rafters and hip jacks are installed next.

The common rafters in a hip roof do not require plumbing. If the hip rafters are correctly cut, installing the hip rafters and the common rafter which projects from the end of the ridge board to the end wall will bring the common rafters plumb.

Hip rafters are toenailed to plate corners with 10d nails, two to each side. At the ridge board, they are toenailed with four 8d nails. After the hip rafters are fastened in place, partially drive a nail in the center of the top edge of the hip rafter at the ridge end and at the plate end. Pull a line taut between these nails and as the hip jacks are nailed to the hip rafter, keep the string centered on the top edge of the hip rafter to insure a straight hip line.

The hip jacks should be nailed in pairs, one opposite the other. *Do not nail* all the jacks on one side of the hip and then all the jacks on the opposite side as this would push the hip out of alignment and cause a bow. Hip jacks are toenailed to hip rafters with 10d nails, three to each jack, and to the plate with 10d nails, two to each side.

Additions and Dormers

For an addition or dormer the valley rafters are usually erected first. Valley rafters are toenailed to plates with 10d nails, two to each side, and to ridge pieces and headers with three 10d nails. Ridge pieces and ridge-end common rafters are erected next, then other addition common rafters, and last, valley and cripple jacks. As with hip rafters, pull a line along the top edge of the valley rafter and nail the jacks in pairs. A valley jack should be held in position for nailing as shown in Fig. 21-21. When properly nailed, the end of a straight-edge laid along the top edge of the jack should contact the center line of the valley rafter as shown.

COLLAR BEAM FRAMING

Gable or double-pitch roof rafters are often reinforced by horizontal members called *collar*

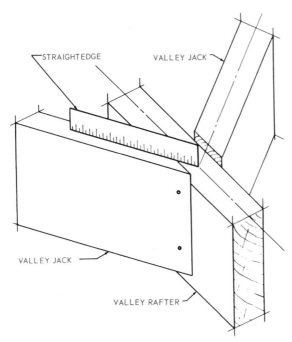

21-21. *Correct position for nailing a valley jack rafter.*

unit rise of a common rafter in the roof is 10. Forty-two divided by 10 is 4.2, and twice 4.2 is 8.4. This is subtracted from the span of the building: $16 - 8.4 = 7.6'$, or about $7'7\frac{3}{16}''$, which is the theoretical length of the beam.

To bring the ends of the collar beam flush with the upper edges of the common rafters, you must add to the theoretical length of the beam, at each end, an amount equal to the *level width* of a rafter minus the width of the rafter seat cut. The level width is obtained by setting the square on the rafter to the cut of the roof, drawing a level line from edge to edge, and measuring the length of this line.

Lay out the end cuts on a collar beam by setting the framing square on the beam to the cut of the roof. Fig. 21-23.

Collar beams are nailed to common rafters with four 8d nails to each end of a one-inch beam. If two-inch material is used for the beams, they are nailed with three 16d nails at each end.

beams. In a finished attic the collar beams also function as ceiling joists.

The length of a collar beam is calculated on the basis of the height of the beam above the level of the side-wall top plates. The theoretical length of a beam in feet is found by dividing this height in inches by the unit rise of a common rafter in the roof, and subtracting twice the result from the span of the building. For example, in the roof shown in Fig. 21-22, the collar beam is $3'6''$, or $42''$, above the rafter plate. The

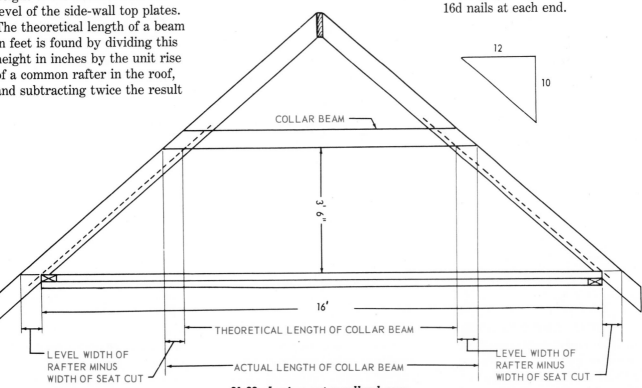

21-22. *Laying out a collar beam.*

187

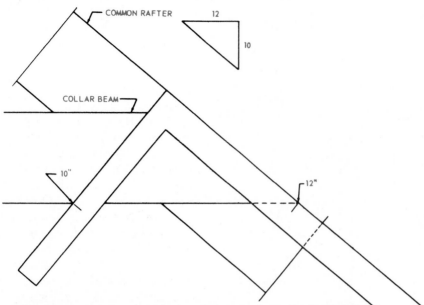

21-23. *Laying out the end cut on a collar beam for a roof with a unit rise of 10".*

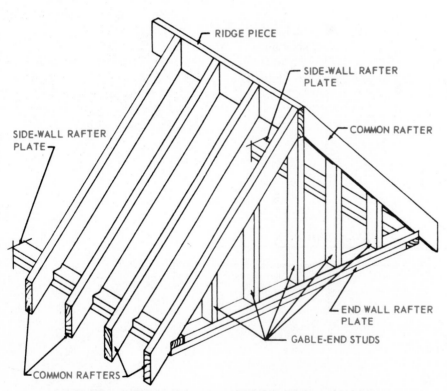

21-24a. *Gable roof framing without a gable overhang.*

GABLE-END FRAMING

Gable-end studs are members which rest on the top plate and extend to the rafter line in the ends of a gable roof. Figs. 21-24a & b. They may be placed with the edge of the stud even with the outside wall and the top notched to fit the rafter (Fig. 21-24c), or they may be installed flatwise with a cut on the top of the stud to fit the slope of the rafter.

The position of the first gable-end stud is located by making a mark on the double plate directly above the wall stud nearest the ridge line. A, Fig. 21-25. Plumb the gable-end stud on this mark and mark the stud where it hits the bottom of the rafter. B, Fig. 21-25. Mark the cut of the roof across the edge of the gable stud and notch the stud to a depth equal to the thickness of the rafter. C, Fig. 21-25.

The lengths of the other gable studs will depend on the spacing. For studs 24" on center, the line DE in Fig. 21-25 represents 2 units of run (one unit is 12"). For a roof with a unit rise of 6" and studs 24" O.C., the second gable stud will be 12" shorter.

The common difference in the length of the gable studs may be figured by the following method:

$$\frac{24'' \text{ (O.C. spacing)}}{12'' \text{ (unit run)}} = 2$$

$2 \times 6''$ (unit rise) = 12" (common difference)

A common difference of 12" means that each stud will be 12" shorter than the first, the third stud 24" shorter than the first, the fourth stud 36" shorter, and so on. If the studs are spaced 16" O.C. for the same roof, the common difference in length is 8":

$$\frac{16'' \text{ (O.C. spacing)}}{12'' \text{ (unit run)}} = 1\frac{1}{3}$$

$1\frac{1}{3} \times 6''$ (unit rise) = 8" (common difference)

The common difference in the length of the gable studs may also be laid out directly with the framing square. Fig. 21-26. Place the framing square on the stud to the cut of the roof (6 and 12 for this example). Draw a line along the blade at A. Slide the square along

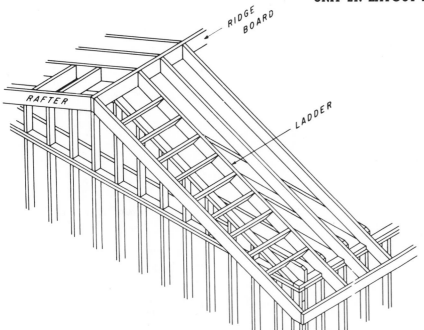

21-24b. *Roof framing for overhang at gable end.*

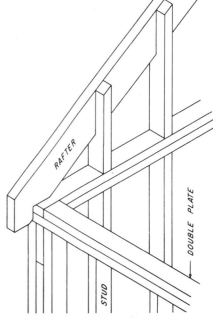

21-24c. *Gable-end studs notched to fit the rafter.*

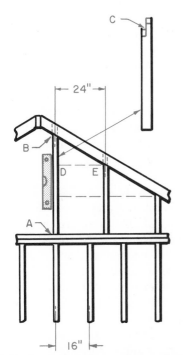

21-25. *Locating the position of the gable-end studs and determining the common difference in length.*

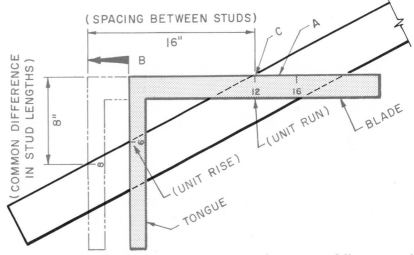

21-26. *Using the framing square to determine the common difference in the length of gable-end studs.*

this line in the direction of the arrow at B until the spacing desired between the studs (16 for this example) is at the intersection of the line drawn at A and the edge of the stud. C, Fig. 21-26. Read the dimension on the tongue which is aligned with the same edge of the stud. This is the common difference (8″ for this example) between the gable studs.

Toenail the studs to the plate with two 8d nails from each side. Fig. 21-8. As the studs are nailed in place, care must be taken not to force a crown into the top of the rafter.

FRAMING A GAMBREL ROOF

The gambrel roof is a gable roof with two slopes. It has the advantage of providing additional space for rooms in the attic area. It also

189

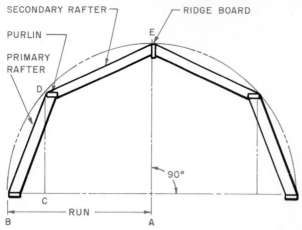

SECONDARY RAFTER ⌐ ⌐ RIDGE BOARD

PURLIN ⌐

PRIMARY
RAFTER

E

D

90°

C

B   RUN   A

21-27. *The patterns for the rafters in a gambrel roof may be made by laying the roof out full size on the subfloor.*

21-29. *Framing a shed dormer.*

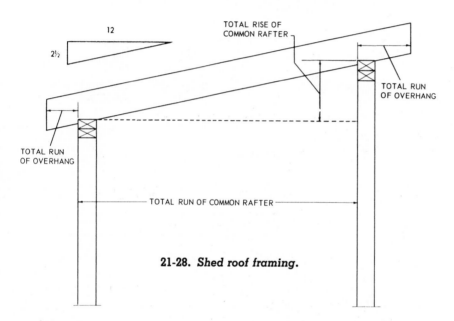

12

2½

TOTAL RISE OF
COMMON RAFTER

TOTAL RUN
OF OVERHANG

TOTAL RUN
OF OVERHANG

TOTAL RUN OF COMMON RAFTER

21-28. *Shed roof framing.*

minimizes the roof area exposed to snow loads. The framing of this roof style is simply a combination of two common rafters, the lower one having a steep pitch and the upper one a low pitch. If the pitches are known, the rafters may be laid out in the same manner as any common rafter.

The roof may also be laid out full size on the subfloor. Use the run of the building (AB) as a radius and draw a semicircle. Fig. 21-27. Draw a perpendicular line from point A to intersect the semicircle at E. This locates the ridge line. Find the height of the partition walls from the plans. Draw a

perpendicular line this length, between the plate and the semicircle. Line CD, Fig. 21-27. Connect the points B and D and the points D and E. This gives the location and pitch of the primary rafter BD and the secondary rafter DE. From this layout the rafter patterns can be made and cut for trial on the building.

FRAMING A SHED ROOF

A shed roof is essentially one-half of a gable roof. Like the full-length rafters in a gable roof, the full-length rafters in a shed roof are common rafters. However, the total run of a shed roof com-

mon rafter is equal to the span of the building *minus the width of the top plate on the higher rafter-end wall.* Fig. 21-28. Also, the run of the overhang on the higher wall is measured from the *inner edge* of the top plate. With these exceptions, shed roof common rafters are laid out like gable roof common rafters. A shed roof common rafter has two bird's-mouths, but they are laid out just like the bird's-mouth on a gable roof common rafter.

For a shed roof, the height of the higher rafter-end wall must exceed the height of the lower by an amount equal to the total rise of a common rafter.

FRAMING A SHED DORMER

When framing a shed dormer (Fig. 21-29), there are three layout problems to be solved:
• Determining the total run of a dormer rafter.
• Determining the angle of cut on the inboard ends of the dormer rafters.
• Determining the lengths of the dormer side-wall studs.

To determine the total run of a dormer rafter, divide the height of the dormer end wall, in inches, by the difference between the unit rise of the dormer roof and the unit rise of the main roof. For example, suppose the height of the dormer end wall is 9′, or 108″. A, Fig. 21-30. The unit rise of the

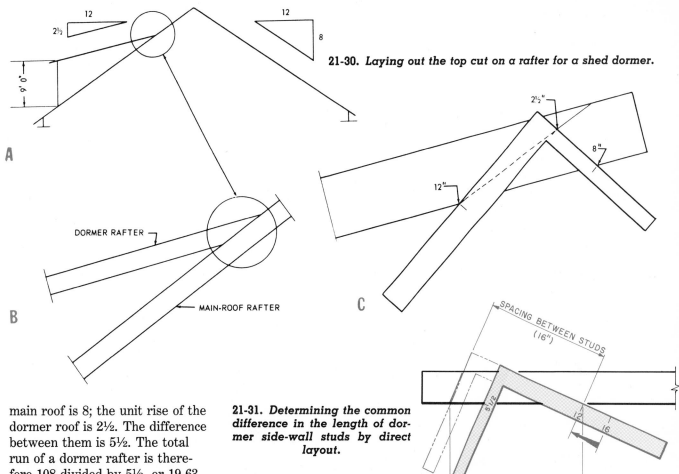

21-30. *Laying out the top cut on a rafter for a shed dormer.*

main roof is 8; the unit rise of the dormer roof is 2½. The difference between them is 5½. The total run of a dormer rafter is therefore 108 divided by 5½, or 19.63 feet. Knowing the total run and the unit rise, you can figure the length of a dormer rafter by any of the methods already described.

The inboard ends of the dormer rafters must be cut to fit the slope of the main roof. B, Fig. 21-30. To get the angle of this cut, set the square on the rafter to the cut of the main roof. C, Fig. 21-30. Measure off the unit rise of the dormer roof from the heel of the square along the tongue. Make a mark at this point and draw the cut off line through this mark from the 12″ mark.

The lengths of the side-wall studs on a shed dormer are determined as follows: Suppose a dormer rafter rises 2½″ for every 12″ of run and a main roof common rafter rises 8″ for every 12″ of run. A, Fig. 21-30. If the studs were spaced 12″ O.C., the length of the shortest stud (which is also the common difference of studs)

21-31. *Determining the common difference in the length of dormer side-wall studs by direct layout.*

would be the difference between 8″ and 2½″, or 5½″. This being the case, if the stud spacing is 16″, the length of the shortest stud is the value of x in the proportional equation 12:5½ :: 16:x. Thus $x = 7\frac{5}{16}$. The shortest stud will be 7⁵⁄₁₆″ long. The next stud will be 2 × 7⁵⁄₁₆″ long, or 14⁵⁄₈″, and so on.

A second method of determining the length of the shortest stud (the common difference of the studs) is to make the layout directly on a stud with the framing square. Fig. 21-31. The difference in the rise of the two roofs is 5½″. Find the 5½″ mark on the tongue of the square and place it on the edge of a stud. Place the 12″ mark of the body of the square on the same edge of the stud. Draw a line along the body of the square onto the stud. Slide the square along this line until the 16″ mark (the on-center spacing between

the studs) is over the point of the 12″ mark. Draw a line along the tongue of the square. This completes the layout for the shortest stud; the second stud will be twice as long, and so on.

To get the lower-end cut off angle for studs, set the square on the stud to the cut of the main roof. To get the upper-end cut off angle, set the square to the cut of the dormer roof.

FRAMING A FLAT OR LOW-PITCH ROOF

The two basic types of roofs—flat and pitched—have numerous variations. The so-called flat roof may actually have some slope for drainage. As discussed earlier, the slope is generally expressed as the inches of vertical rise in 12

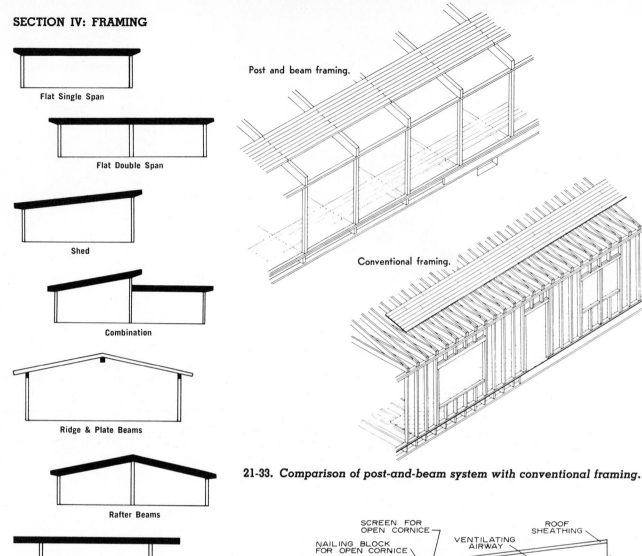

Flat Single Span

Flat Double Span

Shed

Combination

Ridge & Plate Beams

Rafter Beams

Cantilever Beam

21-32. *Variations of flat roof styles.*

Post and beam framing.

Conventional framing.

21-33. *Comparison of post-and-beam system with conventional framing.*

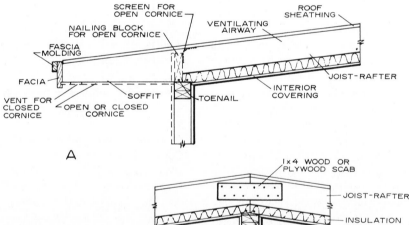

SCREEN FOR OPEN CORNICE · ROOF SHEATHING · NAILING BLOCK FOR OPEN CORNICE · VENTILATING AIRWAY · FASCIA MOLDING · FACIA · JOIST-RAFTER · VENT FOR CLOSED CORNICE · OPEN OR CLOSED CORNICE · SOFFIT · TOENAIL · INTERIOR COVERING

A

1 x 4 WOOD OR PLYWOOD SCAB · JOIST-RAFTER · INSULATION · TOENAIL · INTERIOR WALL OR BEAM

B

21-34. *Rafter-joist construction for a flat roof: A. Detail at exterior wall. B. Detail at interior wall.*

inches of horizontal run. For purposes of definition, flat roofs might be classed as those having less than a 3-in-12 slope. Fig. 21-32.

Post-and-beam construction is frequently used with flat or low-slope roofs. Fig. 21-33. In conventional stud wall framing for buildings with flat or low-slope roofs, the rafters or roof joists usually serve as ceiling joists for the space below.

The flat or low-slope roof sometimes combines ceiling and roof elements in one system. This system serves as an interior finish, or as a fastening surface for the finish, and as an outer surface for application of the roofing. Fig. 21-34. In mild climates flat or low-pitch roofs may be built with 2″ matched planks for roof sheathing supported on large beams spaced about 6′ apart. The planking and beams are exposed on the underside. Fig. 21-35. The exposed ma-

21-35. *In this home the ceiling is extended beyond the exterior wall to become the overhang. The roofing is applied on top and a finish is applied to the bottom. The single system serves as a ceiling and a roof.*

terial may be dressed smooth and finished with varnish or otherwise decorated.

The structural elements can be arranged in several ways by the use of ceiling beams or thick roof decking which spans from the exterior walls to the ridge beam or center bearing partition. Fig. 21-36. The roof is generally covered with a fiberboard insulation, and this in turn with a composition roof.

Roof joists for flat roofs are commonly laid level, with roof sheathing and roofing on top and with the underside utilized to support the ceiling. Sometimes a slight roof slope may be provided for roof drainage by tapering the joist or adding a cant strip (a triangular piece of lumber) to the top. Insulation may be added just above the ceiling, and the space above the insulation should be ventilated to remove hot air in the summer and to provide protection against condensation in the winter.

Flat and low-pitch roofs generally require larger-sized rafters than pitched roofs, but the total amount of framing lumber required is usually less. In flat roof

construction where rafters also serve as ceiling joists, the size of the rafters is based on both roof and ceiling loads. The size is given on the plans or determined from rafter span tables.

When there is an overhang on

all sides of the house, lookout rafters are ordinarily used. Fig. 21-37. The lookout rafters are nailed to a double header and toenailed to the wall plate. The distance from the double header to the wall line is usually twice the overhang. Rafter ends may be finished with an outside header, which will serve as a nailing surface for trim.

ROOF OPENINGS

Roof openings are those which require interruption of the normal run of rafters or other roof framing. Such openings may be required for a ventilator, chimney, skylight, or for dormer windows. Fig. 21-38.

Roof openings, like floor openings, are framed by headers and trimmers. Double headers are used at right angles to the rafters, which are set into the headers in the same manner as joists in floor opening construction. Just

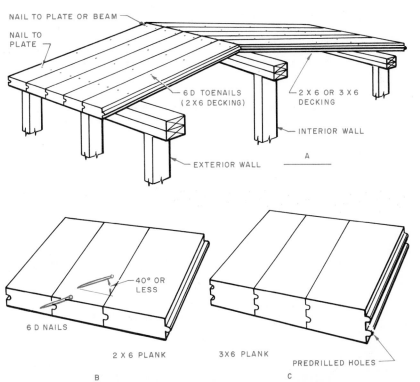

21-36. *Wood-deck construction: A. Installing wood decking. B. Toenailing horizontal joint. C. Edge-nailing 3″ × 6″ solid decking.*

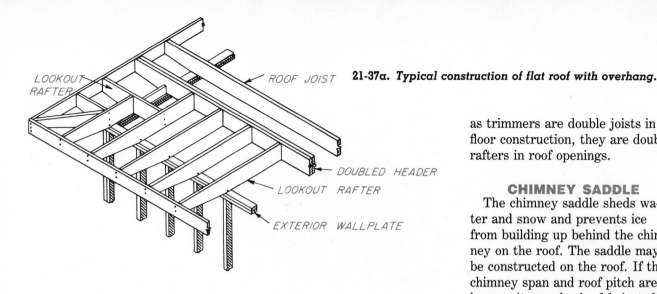

LOOKOUT RAFTER

ROOF JOIST

21-37a. Typical construction of flat roof with overhang.

DOUBLED HEADER

LOOKOUT RAFTER

EXTERIOR WALLPLATE

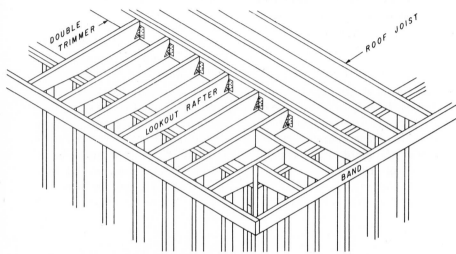

DOUBLE TRIMMER

ROOF JOIST

LOOKOUT RAFTER

BAND

21-37b. Corner framing for flat roofs. Note the use of metal brackets to fasten the lookout rafters to the main roof rafter joist.

21-38. Roof framing around the chimney. The top edges of the headers are kept below the top edge of the rafter (arrow #1). The lower edges of the headers are kept even with the top edge of the rafter (arrow #2).

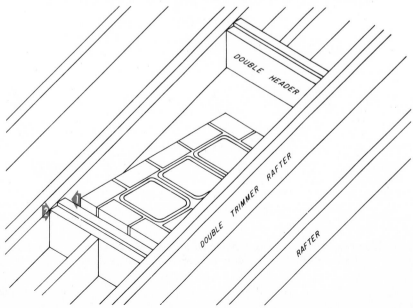

DOUBLE HEADER

DOUBLE TRIMMER RAFTER

RAFTER

as trimmers are double joists in floor construction, they are double rafters in roof openings.

CHIMNEY SADDLE

The chimney saddle sheds water and snow and prevents ice from building up behind the chimney on the roof. The saddle may be constructed on the roof. If the chimney span and roof pitch are known, it can also be fabricated on the ground and the completed assembly nailed to the roof framing. Fig. 21-39.

The valley strips are 1″ × 4″ or 1″ × 6″. The length is determined in the same way as for a valley rafter. Use the framing square to lay out the valley strips. Lay out the top and bottom cuts along the tongue of the square. When measuring off the length of the strip, use the unit length of a common rafter from the roof on which the saddle is to be framed.

For example, a roof with a unit rise of 5″ has a unit length of 13″. Fig. 18-10b. To lay out the valley strip, position the square with the 13″ mark of the tongue and the 12″ mark of the blade on the edge of the strip. Draw a line along the tongue for the top cut. Fig. 21-40. Measure and lay off the length of the valley strip. With the square set the same as for the top cut, place the edge of the blade on the length mark. Then, draw a line along the blade to show the bottom cut.

The end of the ridge rests on the valley strips. A, Fig. 21-39. This cut is the same as the seat cut for a common rafter in the main roof. Place the square on the ridge board for the cut of the roof (5″ on the tongue and 12″ on the blade for the example), and draw a line along the blade. The length of the ridge is equal to the

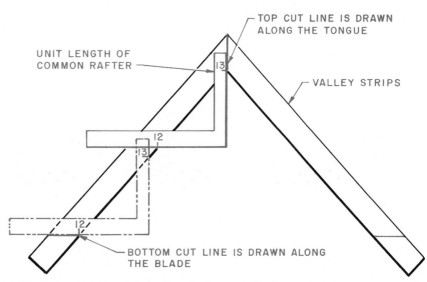

21-40. *Using the framing square to lay out the top and bottom cuts on the valley strips.*

21-39. *The saddle span is less than the chimney width, as shown at "B" in the drawing. This distance ("B") must be subtracted from each side of the chimney width to obtain the actual saddle span. When the sheathing is applied to the saddle rafters, it will project beyond the valley strip.*

run of the common rafter in the saddle span minus the allowance for the drop of the ridge, which is approximately ¾".

The length of the longest rafter is determined by multiplying the saddle run (half the saddle span) by the unit length of the common rafter. Fig. 21-39. Deduct the ridge shortening allowance to obtain the actual length. The top and bottom cuts are the same as for a common rafter in the main roof. However, there is a side cut on the bottom where the rafter rests on the valley strip. This cut is the same as for regular valley jacks. On the rafter table of the framing square, the side cut figure for a valley jack in a roof with a 5" unit rise is 11½". Fig. 18-10b. Lay out and make the cut as described for jack rafters.

The cuts are the same for all the rafters in the chimney saddle. However, the rafter lengths differ. The difference in the length of the rafters can be found on the rafter table of the framing square under "Difference in Length of Jacks." For rafters 16" on center in a roof with a unit rise of 5", the second rafter will be 17⁵⁄₁₆" shorter than the first rafter. Fig. 18-10b. The third rafter will be 34⁵⁄₈" (2 × 17⁵⁄₁₆) shorter than the first rafter, and so on. When the saddle framing is complete, the sheathing, flashing (to prevent water seepage), and roofing are applied.

QUESTIONS

1. Why is the ridge board for a hip roof shorter in length than for a gable roof?

2. Why is it necessary that the rafter locations on the ridge board be laid out exactly from the top plate?

3. Why is it best to erect the ridge board in its proper position before beginning the installation of the rafters?

4. When nailing the common rafters in place, why must the rafter be nailed at the bird's-mouth first?

5. When erecting the rafters for a hip roof, which rafters are erected first?

6. Why must hip jack rafters be installed in pairs?

7. What special treatment is required around roof openings?

8. What is a collar beam?

9. Describe the two methods of installing gable end studs.

10. How does a gambrel roof differ from a gable roof?

Roof Trusses

22

Much modern roof framing of residential and commercial buildings is done with roof trusses. The *simple truss*, or *trussed rafter*, is an assembly of members forming a rigid framework of triangular shapes. These members are usually connected at the joints by gussets. *Gussets* are flat wood, plywood, or similar type members. They are fastened to the truss by nails, screws, bolts, or adhesives. The roof truss is capable of supporting loads over long spans without intermediate support. Fig. 22–1. It has been greatly refined during its development over the years, and the gusset and other preassembled types of roof trusses are being used extensively in the housing field. Figs. 22-2 & 22-3.

Roof trusses save material and on-site labor costs. It is estimated that a material savings of about 30% is made on roof members and ceiling joists. The double top plate on interior partition walls and the double floor joists under interior bearing paritions are not necessary. Roof trusses also eliminate interior bearing paritions because trusses are self-supporting.

Trusses can be erected quickly, and therefore the house can be enclosed in a short time. The roof frame can be ready for sheathing in less than an hour. A long boom mobile crane with a spreader bar can be used to lift the trusses up to the top plates. The trusses are hung on the bar at 24″ centers while on the ground. Six to ten trusses are held in position on the bar by special blocking or by 1″ × 4″ ribbons nailed to the top chord. The assembly is swung to the top wall plates, and a worker on each wall plate positions the trusses and nails them. The bar is then removed and returned for another load. Fig. 22-4.

Trusses are usually designed to span from one exterior wall to the other with lengths of 20′ to 32′ or more. Because no interior bearing walls are required, the interior of the building becomes one large workroom. This allows increased flexibility for interior planning, since partitions can be placed without regard to structural requirements. Fig. 22-5.

Most trusses are fabricated in a shop and then delivered to the job site. Fig. 22-6. Some, however, are constructed at the job site.

The following wood trusses are most commonly used for houses. Fig. 22-7.
- King-post.
- W-type.
- Scissors.

These and similar trusses are most adaptable to rectangular houses because the constant width requires only one type of truss. However, trusses can also be used for L-shaped houses. For hip roofs, hip trusses can be provided for each end and valley area. Fig. 22-8.

Trusses are commonly designed for 24″ spacing. This spacing requires somewhat thicker interior and exterior sheathing or finish material than is needed for con-

22-1. Through the use of trusses, this barn has a clear span of 80′.

22-2. Testing a nail-glued king-post truss.

196

22-3. *Both roof and floor trusses are used in this southern California home.*

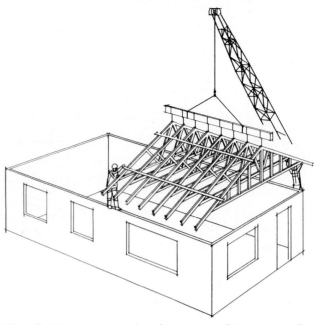

22-4. *Setting a series of **eight prespaced trusses** on the exterior walls with a long boom mobile crane.*

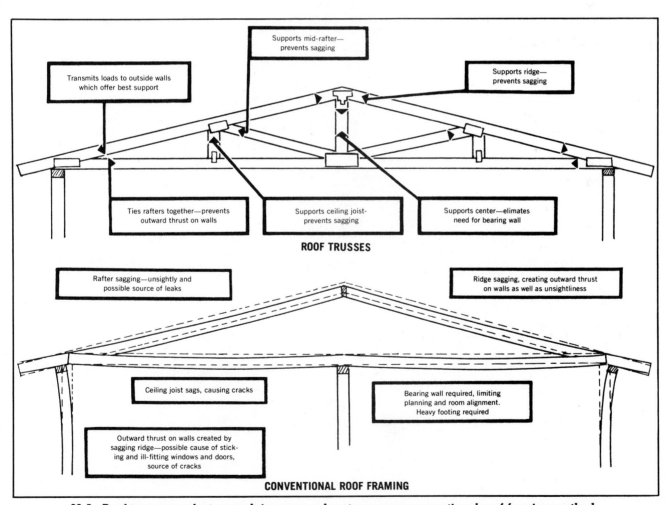

Supports mid-rafter— prevents sagging

Transmits loads to outside walls which offer best support

Supports ridge— prevents sagging

Ties rafters together—prevents outward thrust on walls

Supports ceiling joist- prevents sagging

Supports center—elimates need for bearing wall

ROOF TRUSSES

Rafter sagging—unsightly and possible source of leaks

Ridge sagging, creating outward thrust on walls as well as unsightliness

Ceiling joist sags, causing cracks

Bearing wall required, limiting planning and room alignment. Heavy footing required

Outward thrust on walls created by sagging ridge—possible cause of sticking and ill-fitting windows and doors, source of cracks

CONVENTIONAL ROOF FRAMING

22-5. *Roof truss manufacturers claim many advantages over conventional roof framing methods.*

22-6. *These trusses are bundled in sets with ¾″ steel tape for transporting to the building site.*

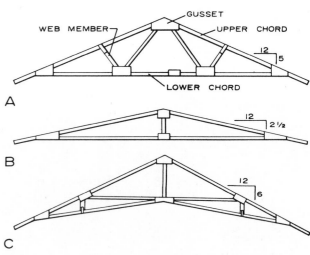

22-7. *Light wood trusses: A. W-type. B. King-post. C. Scissors.*

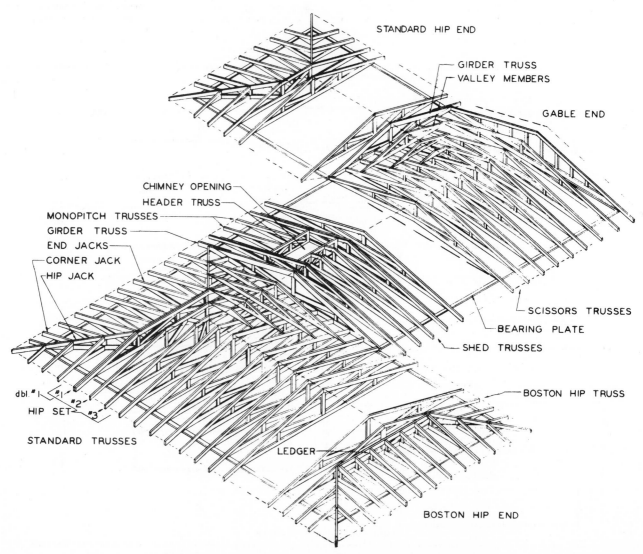

22-8. *Special trusses are available for hip and valley areas.*

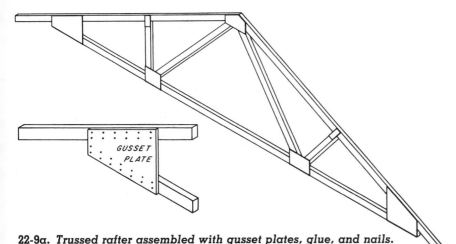

22-9a. *Trussed rafter assembled with gusset plates, glue, and nails.*

22-9b. *Metal gusset plate.*

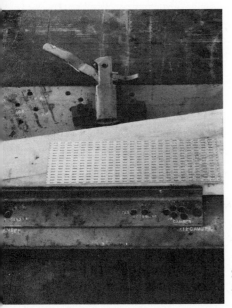

22-9c. *This jig holds the truss firmly in place during application of the metal gusset plate.*

ventional joist and rafter construction using 16″ spacing. Truss designs, lumber grades, and construction details are available from several sources, including the American Plywood Association.

KING-POST TRUSS

The king-post is the simplest form of truss used for houses. It consists of upper and lower chords and a center vertical post. B, Fig. 22-7. Allowable spans are somewhat less than for the W-truss when the same size members are used because of the unsupported length of the upper chord. For example, under the same conditions, a plywood gusset king-post truss with 4-in-12 slope and 2″ spacing is limited to about a 26′ span for 2″ × 4″ members. A W-truss with the same size members and spacing could be used for a 32′ span. Furthermore, the grades of lumber used for the two types might also vary.

For short and medium spans, the king-post truss is probably more economical than other types because it has fewer pieces and

can be fabricated faster. However, local prices and design load requirements (for snow, wind, etc.) as well as the span should govern the type of truss to be used.

W-TYPE TRUSS

The W-type truss is perhaps the most popular and most widely used of the light wood trusses. A, Fig. 22-7. Its design includes the use of three more members than the king-post truss, but distances between connections are less. This usually allows the use of lower grade lumber and somewhat greater spans for the same member size.

SCISSORS TRUSS

The scissors truss is used for houses with a sloping living room ceiling. C, Fig. 22-7. Somewhat more complicated than the W-type truss, it provides good roof construction for a "cathedral" ceiling with a saving in materials over conventional framing methods.

DESIGN

Plans for the fabrication of trussed rafters must include the preparation of an engineered design. For use on FHA-insured projects, the design must be approved by the FHA. Some building codes require approval and a certificate of inspection on trussed rafters.

The design of a truss includes consideration of not only snow and wind loads but the weight of the roof itself. Design also takes into account the slope of the roof. Generally, the flatter the slope, the greater the stresses. Flatter slopes therefore require larger members and stronger connections in roof trusses.

A great majority of the trusses used are fabricated with gussets of plywood (nailed, glued, or

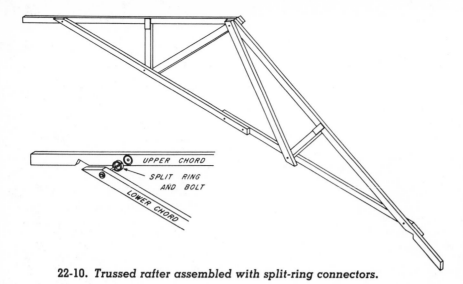

bolted in place) or with metal gusset plates. Fig. 22-9. Others are assembled with split-ring connectors. Fig. 22-10. Some trusses are designed with a 2″ × 4″ soffit return at the end of each upper chord to provide nailing for the soffit of a wide box cornice.

Designs for standard W-type and king-post trusses with plywood gussets are usually available through the American Plywood Association or a local lumber dealer. Fig. 22-11. Many lumber dealers are able to provide the

22-10. *Trussed rafter assembled with split-ring connectors.*

22-11. *An example of a nail-glued truss plan for a king-post truss with a 4-in-12 slope. This is one of the designs available from the American Plywood Association.*

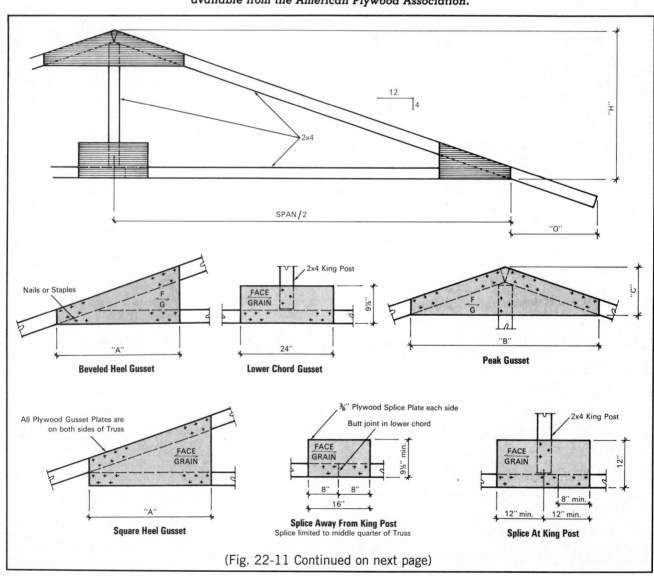

(Fig. 22-11 Continued on next page)

SELECTING THE TRUSS DESIGN

☐ The first step in selecting a design from this plan is to determine the design load used in your area. Consult your local building official for this information; or the nearest FHA office, for FHA-insured construction.

☐ Two loading conditions are given—30 and 40 lbs. per sq. ft. total roof load. Allowance has been made for ceiling and attic loads in addition to these roof loads. Enter the TRUSS DIMENSION TABLE with the design load for your area.

☐ These designs are based on a 24″ O.C. spacing. Where trusses are spaced 16″ O.C. they can carry greater loads. For instance, on the 16″ spacing, total allowable roof loads are 45 lbs. per sq. ft. for the tabulated 30 lbs. per sq. ft. loading condition, and 60 lbs. per sq. ft. for the tabulated 40 lbs. per sq. ft. loading condition.

☐ A choice of two heel gusset arrangements is offered. The beveled heel gusset provides the lowest roof line. The square heel joint offers the most economical fabrication.

☐ With the loading condition and heel gusset arrangement determined, gusset plate sizes and truss dimensions may be selected for the applicable span.

☐ Lumber grades may be chosen from the CHORD CODE TABLE. For each span in the TRUSS DIMENSION TABLE the lumber species and grades for the upper and lower chord members may be picked from those listed under the corresponding code in the CHORD CODE TABLE.

GENERAL NOTES

1. See page 200 for detailed material and fabrication recommendations.

2. Use ⅜″ minimum thickness plywood gusset plates nail-glued in accordance with fabrication recommendations. All plywood shall bear the APA grade-trademark of the American Plywood Association.

3. Use stress grade lumber for chord members from the CHORD CODE TABLE. The king-post may be cut from Construction grade Douglas fir, larch, hemlock, or No. 2 Southern pine.

4. Trusses with pitches over 4 in 12 and not more than 5 in 12, require the same gusset plate sizes and chord lumber grades as specified for the trusses on this sheet.

5. Trusses for spans intermediate between those listed in the TRUSS DIMENSION TABLE, require gusset plate sizes and chord codes for the next longer span. For example, a 27′ span truss would require the gusset plate sizes and chord codes listed for the 28′8″ design.

Nail-Glued Truss Plans

Chords			King Post KP-2
Upper	Lower	Pitch	
2 x 4	2 x 4	4 in 12	

MATERIALS

Plywood

Use only plywood bearing the APA grade-trademark of the American Plywood Association, in the thicknesses specified on the drawings. In normal situations where the moisture content of the trusses in service will not exceed 18%, use the regular Interior-APA-PlyScord or WSP-1 or WSP-2 CD sheathing grades.

The above grades manufactured with Exterior glue or Exterior-APA type plywood may be used for added assurance of durability. These premium grades manufactured with Exterior glue are required in FHA insured construction.

Where the moisture content of the wood is likely to exceed 18% in service, use only Exterior-APA grades.

Plywood for gusset plates shall have a moisture content of 16% or less. Normally, plywood may be used as received unless it has been stored out of doors. Surfaces to be

(Fig. 2-11 Continued on next page)

(Fig. 22-11 Continued)

glued must be clean and free from oil, dust, and paper tape.

Lumber

Lumber must be of the stress grade called for in the CHORD CODE TABLE, as indicated by an approved grading agency. At time of gluing it should be conditioned to a moisture content approximately that which it will attain in service, but in any case between 7% and 16%. Surfaces to be glued should be clean and free from oil, dust, and other foreign matter. Each piece should be machine finished, but not sanded.

Use no lumber which has in the area of the gussets any roughness, cup, or twist which might prevent good contact between gusset and lumber.

Keep surfaces of intersecting lumber members flush within $1/32''$.

Glue

Use casein type, conforming with Federal Specification MMM-A-125, Type II for dry, indoor exposures. For wet conditions, or if any glue joint is exposed, even at a soffit, use resorcinol-type glue, conforming with Military Specification MIL-A-46051.

KP-2 Designs (when using STANDARD sheathing or C-C EXT-APA grade plywood)

Truss Dimension Table KP-2

Loading Condition Total Roof Load (lbs. per sq. ft.)	Span	Beveled Heel Gusset							Square Heel Gusset						
		Dimensions in Inches					Chord Code		Dimensions in Inches					Chord Code	
		A	B	C	H	O	Upper	Lower	A	B	C	H	O	Upper	Lower
30 (a) Meets FHA requirements	20'8"	32	48	12	45⅛	44	2	3	19	32	12	48¾	44	2	3
	22'8"	32	48	12	49⅛	48	1	2	19	32	12	52¾	48	1	3
	24'8"	48	60	16	53⅛	48	1	2	24	48	12	56¾	48	1	3
	26'8"	48	72	16	57⅛	48	1	2	32	60	16	60¾	48	1	2
40 (b)	20'8"	32	48	12	45⅛	43	8	9	19	32	12	48¾	48	7	9
	22'8"	32	60	16	49⅛	48	7	8	19	48	12	52¾	48	7	9
	24'8"								32	60	16	56¾	48	7	9

(a) 30 psf (20 psf live load, 10 psf dead load) on upper chord and 10 psf dead load on lower chord
(b) 40 psf (30 psf live load, 10 psf dead load) on upper chord and 10 psf dead load on lower chord

Chord Code Table

Chord Code	Size	Grade and Species meeting Stress requirements	Grading Rules	f	t//	c//
1	2 x 4	Select Structural Light Framing WCDF No. 1 Dense Kiln Dried Southern Pine 1.8E	WCLIB SPIB WWPA	1950 2000 2100	1700 2000 1700	1400 1700 1700
2	2 x 4	1500f Industrial Light Framing WCDF 1500f Industrial Light Framing WCH No. 1 2" Dimension Southern Pine 1.4E	WCLIB WCLIB SPIB WWPA	1500 1450 1450 1500	1300 1250 1450 1200	1200 1100 1350 1200
3	2 x 4	1200f Industrial Light Framing WCDF 1200f Industrial Light Framing WCH No. 2 2" Dimension Southern Pine	WCLIB WCLIB SPIB	1200 1150 1200	1100 1000 1200	1000 900 900

(Fig. 22-11 Continued on next page)

Chord Code	Size	Grade and Species meeting Stress requirements	Grading Rules	f	t//	c//
		(Fig. 22-11 Continued)				
4	2 x 6	Select Structural J & P WCDF	WCLIB	1950	1700	1600
		Select Structural J & P Western Larch	WWPA	1900	1600	1500
		No. 1 Dense Kiln Dried Southern Pine	SPIB	2000	2000	1700
		1.8E	WWPA	2100	1700	1700
5	2 x 6	Construction Grade J & P WCDF	WCLIB	1450	1300	1200
		Construction Grade J & P WCH	WCLIB	1450	1250	1150
		Structural J & P Western Larch	WWPA	1450	1300	1200
		No. 1 2" Dimension Southern Pine	SPIB	1450	1450	1350
		1.4E	WWPA	1500	1200	1200
6	2 x 6	Standard Grade J & P WCDF	WCLIB	1200	1100	1050
		Standard Grade J & P WCH	WCLIB	1150	1000	950
		Standard Structural Western Larch	WWPA	1200	1100	1050
		No. 2 2" Dimension Southern Pine	SPIB	1200	1200	900
7	2 x 4	Select Structural Light Framing WCDF	WCLIB	1900	1900	1400
		Select Structural Light Framing Western Larch	WWPA	1900	1900	1400
		No. 1 Dense Kiln Dried Southern Pine	SPIB	2050	2050	1750
		1.8E	WWPA	2100	1700	1700
8	2 x 4	1500f Industrial Light Framing WCDF	WCLIB	1500	1500	1200
		Select Structural Light Framing WCH	WCLIB	1600	1600	1100
		Select Structural Light Framing WH	WWPA	1600	1600	1100
		1500f Industrial Light Framing Western Larch	WWPA	1500	1500	1200
		No. 1 2" Dimension Southern Pine	SPIB	1500	1500	1350
		1.4E	WWPA	1500	1200	1200
9	2 x 4	1200f Industrial Light Framing WCDF	WCLIB	1200	1200	1000
		1200f Industrial Light Framing Western Larch	WWPA	1200	1200	1000
		1500f Industrial Light Framing WCH	WCLIB	1500	1500	1000
		1500f Industrial Light Framing WH	WWPA	1500	1500	1000
		No. 2 2" Dimension Southern Pine	SPIB	1200	1200	900
10	2 x 6	Select Structural J & P WCDF	WCLIB	1900	1900	1500
		Select Structural J & P Western Larch	WWPA	1900	1900	1500
		No. 1 Dense Kiln Dried Southern Pine	SPIB	2050	2050	1750
		1.8E	WWPA	2100	1700	1700
11	2 x 6	Construction Grade J & P WCDF	WCLIB	1500	1500	1200
		Construction Grade J & P Western Larch	WWPA	1500	1500	1200
		Select Structural J & P WCH	WCLIB	1600	1600	1200
		Select Structural J & P WH	WWPA	1600	1600	1200
		No. 1 2" Dimension Southern Pine	SPIB	1500	1500	1350
		1.4E	WWPA	1500	1200	1200
12	2 x 6	Standard Grade J & P WCDF	WCLIB	1200	1200	1000
		Standard Grade J & P Western Larch	WWPA	1200	1200	1000
		Standard Grade J & P WCH	WCLIB	1200	1200	1000
		Standard Grade J & P WH	WWPA	1200	1200	1000
		No. 2 2" Dimension Dense Southern Pine	SPIB	1400	1400	1050

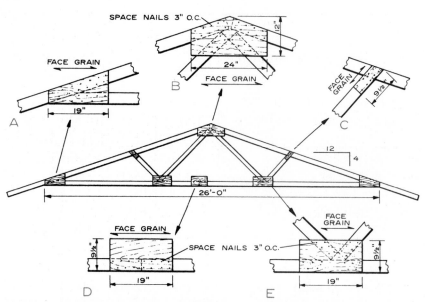

22-12. *Construction details of a 26' W-truss: A. Beveled heel gusset. B. Peak gusset. C. Upper chord intermediate gusset. D. Splice of lower chord. E. Lower chord intermediate gusset.*

builder with completed trusses ready for erection.

To illustrate the design and construction of a typical wood W-truss more clearly, the following example is given:

The span for the gusset truss is 26", the slope 4 in 12, and the spacing 24". Fig. 22-12. The gussets are nail-glued; that is, nails or staples are used to supply pressure while the glue sets. Total roof load is 40 pounds per square foot, which is usually sufficient for areas with moderate to heavy snow. The upper and lower chords can be 2" × 4" in size, but the upper chord requires a slightly higher grade of material. It is often desirable to use dimension material with a moisture content of about 15%. The moisture content should not exceed 19%.

FABRICATION

Applying Gussets

Plywood gussets can be made from ⅜" or ½" standard plywood with exterior glueline or exterior sheathing grade plywood. The cut-out size of the gussets and the general nailing pattern for nail-gluing are shown in Figs. 22-12 & 22-13. More specifically, 4d nails should be used for plywood gussets up to ⅜" thick and 6d for plywood ½" to ⅞" thick. Three-inch spacing should be used when plywood is no more than ⅜" thick and 4" spacing should be used for thicker plywood. When wood truss members are a nominal 4" wide, use two rows of nails with a ¾" edge distance. Use three rows of nails when truss members are 6" wide. Gussets are used on both sides of the truss. Fig. 22-14.

For normal conditions and where relative humidities in the attic area tend to be high, such as might occur in the southern and southeastern United States, resorcinol glue should be used for the gussets. In dry and arid areas where conditions are more favorable, a casein or similar glue might be considered. For estimating purposes, approximately ¹⁄₁₀ pound of glue is required per square foot of gusset.

Glue should be spread on the clean surfaces of the gusset, truss member, or both. When mixing and using the glue, be sure to follow the glue manufacturer's recommendations. If the glue is to be spread on the lumber, it is a good idea to mark the ends of the gussets on the jig table so that the glue will be spread only over the area to be covered by the gusset. To prevent interference with the sheathing or ceiling, the edge of the gussets may be held back ⅛" from the edge of the lumber.

Either nails or staples might be used to supply pressure until the glue has set, although only nails are recommended for plywood ½" and thicker. Use the nail spacing previously outlined. Closer or intermediate spacing may be used to insure "squeezeout" at all visible edges.

Gluing should be done under closely controlled temperature conditions. This is especially true if using the resorcinol adhesives. Follow the assembly temperatures recommended by the manufacturer.

The complete truss should be set aside immediately after assembly. It should not be disturbed until the glue has set. A holding frame which stores the trusses in an upright, inverted position is a convenient way to store the trusses during curing.

Assembly

Plywood trusses can be built without expensive equipment or facilities. The use of trusses is therefore possible for small and medium operations or where a special truss design is needed. A glue spreader or roller and hammers are all the equipment required. Power nailers or staplers are often used. Jig tables are more convenient for hand or power nailing. Fig. 22-15. It is possible, however, to lay out the truss on the subfloor and then cut

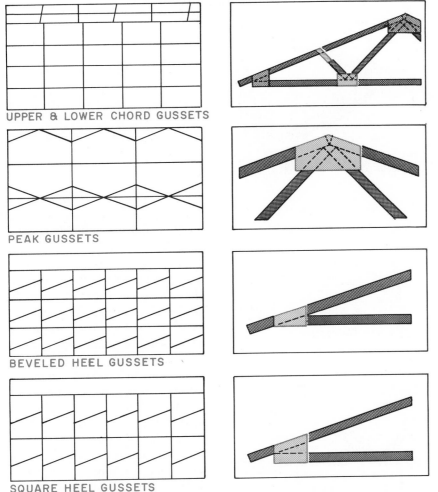

UPPER & LOWER CHORD GUSSETS

PEAK GUSSETS

BEVELED HEEL GUSSETS

SQUARE HEEL GUSSETS

22-13. *Plywood gusset cutting layouts. Make sure the grain of the plywood runs parallel with the lower chord, except for the upper chord intermediate gusset. For this gusset the grain should run parallel to the compression web.*

Plywood Thickness	Nails (1)		Staples (2)	
	Size	Spacing (3)	Size	Spacing (3)
⅜″	4d	3″ o.c.	1⅛″	3″ o.c.
½″–¾″	6d	4″ o.c.	Not recommended	

(1) Nails—Box, common, cement coated or T-nails.
(2) Staples—16 gauge with ⁷/₁₆″ crown width.
(3) Use two rows nails or staples for 4″ wide lumber, three rows for 6″ wide lumber and set in ¾″ from lumber edge. Stagger nails from opposite sides. Nails or staples provide gluing pressure.

22-14. *Nail and staple schedule.*

resorcinol-resin glue are used, a 70° F temperature should be maintained during the assembly and cure. Therefore an area at least big enough to construct the largest truss, with sufficient room for turning and storage after assembly, is recommended.

A simple jig can be made by nailing blocks to a table. Lay out the overall truss dimensions on the table with a chalk line. Position the upper and lower chords. Fig. 22-16. For best appearance don't forget to provide for camber (a slight curvature). Camber of ¼″ for a 30′ truss with proportional amounts for other lengths is recommended.

For the W-type trusses lay out the long diagonals with the diagonal centerline intersecting the lower chord centerline at ⅓ the span of the lower chord. Fig. 22-17. The short diagonals may then be positioned with the upper end centered midway between the clear distance from the peak to

and assemble the trusses before the wall is erected.

To insure proper temperature control, nail-glued trusses should be fabricated in an enclosed build-ing. For normal truss assembly with moisture-resistant glues such as casein, a minimum room temperature of 50° F is needed. Where exterior glues such as

22-15. *A truss jig table designed to allow the worker to walk through while setting the pieces in place for assembly.*

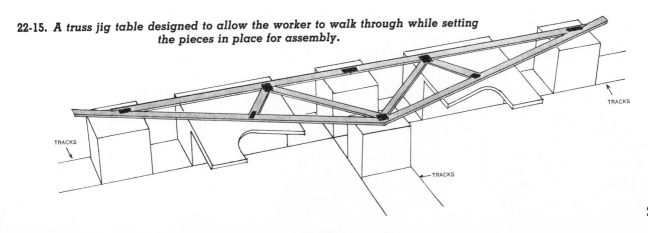

22-16. *A king-post truss laid out and blocked on an assembly table.*

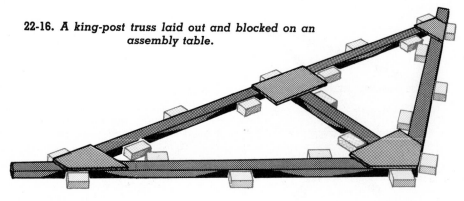

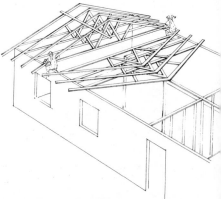

22-19b. *Roof trusses may be tipped up as they are nailed in place.*

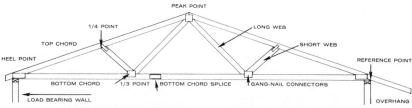

22-17. *Laying out the web locations on a W-type truss.*

22-18. *Nail the holding blocks into position on the jig table. This is a W-type truss.*

22-19a. *Fabricated roof trusses are lightweight and may be placed by hand on the exterior walls.*

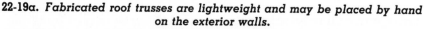

the heel gusset. Another way to determine the points at which the diagonals intersect the upper chord is to divide the bottom chord into four equal parts. Perpendiculars extended from these divisions will intersect the upper chord at the locations for the diagonals (the "¼ point" in Fig. 22-17).

For the king-post truss, simply place the king post in position, and recheck alignment of all members to insure accurate dimensions. When all members of a truss are in position, nail the holding blocks into place. Fig. 22-18.

Handling

In handling and storage of completed trusses, avoid placing unusual stresses on them. They were designed to carry roof loads in a vertical position, and it is important that they be lifted and stored upright. If they must be handled in a flat position, enough support should be used along their length to minimize bending deflections. Never support the trusses only at the center or only at each end when they are in a flat position.

TRUSS ERECTION

Completed trusses can be raised in place with a mobile

206

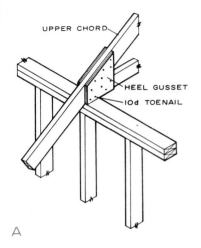

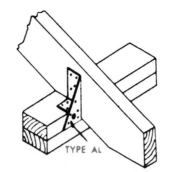

22-21a. Sheet-metal brackets are ideal for attaching the rafter truss to the wall plate.

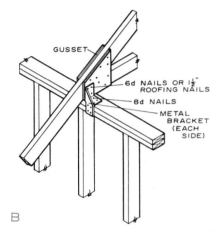

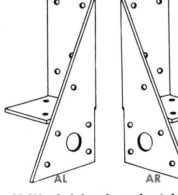

22-21b. Left-hand and right-hand sheet-metal brackets are available for installation on both sides of the rafter truss.

22-20. Fastening trusses to the wall plate: A. Toenailing. B. Metal bracket.

nailing is sometimes done, but this is not always the most satisfactory method. The heel gusset and a plywood gusset or metal gusset plate are located at the wall plate and make toenailing difficult. However, two 10d nails on each side of the truss can be used in nailing the lower chord to the plate. Fig. 22-20. Predrilling may be necessary to prevent splitting. Because of the single-member thickness of the truss and the presence of gussets at the wall plates, it is usually a good idea to use some type of metal connector to supplement the toe-nailings.

A better system of fastening trusses involves the use of a metal connector or bracket. These brackets are available commercially or can be formed from sheet metal. Fig. 22-21. The brackets are nailed to the wall plates at side and top with 8d nails and to the lower chords of the truss with 6d or 1½″ roofing nails. Fig. 22-20.

INTERIOR PARTITION INSTALLATION

Where partitions run parallel to but between the bottom truss chords and the partitions are erected *before* the ceiling finish is applied, install 2″ × 4″ blocking between the lower chords. Fig. 22-22. This blocking should be spaced not over 4′ on center. Nail the blocking to the chords with two 16d nails in each end. To provide nailing for lath or wallboard, nail a 1″ × 6″ or 2″ × 6″ continuous backer to the blocking. Set

crane after delivery to the building site. They can also be placed by hand over the exterior walls in an inverted position and then rotated into an upright position. Fig. 22-19.

The top plates of the two side walls should be marked for the location of each set of trusses. The trusses are fastened to the outside walls and two 1″ × 4″ or 1″ × 6″ temporary horizontal braces

(ribbons) are located near the ridge line to space and align them until the roof sheathing has been applied.

When fastening trusses, resistance to uplift stresses as well as thrust must be considered. Trusses are fastened to the outside walls with nails or framing anchors. The ring-shank nail provides a simple connection which will resist wind uplift forces. Toe-

22-22. Construction details for partitions that run parallel to the roof truss.

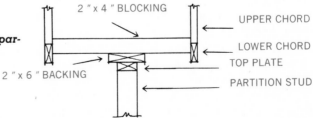

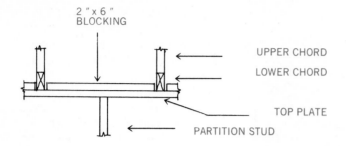

2 " x 6 "
BLOCKING

UPPER CHORD

LOWER CHORD

TOP PLATE

PARTITION STUD

22-23. *Construction details for partitions that run at right angles to the roof truss.*

the bottom face level with the bottom of the lower truss chords.

When partitions are erected *after* the ceiling finish is applied, 2″ × 4″ blocking is set with the bot-tom edge level with the bottom of the truss chords. Nail the block-ing with two 16d nails in each end.

If the partitions run at right angles to the bottom of the truss chords, the partitions are nailed directly to lower chord members. A 2″ × 6″ backer for the applica-tion of the ceiling finish is nailed on top of the partition plates be-tween the trusses. Fig. 22-23.

QUESTIONS

1. What are some of the advantages of using roof trusses in residential construction?

2. What three types of trusses are most com-monly used for residential building?

3. Which of the three types of trusses used in building construction is the most popular and ex-tensively used?

4. What are gussets usually made from?

5. What is a truss plate?

6. What is the best method of attaching the truss to the wall plates?

Roof Sheathing 23

Roof sheathing covers the raf-ters or roof joists. The roof shea-thing, like the wall sheathing and the subflooring, is a structural element. Therefore it is a part of the framing. Sheathing provides a nailing base for the finish roof covering and gives rigidity and strength to the roof framing. Ply-wood or lumber roof sheathing is most commonly used for pitched roofs. Lumber or laminated roof decking is sometimes used in homes with exposed ceilings. Fig. 23-1. A manufactured wood fiber roof decking is also adaptable to exposed ceiling applications.

LUMBER ROOF SHEATHING

Roof sheathing boards are gen-erally No. 3 common or better. The species used are the pines—Douglas fir, redwood, hemlocks, western larch, the firs, and the spruces. If the roof is to be cov-ered with asphalt shingles, it is important that thoroughly sea-soned material be used for the sheathing. Unseasoned wood will dry out and shrink in width. This shrinkage will cause buckling or lifts of the shingles which may ex-tend along the full length of the board.

Nominal 1″ boards are used for both flat and pitched roofs. Where flat roofs are to be used for a deck or a balcony, thicker shea-thing boards will be required.

Board roof sheathing, like board wall sheathing and subflooring, may be laid either horizontally or diagonally. Horizontal board sheathing may be either closed (laid with no spaces between the courses) or open (laid with spaces between the courses). In areas where wind-driven snow conditions prevail, a solid roof deck is recommended.

Installation

CLOSED SHEATHING

Roof boards used for sheathing under materials that require solid and continuous support—such as asphalt shingles, composition roofing, and sheet-metal roofing—must be laid closed. Fig. 23-2. Closed roof sheathing may also be used for wood shingles. The boards are a nominal 1″ × 8″ and may be square-edged, dressed and matched, or shiplapped.

OPEN SHEATHING

Open sheathing is used under wood shingles or shakes as a roof covering in blizzard-free areas or damp climates. Open sheathing usually consists of 1″ × 4″ strips with the on-center spacing equal to the shingle weather exposure but not over 10″. (A 10″ shingle

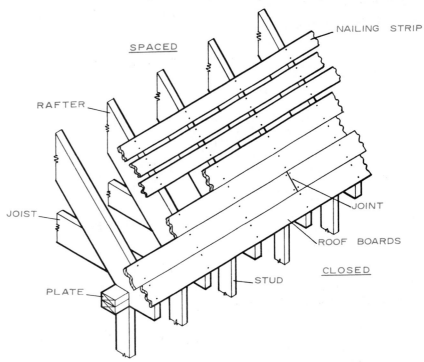

23-2. Installation of board roof sheathing, showing both closed and spaced types.

which is lapped 4″ by the shingle above it is said to be laid 6″ to the weather.) When applying open sheathing, the boards should be laid up without spacing to a point on the roof above the overhang. Fig. 23-2.

Nailing

Lumber roof sheathing is nailed to each rafter with two 8d nails. Joints must be made on the rafters, just as wall sheathing joints must be made over studs. When end-matched boards are used, joints may be made between rafters, but in no case should the joints of adjoining boards be made over the same rafter space. Each board should bear on at least two rafters.

PLYWOOD ROOF SHEATHING

Many different roof forms are possible with plywood construc-

23-1. Lumber roof decking is available in a variety of patterns and is used in homes with exposed ceilings.

23-3. Both the roof and sidewalls of this large home are sheathed with plywood.

Plywood roof sheathing[1][2][3] / (Plywood continuous over two or more spans; grain of face plies across supports)

Panel Identification Index	Plywood Thickness (inch)	Max. Span (inches)[4]	Unsupported Edge-Max. Length (inches)[5]	Allowable Roof Loads (psf)[6][7]										
				Spacing of Supports (inches center to center)										
				12	16	20	24	30	32	36	42	48	60	72
12/0	5/16	12	12	100 (130)										
16/0	5/16, 3/8	16	16	130 (170)	55 (75)									
20/0	5/16, 3/8	20	20		85 (110)	45 (55)								
24/0	3/8, 1/2	24	24		150 (160)	75 (100)	45 (60)							
30/12	5/8	30	26			145 (165)	85 (110)	40 (55)						
32/16	1/2, 5/8	32	28				90 (105)	45 (60)	40 (50)					
36/16	3/4	36	30				125 (145)	65 (85)	55 (70)	35 (50)				
42/20	5/8, 3/4, 7/8	42	32					80 (105)	65 (90)	45 (60)	35 (40)			
48/24	3/4, 7/8	48	36						105 (115)	75 (90)	55 (55)	40 (40)		
2-4-1	1 1/8	72	48							175 (175)	105 (105)	80 (80)	50 (50)	30 (35)
1 1/8 G1&2	1 1/8	72	48							145 (145)	85 (85)	65 (65)	40 (40)	30 (30)
1 1/4 G3&4	1 1/4	72	48							160 (165)	95 (95)	75 (75)	45 (45)	25 (35)

Notes: (1) Applies to Standard, Structural I and II and C-C grades only.

(2) For applications where the roofing is to be guaranteed by a performance bond, recommendations may differ somewhat from these values. Contact American Plywood Association for bonded roof recommendations.

(3) Use 6d common smooth, ring-shank or spiral thread nails for 1/2" thick or less, and 8d common smooth, ring-shank or spiral thread for plywood 1" thick or less (if ring-shank or spiral thread nails same diameter as common). Use 8d ring-shank or spiral thread or 10d common smooth shank nails for 2.4.1, 1 1/8" and 1 1/4" panels. Space nails 6" at panel edges and 12" at intermediate supports except that where spans are 48" or more, nails shall be 6" at all supports.

(4) These spans shall not be exceeded for any load conditions.

(5) Provide adequate blocking, tongue and grooved edges or other suitable edge support such as PlyClips when spans exceed indicated value. Use two PlyClips for 48" or greater spans and one for lesser spans.

(6) Uniform load deflection limitation: 1/180th of the span under live load plus dead load, 1/240th under live load only. Allowable live load shown in boldface type and allowable total load shown within parenthesis.

(7) Allowable roof loads were established by laboratory test and calculations assuming evenly distributed loads.

23-4. *Plywood roof sheathing application specifications.*

tion. Plywood offers flexibility in design, ease of construction, economy, and durability. Fig. 23-3.

It can be installed quickly over large areas and provides a smooth, solid base with a minimum of joints. A plywood deck is also equally effective under any type of shingles or built-up roofing. Waste is minimal, contribut-ing to the low in-place cost. It is frequently possible to cut costs still further by using fewer rafters with a somewhat thicker panel for the decking, for example, 3/4" plywood over framing 4' on center. Plywood and trusses are often combined in this manner. For recommended spans and plywood grades, see Fig. 23-4.

Installation

Plywood roof sheathing should be laid with the face grain perpendicular to the rafters. Fig. 23-5. Sheathing grade (unsanded) plywood is ordinarily used. Joints should be made over the centers of the rafters.

For wood or asphalt shingles with a rafter spacing of 16", 5/16"

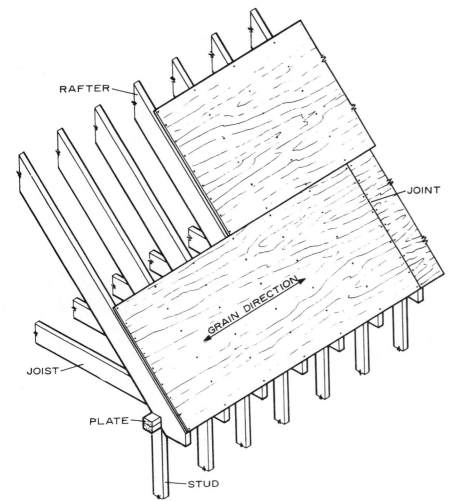

23-5a. *The grain of plywood sheathing should be at right angles to the supporting members.*

a supply of plywood sheathing on the roof that is readily accessible to the worker. Another aid which may be constructed is a plywood ladder. It will eliminate the need for a second person to hand the plywood up to the one on the roof. Fig. 23-7. This ladder is leaned against the building in the normal position of a ladder. The plywood sheets are then set on end on the ⅜″ plywood gusset. From there they can be pulled onto the roof as needed.

NAILING

Nails should be spaced 6″ at the panel edges and 12″ at intermediate supports except where the spans are 48″ or more. Then the nails should be spaced 6″ on all supports. Use 6d common smooth, ring-shank, or spiral thread nails for plywood ½″ thick or less. For plywood 1″ thick or less use 8d common smooth, ring-shank, or spiral thread nails. Fig. 23-4.

DECKING OR PLANKING

Roof decking provides a solid

plywood may be used. For a 24″ span ⅜″ plywood should be used. For slate, tile, and asbestos cement shingles ½″ plywood is recommended for 16″ rafter spacing, and ⅝″ plywood for 24″ spacing. Fig. 23-4. If wood shingles are used and the plywood sheathing is less than ½″ thick, 1″ × 2″ nailing strips spaced according to the shingle exposure should be nailed to the plywood.

Plywood roof sheathing, unless it is of the exterior type, should have no surface or edge exposed to the weather. To reduce handling costs and help the worker apply plywood roof sheathing, a roof platform may be constructed. Fig. 23-6. This platform supports

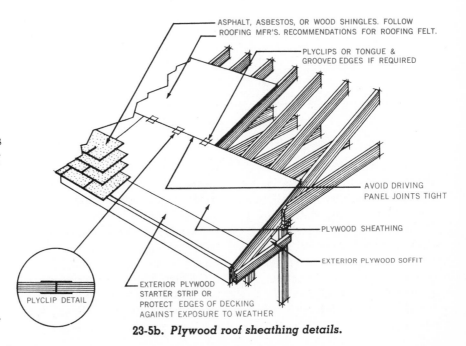

23-5b. *Plywood roof sheathing details.*

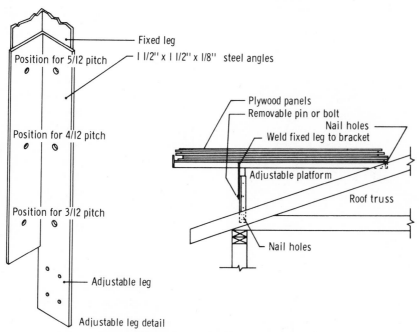

23-6. Details for adjustable roof platform.

23-8. Decking with tongue-and-groove edges and decorative face patterns provides a durable roof and an attractive ceiling for residential and commercial buildings.

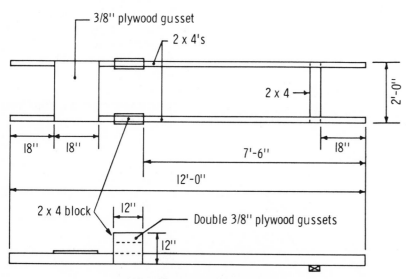

23-7. Plywood ladder detail.

permanent roof deck and an attractive ready-to-finish interior ceiling. It serves as an excellent base for any roofing material. Decking, with tongue-and-groove edges and decorative face patterns, is a standard building product for residential, commercial, and institutional construction. Fig. 23-8. Although known and used as roof decking, its load bearing capacities also make it useful as floor decking and for solid sidewall construction. This material is available in grades, patterns, and sizes suitable for both residential and commercial construction.

Grades. The *commercial* grade is designed for use in buildings where appearance and strength requirements are not a prime fac-

tor. The *select* grade decking is ideally suited for homes, schools, churches, motels, and restaurants, or wherever an attractive surface appearance is important.

Patterns. Lumber roof decking with a double tongue and groove is available in several patterns. Some of the more common are the regular V-joint, grooved, striated, and eased joint (bullnosed) patterns. Single tongue-and-groove decking in nominal 2″ × 6″ and 2″ × 8″ sizes is available with the V-joint pattern only. Fig. 23-9.

Sizes. Decking comes in nominal widths of 4″ to 12″ and in nominal thicknesses of 2″ to 4″. The 3″ and 4″ roof decking is available in random lengths of 6′ to 20′ or longer (odd and even).

Decking is also available laminated. It comes in six different species of softwood lumber: Idaho white pine, inland red cedar, Idaho white fir, ponderosa pine, Douglas fir, larch, and southern pine. Because of the laminating feature, this material may have a

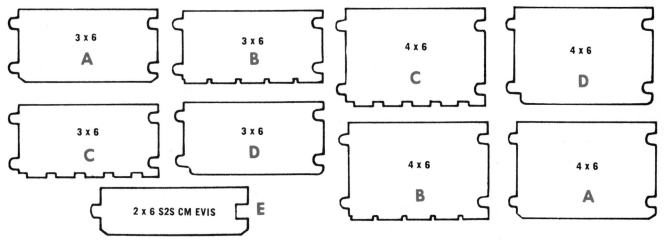

23-9. *Some lumber roof decking patterns and sizes: A. Regular V-jointed. B. Striated. C. Grooved. D. Eased joint (bullnosed). E. Single tongue-and-groove V-joint.*

facing of one wood species and back and interior laminations of different woods. It is also available with all laminations of the same species. For all types of decking make sure the material is the correct thickness for the span by checking the manufacturer's recommendations.

Installation

Roof decking that is to be applied to a flat roof should be installed with the tongue away from the worker. Roof decking that is being applied to a sloping roof should be installed with the tongues up. The butt ends of the pieces are cut at approximately a 2° angle. This provides a bevel cut from the face to the back to insure a tight face butt joint when the decking is laid in a random length pattern. Fig. 23-10. If there are three or more supports for the decking, a controlled random laying pattern may be used. Fig. 23-11. This is an economical pattern because it makes use of random plank lengths, but the following rules must be observed:
● Stagger the end joints in adjacent planks as widely as possible and not less than 2′.
● Separate the joints in the same

general line by at least two courses.
● Minimize joints in the middle one-third of all spans, make each plank bear on at least one support, and minimize the joints in the end span.

The ability of the decking to support specific loads depends on the support spacing, plank thickness, and span arrangement. Although two-span continuous layout offers structural efficiency, use of random-length planks is the most economical. Random-length double tongue-and-groove decking is used when there are three or more spans. It is not intended for use over single spans and it is not recommended for use over double spans. Fig. 23-11. Each piece should bear on at least one support.

NAILING

Fasten the decking with common nails twice as long as the nominal plank thickness. For widths 6″ or less, toenail once and face-nail once at each support. For widths over 6″, toenail once and face-nail twice. Decking 3″ and 4″ thick must be predrilled and toenailed with 8″ spikes. Fig. 23-12. Some manufacturers pro-

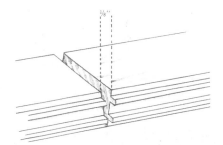

23-10. *The ends of lumber decking are cut at a 2° angle to insure a tight face joint on the exposed ceiling.*

vide the 3″ and 4″ thick roof decking with predrilled nail holes on 30″ centers. Bright common nails may be used but dipped galvanized common nails have better holding power and reduce the possiblity of rust streaks. End joints not over a support should be side-nailed within 10″ of each plank end. Metal splines are recommended on end joints of 3″ and 4″ material for better alignment, appearance, and strength.

WOOD FIBER ROOF DECKING

The all-wood fiber roof decking combines strength and insulation advantages that make possible quality construction with economy. This type of decking is

A — SIMPLE SPAN   B — TWO SPAN CONTINUOUS   C — MULTIPLE SPAN (A & B)

D — CANTILEVERED INTERMIXED SPAN   E — CONTROLLED RANDOM PATTERN

23-11. *Lumber decking span arrangements.*

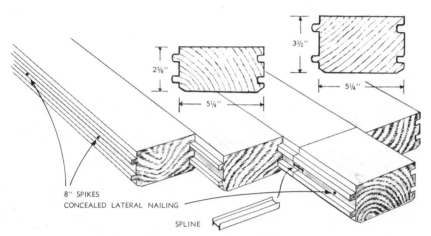

8" SPIKES
CONCEALED LATERAL NAILING

SPLINE

23-12. *Nailing details for lumber decking.*

23-14. *Butt the adjacent pieces together at the calking.*

weatherproof and protected against termites and rot. It is ideally suited for built-up roofing, as well as for asphalt and wood shingles on all types of buildings. It is available in four thicknesses: 2⅜", 1⅞", 1⅜", and ¹⁵⁄₁₆". The standard panels are 2' × 8' with tongue-and-groove edges and square ends. Fig. 23-13. The surfaces are coated on one or both sides at the factory in a variety of colors.

Installation

Wood fiber roof decking is laid with the tongue-and-groove joint at right angles to the support members. Begin laying at the eave line with the groove edge away from the applicator. Staple wax paper in position over the

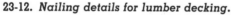

Thickness	Rafter spacing, with allowable live load* of		Size	Finish	Attachment
	40 psf.	50 psf.			To wood rafters with galvanized common nail
15/16"	24" o.c.	16" o.c.	Standard panel 2' x 8' (nominal). Actual is 23⅛" x 95⅞"	Ends: square Long edges; T&G and V-jointed on ceiling side; Special order: All 4 edges T&G up to 4' x 8'	6d.
1⅜"	32" o.c.	24" o.c.			8d.
1⅞"	48" o.c.	32" o.c.	Special order; Widths to 4' Lengths to 12'		10d.
2⅜"	60" o.c. +	48" o.c.			16d.

*Based on spacing and thickness indicated.

23-13. *In laying wood fiber roof decking, the following specifications should be consulted.*

rafter before installing the roof deck. The wax paper protects the exposed interior finish of the decking when the beams are stained later. Calk the end joints with a nonstaining calking compound. Butt the adjacent piece up

against the calked joint. Fig. 23-14. Drive the tongue-and-groove edges of each of the units firmly together with a wood block cut to fit the grooved edge of the decking. Fig. 23-15. End joints must be made over a support member.

RAFTER

ROOF BOARDS
OR PLYWOOD

PROJECTION

SHEATHING

FLUSH BOARDS

STUD

23-16. *Sheathing at the ends of a gable roof.*

NAILING

These panels are tongued and grooved but are nailed through the face into the wood, rafters, or trusses. Face-nail 6″ on center with 6d nails for $^{15}/_{16}$″, 8d for $1^{3}/_{8}$″, 10d for $1^{7}/_{8}$″, and 16d for $2^{3}/_{8}$″ thickness.

SHEATHING AT THE ENDS OF THE ROOF

Where the gable ends of the roof have little or no extension other than the molding and trim,

23-15. *Protect the edges of the roof decking with a wood block when driving the joint up tightly.*

RAFTER

ROOF BOARDS
OR PLYWOOD

CHIMNEY
OPENING

VALLEY

HEADER

RIDGE
BOARD

PLATE

STUD

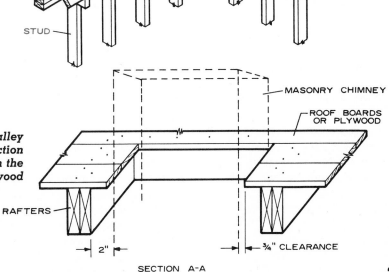

23-17. *Sheathing details at the valley and at the chimney opening. Section A-A shows the clearance between the masonry chimney and the wood structure.*

MASONRY CHIMNEY

ROOF BOARDS
OR PLYWOOD

RAFTERS

2″

¾″ CLEARANCE

SECTION A-A

215

the roof sheathing is usually sawed flush with the outer face of the side wall sheathing. Cuts should be made even so that the trim and molding can be properly installed. See Unit 25, "Roof Trim." Roof sheathing that projects beyond the end walls should span not less than three rafter spaces to insure proper anchorage to the rafters and to prevent sagging. Fig. 23-16. In general, it is desirable to use the longest boards at overhangs to secure good anchorage.

SHEATHING DETAILS AT CHIMNEY OPENINGS

Where chimney openings occur in the roof structure, the roof sheathing should have a clearance of ¾″ from the finished masonry on all sides. Fig. 23-17. Framing members should have a 2″ clearance for fire protection. The sheathing should be securely nailed to the rafters and to the headers around the opening.

SHEATHING AT VALLEYS AND HIPS

The sheathing at the valleys and hips should be fitted to give a tight joint. It should be securely nailed to the valley or the hip rafter. Fig. 23-17. This will give a solid and smooth base for the flashing.

QUESTIONS

1. What are the most commonly used sheathing materials for pitched roofs?

2. List several materials that are commonly used for roof sheathing.

3. Why is it important that lumber roof sheathing be thoroughly seasoned when used with asphalt shingles?

4. When installing plywood roof sheathing, in which direction should the grain run in relation to the rafters?

5. What is the minimum thickness of plywood that may be used for roof sheathing under asphalt shingles with a rafter spacing of 16″?

6. What are some of the advantages of decking or planking in contemporary architecture?

7. What are some of the advantages of wood fiber roof decking?

8. What clearance is recommended between roof sheathing and finished masonry, such as a chimney?

Roof Coverings* 24

The roof covering, or roofing, is a part of the exterior finish. It should provide long-lived waterproof protection for the building and its contents from rain, snow, wind and, to some extent, heat and cold. Fig. 24-1. Materials used for pitched roofs include shingles of wood, asphalt, and asbestos. Tile and slate are also popular. Sheet materials such as roll roofing, galvanized iron, aluminum, copper, and tin are sometimes used. For flat or low-pitched roofs, composition or built-up roofing with a gravel topping or cap sheet are frequent combinations. Built-up roofing consists of a number of layers of asphalt-saturated felt mopped down with hot asphalt or tar. Metal roofs are sometimes used on flat decks of dormers, porches, or entryways.

The choice of materials and method of application is influenced by cost, roof slope, expected service life of the roofing, wind resistance, fire resistance, and local climate. Due to the large amount of exposed surface, appearance is also important. Shingles, for example, add color, texture, and pattern to the roof surface. All shingles are applied to roof sur-

*Some of the material from this unit was adapted from *Construction: Principles, Materials & Methods* by courtesy of the American Savings & Loan Institute Press.

24-1. Applying wood shake roof covering.

faces in some overlapping fashion to shed water. Therefore they are suitable for any roof with enough slope to insure good drainage.

ROOFING TERMINOLOGY

Square. Roofing is estimated and sold by the square. A square of roofing is the amount required to cover 100 sq. ft. of roof surface.

Coverage. This term indicates the amount of weather protection provided by the overlapping of shingles. Depending on the kind of shingle and method of application, shingles may furnish one (single coverage), two (double coverage), or three (triple coverage) thicknesses of material over the surface of the roof.

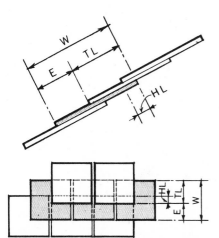

24-2. Terminology used in roofing: E = exposure, TL = toplap, W = width of strip shingles or length of individual shingles.

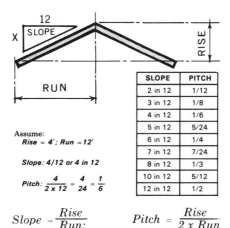

Assume:
$Rise = 4'; Run = 12'$

$Slope: 4/12$ or 4 in 12

$Pitch: \dfrac{4}{2 \times 12} = \dfrac{4}{24} = \dfrac{1}{6}$

SLOPE	PITCH
2 in 12	1/12
3 in 12	1/8
4 in 12	1/6
5 in 12	5/24
6 in 12	1/4
7 in 12	7/24
8 in 12	1/3
10 in 12	5/12
12 in 12	1/2

$$Slope = \frac{Rise}{Run}; \qquad Pitch = \frac{Rise}{2 \times Run}$$

24-3. Slope and pitch.

Shingles providing single coverage are suitable for reroofing over existing roofs. Shingles providing double and triple coverage are used for new construction, both having increased weather resistance and a longer service life.

Exposure. The shortest distance in inches between exposed edges of overlapping shingles. Fig. 24-2.

Toplap. The shortest distance in inches from the lower edge of an overlapping shingle or sheet to the upper edge of the lapped unit in the first course below (that is, the width of the shingle minus the exposure). Fig. 24-2.

Headlap. The shortest distance in inches from the lower edges of an overlapping shingle or sheet to the upper edge of the unit in the second course below. Fig. 24-2.

Side- or Endlap. The shortest distance in inches by which adjacent shingles or sheets horizontally overlap each other. Fig. 24-2.

Shingle Butt. The lower exposed edge of the shingle.

Slope and Pitch

These terms are often incorrectly used synonymously when referring to the incline of a sloped roof. Both are defined on the next page. Figure 24-3 also compares

some common roof slopes to corresponding roof pitches.

Slope. Slope indicates the incline of a roof as a ratio of vertical rise to horizontal run. It is expressed sometimes as a fraction but typically as X in 12. For example, a roof that rises at the rate of 4″ for each foot (12″) of run, is designated as having a 4-in-12 slope. The triangular symbol above the roof in Fig. 24-3 conveys this information.

Pitch. Pitch indicates the incline of a roof as a ratio of the vertical rise to *twice* the horizontal run. It is expressed as a fraction. For example, if the rise of a roof is 4′ and the run 12′, the roof is designated as having a pitch of ⅙ ($4/24 = ⅙$).

ROOFING ACCESSORIES

In addition to the shingles, many accessory materials are required to prepare the roof deck and to apply the shingles. These accessories include: underlayment, flashing, roofing cements, eaves flashing, drip edge, and roofing nails or fasteners. With some kinds of shingles, other accessories may be required, such as starter shingles and hip and ridge units. Regardless of the type of shingle to be installed, always check the instructions and recommendations of the shingle manufacturer to insure proper performance.

When applying shingles, the exposure distance is important. This distance depends mostly on roof slope and shingle type. Fig. 24-4. The minimum slope on main roofs is 4 in 12 for wood, asphalt, asbestos, and slate shingles. For built-up roofs the maximum slope is 3 in 12.

Underlayment

Underlayment is normally required for asphalt, asbestos, and slate shingles and for tile roofing,

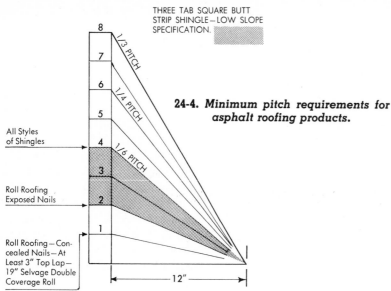

THREE TAB SQUARE BUTT STRIP SHINGLE—LOW SLOPE SPECIFICATION.

1/3 PITCH

1/4 PITCH

1/6 PITCH

All Styles of Shingles

Roll Roofing Exposed Nails

Roll Roofing—Concealed Nails—At Least 3" Top Lap—19" Selvage Double Coverage Roll

12"

24-4. Minimum pitch requirements for asphalt roofing products.

INSTALLING UNDERLAYMENT

Apply the underlayment as soon as the roof sheathing has been completed. For single underlay, start at the eave line with the 15-pound felt. Roll across the roof with a toplap of at least 2" at all horizontal joints and a 4" sidelap at all end joints. Fig. 24-7. Lap the underlayment over all hips and ridges 6" on each side. A double underlay can be started with two layers at the eave line,

but it may be omitted for wood shingles. In areas where snow is common and ice dams occur (melting snow freezes at the eave line), it is a good practice to apply one course of 55-pound smooth-surfaced roll roofing at the eaves. Fig. 24-5. Roof underlayment generally has three purposes:

• It protects the sheathing from moisture absorption until the shingles can be applied.

• It provides important additional weather protection by preventing the entrance of wind-driven rain below the shingles onto the sheathing or into the structure.

• In the case of asphalt shingles, it prevents direct contact between the shingles and resinous areas in wood sheathing which may be damaging to the shingles because of chemical incompatibility.

Underlayment should be a material with low vapor resistance, such as asphalt-saturated felt. Do not use materials such as coated felts or laminated waterproof papers which act as a vapor barrier. These allow moisture or frost to accumulate between the underlayment and the roof sheathing. Underlayment requirements for different kinds of shingles and various roof slopes are shown in Fig. 24-6.

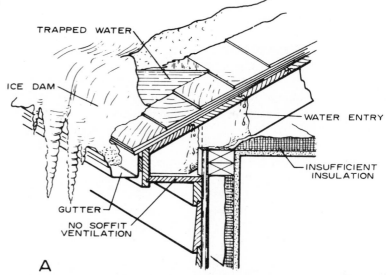

TRAPPED WATER

ICE DAM

WATER ENTRY

INSUFFICIENT INSULATION

GUTTER

NO SOFFIT VENTILATION

A

24-5. A. Snow and ice dams can build up on the overhang of roofs and gutters, causing melting snow to back up under shingles and under the fascia board of closed cornices. Damage to interior ceilings and walls and to exterior paint results from this water seepage. B. Protection from snow and ice dams is provided by eaves flashing. Ventilation of the cornice by means of vents in the soffit and sufficient insulation will minimize the melting.

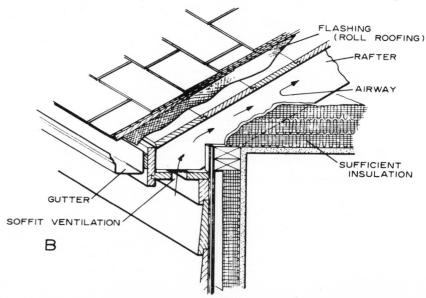

FLASHING (ROLL ROOFING)

RAFTER

AIRWAY

SUFFICIENT INSULATION

GUTTER

SOFFIT VENTILATION

B

Type of Roofing	Sheathing	Type of Underlayment	Normal Slope		Low Slope	
Asbestos-Cement Shingles	Solid	No. 15 asphalt saturated asbestos (inorganic) felt, OR No. 30 asphalt saturated felt	5/12 and up	Single layer over entire roof	3/12 to 5/12	Double layer over entire roof[1]
Asphalt Shingles	Solid	No. 15 asphalt saturated felt	4/12 and up	Single layer over entire roof	2/12 to 4/12	Double layer over entire roof[2]
Wood Shakes	Spaced	No. 30 asphalt saturated felt (interlayment)	4/12 and up	Underlayment starter course; interlayment over entire roof	Shakes not recommended on slopes less than 4/12 with spaced sheathing	
	Solid[3,5]	No. 30 asphalt saturated felt (interlayment)	4/12 and up	Underlayment starter course; interlayment over entire roof	3/12 to 4/12[4]	Single layer underlayment over entire roof; interlayment over entire roof
Wood Shingles	Spaced	None required.	5/12 and up	None required	3/12 to 5/12[4]	None required
	Solid[5]	No. 15 asphalt saturated felt	5/12 and up	None required[6]	3/12 to 5/12[4]	None required[6]

1. May be single layer on 4 in 12 slope in areas where outside design temperature is warmer than 0° F.
2. Square-Butt Strip shingles only; requires Wind Resistant shingles or cemented tabs.
3. Recommended in areas subject to wind driven snow.
4. Requires reduced weather exposure.
5. May be desirable for added insulation and to minimize air infiltration.
6. May be desirable for protection of sheathing.

24-6. *Shingle roof recommendations.*

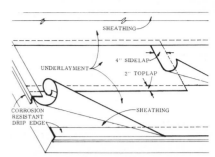

24-7. *Details for applying the underlayment for single coverage.*

valley of a roof, around chimneys, and at the point where a wall intersects a roof.

SOIL STACKS
Apply the roofing up to the stack, cutting it to fit. Fig. 24-9. Then install a corrosion-resistant metal sleeve which slips over the stack and has an adjustable flange to fit any roof slope. Figs. 24-10. & 24-11. A piece of 55-pound roll roofing can also be used. For roll

flush with the fascia board or molding. The second and remaining strips have 19″ headlaps with 17″ exposures. Fig. 24-8. Cover the entire roof in this manner, making sure that all surfaces have double coverage. Use only enough fasteners to hold the underlayment in place until the shingles are applied. Do not apply shingles over wet underlayment.

Flashings
Flashing is a special construction of sheet metal or other material used to protect the building from water seepage. Flashing must be made watertight and water-shedding. Metal used for flashing must be corrosion-resistant. It should be galvanized steel (at least 26-gauge), 0.019″ thick aluminum, or 16 oz. copper.

Flashing is required at the point of intersection between roof and soil stack or ventilator, in the

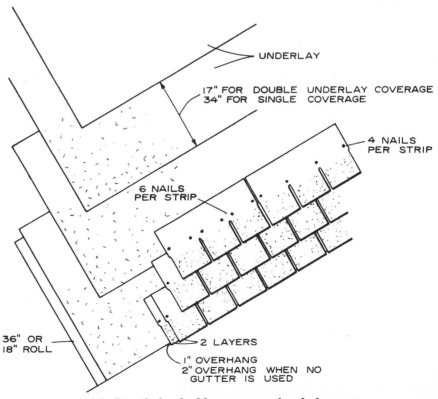

24-8. *Details for double coverage of underlayment.*

219

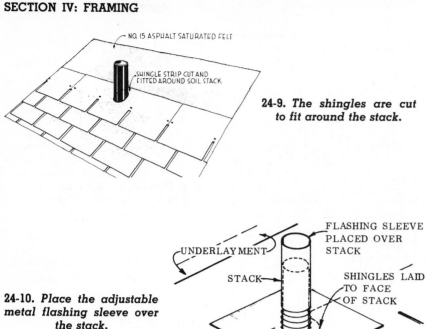

24-9. *The shingles are cut to fit around the stack.*

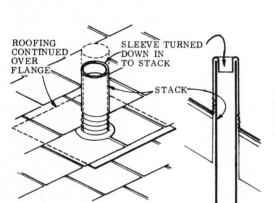

24-10. *Place the adjustable metal flashing sleeve over the stack.*

24-11. *Lay the shingles over the flange. Turn the top of the sleeve down into the stack to complete the installation.*

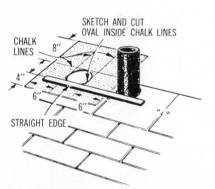

24-12. *Laying out a flange of roll roofing for a soil stack.*

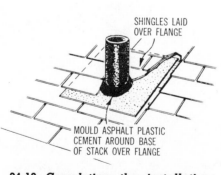

24-13. *Completing the installation of the flange. Lay shingles over the flange, fit them around the stack, and press the shingles firmly into the cement.*

roofing, lay out and cut an opening for the stack as shown in Fig. 24-12. Slip this flange in place over the stack and apply a roof cement 2″ up on the stack and 2″ out on the flange. Continue shingling over the flange. Cut the shingles to fit around the stack, pressing them firmly into the cement. Fig. 24-13.

VALLEYS

The open or closed method may be used to construct valley flashing. A valley underlayment strip of No. 15 asphalt-saturated felt, 36″ wide, is applied first. Fig. 24-14. The strip is centered in the valley and secured with enough nails to hold it in place. The horizontal courses of underlayment are cut to overlap this valley strip a minimum of 6″. Where eaves flashing is required, it is applied over the valley underlayment strip.

Open Valleys. Open valleys may be flashed with metal or with 90 lb. mineral-surfaced asphalt roll roofing in a color to match or to contrast with the roof shingles. The method is illustrated in Fig. 24-15. An 18″ wide strip of mineral-surfaced roll roofing is placed over the valley underlayment. It is centered in the valley with the surfaced side down and the lower edge cut to conform to and be flush with the eaves flashing. When it is necessary to splice the material, the ends of the upper segments are laid to overlap the lower segments 12″ and are secured with asphalt plastic cement. Only enough nails are used in rows 1″ in from each edge to hold the strip smoothly in place.

Another strip, 36″ wide, is placed over the first strip. It is centered in the valley with the surfaced side up and secured with nails. It is lapped if necessary the same way as the underlying 18″ strip.

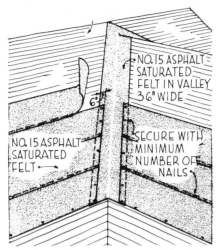

24-14. Applying the underlayment in the valley.

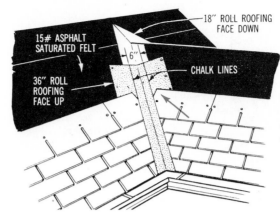

24-16. Begin laying each course at the chalk line in the valley. Clip the top corners of the shingles as shown at the arrow to prevent water penetration.

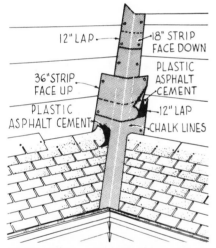

24-15. Open valley flashing details using roll roofing.

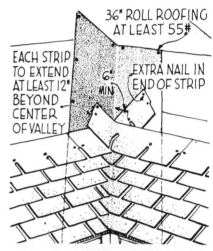

24-17. Closed valley flashing using woven strip shingles.

Before shingles are applied, a chalk line is snapped on each side of the valley for its full length. The lines should start 6″ apart at the ridge and spread wider apart (at the rate of ⅛″ per foot) to the eave. The chalk lines serve as a guide in trimming the shingle units to fit the valley, insuring a clean, sharp edge. The upper corner of each end shingle is clipped to direct water into the valley and prevent water penetration between courses. Fig. 24-16. Each shingle is cemented to the valley

lining with asphalt cement to insure a tight seal. No exposed nails should appear along the valley flashing.

Closed Valleys. Closed (woven) valleys can be used only with strip shingles. This method has the advantage of doubling the coverage of the shingles throughout the length of the valley, increasing weather resistance at this vulnerable point. A valley lining made from a 36″ wide strip of 55-pound (or heavier) roll roofing is placed over the valley underlayment and centered in the valley. Fig. 24-17.

Valley shingles are laid over the lining by either of two methods:
• They may be applied on both roof surfaces at the same time, with each course in turn woven over the valley.
• Each surface may be covered to the point approximately 36″ from the center of the valley and the valley shingles woven in place later.

In either case, the first course at the valley is laid along the eaves of one surface over the valley lining and extended along the adjoining roof surface for a distance of at least 12″. The first course of the adjoining roof surface is then carried over the valley on top of the previously applied shingle. Succeeding courses are then laid alternately, weaving

the valley shingles over each other.

The shingles are pressed tightly into the valley and nailed in the usual manner (see "Fastening Strip Shingles," page 225) except that no nail should be located closer than 6″ to the valley center line and two nails are used at the end of each terminal strip. Fig. 24-17.

CHIMNEYS

Apply the shingles over the felt up to the chimney face. If 90-pound roll roofing is to be used for flashing, cut wood cant strips and install them above and at the sides of the chimney. Fig. 24-18. The roll roofing flashing should be

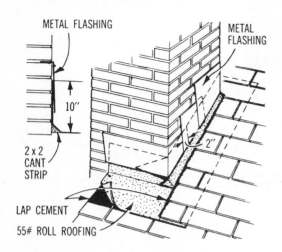

24-18. *How to flash the chimney.*

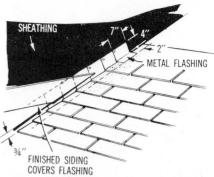

24-19. *Flashing the intersection of a roof and wall.*

cut to run 10″ up the chimney. Working from the bottom up, fit metal flashing over the base flashing and insert it 1½″ into the mortar joints. Refill the joints with mortar or roofing cement.

WALL INTERSECTIONS

Start at the eave line and work upward. Apply the metal flashing over the felt and up onto the wall sheathing but under the roofing and the siding. The siding should be cut so that it clears the roof by at least ¾″. Fig. 24-19.

Roofing Cements

Roofing cements are used for installing eaves flashing, for flashing assemblies, for cementing tabs of asphalt shingles and laps in sheet material, and for roof repairing. There are several types of cement, including plastic as-

phalt cements, lap cements, quick-setting asphalt adhesives, roof coatings, and primers. The type and quality of materials and methods of application on a shingle roof should be those recommended by the manufacturer of the shingle roofing.

Eaves Flashing

Eaves flashing is recommended in areas where the temperature goes below 0° F, or wherever there is a possibility of ice form-

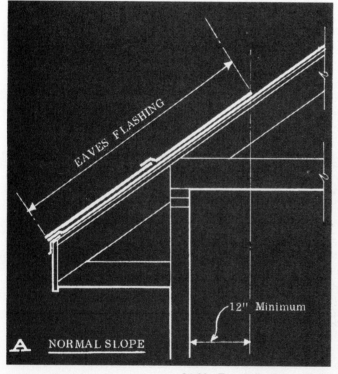

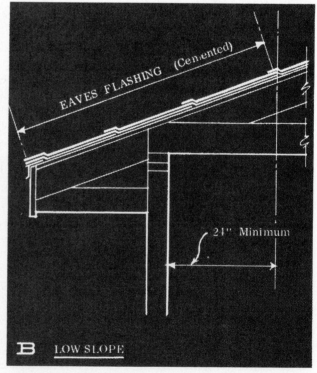

24-20. *Eaves flashing: A. Normal slope. B. Low slope.*

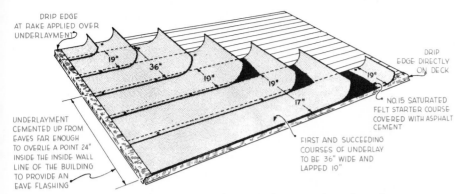

24-21. *Details for applying the underlayment for a low slope.*

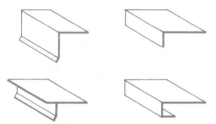

24-22. *Various drip edge shapes.*

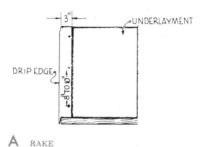

A RAKE

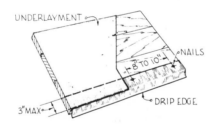

B EAVE

24-23. *Drip edge application: A. Along the rake. B. At the eave. Note that the underlayment goes under the drip edge at the rake and over the drip edge at the eave.*

COMMON WIRE

BOX

ANNULAR THREADED

SPIRAL THREADED

SCREW THREADED

24-24. *Types of smooth and threaded shank nails recommended for the application of shingle roofing.*

ing along the eaves. This ice forms a dam which allows water to back up under the shingles.

Eaves flashing is usually formed by an additional course of underlayment or roll roofing applied over the underlayment. For normal slopes this extends up the roof to cover a point at least 12″ inside the interior wall line of the building. A, Fig. 24-20.

For low slopes or in areas subject to severe icing, eaves flashing is formed by cementing an additional course of underlayment over the first underlayment as for normal conditions. However, it extends up the roof to cover a point at least 24″ inside the interior wall line of the building. B, Fig. 24-20 & Fig. 24-21.

Drip Edge

Drip edges are designed and installed to protect the edges of the roof. They prevent leaks at this point by causing water to drip free of underlying eave and cornice construction. Some shapes of

preformed drip edges are shown in Fig. 24-22. A drip edge is recommended for most shingle roofs. It is applied to the sheathing and under the underlayment at the eaves, but over the underlayment up the rake. Fig. 24-23.

Roofing Nails

No single step in applying roof shingles is more important than proper nailing. Suitability is dependent on several factors:
• Selecting the correct nail for the kind of shingle and type of roof sheathing. Fig. 24-24.
• Using the correct number of nails.
• Locating them in the shingle correctly.
• Choosing nail metal compatible with metal used for flashing.

Roofing nails should be long enough to penetrate through the shingle and through the roof sheathing. They should penetrate at least 1″ into plank decking. Nails for applying shingles over plywood sheathing should have threaded shanks.

Specific recommendations for the type, size, number, and spacing of roofing nails are given later in conjunction with the information on asphalt and wood shingles and wood shakes.

ASPHALT SHINGLES

Asphalt roof shingles are manufactured in three basic kinds of units. Fig. 24-25.
• Strip shingles of the square-butt or hexagonal type.
• Individual shingles of the interlocking or staple-down type.
• Giant individual shingles for application by either the American or Dutch lap methods.

In areas where high winds prevail, wind resistant strip shingles with factory-applied adhesive or integral locking tabs are recommended.

Shingles are laid so that they

SHINGLE TYPE*		SHIPPING WEIGHT PER SQUARE	PACKAGES PER SQUARE	LENGTH	WIDTH	UNITS PER SQUARE	SIDE-LAP	TOP-LAP	HEAD-LAP	EXPOSURE
STRIP SHINGLES	2 & 3 TAB SQUARE BUTT	235 Lb	3	36″	12″	80		7″	2″	5″
	2 & 3 TAB HEXAGONAL	195 Lb	3	36″	11⅓″	86		2″	2″	5″
INDIVIDUAL	STAPLE LOCK	145 Lb	2	16″	16″	80	2½″			
GIANT INDIVIDUAL	AMERICAN	330 Lb	4	16″	12″	226		11″	6″	5″
	DUTCH LAP	165 Lb	2	16″	12″	113	3″	2″		10″

24-25. *Asphalt roof shingles.*

overlap and cover each other to shed water. Before applying shingles make sure that:

• The underlayment, drip edge, and flashings are in place.

• The roof deck is tight and provides a suitable nailing base.

• The chimney is completed and the counter flashing installed.

• Stacks and other equipment requiring openings in the roof are in place with counter flashing where necessary.

Strip Shingles

On small roofs strip shingles may be laid from either rake. Fig. 24-26. On roofs 30′ and longer, shingles should be started at the center and applied both ways from a vertical line. This will assure more accurate vertical alignment and will provide for meeting and matching above a projection such as a dormer or chimney. To assure accurate alignment of shingles, use horizontal and vertical chalk lines.

The first course of shingles, called the *starter* course, is applied over the eaves flashing strip and even with its lower edge along the eave. The starter course may be a 9″ wide (or wider) starter strip of mineral-surfaced roll roofing of a color to match the shingles. A row of inverted shingles may also be used for the starter course. Fig. 24-27. Fasten the starter strip with roofing nails placed about 3″ or 4″ above the eave edge and spaced so that the nailheads will not be

exposed at the cutouts between the tabs on the first course. If square-butt strip shingles are used as a starter strip, cut off 3″ of the first inverted starter course shingle to be laid at the rake. Then the first course laid right side up is started with a full shingle. Succeeding courses are started with full or cut strips depending upon the pattern desired. Three variations for laying square-butt strip shingles are as follows:

• *Cutouts breaking joints on*

24-26. *Shingles laid out on a roof. Some of them have been opened up in preparation for the roofer, who is stapling the shingles in place. The helper is cutting the starter shingles on the edge of the roof.*

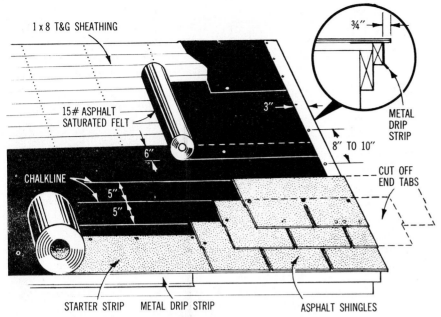

24-27. *Laying asphalt strip shingles.*

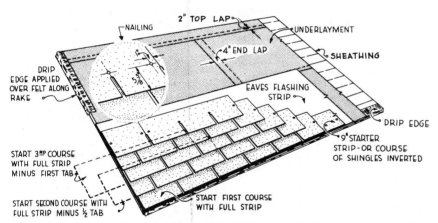

24-28. *Laying asphalt square-butt strip shingles with the cutouts breaking joints on halves.*

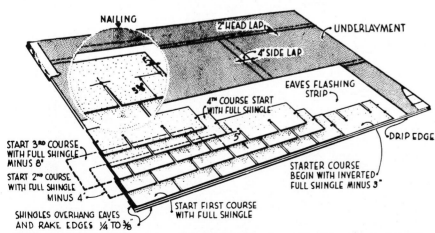

24-29. *Laying asphalt square-butt strip shingles with the cutouts breaking joints on thirds.*

halves. When the cutouts break the joints on halves, the second course is started with a full shingle cut 6″ short (half a tab). The third is started with a full shingle minus the entire first tab. The fourth is started with 1½ tabs cut from the shingle, and so on, causing the cutouts to be centered on the tabs of the course below. Fig. 24-28. This pattern can also be made with less cutting by starting a third course with a full shingle.

● *Cutouts breaking joints on thirds.* When the cutouts break the joints on thirds, the second course is started with a full shingle cut 4″ short (⅓ of a tab). The third course is started with a full shingle cut 8″ short (⅔ of a tab), and the fourth with a full shingle. Fig. 24-29.

● *Random spacing* is achieved by removing different amounts from the starting tab of succeeding courses in accordance with the following general principles:

1. The width of any starting tab should be at least 3″.

2. Cutout centerlines of any course should be located at least 3″ laterally from the cutout centerlines in both the course above and the course below.

3. Starting tab widths should be varied sufficiently so that the eye will not follow a cutout alignment. Fig. 24-30.

Regardless of the laying pattern, each succeeding course of shingles is placed so that the lower edges of the butts are aligned with the top of the cutouts on the underlying course.

Fastening Strip Shingles

Nails for applying asphalt roofing should be corrosion-resistant. Hot-dipped galvanized steel or aluminum nails with sharp points and flat heads that are ⅜″ to ⁷⁄₁₆″ in diameter are recommended. Shanks should be 10- to 12-gauge wire and may be smooth or

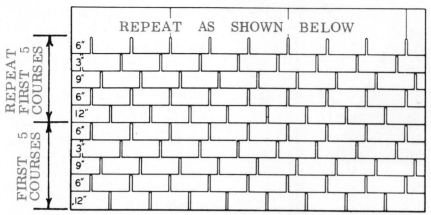

24-30. *Random spacing of asphalt square-butt strip shingles. The first course was started with a full-length strip.*

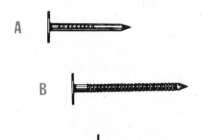

24-31. *Asphalt shingle nails: A. Smooth. B. Annular threaded. C. Screw threaded.*

threaded. Threaded nails are preferred because of their increased holding power. Aluminum nails should have screw threads with approximately a 12½° thread angle. Fig. 24-31. Galvanized steel nails, if threaded, should have annular threads. Nail lengths typically required are given in Fig. 24-32.

The number and the placement of nails are important for good roof application. Nailing should start at the end nearest the shingle last applied and proceed to the opposite end. To prevent buckling be sure each shingle is in perfect alignment before driving any nail. Drive the nail straight to avoid cutting the shingle with the edge of the nailhead. The nailhead should be driven flush, not

sunk below the surface of the shingle.

If the shingles are laid in windy areas, they will require additional protection. A spot of quick-setting cement about 1″ square for each

tab is applied on the underlying shingle with a putty knife or calking gun. The free tab is then pressed against the cement. Fig. 24-33.

Three-tab square-butt shingles require four nails for each strip. When the shingles are applied with a 5″ exposure, the four nails are placed ⅝″ above the top of the cutouts and located horizontally with one nail 1″ back from each end, and one nail on the center line of each cutout. Fig. 24-28. Two-tab square-butt shingles are nailed in a similar manner.

HEXAGONAL STRIP SHINGLES

Hexagonal strip shingles permit no spacing variations as do square-butt strips. Application begins with a roll roofing starter course or with inverted shingles. The

Application	1″ Sheathing	⅜″ Plywood
Strip or Individual Shingle (new construction)	1¼″	⅞″
Over Asphalt Roofing (reroofing)	1½″	1″
Over Wood Shingles (reroofing)	1¾″	———

24-32. Asphalt Shingle Nail Lengths

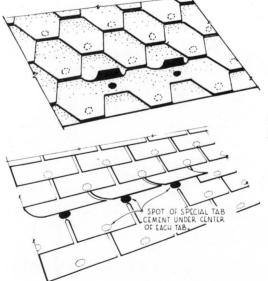

24-33. *Shingle tabs are cemented down for wind protection.*

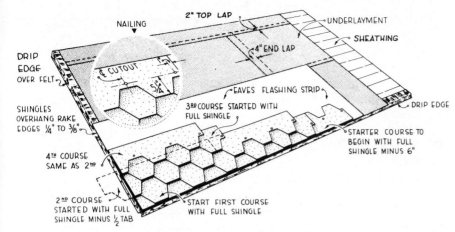

24-34a. *Laying two-tab hexagonal strip shingles.*

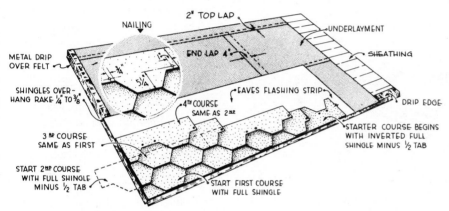

24-34b. *Laying three-tab hexagonal strip shingles.*

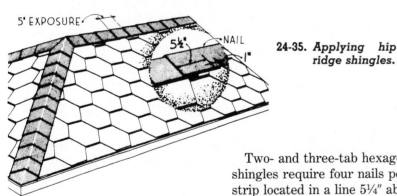

24-35. *Applying hip and ridge shingles.*

first course starts with a full strip. The remaining courses begin alternately with a full strip minus ½ tab and then a full strip. Fig. 24-34. Each course is applied so that the lower edge of the tabs is aligned with the top of the cutouts on the preceding course.

Two- and three-tab hexagonal shingles require four nails per strip located in a line 5¼″ above the exposed butt-edge and horizontally as follows:
• For two-tab shingles, one nail 1″ back from each end of the strip and one nail ¾″ back from each angle of the cutouts. Fig. 24-34a.
• For three-tab shingles, one nail 1″ back from each end and one nail centered above each cutout. Fig. 24-34b.

HIPS AND RIDGES

Hips and ridges may be finished by using hip and ridge shingles furnished by the manufacturer or by cutting pieces at least 9″ × 12″ either from 12″ × 36″ square-butt shingle strips or from mineral-surfaced roll roofing of a color to match the shingles. They are applied by bending each shingle lengthwise down the center with an equal amount on each side of the hip or ridge. Proper alignment can best be maintained by snapping a chalk line down one side of the ridge on which the edge of the shingle is aligned as it is nailed in place.

Apply the hip and ridge shingles by beginning at the bottom of a hip or one end of the ridge. Use a 5″ exposure. Each shingle is secured with one nail at each side 5½″ back from the exposed end and 1″ up from the edge. Fig. 24-35. When laying the shingles on the ridge, always lay the exposed edge away from the prevailing winds.

VALLEYS

Valley treatment may be open or closed. See pages 220-221 for details about shingle application at the valleys.

Strip Shingles on a Low-Pitch Roof

Square-tab strip shingles may be used on roof slopes less than 4 in 12 but not less than 2 in 12. Low-slope application requires:
• Double underlayment.
• Cemented eaves flashing strip.
• Shingles provided with factory-applied adhesives. Or, each free tab of square-butt shingles should be cemented. Fig. 24-33. Application of strip shingles over double underlayment and cemented eaves flashing is shown in Fig. 24-36. Any shingle laying pattern described under normal slope application may be used.

227

Interlocking Shingles

Interlocking (lock-down) shingles are designed to provide resistance to strong winds. They have integral locking devices that vary in detail but which can be classified into five general groups. A, Fig. 24-37. Types 1, 2, 3, and 4 are individual shingles while Type 5 is a strip shingle usually having two tabs per strip. Interlocking shingles do not require use of adhesives although cement may occasionally be needed along rakes and eaves where the locking devices may have to be removed. The roof slope should not be less than the minimum specified by the shingle manufacturer. The individual shingles (Types 1 through 4) are intended for roof slopes of 4 in 12 and greater.

INSTALLING INTERLOCKING SHINGLES

Due to the number of designs available, the manufacturer's instructions should be studied carefully. Interlocking shingles are self-aligning but are sufficiently flexible to allow for a limited amount of adjustment to save time, especially on long roofs. It is recommended that the roofs be laid out with horizontal chalk lines to provide guides for the meeting, matching, and locking of courses above dormers and other projections through the roof. The locking devices should be engaged correctly. B, Fig. 24-37.

The proper location of nails is essential to the performance of the locking device. The shingle manufacturer's instructions will specify where the nail should be located to insure the best results.

The procedure for finishing hips and ridges is the same as for strip shingles. Fig. 24-35.

WOOD SHAKES

There are three types of wood shakes: handsplit-and-resawn, tapersplit, and straightsplit. Shakes are produced in three lengths: 18″, 24″, and 32″. Fig. 24-38a. The maximum exposure recommended for double coverage on a roof is 13″ for 32″ shakes, 10″ for 24″ shakes, and 7½″ for 18″ shakes. A triple coverage roof can be achieved by reducing these exposures to 10″ for 32″ shakes, 7½″ for 24″ shakes, and 5½″ for 18″ shakes. Fig. 24-38b.

Shakes are recommended on slopes of 4 in 12 or steeper. By taking special precautions, installations may be made on slopes as low as 3 in 12. These precautions are:
• Reduce the exposure to provide triple coverage.
• Use solid sheathing with an underlayment of No. 30 asphalt-saturated felt applied over the entire roof with a No. 30 asphalt-saturated felt interlayment between each course. Shakes may be applied over either spaced or solid sheathing depending on climate conditions. See "Roof Sheathing," Unit 23.

24-36. *Applying strip shingles on a low slope over a double underlayment.*

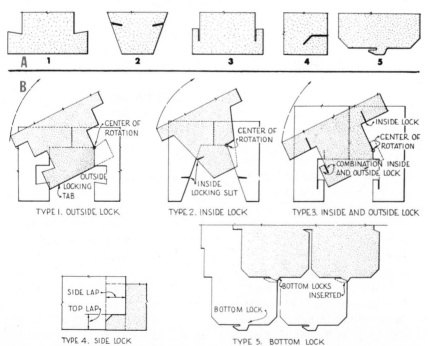

24-37. *Interlocking shingles: A. Various styles of interlocking shingle tabs. B. Methods of locking shingles.*

Grade	Length and Thickness	Bundles Per Square	Weight (lbs. per square)	Description
No. 1 Handsplit & Resawn	18″ x ½″ to ¾″ 18″ x ¾″ to 1¼″ 24″ x ½″ to ¾″ 24″ x ¾″ to 1¼″ 32″ x ¾″ to 1¼″	4 5 4 5 6	220 250 280 350 450	These shakes have split faces and sawn backs. Cedar blanks or boards are split from logs and then run diagonally through a bandsaw to produce two tapered shakes from each.
No. 1 Tapersplit	24″ x ½″ to ⅝″	4	260	 Produced largely by hand, using a sharp-bladed steel froe and a wooden mallet. The natural shingle-like taper is achieved by reversing the block, end-for-end, with each split.
No. 1 Straight-Split (Barn)	18″ x ⅜″ 24″ x ⅜″	5 5	200 260	 Produced in the same manner as taper-split shakes except that by splitting from the same end of the block, the shapes acquire the same thickness throughout.

24-38a. *Types of red cedar shakes.*

24-38b.
Roof Coverage Of Shakes At Varying Weather Exposures

Length and Thickness	Type of Shake	No. of Bundles	Approximate Coverage (sq. ft.)								
			Weather Exposures								
			5½″	6½″	7″	7½″	8″	8½″	10″	11½″	13″
18″ x ½″ to ¾″	Handsplit-and-Resawn	4	55*	65	70	75**	—	—	—	—	—
18″ x ¾″ to 1¼″	Handsplit-and-Resawn	5	55*	65	70	75**	—	—	—	—	—
24″ x ½″ to ¾″	Handsplit-and-Resawn	4	—	65	70	75*	80	85	100**	—	—
24″ x ¾″ to 1¼″	Handsplit-and-Resawn	5	—	65	70	75*	80	85	100**	—	—
32″ x ¾″ to 1¼″	Handsplit-and-Resawn	6	—	—	—	—	—	—	100**	115	130**
24″ x ½″ to ⅝″	Tapersplit	4	—	65	70	75*	80	85	100**	—	—
18″ x ⅜″	Straight-Split	5	65*	—	—	—	—	—	—	—	—
24″ x ⅜″	Straight-Split	5	—	65	70	75*	—	—	—	—	—
15″ Starter-Finish Course			Use supplementary with shakes applied not over 10″ exposure								

*Recommended maximum weather exposure for 3-ply roof construction.

**Recommended maximum weather exposure for 2-ply roof construction.

Eaves Flashing

In areas where the outside design temperature is 0° F or colder, or where there is a possibility of ice forming along the eaves and causing a backup of water, eaves flashing is recommended. Fig. 24-39. In these areas shakes should be applied over solid sheathing.

On slopes 4 in 12 or steeper, eaves flashing is formed by applying an additional course of No. 30 asphalt-saturated felt over the underlayment starting course at the eaves. The eaves flashing should extend up the roof to cover a point at least 24″ inside the exterior wall line of the building. When the eave overhang requires flashing to be wider than 36″, the necessary horizontal joint is cemented and located outside the exterior wall line of the building. A, Fig. 24-20.

For slopes 3 in 12 to 4 in 12 or in areas subject to severe icing, eaves flashing may be formed as described for 4 in 12 slopes except that a double layer of No. 30 asphalt-saturated felt underlayment is cemented together with a continuous layer of plastic asphalt cement. Cement is also applied to the 19″ underlying portion of each succeeding course which lies within the eaves flashing area, before applying the next course of asphalt felt.

229

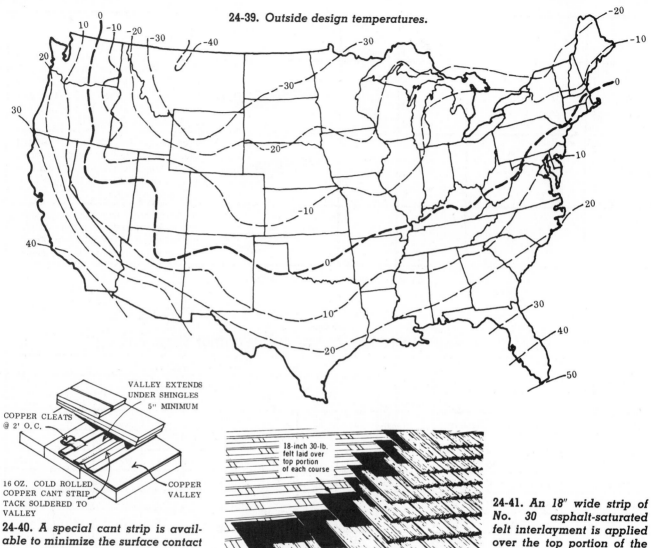

24-39. *Outside design temperatures.*

24-40. *A special cant strip is available to minimize the surface contact of copper with red cedar.*

COPPER CLEATS @ 2' O.C.

VALLEY EXTENDS UNDER SHINGLES 5" MINIMUM

16 OZ. COLD ROLLED COPPER CANT STRIP TACK SOLDERED TO VALLEY

COPPER VALLEY

18-inch 30-lb. felt laid over top portion of each course

Double starter course

24-41. *An 18" wide strip of No. 30 asphalt-saturated felt interlayment is applied over the top portion of the shakes between each course.*

Drip Edge

Wood shakes should extend out over the eave and rake a distance of 1″ to 1½″ to form a drip. To align this drip edge, nail a shingle at each end of the roof line and allow it to project. Attach a chalk line to the butt edge and pull it taut between the two shingles. The butt edge of each wood shingle applied as a drip edge can then be aligned to the chalk line.

Flashing

Unless special precautions are taken, copper flashing materials

non-skid head

lightweight handle

sliding gauge

24-42. *Shingler's hatchet with an adjustable gauge for setting the exposure.*

sharp blade and heel

are not recommended for use with red cedar shakes. Fig. 24-40.

For valley flashing the *open method* (roofing, felt, and sheet metal) or the *closed method* (hand-fitted shakes) may be used. The open method is highly recommended for longer service life. Open valleys are first covered with a valley underlayment strip of No. 30 asphalt-saturated felt at least 20″ wide. The strip is centered in the valley and secured with enough nails to hold it in place. Metal flashing strips 20″ wide are then nailed over the underlayment. If the flashing is galvanized steel, it should be preferably 18-gauge, but not less than 26-gauge. Flashing that is center crimped and painted on both surfaces is preferred. The valley edges should be edge crimped to provide an additional water stop. This is done by turning the edges up and back approximately ½″ toward the valley center line. The shakes laid to finish at the valley are trimmed parallel with the valley to form a 6″ wide gutter.

Closed valleys are first covered with a 1″ × 6″ wood strip. This strip is nailed flat into the saddle and covered with roofing felt as specified previously for open valleys. Shakes in each course are edge trimmed to fit into the valley, then laid across the valley with an undercourse of metal flashing having a 2″ headlap and extending 10″ under the shakes on each side of the saddle.

Application of Shakes

Apply a starter strip of No. 30 asphalt-saturated felt underlayment 36″ wide over the sheathing. The starter course of shakes at the eave line should be doubled, using an undercourse of 24″, 18″, or 15″ shakes. The latter is made expressly for this purpose.

After each course of shakes is applied, an 18″ wide strip of No.

30 asphalt-saturated felt interlayment should be applied over the top portion of the shakes extending onto the sheathing. Fig. 24-41. The bottom edge of the interlayment should be positioned at a distance above the butt edge of the shake equal to twice the exposure. For example, if 24″ shakes are being laid at 10″ exposure, the bottom edge of the felt should be positioned at a distance 20″ above the shake butts. The 18″ felt strip will then cover the top 4″ of the shakes and extend 14″ onto the sheathing.

Individual shakes should be spaced approximately ¼″ to ⅜″ apart to allow for possible expansion due to moisture absorption. The joints between shakes should be offset at least 1½″ in adjacent courses. The joints in alternate courses also should be kept out of direct alignment when a three-ply roof is being built.

When straightsplit shakes, which are of equal thickness throughout, are applied, the froe-end of the shake (the smoother end from which it has been split) should be laid undermost. The application of shakes can be speeded by use of a shingler's or lather's hatchet. Fig. 24-42. The hatchet is used for nailing, carries an exposure gauge, and has a sharpened heel which can be used for trimming.

NAILING

Only two nails should be used to apply each shake, regardless of its width. The nails should be placed approximately 1″ in from each edge and from 1″ to 2″ above the butt line of the succeeding course. Nails should be driven until their heads meet the shake surface but no farther. Fig. 24-43.

Nail lengths typically required are 6d (2″) for shakes with ¾″ to 1¼″ butt thickness and 5d (1¾″) for shakes with ½″ to ¾″ butt thickness. Nails two-penny (2d) sizes larger should be used to apply hip and ridge units. The nail should always be long enough to penetrate through the sheathing.

HIPS AND RIDGES

The final shake course at the ridge line, as well as shakes that terminate at hips, should be secured with additional nails. This final shake course should also be composed of smoother textured shakes. A strip of No. 30 asphalt-saturated felt, at least 12″ wide, should be applied over the crown of all hips and ridges, with an equal exposure of 6″ on each side.

Prefabricated hip and ridge units can be used, or the hips and

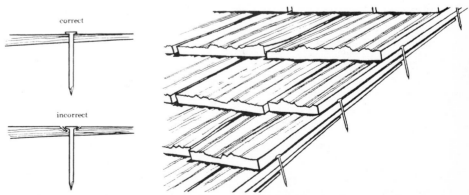

24-43. *Nails should be driven flush with the surface of the wood shake but no farther. Place the nails from 1″ to 2″ above the butt line of the succeeding course.*

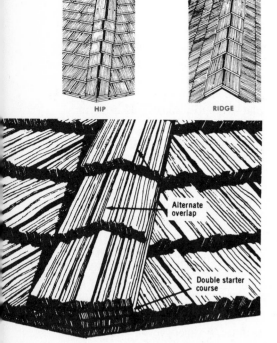

HIP   RIDGE

24-44. *Hip and ridge construction using wood shakes.*

ridges can be cut and applied on the site. In site-construction of hips, shakes approximately 6″ wide are sorted out. Two wooden straight-edges are tacked on the roof 6″ from the center line of the hip, one on each side. The starting course of shakes should be doubled. The first shake on the hip is nailed in place with one edge resting against the guide strip. The edge of the shake projecting over the center of the hip is cut back on a bevel. The shake on the opposite side is then applied and the projecting edge cut

back to fit. Shakes in the following courses are applied alternately in reverse order. Fig. 24-44.

Ridges are constructed in a similar manner. Exposure of the hip and ridge shakes normally is the same as the shakes on the roof. Ridge shakes are laid along an unbroken ridge that terminates in a gable at each end. They should be started at each gable end and terminate in the middle of the ridge. At that point, a small saddle is face-nailed to splice the two lines. The first course of shakes should always be doubled at each end of the ridge.

GABLE RAKES

Dripping water may be eliminated by inserting a single strip of bevel siding the full length of each gable rake with the thick edge flush with the sheathing edge. The inward pitch of the roof surface will then divert the water away from the gable edge.

ALTERNATE APPLICATION METHODS

• **Graduated exposures.** By reducing the exposure of each course from eaves to ridge, a variation in roof appearance may

be achieved with handsplit shakes. This requires shakes of several lengths. Fig. 24-45.

• **Staggered lines.** Irregular and random roof patterns can be achieved by laying shakes with butts placed slightly above or below the horizontal lines governing each course. Fig. 24-46. For an extremely irregular pattern, longer shakes may be interspersed over the roof with their butts several inches lower than the course lines.

WOOD SHINGLES

Wood shingles are manufactured in 24″, 18″, and 16″ lengths conforming to three grades: #1, #2, and #3. Fig. 24-47. Preformed, factory-built hip and ridge units are available.

The exposure of wood shingles is dependent on the slope of the roof. Standard exposures of 5″, 5½″, and 7½″ for shingle lengths of 16″, 18″, and 24″ respectively are used on slopes 5 in 12 or greater. On 4-in-12 slopes, exposures should be reduced to 4½″, 5″, and 6¾″ for 16″, 18″, and 24″ shingles respectively. On 3-in-12 slopes they should be reduced further to 3¾″, 4¼″, 5¾″ for 16″, 18″, and 24″ shingles respectively. This will assure four layers of shingles over the roof area. Fig. 24-48. Wood shingles are not recommended on slopes less than 3 in 12.

Underlayment

Underlayment is not usually used between shingles on spaced or solid sheathing. However, it

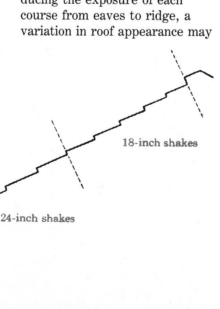

18-inch shakes

24-inch shakes

32-inch shakes

24-45. *A roof with graduated exposure is created by using all three shake sizes.*

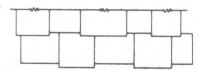

24-46. *Irregular roof patterns can be made by laying the shakes with butts slightly above or below the horizontal lines.*

Grade	Size	Bundles Per Square	Weight Per Square		Description
No. 1	24″ 18″ 16″	4 bdls. 4 bdls. 4 bdls.	192 lbs. 158 lbs. 144 lbs.		The premium grade of shingles for roofs and side-walls. These shingles are 100% heartwood, 100% clear and 100% edge-grain.
No. 2	24″ 18″ 16″	4 bdls. 4 bdls. 4 bdls.	192 lbs. 158 lbs. 144 lbs.		A good grade for all applications. Not less than 10″ clear on 16″ shingles, 11″ clear on 18″ shingles and 16″ clear on 24″ shingles. Flat grain and limited sapwood are permitted.
No. 3	24″ 18″ 16″	4 bdls. 4 bdls. 4 bdls.	192 lbs. 158 lbs. 144 lbs.		A utility grade for economy applications and secondary buildings. Guaranteed 6″ clear on 16″ and 18″ shingles, 10″ clear on 24″ shingles.

24-47. *Types of red cedar shingles.*

Length and Thickness*	Approximate Coverage (sq. ft.) of Four Bundles								
	Weather Exposures								
	3½″	4″	4½″	5″	5½″	6″	6½″	7″	7½″
16′ x 5/2″	70	80	90	100**	——	——	——	——	——
18″ x 5/2¼″	——	72½	81½	90½	100**	——	——	——	——
24″ x 4/2″	——	——	——	——	——	80	86½	93	100**

*Sum of the thickness e.g. 5/2″ means 5 butts = 2″
**Maximum exposure recommended for roofing

24-48. *Roof coverage of wood shingles at varying exposures.*

may be desirable for the protection of sheathing and to insure against air infiltration. For underlayment, No. 15 asphalt-saturated felt may be used.

Drip Edge

Wood shingles should extend out over the eave and rake a distance of 1″ to 1½″ to form a drip. This edge is aligned in the same manner as described for wood shakes.

Eaves Flashing

In areas where the outside design temperature is 0° F or colder, or where there is possibility of ice forming along the eaves and causing a backup of water, eaves flashing is recommended. Fig. 24-39. Sheathing should be applied solidly above the eave line to cover a point at least 24″ inside the interior wall line of the building. Fig. 24-20.

For 4-in-12 slopes, the eaves flashing is formed by applying a double layer of No. 15 asphalt-saturated felt to cover this section of solid sheathing. When the eave overhang requires the flashing to be wider than 36″, the necessary horizontal joint between the felt strips is cemented and located outside the exterior wall line.

For slopes from 3 in 12 up to 4 in 12, or in areas subject to severe icing, eaves flashing may be formed as described 4-in-12 slopes, except that the double layer of No. 15 asphalt-saturated underlayment is cemented. The eaves flashing is formed by applying a continuous layer of plastic asphalt cement, at the rate of 2 gals. per 100 sq. ft., to the surface of the underlayment starter course before the second layer of underlayment is applied.

Cement is also applied to the 19″ underlying portion of each succeeding course which lies within the eaves flashing area, be-

fore placing the next course. It is important to apply the cement uniformly with a comb trowel, so that at no point does underlayment touch underlayment when the application is completed. The overlying sheet is pressed firmly into the entire cemented area. Fig. 24-21.

Flashing

If copper flashing is used with wood shingles, take special precautions. Fig. 24-40. Premature deterioriation of the copper may occur when the metal and wood are in intimate contact in the presence of moisture.

Only the open method should be used to construct valley flashing. A closed valley is not recommended.

On slopes up to 12 in 12, metal valley sheets should be wide enough to extend at least 10″ on each side of the valley center line. Fig. 24-49. On roofs of steeper

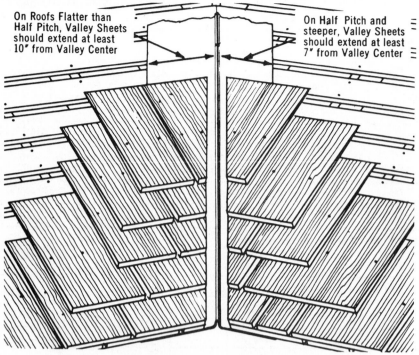

On Roofs Flatter than Half Pitch, Valley Sheets should extend at least 10" from Valley Center

On Half Pitch and steeper, Valley Sheets should extend at least 7" from Valley Center

24-49. Open valley flashing construction with wood shingles.

slope, narrower sheets may be used extending on each side of the valley center line for a distance of at least 7". The open portion of the valley should be at least 4" wide, but valleys may begin 2" wide and increase at the rate of ½" per 8' of length as they descend.

In areas where the outside design temperature is 0° F or colder, underlayment should be installed under metal valley sheets.

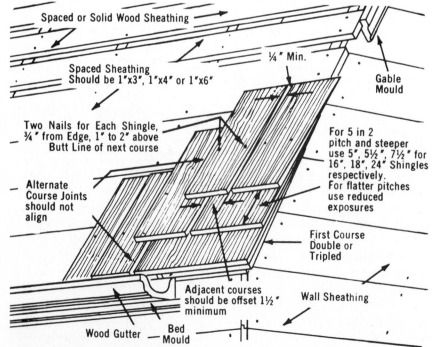

Spaced or Solid Wood Sheathing

Spaced Sheathing Should be 1"x3", 1"x4" or 1"x6"

¼" Min.

Gable Mould

Two Nails for Each Shingle, ¾" from Edge, 1" to 2" above Butt Line of next course

For 5 in 2 pitch and steeper use 5", 5½", 7½" for 16", 18", 24" Shingles respectively. For flatter pitches use reduced exposures

Alternate Course Joints should not align

First Course Double or Tripled

Adjacent courses should be offset 1½" minimum

Wall Sheathing

Wood Gutter — Bed Mould

24-50. Details for applying wood shingles over spaced or solid sheathing.

Application of Wood Shingles

The fist course of shingles at the eaves should be doubled or tripled. It should project 1" to 1½" beyond the eaves to provide a drip.

The second layer of shingles in the first course should be nailed over the first layer to provide a minimum sidelap of at least 1½" between joints. Fig. 24-50. If possible, joints should be "broken" by a greater margin. A triple layer of shingles in the first course provides additional insurance against leaks at the cornice. No. 3 grade shingles frequently are used for the starter course.

Shingles should be spaced at least ¼" apart to provide for expansion. Joints between shingles in any course should be separated not less than 1½" from joints in the adjacent course above or below. Joints in alternate courses should not be in direct alignment. When shingles are laid with the recommended exposure, triple coverage of the roof results. Fig. 24-50.

When the roof terminates in a valley, the shingles for the valley should be carefully cut to the proper miter at the exposed butts. These shingles should be nailed in place first so that the direction of shingle application is away from the valley. This permits valley shingles to be carefully selected and insures shingle joints will not break over the valley flashing.

NAILING

To insure that shingles will lie flat and give maximum service, only two nails should be used to secure each shingle. Nails should be placed not more than ¾" from the side edge of shingles, at a distance of not more than 1" above the exposure line. Fig. 24-50. Nails should be driven flush but

Size	Length	Gauge	Head	Shingles
3d*	1¼″	14½	⁷/₃₂″	16″ & 18″
4d*	1½″	14	⁷/₃₂″	24″
5d**	1¾″	14	⁷/₃₂″	16″ & 18″
6d**	2″	13	⁷/₃₂″	24″

*3d and 4d nails are used for new construction.
**5d and 6d nails are used for reroofing.

24-51.
Nail Sizes Recommended For Application Of Wood Shingles

not so that the nailhead crushes the wood. Fig. 24-43. The recommended nail sizes for the application of wood shingles are shown in Fig. 24-51. As with shakes, the application of wood shingles can be speeded by the use of a shingler's or lather's hatchet. Fig. 24-42.

HIPS AND RIDGES

Hips and ridges should be of the modified "Boston" type with protected nailing. Fig. 24-52a. Nails at least two sizes larger than the nails used to apply the shingles are required.

Hips and ridges should begin with a double starter course. Either site-applied or preformed factory-constructed hip and ridge units may be used. Fig. 24-52b.

GABLE RAKES

Shingles should project 1″ to 1½″ over the rake. End shingles may be canted to eliminate drips as discussed in the application of wood shakes.

ALTERNATE APPLICATION METHODS

Several alternate methods of applying shingles, giving a different appearance to the roof, are illustrated in Fig. 24-53.

ROOF ROOFING

When economy is a factor in construction, the use of mineral-surfaced roll roofing might be considered. While this type of roofing will not be as attractive as an asphalt shingle roof and perhaps not as durable, it may cost up to 15%

less than standard asphalt shingles. Roll roofing is excellent over old roofs as well as for new decks. The 19″ selvage, double coverage rolls (65 pounds minimum weight with a mineral surface) are designed for flat decks with a slope of 1″ or more per foot.

Roll roofing should be installed over a double coverage underlay. First nail the metal drip edges at

eaves and rakes. Use the 19″ selvage cut, trimmed out of a full sheet, as a starter strip. Save the remaining 17″ strip for the last course. Nail the strip so that it overhangs eaves and rakes ¼″ to ⅜″. With galvanized roofing nails, apply the first course. Fig. 24-54. Apply the second and following courses with a full 19″ overlap, leaving just the mineral surface exposed.

Next, lift the mineral surface of each course and apply a quick-setting lap cement to the underlying sheet to within ¼″ of the exposed edge. Apply firm pressure over

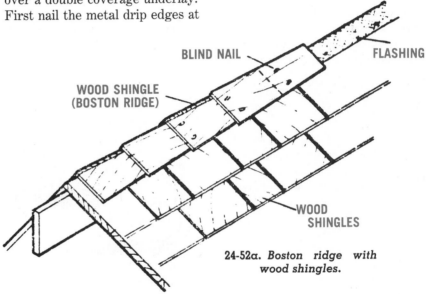

24-52a. Boston ridge with wood shingles.

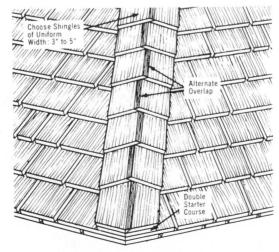

24-52b. Wood shingle hip and ridge construction.

the entire cemented area, using a light roller or broom.

After you have finished the whole roof, check for any loose laps, and re-cement to insure complete bond. The ridge can be finished with a Boston-type covering or by 12″ wide strips of the roll roofing, using at least 6″ on each side.

BUILT-UP ROOFS

Built-up roof coverings are intalled by roofing companies that specialize in this field. Roofs of this type may have three, four, or five layers of roofer's felt, each mopped down with tar or asphalt. The final surface may be coated with asphalt and covered with gravel embedded in asphalt or tar, or covered with a cap sheet. For convenience, it is customary to refer to built-up roofs as 10-year, 15-year, or 20-year roofs, depending upon the method of application. Fig. 24-55.

For example, a 15-year roof over a wood deck may have a base layer of 30-pound saturated roofer's felt laid dry, with edges lapped and held down with roofing nails. All nailing is done with either roofing nails driven through flat tin caps or with 10-gauge roofing nails having heads of not less than ⅝″ diameter. The dry sheet is intended to prevent tar or asphalt from entering the rafter spaces. Three layers of 15-pound saturated felt follow, each of which is mopped on with hot tar rather than being nailed. The final coat of tar or asphalt may be covered with roofing gravel or a cap sheet of roll roofing. Fig. 24-56.

The cornice or eave line of projecting roofs is usually finished with metal edging or flashing, which acts as a drip edge. A metal gravel strip is used in conjunction with the flashing at the eaves when the roof is covered

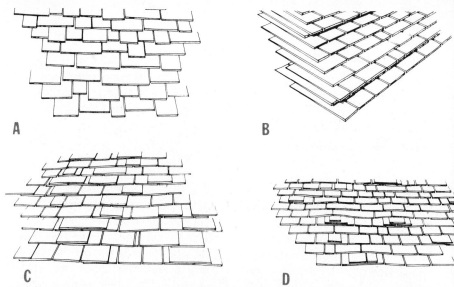

A   B

C   D

24-53. *Alternate application methods for wood shingles: A. Thatch. Shingles are positioned above and below a hypothetical course line, with deviation from the line not to exceed 1″. B. Serrated. Courses are doubled every 3rd, 4th, 5th, or 6th course. Doubled courses can be laid butt-edge flush or with a slight overhang. C. Dutch weave. Shingles are doubled or superimposed at random throughout the roof area. D. Pyramid. Two extra shingles, narrow shingle over a wide one, are superimposed at random.*

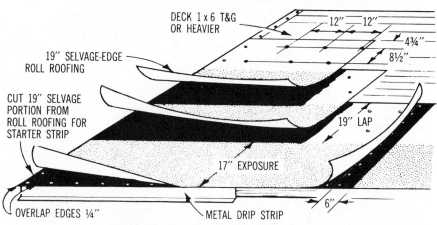

DECK 1 x 6 T&G OR HEAVIER

19″ SELVAGE-EDGE ROLL ROOFING

CUT 19″ SELVAGE PORTION FROM ROLL ROOFING FOR STARTER STRIP

12″   12″

4¾″

8½″

19″ LAP

17″ EXPOSURE

OVERLAP EDGES ¼″

METAL DRIP STRIP

6″

24-54. *Application details for roll roofing.*

24-55. *Cross section of a 20-year, 5-ply built-up roof.*

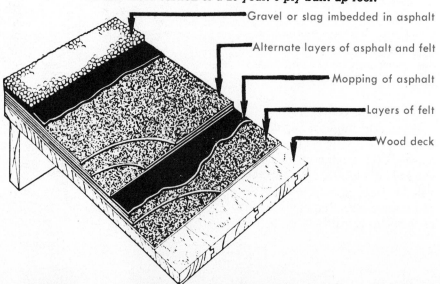

Gravel or slag imbedded in asphalt

Alternate layers of asphalt and felt

Mopping of asphalt

Layers of felt

Wood deck

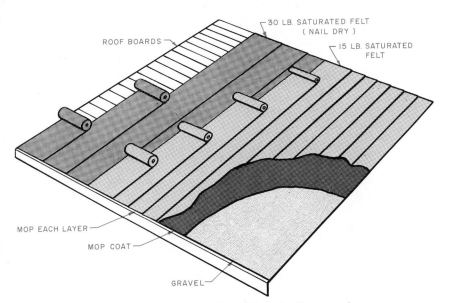

24-56. *Application detail for a built-up roof.*

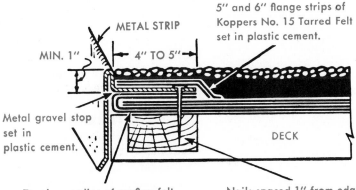

24-57a. *A cross section of a built-up roof without insulation showing the metal gravel strip at the roof edge.*

24-57b. *A cross section of a built-up roof with insulation showing the metal gravel strip at the roof edge.*

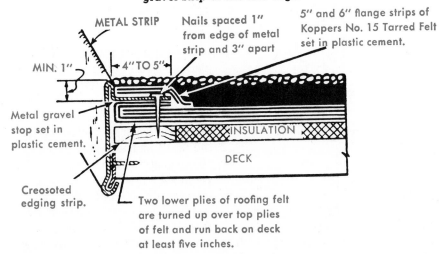

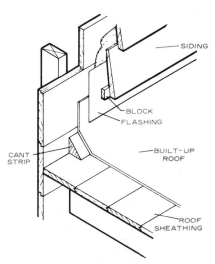

24-58. *The intersection of a flat roof with a wall showing the cant strip and flashing.*

with gravel. Fig. 24-57. Where built-up roofing is finished against another wall, the roofing is turned up on the wall sheathing over a cant strip and is flashed with metal. Fig. 24-58. This flashing is generally extended up about 4″ above the bottom of the siding.

METAL ROOFING

Corrugated metal roofing comes in widths up to 4′ and lengths up to 24′ and therefore covers large areas quickly. It is ideal for utility buildings such as garages and sheds. It can be used on slopes as low as 4 in 12, or as low as 3 in 12 if a single panel will cover from eave to ridge.

Apply purlins across the roof rafters to support the roofing, spacing according to roofing manufacturer's instructions. No sheathing is required. Set closure strips (they come with the roofing) at eave. Cut panels to length with tin snips, the length being the dimension from ridge to eave plus 2½″. Nail through tops of ribs, using 1¾″ screwshank nails with neoprene washers to prevent leaks. Next, set closure strips (inverted) at ridge and apply metal cap. Finally, flash at eaves as shown in Fig. 24-59.

Sheet-metal roofing must be

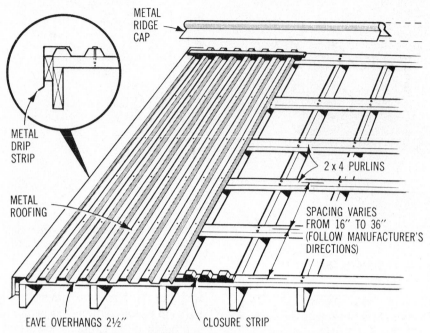

24-59. *Installation details for corrugated roofing.*

METAL RIDGE CAP

METAL DRIP STRIP

METAL ROOFING

2 x 4 PURLINS

SPACING VARIES FROM 16" TO 36" (FOLLOW MANUFACTURER'S DIRECTIONS)

EAVE OVERHANGS 2½"

CLOSURE STRIP

24-61a. *Gutters: A. Formed. B. Half-round.*

24-61b. *Corrugated downspouts: A. Rectangular. B. Round.*

tin roofs, where steel nails may be used. All exposed nailheads on tin roofs should be soldered.

GUTTERS AND DOWNSPOUTS

Various types of gutters are on the market. Fig. 24-60. The two general types are the formed metal and the half-round. Fig. 24-61a. Downspouts or leaders are rectangular or round, with the round leader being ordinarily used with the half-round gutter. Both the round and the rectangular leaders are usually corrugated for added strength. Fig. 24-61b. The corrugated patterns are less likely to burst when plugged with ice.

Wood gutters may be used in place of metal gutters and are usually fastened by means of rust-resistant screws or nails. Fig. 24-62. Nailing or spacing blocks are placed between the gutter and the fascia or frieze board about 16" on center. Wood gutters are given very little pitch because they usually are part of the architectural treatment. Joints in wood gutters are best made by dowels or splines. The joints are covered

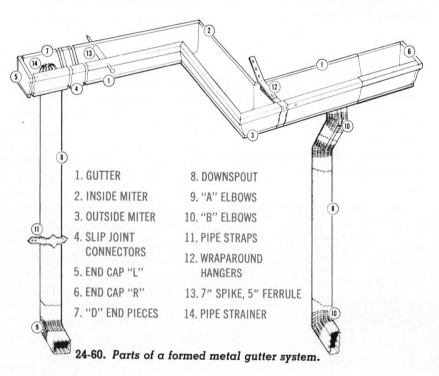

1. GUTTER
2. INSIDE MITER
3. OUTSIDE MITER
4. SLIP JOINT CONNECTORS
5. END CAP "L"
6. END CAP "R"
7. "D" END PIECES
8. DOWNSPOUT
9. "A" ELBOWS
10. "B" ELBOWS
11. PIPE STRAPS
12. WRAPAROUND HANGERS
13. 7" SPIKE, 5" FERRULE
14. PIPE STRAINER

24-60. *Parts of a formed metal gutter system.*

laid over sheathing. The joints should be watertight, and the deck should be properly flashed

where it joins with a wall. Nails should be of the same metal as that used on the roof, except with

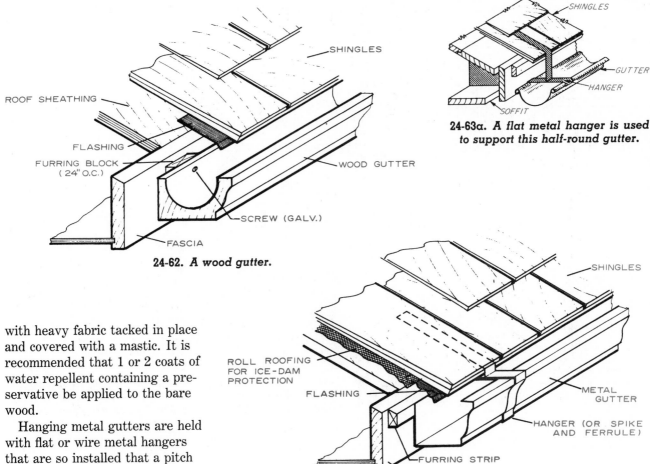

24-63a. *A flat metal hanger is used to support this half-round gutter.*

24-62. *A wood gutter.*

with heavy fabric tacked in place and covered with a mastic. It is recommended that 1 or 2 coats of water repellent containing a preservative be applied to the bare wood.

Hanging metal gutters are held with flat or wire metal hangers that are so installed that a pitch is formed for drainage. Fig. 24-63. Joints in metal gutters and downspouts should be soldered. Gutters should be mounted so that the shingle extension is over the center of the gutter. Hangers should be spaced 3' to 4' on center.

Another type of formed metal gutter has an extension or flashing strip that is fastened to the roof boards. Fig. 24-64. These gutters are usually made so that the back varies in height to allow a pitch for drainage. Hangers are used at the outer rim to add stiffness.

Metal gutters are placed to drain toward the downspouts. With a slope of approximately 1" in 10', the maximum run between the high point and the downspout should not ordinarily exceed 25'. A gooseneck is used to bring the downspout in line with the wall.

24-63b. *A flat metal hanger used to support a formed gutter.*

The form of this gooseneck will vary according to the extent of the cornice overhang.

Downspouts are fastened to the wall by means of leader straps or hooks. Fig. 24-65. Many patterns of straps are made to allow a space between the wall and the downspout. A minimum of two straps should be used in an 8' length of leader: one at the gooseneck and one at the bottom elbow. The elbow is used to lead the water to a splash block that carries the water away from the foundation. The minimum length of the splash block should be 3'. It is the practice, in some areas, to carry the water to a storm

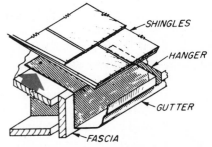

24-64. *This formed metal gutter has an extension which also serves as a flashing strip. It is used to fasten the gutter to the roof boards.*

sewer by means of tile lines. Fig. 24-66. In final grading the slope should be such as to insure positive drainage of water away from the foundation walls.

239

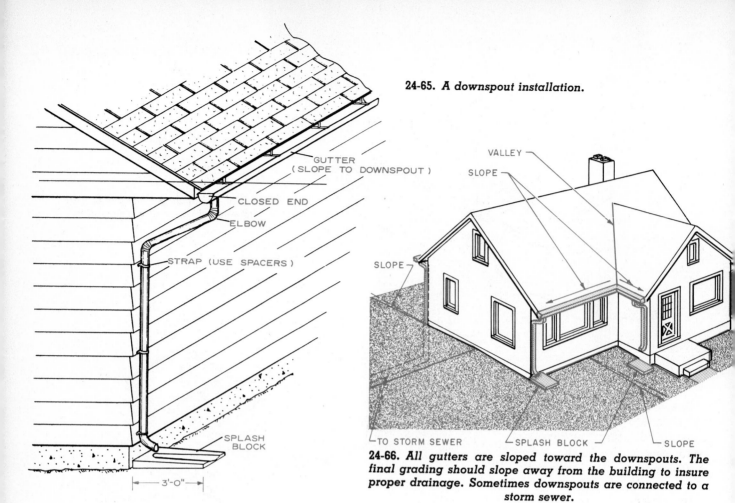

24-65. *A downspout installation.*

GUTTER
(SLOPE TO DOWNSPOUT)

CLOSED END

ELBOW

STRAP (USE SPACERS)

SPLASH
BLOCK

3'-0"

VALLEY

SLOPE

SLOPE

SLOPE

TO STORM SEWER

SPLASH BLOCK

SLOPE

24-66. *All gutters are sloped toward the downspouts. The final grading should slope away from the building to insure proper drainage. Sometimes downspouts are connected to a storm sewer.*

QUESTIONS

1. List several materials used for covering pitched roofs.

2. List several materials which are considered roofing accessories.

3. What is the minimum slope on main roofs for the application of wood, asphalt, asbestos, and slate shingles?

4. What is the maximum slope recommended for built-up roofs?

5. Why is roof underlayment used under shingles?

6. What is flashing?

7. On roofs longer than 30', why are strip shingles started at the center and applied toward the ends?

8. List 3 variations for laying square-butt strip shingles.

9. When laying shingles on the ridge, in what direction should the exposed edge be laid?

10. What are the two methods used to construct valley flashing? Describe the difference.

11. What special precautions should be taken when laying strip shingles on a low-pitch roof?

12. What is the main advantage of interlocking shingles?

13. What is the difference between wood shakes and wood shingles?

14. What is considered the least expensive type of roofing?

15. How is the quality of a built-up roof designated?

16. What is the recommended slope for metal gutters?

17. Describe one method for determining the area of a plain gable roof. For determining the area of a plain hip roof.

SECTION V

Completing the Exterior

Roof Trim

25

As explained in Unit 17, the rafter-end overhangs of a roof are called the *eaves*. For example, on a hip roof, all four edges—the sides and ends—of the roof have eaves. A gable roof, however, has eaves only on the side-wall edges. The gable-end (end-wall) edges are called *rakes*. The exterior finish at and just below the eaves is called the *cornice*.

The cornice work may be done as soon as the roof has been framed. It may also, with the exception of the fascia, be done after the roofing has been applied. In most geographical areas, workers will put on the roof covering first to protect the structure from the weather. The rake molding on a gable roof, however, must be installed before the roofing.

TYPES OF CORNICES

The type of cornice required for a particular structure is shown on the wall sections of the house plans, and there are usually cornice detail drawings as well. Basically, there are three types of cornices:

- Close.
- Open.
- Box.

A roof with no rafter overhang normally has a close cornice. Fig. 25-1. This cornice consists of a single strip called a *frieze*. The frieze is beveled on its upper edge to fit close under the overhang of the eaves and rabbeted on its lower edge to overlap the upper edge of the top siding course. If trim is used, it usually consists of molding installed as shown in Fig. 25-1. Molding trim in this position is called *crown* or *shingle molding*.

A roof with a rafter overhang may have either an open cornice or a box cornice. The simplest type of open cornice consists of only a frieze, which must be notched to fit around the rafters. Fig. 25-2. If trim is used, it usually consists of molding cut to fit between the rafters. Molding in this position is called *bed molding*.

Another type of open cornice consists of a frieze and a fascia. Fig. 25-3. A *fascia* is a strip nailed to the tail plumb cuts of

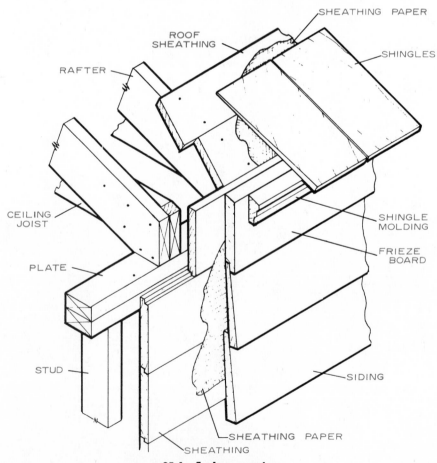

25-1. *A close cornice.*

242

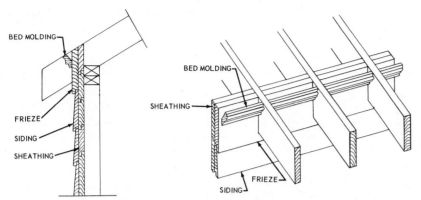

25-2. *Simplest type of open cornice.*

the rafters. Shingle molding can be attached to the top of the fascia, but it is seldom used.

With a box cornice, the rafter overhang is entirely boxed in by the roof covering, the fascia, and a bottom strip called a *plancier*, or *soffit*. Figs. 25-4 & 25-5. The soffit can be nailed to the rafters. It can also be nailed to lookouts. The *lookouts* are a series of horizontal members which are nailed

to the rafters and extend from the rafter ends to the face of the sheathing. Fig. 25-5b. The frieze, if any, is set just below the lookouts. If trim is used, it is placed at the intersection of the frieze and the soffit.

WOOD CORNICE CONSTRUCTION

If building paper is used on the sidewalls, the top course of paper

must be applied before beginning work on the cornice. For an open or a box cornice the paper must be slit to fit around the rafters.

Open Cornice

One method of constructing an open cornice is to measure the distance between the rafter at either the top or bottom edge. Fig. 25-6. Cut material to this length, making sure that both ends are cut square. Do this for each rafter spacing. Nail the material in position, as shown at A in Fig. 25-7. The nails can be driven through the side of the rafter into the end of the block on one side. Nails will have to be toenailed on the other side, as the block installed previously gets in the way for nailing. Fig. 25-8 (page 246).

If vents are needed, determine their location and bore the necessary holes. Staple or tack a piece of window screen on the back of the vent openings. Fig. 25-6.

If tongued and grooved mate-

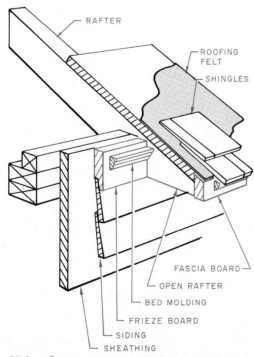

25-3a. *An open cornice with a fascia board.*

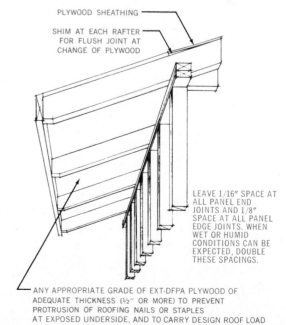

25-3b. *An open cornice, sometimes referred to as an open soffit, with plywood roof sheathing.*

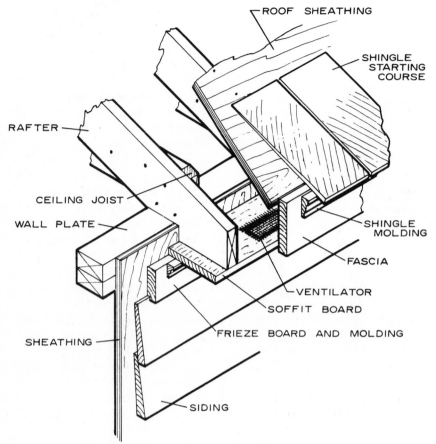

25-4. A narrow box cornice.

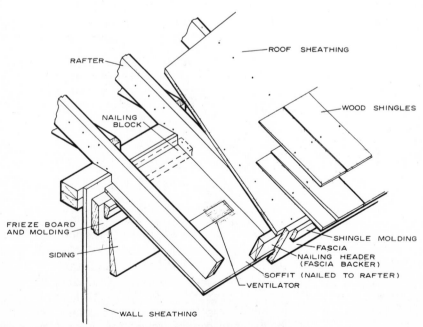

25-5a. A box cornice with a sloping soffit.

rial is to be used on the roof, nail the material on the rafters with the good surface down since it will be visible from below. Fig. 25-9. Remove the groove from the starter board. Bevel the edge if desired. Place the starter board in position along the top of the rafter tail and align it before nailing it in place. B, Fig. 25-7. The type of sheathing used above the blocking will depend on the kind of roof covering. Plywood or lumber sheathing may be chosen.

All joints in the construction of an open cornice should be planed smooth and fitted together tightly. All moldings must be mitered for joining on outside corners and mitered or coped (See page 453) for joining on inside corners. Care should be taken in this type of cornice construction because the workmanship is readily visible from the ground.

Box Cornice

Before adding a box cornice, check the plumb cuts on the rafter tails to make certain they are all in line. This can be done by stretching a line along the top ends of the rafters from one corner of the building to the other. Many carpenters do not make the plumb cut on the rafter tails when the rafter is cut. Instead, the rafters are nailed in place with the tails projecting beyond the exterior wall at various lengths. Then the point at which the tails are to be cut off is determined, and a chalk line is snapped along the top edge of the rafter tails. The plumb line is drawn down the side of each rafter from this line. Each rafter tail is then cut off along this plumb line. Fig. 25-10.

LOOKOUTS

1. Use a piece of 1″ × 4″ material to serve as a ledger and nail it temporarily against the exterior wall tight up under the rafters

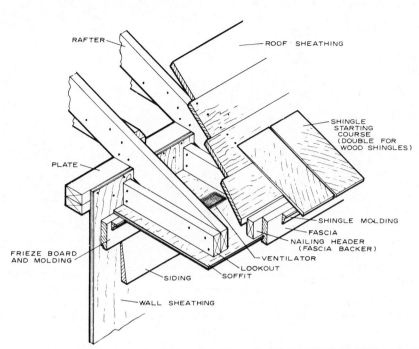

RAFTER

ROOF SHEATHING

SHINGLE STARTING COURSE (DOUBLE FOR WOOD SHINGLES)

PLATE

SHINGLE MOLDING

FASCIA

NAILING HEADER (FASCIA BACKER)

FRIEZE BOARD AND MOLDING

VENTILATOR

LOOKOUT

SIDING

SOFFIT

WALL SHEATHING

25-5b. A box cornice with a flat soffit. Note that the soffit is nailed to lookouts.

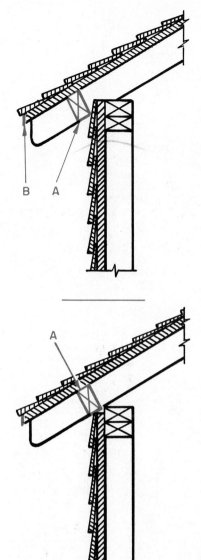

25-7. The location of the blocking used in the open cornice is determined by the bird's-mouth in the rafter as shown at A in both drawings. Notice that the roof board at B in the first drawing has been beveled on the edge to conform to the roof pitch.

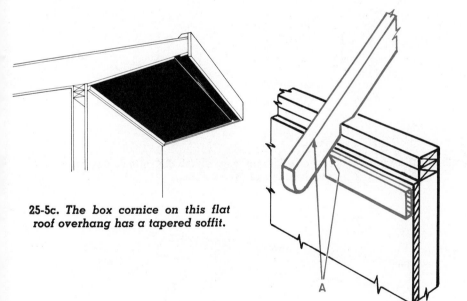

25-5c. The box cornice on this flat roof overhang has a tapered soffit.

25-6. Determining the length of blocking needed for an open cornice. Measuring may be done at either the top or bottom edge of the rafter, as shown by the arrows at A. For ventilation, the blocking should be drilled as shown at B and then screening stapled behind the holes as shown at C.

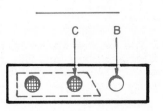

and aligned with the inside edge of the first rafter. Fig. 25-11. With a straightedge against the side of the rafter, make a line on the surface of the ledger. Place an X on the side of the line away from the underside of the rafter

25-8. *Nailing the blocking in place on an open cornice. Notice the vent holes which have been drilled in some of the blocks.*

25-9. *This open cornice was extended to serve as a roof for the porch area. Notice that the roof board material needs to be carefully selected and applied because it is clearly visible from below.*

25-10. *Cutting off the rafter tails after the rafters have been nailed in place.*

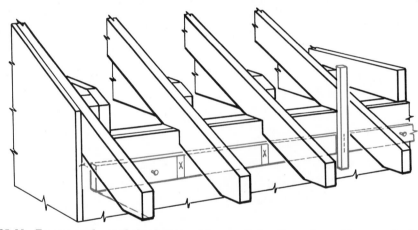

25-11. *Temporarily nail the ledger strip up under the rafters. Then mark the location of the lookouts.*

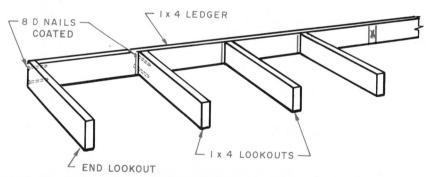

25-12. *The lookouts are nailed to the ledger strip next to the line and over the X made earlier. Note that the end lookout is nailed into the end of the ledger strip. This means that the end lookout has to be of the same thickness as the rafter and longer than the rest of the lookouts. It will have to be cut to fit under the rafter tail.*

to indicate the location of the lookout. Do this along the entire length of the building.

2. Determine the length of the lookouts. Measure on a level line from the plumb cut on the rafter tail to the wall of the building. Subtract ¾″ from this measurement to allow for the thickness of the nominal 1″ × 4″ ledger to which the lookouts will be nailed. Subtract another ¾″ to make sure that the lookouts do not project

246

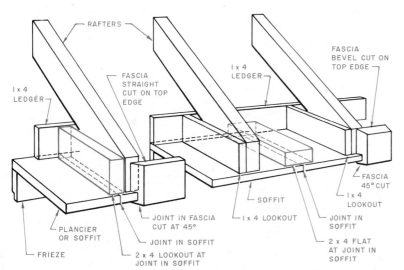

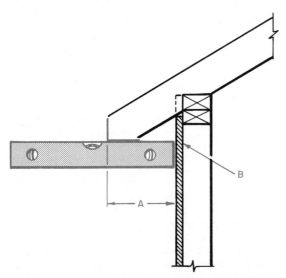

25-13. *The soffit material must be supported wherever it is joined together. If possible, the joints are usually located under a lookout. However, rather than cut the soffit off, nail a 2 × 4 laid flat from the ledger strip to the fascia over the soffit joint.*

25-14. *To locate the position of the ledger strip on the side wall of the building (point B), level a line in from the rafter tail as shown at A. The rafters are usually cut off before the ledger strip is located on the building. However, in some cases, point B is located on the building first. The cutoff line on the bottom of the rafter tail is then marked by leveling out from point B.*

beyond the end of the rafters. Otherwise, if there is any deviation in the alignment of the exterior wall, such as a slight bow or crooked stud, the lookout may extend beyond the end of the rafter tail. This will interfere later with the proper installation and alignment of the fascia board.

3. After the lookouts have

been cut to length, remove the ledger from its temporary nailing and nail the lookouts to the ledger over the Xs. Nail through the back of the ledger into the end of the lookout with two 8d coated nails. Nail the last lookout into the end of the strip. Fig. 25-12. The lookouts may be made from either 2 × 4s or 1 × 4s. If 1 ×

4s are used, place a 2 × 4 for additional nailing surface wherever the soffit pieces must be joined. Fig. 25-13.

4. Locate the position of the ledger on the exterior wall by leveling from the rafter tail in toward the wall and placing a mark on the sheathing. Point B, Fig. 25-14. Do this at each end of the building. Then snap a chalk line along the full length of the building on the sheathing.

5. Place bottom edge of ledger on this line. Nail it to studs through sheathing. Nail each lookout to side of rafter tail, except end lookout, which is nailed under rafter. Level each lookout as it is nailed.

FASCIA BOARD

Lay out, rip (if necessary), and groove the fascia board. The groove is made in the fascia board to receive the soffit. It should be cut about ⅜″ up from the bottom edge of the fascia. This is done to provide a drip edge which prevents water from backing up into the groove.

Nail the fascia board in position along the ends of the rafter tails with the top of the groove even with the bottom edge of the lookouts. Fig. 25-15. If the fascia board must be spliced, it should be done with the joint on the end of a rafter tail, and the joint should be mitered. Fig. 25-16. The top edge of the fascia board may be beveled to the same angle as the pitch of the roof. If it is not, make certain that the fascia board is installed with its top outer edge in line with the top surface of the roof sheathing. Fig. 25-15. Make certain that the fascia is straight along its length. If necessary, straighten the fascia by driving shims between the rafter tail ends and the inside of the fascia board.

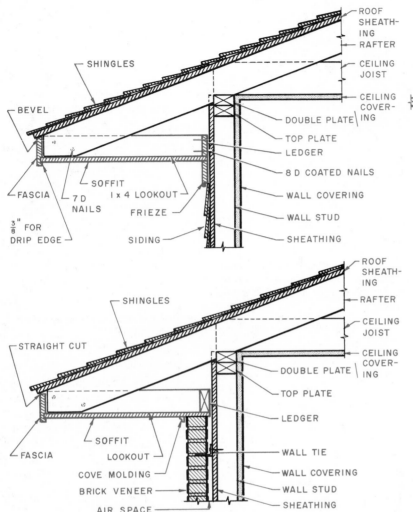

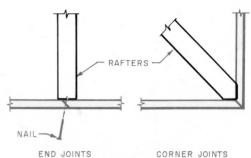

25-16. *On a hip roof the fascia is mitered at the corner on the end of the hip rafter. On any roof where the fascia must be joined, join it on a rafter end. Miter the joint as shown.*

25-15. *The fascia board may be nailed to the ends of the rafters using either of the methods shown here. Bevel the top edge of the fascia board to conform to the roof pitch, as at A. Or rip the fascia board to width so that the outside top corner is in line with the top edge of the rafter, as at B. In either case, when the roof sheathing is applied it must lie flat. Notice also that the top edge of the groove in the fascia must be in line with the bottom edge of the lookout for proper installation of the soffit material.*

SOFFIT

Several materials may be used for the soffit on a box cornice. Because of the popularity of wide overhangs, materials which are available in large sheets are frequently used. These include plywood, gypsum board, and hardboard.

Plywood. This is one of the most popular materials for a box cornice. It simplifies construction and presents a smooth, attractive surface. Plywood also has the advantage of matching other wood surfaces when a stained wood grain finish is desirable.

The recommended spans for box soffits are shown in Fig. 25-17. Exterior plywood should be used wherever the underside of the roof deck is exposed to the weather. Fig. 25-18

To install plywood soffit, rip the soffit to width and slip the outer edge into the fascia groove. Then push the inside edge up against the lookouts and ledger strip. Nail the soffit securely to the ledger and to each lookout with 4d nails. The nails should be spaced about 6″ apart. If the soffit has to be made up of several pieces, join it under a lookout. If this is not possible, a 2 × 4 can be laid flat, toenailed into the ledger, and face-nailed through the fascia. The two pieces of soffit can then be joined under the center of the flat side of the 2 × 4. This will give adequate nail backing for the joint. Fig. 25-13.

Gypsum Board. Gypsum soffit board is a noncombustible product developed for use in soffits, carports, and similar installations where there is no direct exposure to the weather.

This soffit board has a water-resistant core and blue face paper. It is cut and scored instead of sawed, and the joints are taped and finished to provide a smooth surface. Gypsum soffit board is available in 8′ and 12′ lengths, ½″ thick and 4′ wide.

Gypsum soffit boards are installed in the same manner as regular wallboard over framing members spaced a maximum of 24″ on center. Space nails 7″ on center. Use trim around the edges that abut the building and

Exterior plywood soffits (closed) / (Plywood continuous over two or more spans; grain of face plys across supports)

Plywood Thickness (inch)	Closed Soffits			Nail		Nail Spacing (inches)	
	Group	Max. Spacing of Supports c. to c. (inches)	Size	Type		Panel Edges	Intermediate (each support)
⅜	1, 2, 3, or 4	24	6d	Non-corrosive type (galv. or alum.) box or casing.		6 (or one nail each support)	12
⅝		48	8d				12

Use plywood with these typical APA grade-trademarks.	**Sanded Grades** **A-C** GROUP 2 EXTERIOR PS 1-74 000 (APA)	**Specialty Panels** 303 SIDING 16 oc GROUP 3 EXTERIOR PS 1-74 000 (APA)	M.D. OVERLAY GROUP 1 EXTERIOR PS 1-74 000 (APA)

25-17. *This chart indicates the maximum spacing of supports for various thicknesses of box (closed) plywood soffits. It also gives the correct nail sizes and spacing to be used.*

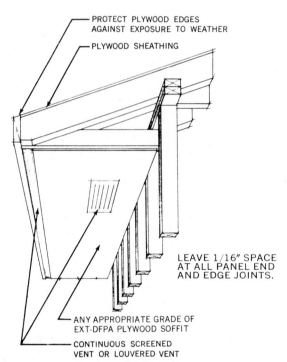

LEAVE 1/16" SPACE AT ALL PANEL END AND EDGE JOINTS.

PROTECT PLYWOOD EDGES AGAINST EXPOSURE TO WEATHER

PLYWOOD SHEATHING

ANY APPROPRIATE GRADE OF EXT-DFPA PLYWOOD SOFFIT

CONTINUOUS SCREENED VENT OR LOUVERED VENT

25-18. *Plywood is frequently used for soffit material on a box soffit, but certain precautions should be taken, as shown here.*

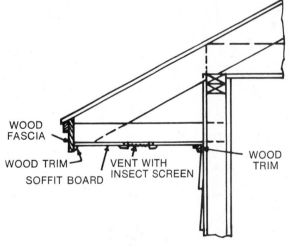

WOOD FASCIA

WOOD TRIM

SOFFIT BOARD

VENT WITH INSECT SCREEN

WOOD TRIM

FRAME WALL

25-19a. *Gypsum board provides a flat, smooth surface for the soffit of a box cornice.*

fascia. The fascia must come at least ⅜" below the level of the soffit board to provide a drip edge. The round-edge joints should be prefilled with a joint compound. The application should then be taped and finished in the conventional manner. (See Unit 31 for detailed information about taping and finishing gypsum board.) Apply two coats of compound over the tape and three coats over the nailheads. Provide adequate ventilation for the space above the soffit (see *FHA Minimum Property Standards*). Install control joints (to allow for expansion and contraction) at a maximum spacing of 40'. Fig. 25-19.

Hardboard. Hardboard panels are frequently used for soffits on the undersides of eaves and the ceilings of porches, breezeways,

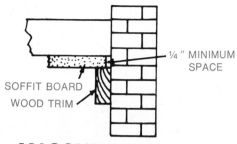

¼" MINIMUM SPACE

SOFFIT BOARD

WOOD TRIM

MASONRY WALL

25-19b. *A gypsum-board soffit on a masonry exterior wall should be supported along its entire length. This support is provided by a piece of wood trim attached to the exterior wall. Also be sure to provide a minimum of ¼" of space behind the soffit for expansion.*

TYPICAL SOFFIT FRAMING

LOOKOUT

fillers required every 4 ft. and at joints

if lookouts are spaced more than 16" apart, use 2 x 2's instead of 1 x 2's.

if soffit is more than 16" wide, one or more intermediate stringers are required. stringers must not be more than 16" apart

CONTINUOUS NAILING BASE

GROOVE IN FASCIA AS SOFFIT SUPPORT

HEADER PROVIDES CONTINUOUS NAILING SUPPORT

AT HOUSE

25-20. *Hardboard soffits should be supported as shown. Nail them 4" on center around the edges and 6" on center at the intermediate supports.*

and carports. If the hardboard does not have a factory-applied primer, the panel should be conditioned prior to use. Follow the manufacturer's recommendations included in each product bundle. The soffit framing must provide continuous support at the edges, ends of panels, and joints.

The installation of hardboard soffits is similar to plywood soffit installation. Fasten the panels with 5d galvanized box, siding, or sinker nails. Space the nails 4" on center around edges and approximately 6" on center at intermediate supports. Never nail closer than ⅜" to the edge. Fig. 25-20. Metal moldings are available for use in installing hardboard soffit material. With these moldings, there are no exposed nailheads on soffits 2' or less in width.

METAL CORNICE CONSTRUCTION

Metal cornice material may be used for box cornices on most roofs, entryways, porches, and carport ceilings. This aluminum system requires little maintenance, is self-ventilating and self-supporting, and will not rust. It is entirely prefinished. The ribbed soffit material may be nonperforated or it may be perforated to give approximately 8% open area. It comes in coils usually 50' long and of various widths. Fig. 25-21.

Installation

It is important to plan the sequence of operation. The soffit material is pulled from coils and fed between the fascia and frieze runner guides into its proper position. Therefore one end of a soffit run cannot be closed off with enclosures, fascia runners, or extensions of frieze runners until the respective runs of soffit material are in place. Fig. 25-21. To illustrate metal cornice construction, this section describes the installation procedure for a box cornice on a hip roof.

FASCIA RUNNERS

Hang the fascia level, using a chalk line for alignment. Do not force the aluminum or metal fascia to conform to the wood to which it is being attached.

Secure the top edge of the fascia runner to the fascia with 1½" spiral-shank aluminum nails. Place the nails no more than 2' apart along the length of the fascia.

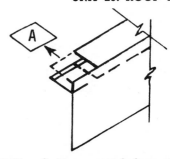

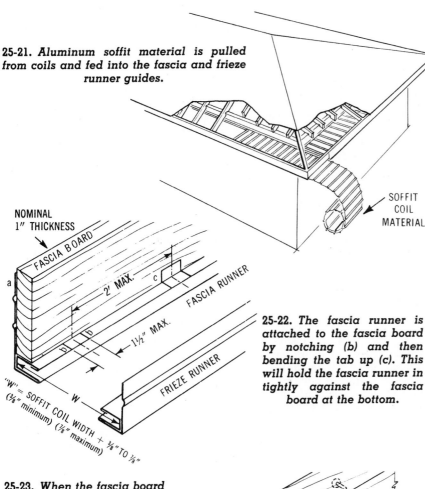

25-21. *Aluminum soffit material is pulled from coils and fed into the fascia and frieze runner guides.*

NOMINAL 1" THICKNESS

FASCIA BOARD

2' MAX.

FASCIA RUNNER

1½" MAX.

FRIEZE RUNNER

"W" = SOFFIT COIL WIDTH + ⅜" TO ⅞"
(⅜" minimum) (⅞" maximum)

a

b

c

SOFFIT COIL MATERIAL

25-22. *The fascia runner is attached to the fascia board by notching (b) and then bending the tab up (c). This will hold the fascia runner in tightly against the fascia board at the bottom.*

25-24a. *Cutting a notch for end lap joints. On the bottom flange of the runner make a cut through to the back of the runner. Then cut through both thicknesses along the back edge of the runner groove back and remove piece A.*

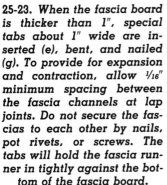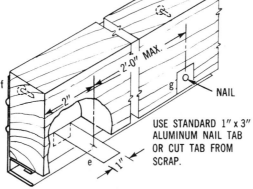

25-23. *When the fascia board is thicker than 1", special tabs about 1" wide are inserted (e), bent, and nailed (g). To provide for expansion and contraction, allow ¹⁄₁₆" minimum spacing between the fascia channels at lap joints. Do not secure the fascias to each other by nails, pot rivets, or screws. The tabs will hold the fascia runner in tightly against the bottom of the fascia board.*

2'-0" MAX.

2"

f

g

NAIL

e

1"

USE STANDARD 1" x 3" ALUMINUM NAIL TAB OR CUT TAB FROM SCRAP.

Make two cuts 1" to 1½" apart and about ⅝" deep into the top flange of the runner's guide. Fig. 25-22 at b. While holding the fascia runner flush against the outside face of the fascia board, bend the 1" to 1½" tab up against the inside face of the fascia board. Fig. 25-22 at c. When bending the tab, make sure that the top flange of the runner's guide is still straight. If not, straighten as necessary. Cut and bend as many of these tabs as necessary to hold the fascia runner in place. Space these tabs not more than 2' apart.

If the fascia board is thicker than 1", the runner will not be wide enough for a tab. In that case, cut some nailing tabs from scrap material about 1" wide. In-sert the nailing tabs into the slot between the fascia and channel runner. Fig. 25-23 at e. Secure the fascia runner as described previously. Bend the tabs up against the inside of the fascia board. Secure with a 1½" spiral-shank aluminum nail. Refer to Fig. 25-23 at g.

Be sure to allow for expansion and contraction between the fascia channels at the lap joints. Do not secure the channels to each other by nailing or any other fastening device and do not nail along the bottom edge of fascia runners. If this is done it will cause buckling on the face when the aluminum expands.

CUTTING AND FITTING FASCIA AND FRIEZE RUNNERS

A 1¼" notch is made in the fascia and frieze runners for end lap joints. To make lap cuts, use a pair of snips to cut the bottom flange of the runner through to the back of the runner. Fig. 25-24a. Then cut through both thicknesses of the runner groove back, removing piece A. On the top flange of the runner, cut to the inside of the runner groove back. Now bend the piece back and forth until it breaks off as shown at B in Fig. 25-24b. Trim the rough edges and reshape the

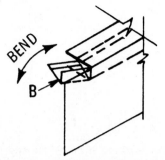

25-24b. *On the top flange of the runner, cut to the inside of the runner groove back and bend piece B back and forth until it breaks.*

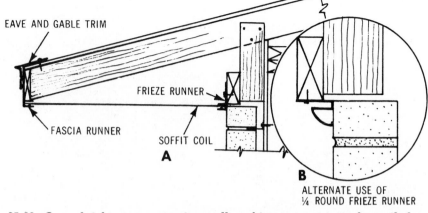

25-26. *On a brick veneer exterior wall, a frieze runner may be nailed to blocking material to bring it out flush with the wall (A). Or a quarter-round frieze runner can be used and blocked out (B).*

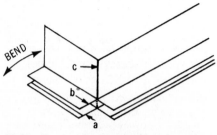

25-25. *To cut the material to length, cut through the flanges at a and b and then bend the piece back and forth at c until the back breaks.*

25-27. *On a flat roof installation where the overhang is wider than 48", a double-channel runner (H molding) may be used to support two widths of the soffit coil material.*

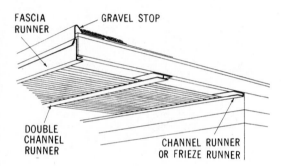

flanges as necessary. To make end cuts, cut through flanges a, b, and c to the runner groove back as shown in Fig. 25-25. Bend the piece back and forth until the groove back breaks. Trim off rough edges as necessary.

If the metal soffit system is to be used on a home which is of brick veneer, block out at the exterior wall above the brick line to permit the frieze runner to be installed flush with the top edge of the brick. Fig. 25-26. (Note in the detail B that a quarter-round frieze runner may also be used for this installation with a slightly different method of blocking.)

Where overhangs exceed the width of the soffit material available, a double channel runner can be used so that two pieces of soffit can be installed. Fig. 25-27.

Trimming the corners of a horizontal soffit is slightly different

from a sloping soffit. In the case of a horizontal soffit such as would be used on a hip roof, miter corner trim is available and is installed as shown in Fig. 25-28a. For a sloping soffit such as

25-28a. *On a horizontal soffit a special miter corner trim can be used to enclose the corner.*

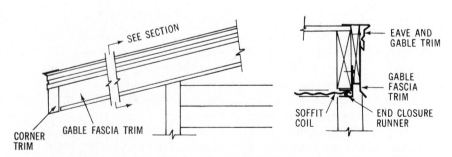

25-28b. *On a sloping soffit for a gable, special corner trim can be applied as shown here.*

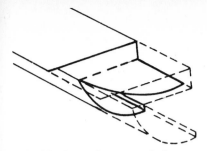

25-29. *To make an end closure on a quarter-round frieze molding, notch as shown. Bend the tab down, then cut the tab to conform to the profile of the quarter-round.*

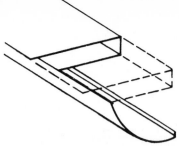

25-30. *When making an end lap, notch the piece as shown. With tin snips, cut away the dotted line area.*

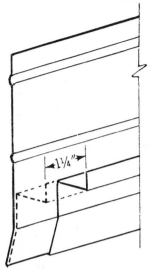

25-32. *Notch the gable fascia trim for end laps as shown by the broken lines.*

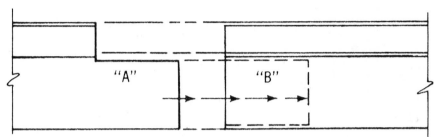

"A" "B"

25-31. *When joining two pieces of quarter-round end to end, the pieces can be cut and telescoped as shown. Part A will fit over B, and the frieze runner will butt together with the cove overlapping.*

the end of a gable, another type of corner trim is available and can be installed as shown in Fig. 25-28b.

CUTTING AND FITTING QUARTER-ROUND FRIEZE MOLDING

End closures are made by cutting away the area shown by the dotted line on the flanges in Fig. 25-29. Bend the flange down over the end and cut the rounded contour with double action snips to complete the closure. Corners are fitted by mitering in the usual manner. End-laps are made by cutting away the area shown by the broken lines in Fig. 25-30 and shoving the ends together. Fig. 25-31. On the gable fascia trim, the end laps are made by notching. Fig. 25-32.

25-33. *Fascia and frieze runners are available either straight or sloping, in a variety of sizes.*

Part Name		Part Description
Straight Fascia and Frieze Runners (Prenotched)		
1″ Leg	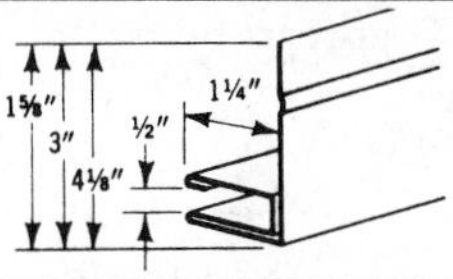	1⅝″ wide x 121¼″ long
3″ Leg		3″ wide x 121¼″ long
4″ Leg		4⅛″ wide x 121¼″ long
6″ Leg	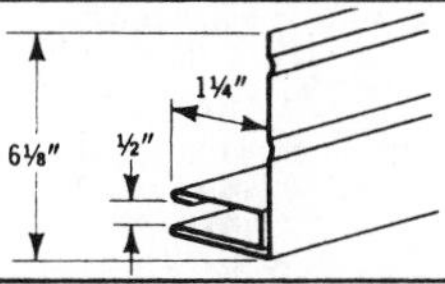	6⅛″ wide x 121¼″ long
8″ Leg	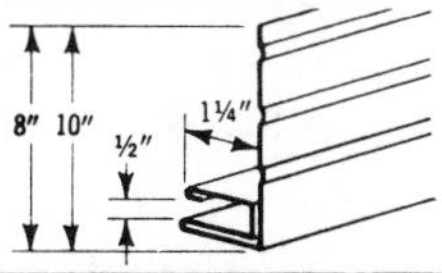	8″ wide x 121¼″ long
10″ Leg		10″ wide x 121¼″ long
Sloping Fascia and Frieze Runners, Type #1 (Prenotched)		
3″ Leg	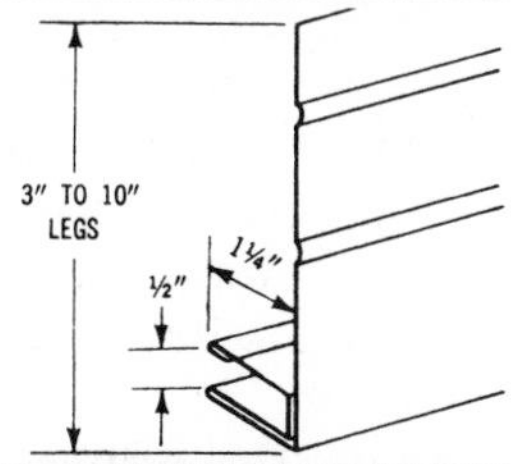	3″ wide x 121¼″ long
4″ Leg		4⅛″ wide x 121¼″ long
6″ Leg		6⅛″ wide x 121¼″ long
8″ Leg		8″ wide x 121¼″ long
10″ Leg		10″ wide x 121¼″ long
Sloping Fascia and Frieze Runners, Type #2 (Prenotched)		
3″ Leg	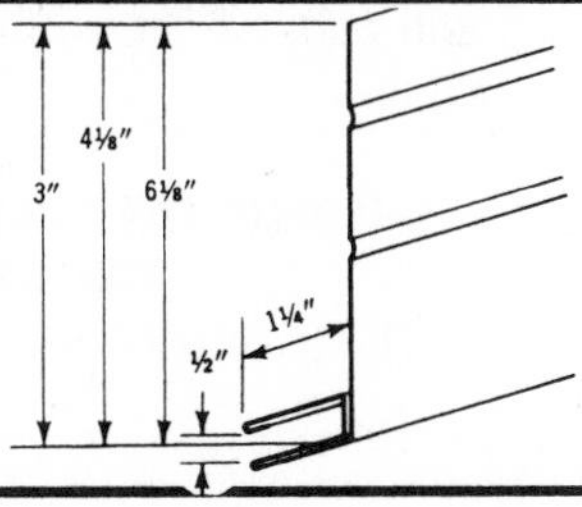	3″ wide x 121¼″ long
4″ Leg		4⅛″ wide x 121¼″ long
6″ Leg		6⅛″ wide x 121¼″ long

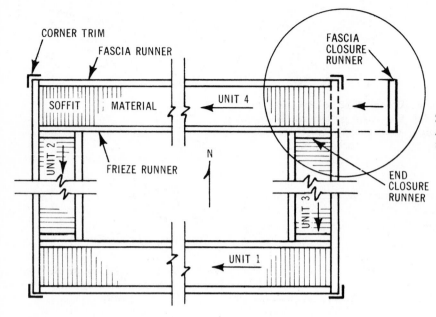

25-34. *This diagram shows the soffit of a hip roof from below. The arrows indicate the direction and sequence in which to pull the soffit coil into place.*

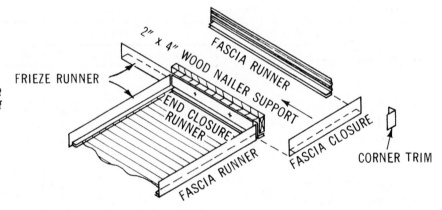

25-35a. *A corner assembly of the soffit on a hip roof showing the method of end and fascia closures.*

INSTALLATION SEQUENCE

1. Fascia and frieze runners are available in a variety of sizes. Fig. 25-33. Select the correct size for the job and apply the fascia and frieze runners for Unit 1 as shown in Fig. 25-34.

2. Pull the soffit coil into place in the direction of the arrow as shown in Fig. 25-34 for Unit 1.

3. Install end closure runners to Units 2 and 3 where they intersect with Unit 1. Fig. 25-35a.

4. Install the fascia and frieze runners of Units 2 and 3. Fig. 25-34. Note that you must leave out one section of fascia runner equal to the soffit width to allow Unit 4 to be installed later.

5. Pull in the soffit coil for Units 2 and 3. Fig. 25-34.

6. Apply end closure runners to Units 2 and 3 where they intersect with Unit 4. Fig. 25-34.

7. Apply the fascia and frieze runners of Unit 4 and pull in the soffit coil.

8. Install the fascia closure runner referred to in Step 4. Figs. 25-34 & 25-35a.

9. Apply the corner trim angles or mitered corner trim. Fig. 25-35b.

10. Use a spline tool to apply a polyethylene spline along all sides and ends of the soffit sheet. Fig. 25-36.

RAKE OR GABLE-END FINISH

The extension of a gable roof beyond the end wall is called the *rake section*. This detail may be classed as being:

• A close rake with little projection.

• A boxed or open extension varying from 6″ to 2′ or more.

When the rake extension is only 6″ to 8″, the fascia and soffit can be nailed to a series of short lookout blocks. A, Fig. 25-37. In addition, the fascia is further secured by nailing through the projecting roof sheathing. A frieze board and appropriate moldings will complete the construction.

In a moderate overhang of up to 20″, both the extending shea-

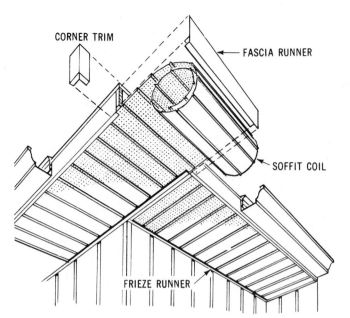

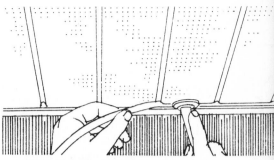

25-36. *Use the spline tool to insert the polyethylene spline along all sides and ends of the soffit sheet.*

25-35b. *After all of the soffit coils have been pulled into place, the last soffit to be pulled in must be closed off by installing the fascia runner and the corner trim.*

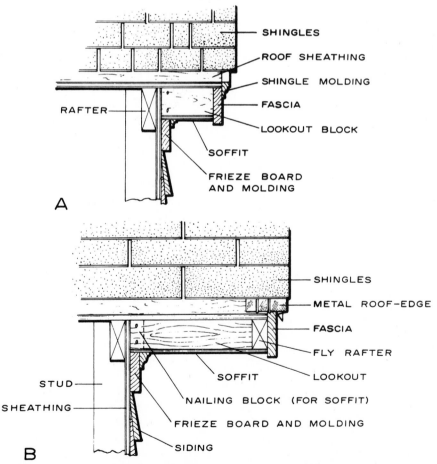

25-37. *Normal gable-end extensions: A. Narrow overhang. B. Moderate overhang.*

thing and a *fly rafter* aid in supporting the rake section. B, Fig. 25-37. The fly rafter extends from the ridge board to the nailing header which connects the ends of the rafters. The roof sheathing boards or the plywood should extend from inner rafters to the end of the gable projection to provide rigidity and strength.

The roof sheathing is nailed to the fly rafter and to the lookout blocks which aid in supporting the rake section and also serve as a nailing area for the soffit. Additional nailing blocks against the sheathing are sometimes required for thinner soffit materials.

Wide gable extensions (2′ or more) require rigid framing to resist roof loads and prevent deflection of the rake section. This is usually accomplished by a series of *purlins* or lookout members nailed to a fly rafter at the outside edge and supported by the end wall and a doubled interior rafter. Fig. 25-38. This framing is often called a "ladder" and may be constructed in place or on the ground or other convenient area and hoisted in place.

When ladder framing is preassembled, it is usually made up with a header rafter on the inside and a fly rafter on the outside. Each is nailed to the ends of the lookouts which bear on the gable-end wall. When the header is the

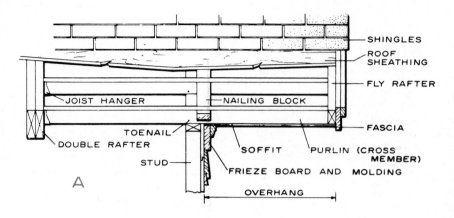

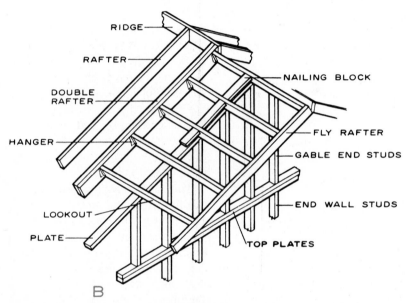

25-38. Wide gable-end extension: A. Wide overhang. B. Ladder framing for a wide overhang.

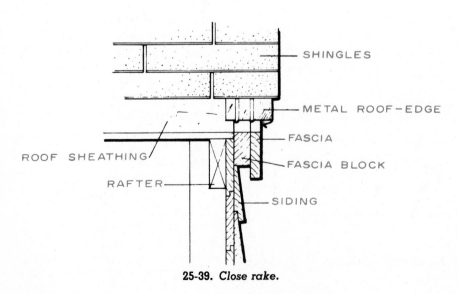

25-39. Close rake.

same size as the rafter, be sure to provide a notch for the wall plates the same as for the regular rafters. In moderate width overhangs, nailing the header and fly rafter to the lookouts with supplemental toenailing is usually sufficiently strong to eliminate the need for the metal hangers shown in B, Fig. 25-38. The header rafters can be face-nailed directly to the end rafters with 12d nails spaced 16″ to 20″ apart.

Other details of soffits, fascia, frieze board, and moldings can be similar to those used for a wide gable overhang. Lookouts should be spaced 16″ to 24″ apart, depending on the thickness of the soffit material.

A close rake has no extension beyond the end wall other than the frieze board and moldings. Some additional protection and overhang can be provided by using a 2″ × 3″ or 2″ × 4″ fascia block over the sheathing. Fig. 25-39. This member acts as a frieze board, as the siding can be butted against it. The fascia, often 1″ × 6″, serves as a trim member. Metal roof edging is often used along the rake section as flashing.

CORNICE RETURN

The cornice return is the end finish of the cornice on a gable roof. On hip roofs and flat roofs, the cornice is usually continuous around the entire house. On a gable roof, however, it must be terminated or joined with the gable ends. The method selected depends to a great extent on the type of cornice and the projection of the gable roof beyond the end wall.

A narrow box cornice, often used in houses with Cape Cod or Colonial details, has a boxed return when the rake section has some projection. A, Fig. 25-40. The fascia board and shingle molding of the cornice are carried

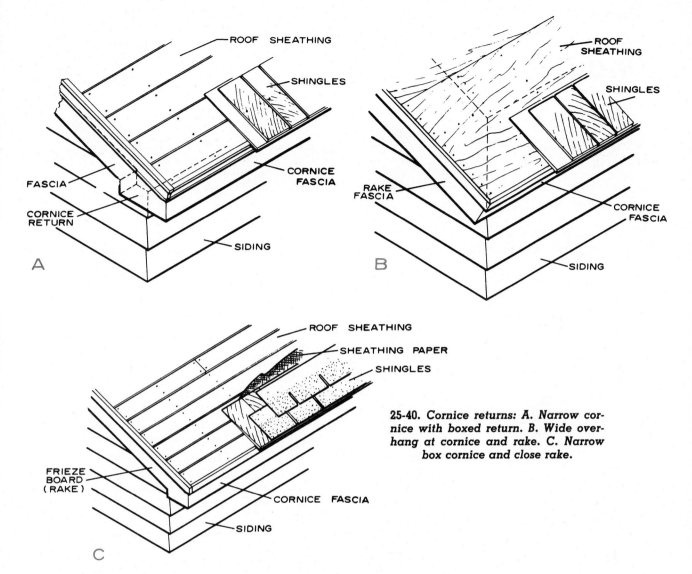

25-40. Cornice returns: A. Narrow cornice with boxed return. B. Wide overhang at cornice and rake. C. Narrow box cornice and close rake.

around the corner of the rake projection.

When a wide box cornice has no horizontal lookout members, the soffit of the gable-end overhang is at the same slope and coincides with the cornice soffit. B, Fig. 25-40. This is a simple system and is often used when there are wide overhangs at both sides and ends of the house.

A close rake (a gable end with little projection) may be used with a narrow box cornice or a close cornice. In this type, the frieze board of the gable end, into which the siding butts, joins the frieze board or fascia of the cornice. C, Fig. 25-40.

While close rakes and cornices with little overhang are lower in cost, the extra material and labor required for good gable and cornice overhangs are usually justified. Better sidewall protection and lower paint maintenance costs are only two of the benefits derived from good roof extensions.

GUTTERS

Wooden gutters are either built into the cornice or prefabricated and attached on the job site. They were once used extensively but are now almost obsolete. Most modern gutters are of prefabricated metal, equipped with metal straps for attaching to the roof boards. See Unit 24.

QUESTIONS

1. List several types of cornices.
2. When constructing a closed cornice, what procedure is used to insure alignment of the plumb cuts on the rafter tails?

3. What is a fascia?
4. List three materials that are commonly used for the soffit on a closed cornice.
5. What is a cornice return?

Windows 26

Windows are millwork items and are usually fully assembled at the factory. Window units often come with the sash fitted and weather-stripped, the frame assembled, and the exterior casing in place. Fig. 26-1. Standard combination storms and screens or separate units can also be included. All wood components are treated with a water-repellent preservative at the factory to provide protection before and after they are placed in the walls.

Besides letting in light and air, windows are an important part of the architectural design. Fig. 26-2. Generally the glass area of a room should be not less than 10% of the floor area. The window area that can be opened for ventilation should be not less than 4% of the floor area unless a complete air conditioning system is used.

There should be a balance of fixed picture windows and operating windows. An operating window can always be closed to seal out unpleasant weather or opened to a cooling breeze, but a fixed window cannot be opened. Local climate and prevailing winds determine the best window placement and the degree of ventilation required.

The type of window specified in the plans will vary with the room requirements. Not every room will need the same size and type of window. In bedrooms, light and ventilation are a necessity, but privacy and wall space for furniture are also important factors. A row of narrow operating windows placed high on two walls of the room will provide light and ventilation as well as privacy and wall space. In the kitchen, windows should provide good ventilation of cooking odors. For the area over the sink and other hard-to-reach spots, a casement window or awning type that opens with a crank or lever action would be a good choice. The living areas are an ideal location for large picture windows which bring in scenic views.

The window style and size should be such that it is convenient to look through the window whether a person is seated or

26-1. *Parts of an assembled double-hung window: 1. Head flashing. 2. Blind stops. 3. Casing. 4. Sash. 5. Counterbalancing unit. 6. Tracks. 7. Weather stripping. 8. Glazing. 9. Grill (installed on the inside when insulating glass is used). 10. Grill (installed between the glass when storm panels are used). 11. Storm panel.*

26-2. **Windows are an important part of the design in this home.**

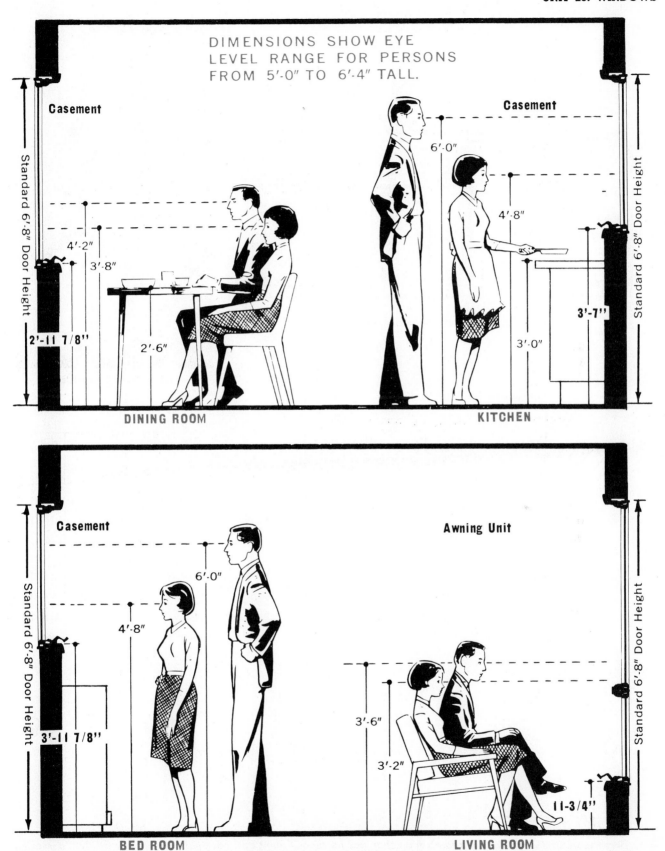

DIMENSIONS SHOW EYE LEVEL RANGE FOR PERSONS FROM 5'-0" TO 6'-4" TALL.

Casement

Casement

Standard 6'-8" Door Height

6'-0"

4'-2"

3'-8"

4'-8"

3'-7"

2'-11 7/8"

2'-6"

3'-0"

Standard 6'-8" Door Height

DINING ROOM

KITCHEN

Casement

Awning Unit

Standard 6'-8" Door Height

6'-0"

4'-8"

3'-11 7/8"

3'-6"

3'-2"

11-3/4"

Standard 6'-8" Door Height

BED ROOM

LIVING ROOM

26-3. *Recommended window heights for various rooms. Note that the header height of the window is standard and corresponds to the 6'8" door height. The sill height, however, will vary with the style and size of the window.*

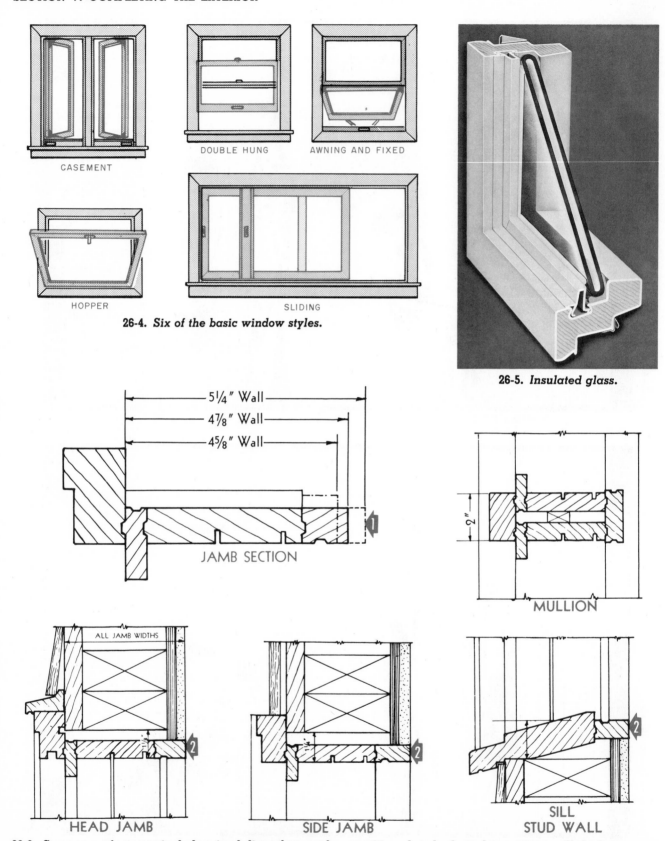

CASEMENT

DOUBLE HUNG

AWNING AND FIXED

HOPPER

SLIDING

26-4. *Six of the basic window styles.*

26-5. *Insulated glass.*

5¼" Wall

4⅞" Wall

4⅝" Wall

JAMB SECTION

2"

MULLION

ALL JAMB WIDTHS

HEAD JAMB

SIDE JAMB

SILL
STUD WALL

26-6. *Some manufacturers include a jamb liner that can be repositioned and adapted to various wall thicknesses as shown at arrow No. 1 in the jamb section. The jamb liner is shown at the No. 2 arrows with the window installed in a framed wall with lath and plaster.*

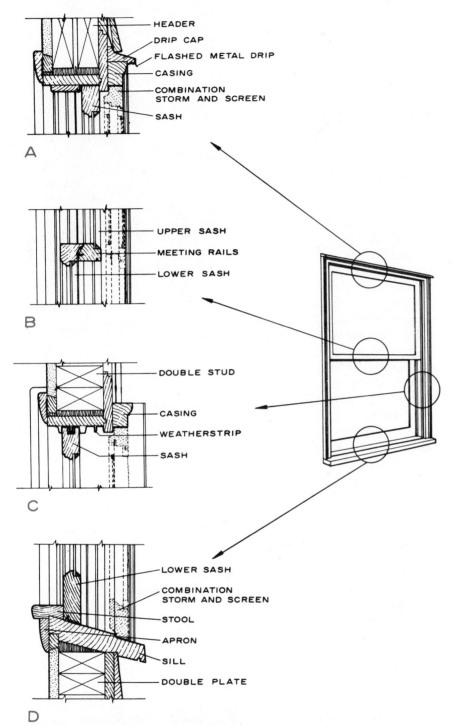

26-7. *Double-hung window. Cross section: A. Head jamb. B. Meeting rails. C. Side jamb. D. Sill.*

- Casement.
- Stationary (fixed).
- Awning.
- Hopper.
- Horizontal-sliding.
- Jalousie.

Glass blocks are sometimes used for admitting light in places where transparency or ventilation is not required.

Windows may be of wood or metal. Heat loss through metal frames and sash is much greater than through similar wood units. On the other hand, metal frames and sash require less maintenance. They also can be made narrower and thus allow larger glass areas.

Wooden window frames and sash should be made from a clear grade of all-heartwood stock of a decay-resistant wood species or from wood which has been given a preservative treatment. Species commonly used include ponderosa and other pines, cedar, cypress, redwood, and spruce. Metal window frames and sash are made of aluminum or steel.

Insulated glass, used for both stationary and movable sash, consists of two or more sheets of glass with air space between the sheets. The edges are hermetically sealed to keep the air in. This type of glass has more resistance to heat loss than a single thickness and is often used without a storm sash. Fig. 26-5.

Window jambs (sides and tops of the frames) must be the same width as the wall section, including the exterior sheathing and the interior finished wall covering. Jambs are made of nominal 1″ lumber; jamb liners are used to adapt the window unit to various wall thicknesses. Fig. 26-6. Sills (bottoms of frame) are made from nominal 2″ lumber and are sloped at about 3 in 12 for good drainage. D, Fig. 26-7. Sash are normally 1⅜″ thick, and wood combi-

standing. Eye-level (seated and standing) height charts for various window types are shown in Fig. 26-3.

TYPES OF WINDOWS

These are the principal types of windows. Fig. 26-4.
- Double-hung.

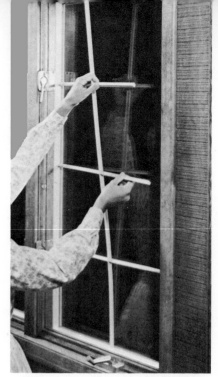

26-8. *Most double-hung windows are designed so that the sash can be easily removed.*

counterbalance. Several types allow the sash to be removed for easy cleaning, painting, or repair. Fig. 26-8.

Sash may be divided into a number of compartments, or lights, by small wood members called muntins. A ranch-type house may provide the best appearance with top and bottom sash divided into two horizontal lights. A Colonial or Cape Cod house usually has each sash divided into six or eight lights. Some manufacturers provide preassembled dividers which snap in place over a single light, dividing it into six or eight lights. This simplifies painting and other maintenance. Fig. 26-9.

Hardware for double-hung windows includes the sash lifts (for opening the window) that are fastened to the bottom rail, although they are sometimes eliminated by providing a finger groove in the rail. Other hardware consists of sash locks or fasteners located at the meeting rails. They not only lock the window, but also draw the sash together to provide a "windtight" fit.

Double-hung windows can be arranged in a number of ways—as a single unit, doubled (or mullion) type, or in groups of three or more. One or two double-hung windows on each side of a large

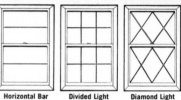

26-9a. *Preassembled dividers of various styles for subdividing window lights are easily snapped into place.*

stationary insulated window are often used to effect a window wall. Fig. 26-10. Such large openings must be framed with headers large enough to carry roof loads.

Casement Windows

Casement windows have side-hinged sash, usually designed to

nation storm and screen windows are usually 1⅛″ thick.

Double-Hung Windows

The double-hung window is perhaps the most familiar window type. It consists of an upper and a lower sash that slide vertically in separate grooves in the side jambs or in full-width metal weather stripping. Fig. 26-7. This type of window provides a maximum face opening for ventilation of one-half the total window area. Each sash is provided with springs, balances, or compression weather stripping to hold it in place in any location. Compression weather stripping, for example, prevents air infiltration, provides tension, and acts as a

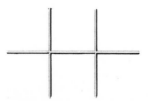

26-9b. *The use of various grill patterns can change the architectural style of a window. A divided-light grill pattern installed in conventional two-light double-hung windows enhances the traditional styling of this house.*

26-10. *Double-hung windows are frequently used in combination with a large stationary window.*

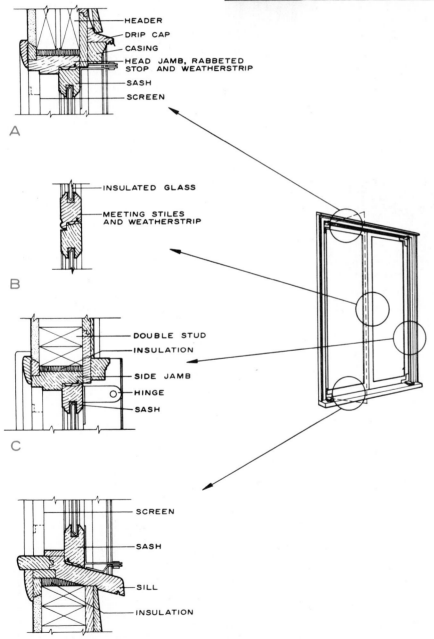

HEADER
DRIP CAP
CASING
HEAD JAMB, RABBETED
STOP AND WEATHERSTRIP
SASH
SCREEN

A

INSULATED GLASS
MEETING STILES
AND WEATHERSTRIP

B

DOUBLE STUD
INSULATION
SIDE JAMB
HINGE
SASH

C

SCREEN
SASH
SILL
INSULATION

D

26-11a. *Casement window. Cross sections: A. Head jamb. B. Meeting stiles. C. Side jamb. D. Sill.*

26-11b. *Wood casement window.*

swing outward because this type can be made more weathertight than the type that swings inward. Fig. 26-11. Screens are located inside these outward-swinging windows, and winter protection is obtained with a storm sash or by using insulated glass in the sash. One advantage of the casement window over the double-hung type is that the entire window area can be opened for ventilation.

Weather stripping is also provided for this type of window, and units are usually received from the factory entirely assem-

26-12a. *Casement window roto-gear operator.*

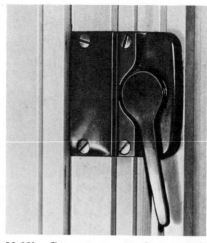

26-12b. *Casement window sash lock.*

bled with hardware in place. Closing hardware consists of a rotary operator and sash lock. Fig. 26-12.

As in the double-hung units, casement sash can be used in a number of ways—as a pair or in combinations of two or more pairs. Style variations are achieved by divided lights. For example, snap-in muntins provide a small multiple-pane appearance for traditional styling.

Stationary Windows

Stationary windows, used alone or in combination with other types of windows, usually consist of a wood sash with a large single light of insulated glass. They are

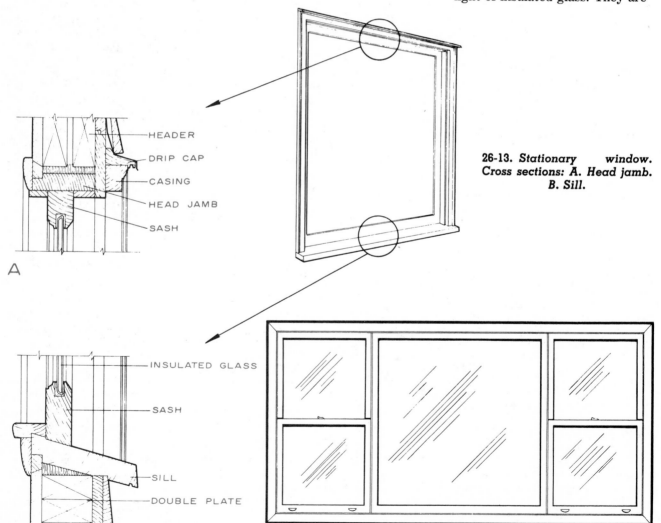

HEADER

DRIP CAP

CASING

HEAD JAMB

SASH

A

INSULATED GLASS

SASH

SILL

DOUBLE PLATE

B

26-13. *Stationary window. Cross sections: A. Head jamb. B. Sill.*

TYPICAL USE IN COMBINATION WITH OTHER TYPES

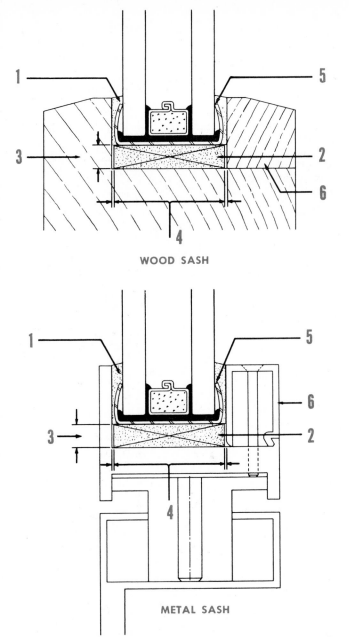

WOOD SASH

METAL SASH

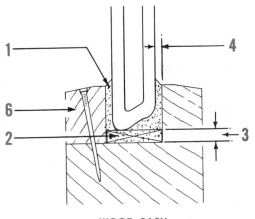

WOOD SASH

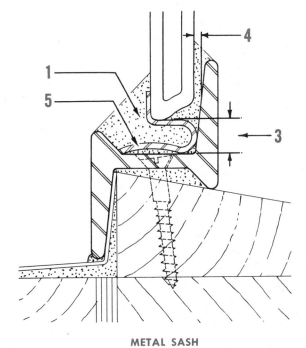

METAL SASH

26-14a. *Putting (glazing) of a metal edge insulated glass unit: 1. Glazing compound. 2. Setting blocks (two, treated wood, 4″ long, located ¼ of the width in from each end of the unit). 3. Edge clearance (⅛″, 3⁄16″, or ¼″ at all edges, depending on the size; distribute proportionately). 4. Lateral clearance (a minimum of 1⁄16″ for glazing compound). 5. Edge coverage (cover the channel with glazing compound for uniform appearance and maximum edge insulation). 6. Glazing stop (must not bear on the unit).*

26-14b. *Glazing of insulated glass with a glass edge: 1. Glazing compound. 2. Setting blocks (two, treated wood, 2″ to 3″ long, located ¼ of the width in from each end of the unit). 3. Edge clearance (avoid glass to metal sash contact, allow ⅛″ at all edges, and distribute proportionately). 4. Lateral clearance (a minimum of 1⁄16″ for glazing compound). 5. Clips (use special clips in metal sash). 6. Glazing stop (do not allow to bear on the unit).*

designed for providing light, as well as for attractive appearance, and are fastened permanently into the frame. Fig. 26-13. Because of their size (sometimes 6 to 8 feet in width) and because of the thickness of the insulating glass, a 1¾″ thick sash is usually used to provide strength.

Stationary windows may also be installed without a sash. The glass is set directly into rabbeted frame members and held in place with stops. As with the window-sash units, back puttying and face puttying of the glass (with or without a stop) will assure moisture-resistance. Fig. 26-14.

265

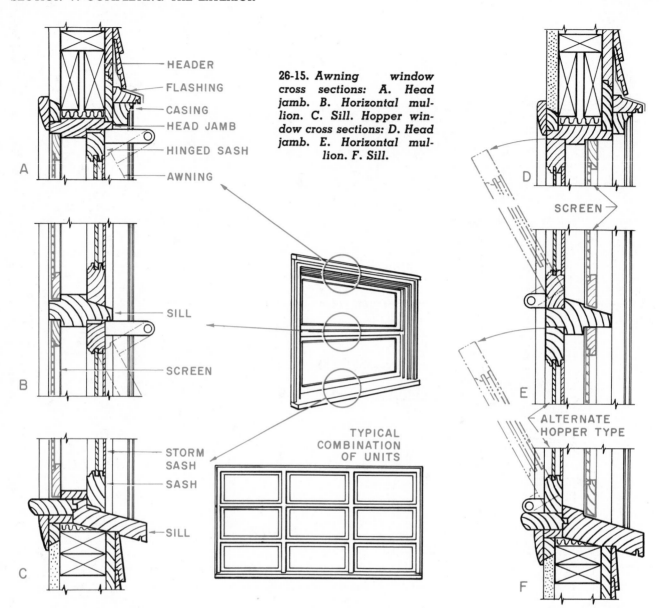

26-15. *Awning window cross sections: A. Head jamb. B. Horizontal mullion. C. Sill. Hopper window cross sections: D. Head jamb. E. Horizontal mullion. F. Sill.*

Labels in figure A: HEADER, FLASHING, CASING, HEAD JAMB, HINGED SASH, AWNING

Labels in figure B: SILL, SCREEN

Labels in figure C: STORM SASH, SASH, SILL

TYPICAL COMBINATION OF UNITS

Labels in figures D, E, F: SCREEN, ALTERNATE HOPPER TYPE

Awning and Hopper Windows

Awning window units have a frame in which one or more operative sash are installed. Fig. 26-15. They often are made up for a large window wall and consist of three or more units in width and height.

Sash of the awning type are made to swing outward at the bottom. A similar unit, the hopper type, is one in which the top of the sash swings inward. Both types provide protection from rain when open.

Jambs are usually 1 1/16" or more thick because they are rabbeted, while the sill is at least 1 5/16" thick when two or more sash are used in a complete frame. Each sash may also be provided with an individual frame so that any combination in width and height can be used. Awning or hopper window units may consist of a combination of one or more fixed sash with the remainder being the operable type. Fig. 26-16. Operable sash are provided with hinges, pivots, and sash supporting arms. Fig. 26-17. There are three types of

operating hardware available for awning windows: the standard push bar, the lever lock, and the rotary gear. Fig. 26-18. Weather stripping and storm sash and screens are usually provided. The storm sash is eliminated when the windows are glazed with insulated glass.

Horizontal-Sliding Window Units

Horizontal-sliding windows appear similar to casement sash. However, the sash (in pairs) slide horizontally in separate tracks or

VERTICAL SECTION

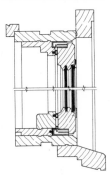

MULLION SECTION

HORIZONTAL SECTION

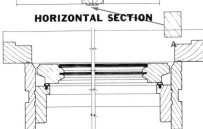

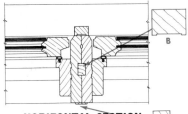

26-16. *In the cross section of the mullion, notice the spline, A, inserted in the groove provided in the side jambs. This aligns the window units when they are used in combination. A mullion strip, B, is then installed on the front edge for trim. In the vertical section, notice that the head and sill jambs also contain grooves in which a spline may be inserted to enable stacking these window units for other possible combinations.*

guides located on the sill and head jamb. Fig. 26-19. Multiple window openings consist of two or more single units and may be used when a window-wall effect is desired. Fig. 26-20. As in most modern window units of all types, weather stripping, water-repellent preservative treatments, and sometimes hardware are included in these fully factory-assembled units.

26-17. *Roto-lock hardware for an awning window. As the window is closed, the bars slide toward the corners of the sash on a track. This provides a tight seal when the sash is closed.*

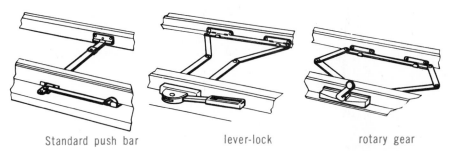

Standard push bar   lever-lock   rotary gear

26-18. *The three types of operating hardware for awning windows.*

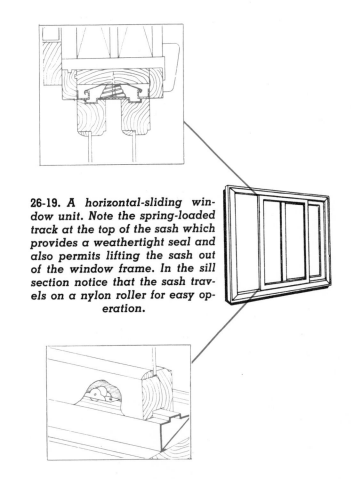

26-19. *A horizontal-sliding window unit. Note the spring-loaded track at the top of the sash which provides a weathertight seal and also permits lifting the sash out of the window frame. In the sill section notice that the sash travels on a nylon roller for easy operation.*

267

26-20. A window wall of horizontal-sliding window units.

26-21. A jalousie window.

26-22. Metal casement sash. Cross sections: A. Head jamb. B. Side jamb. C. Sill. The arrows indicate the wood buck which holds the metal frame in place.

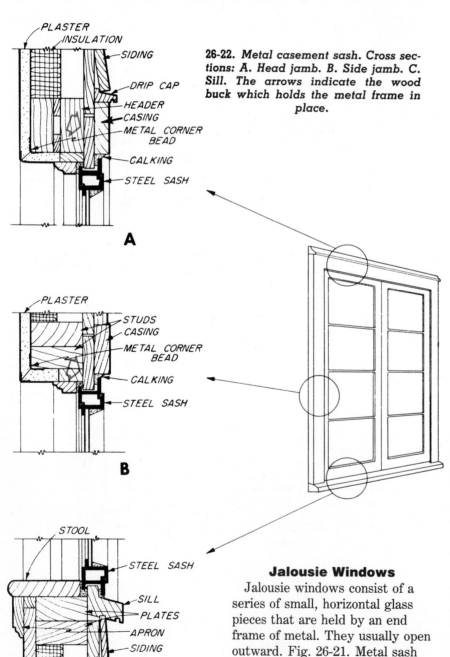

installed carefully to prevent condensation and frosting on the interior surfaces during cold weather. A full storm window unit is sometimes necessary to eliminate this problem in cold climates.

Metal Sash

Metal sash are available in units made of aluminum or steel, and the principal types are casement (Fig. 26-22), double-hung (Fig. 26-23), sliding, and stationary. Lights in the sash are divided in various patterns. The aluminum sash and frames are generally made of solid extruded aluminum alloy, welded at the joints. Steel sash are made of rolled shapes about 1/8" thick, most parts being Z-shaped or T-shaped, with welded butt joints. Steel sash should be treated to make them rust-resistant. Hardware, such as hinges, latches, and operators, is special and provided with the window. Screens should be ordered with the windows.

Actual details for installation vary according to the method of manufacture. It is common practice in frame construction to use a wood buck in the window opening

Jalousie Windows

Jalousie windows consist of a series of small, horizontal glass pieces that are held by an end frame of metal. They usually open outward. Fig. 26-21. Metal sash are sometimes used but, because of low insulating value, should be

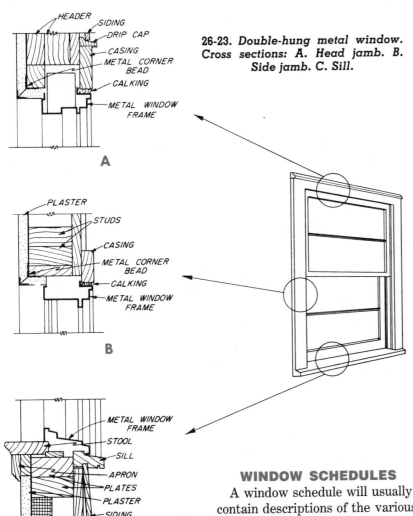

26-23. Double-hung metal window. Cross sections: A. Head jamb. B. Side jamb. C. Sill.

A (labels: HEADER, SIDING, DRIP CAP, CASING, METAL CORNER BEAD, CALKING, METAL WINDOW FRAME)

B (labels: PLASTER, STUDS, CASING, METAL CORNER BEAD, CALKING, METAL WINDOW FRAME)

C (labels: METAL WINDOW FRAME, STOOL, SILL, APRON, PLATES, PLASTER, SIDING, INSULATION)

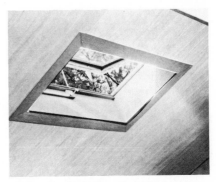

26-24. A skylight installed on a pitched roof.

to hold the metal frame. Fig. 26-22. The space between the metal and the wood is filled with calking compound.

Skylights

Skylights are installed on either pitched or flat roofs for ventilation and light. The dome is double plastic for insulation. Rather than building a dormer on a room directly under a roof and enclosed with interior walls, a skylight can be installed. This is a good solution to the problem of light in a remodeled attic also. Fig. 26-24.

WINDOW SCHEDULES

A window schedule will usually contain descriptions of the various windows, plus sash openings, glass sizes, and sometimes the rough opening sizes. The location of each window in a house is found by matching the number of the window in the window schedule with the corresponding number on the house plan. Fig. 26-25 (page 270).

Figuring Rough Opening Sizes

When the rough opening size is not provided, it will have to be figured by the builder or obtained from the window manufacturer's catalog. Tables showing glass size, sash size, and rough opening size for windows from various manufacturers are available from suppliers. In this book, typical openings for double-hung win-

dows are shown in Unit 14, "Wall Framing."

The rough opening size can also be figured if the glass size is known. Make the rough opening at least 6″ wider and 10″ higher than the window glass size. In specifying a window, the width of the glass is always given first, then the height, then the number of pieces of glass (or lights) and the window style. For example, 28½″ × 24″, 2 lights D. H. means that the glass itself is 28½″ wide and 24″ high and that there are two pieces of glass in a double-hung unit.

To figure the rough opening width, add 6″ to the given width: 28½″ + 6″ = 34½″, or 2′10½″. To obtain the rough opening height, add the upper and lower glass height together, and then add another 10″: 24″ + 24″ + 10″ = 58″, or 4′10″. These allowances are fairly standard and provide for the weights, springs, balances, room for plumbing and squaring, and for the normal adjustments. However, when the window manufacturer is known, use his recommended rough opening sizes.

The rough opening sizes vary slightly among manufacturers, as can be seen by comparing sample tables from two typical manufacturers' catalogs designating the sizes of standard units. Fig. 26-

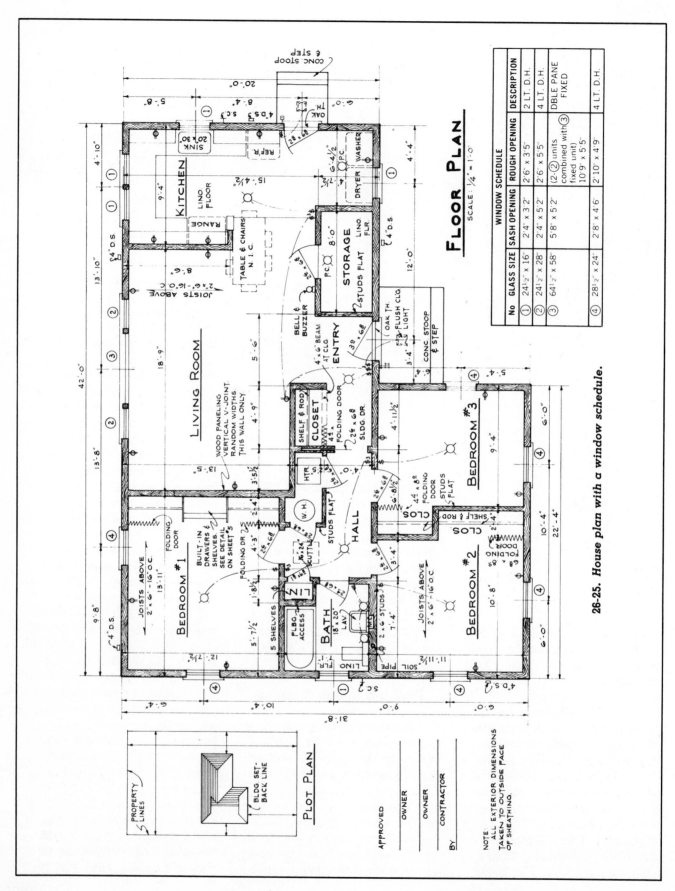

Floor Plan

SCALE: 1/4" = 1'-0"

		WINDOW SCHEDULE		
No	GLASS SIZE	SASH OPENING	ROUGH OPENING	DESCRIPTION
1	24½" x 16"	2'4" x 3'2"	2'6" x 3'5"	2 LT. D.H.
2	24½" x 28"	2'4" x 5'2"	2'6" x 5'5"	4 LT. D.H.
3	64½" x 58"	5'8" x 5'2"	(2-2 units combined with ③ fixed unit) 10'9" x 5'5"	DBLE PANE FIXED
4	28½" x 24"	2'8" x 4'6"	2'10" x 4'9"	4 LT. D.H.

26-25. House plan with a window schedule.

Window Sizes

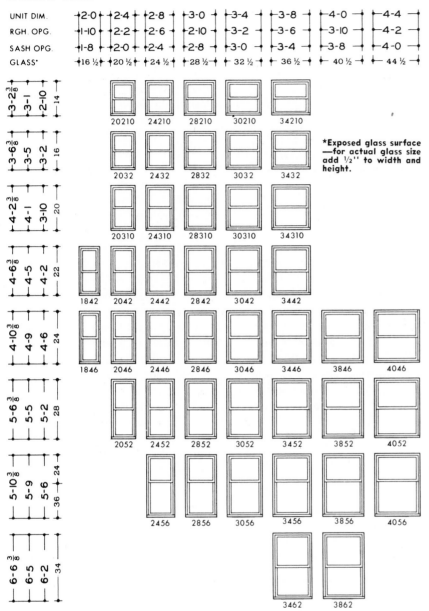

UNIT DIM.	2-0	2-4	2-8	3-0	3-4	3-8	4-0	4-4
RGH. OPG.	1-10	2-2	2-6	2-10	3-2	3-6	3-10	4-2
SASH OPG.	1-8	2-0	2-4	2-8	3-0	3-4	3-8	4-0
GLASS*	16½	20½	24½	28½	32½	36½	40½	44½

*Exposed glass surface —for actual glass size add ½'' to width and height.

26-26. A manufacturer's table for double-hung windows.

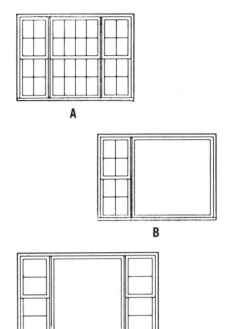

A

B

C

26-27. Typical combinations for contemporary or traditional treatment with grills and mullions.

26. The window schedule in Fig. 26-25 indicates a glass size of 28½″ × 24″ for window No. 4. In Fig. 26-26, read the width of the glass size across the top (28½″) and then follow down that column to the glass length (24″). The window is designated 2846, which is the manufacturer's catalog number. Note, in the case of this particular manufacturer, the number is also the sash opening. Window 2846 has a sash opening width of 2′8″ and a length of 4′6″. The rough opening given on this chart is 2′10″ × 4′9″ for a framed wall. (The unit size, which is the masonry opening for a brick veneer wall, is also given on these tables.) The rough opening figured earlier (2′10½″ × 4′10″) is larger than the manufacturer's recommended opening. Using the method for figuring a rough opening will insure getting the window unit into the opening, but it may require additional blocking or shimming. Therefore, it is always best to use the specific manufacturer's recommendations for a rough opening size.

FIGURING ROUGH OPENINGS FOR COMBINATION UNITS

Many times, window units of various styles and sizes are combined to make up larger units for a particular room and use. These combined units are separated only by vertical piers called mullion strips. Fig. 26-27. Therefore, the rough opening for the combined unit will be less than the total of the rough openings for the units if they were used individually.

In the house plan in Fig. 26-25, note the window schedule on the plan calls for a combination unit in the living room consisting of

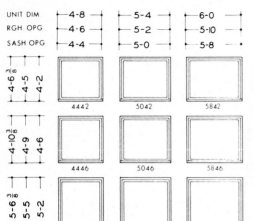

PICTURE WINDOWS

UNIT NO.	1″ GLASS RABBET	1″ GLASS SIZE
4442	49 x 46½	48½ x 46
5042	57 x 46½	56½ x 46⅛
5842	65 x 46½	64½ x 46
4446	49 x 50½	48½ x 50
5046	57 x 50½	56½ x 50
5846	65 x 50½	64½ x 50
4452	49 x 58½	48½ x 58
5052	57 x 58½	56½ x 58⅛
5852	65 x 58½	64½ x 58

26-28. *A manufacturer's table for stationary windows.*

two window units No. 2 and one window unit No. 3. If this unit is to be made up to look like C in Fig. 26-27, figure the width of the multiple opening by adding the individual sash openings to the width of the mullions, plus 2″ for the overall rough opening. In this example the sash opening for the

No. 2 unit in the window schedule is 2′4″. The sash opening for the No. 3 unit is 5′8″. Figs. 26-25 & 26-28. Since there are two No. 2 units, add 2′4″ twice to the 5′8″: 2′4″ + 2′4″ + 5′8″ = 9′16″, or 10′4″. There are two mullions in our example. If each mullion is 1½″ wide, the total mullion width would be 3″. Add this to the sash width and then add 2″ for the

rough opening: 10′4″ + 3″ + 2″ = 10′9″. Note that this is the rough opening width listed for the combination window unit (No. 3) in the window schedule. Fig. 26-25. The rough opening height is figured the same way for combination units as for individual units.

INSTALLATION

Window frames are generally assembled in a mill. The preassembled window frame is easily installed, but care should be taken. Regardless of the quality of the window purchased, it is only as good as its installation. Before actually installing a window, apply a primer coat of paint to all wood members to prevent undue warpage.

The general procedure for the installation of a window frame is very similar regardless of style or manufacturer. Fig. 26-29. However, always refer to the manufacturer's instructions for any specific recommendations. For example, some manufacturers recommend that the sash be removed from the frame to prevent

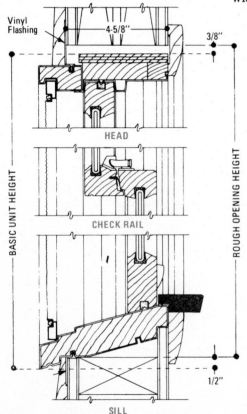

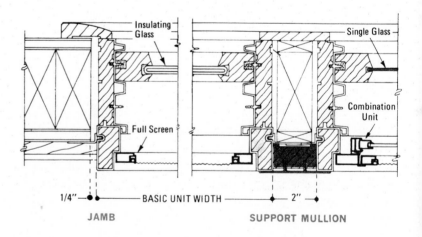

26-29a. *Installation details for double-hung windows.*

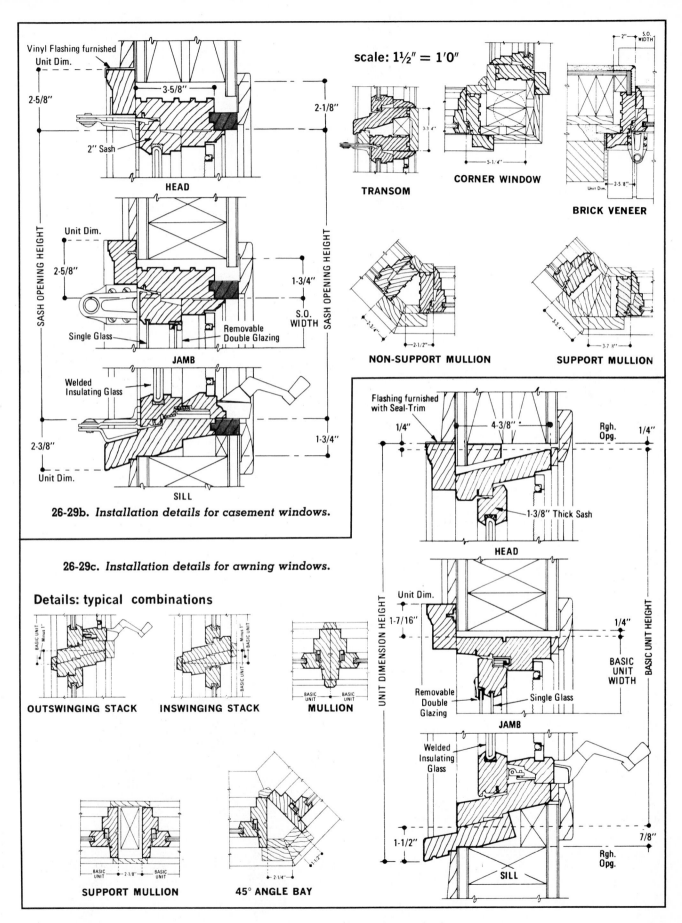

scale: 1½" = 1'0"

HEAD

TRANSOM

CORNER WINDOW

BRICK VENEER

JAMB

NON-SUPPORT MULLION

SUPPORT MULLION

SILL

26-29b. *Installation details for casement windows.*

26-29c. *Installation details for awning windows.*

Details: typical combinations

OUTSWINGING STACK

INSWINGING STACK

MULLION

SUPPORT MULLION

45° ANGLE BAY

HEAD

JAMB

SILL

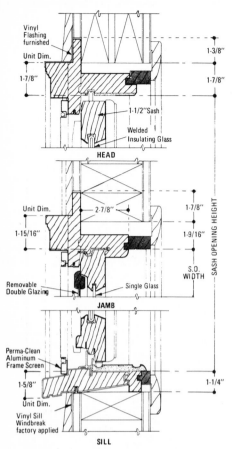

26-29d. *Installation details for horizontal-sliding windows.*

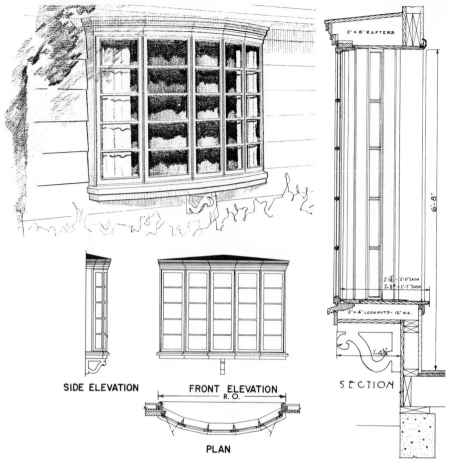

SIDE ELEVATION FRONT ELEVATION
R.O.

PLAN

SECTION

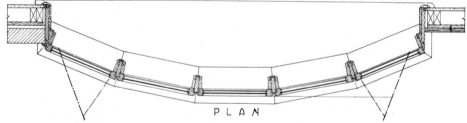

26-29e. *Installation details for a casement bow window.*

26-30. *These windows have been installed with the sash removed for easier handling.*

breakage and provide easier handling of the unit. Fig. 26-30. Others specify not only that the sash be left in the frame but also that diagonal braces and, in some cases, reinforcing blocks be left in place to insure that the frame remains square and in proper alignment.

When a siding material is applied over sheathing, the windows are installed first and the siding applied later. Strips of 15-pound asphalt felt should be put over the sheathing around the openings. Fig. 26-31.

Place the frame in the opening from the outside, allowing the subsill to rest on the rough frame at the bottom, and hold the unit up tightly against the building. Fig. 26-32a. Level and plumb the window frame, then wedge with shingles and tack in place. Fig.

274

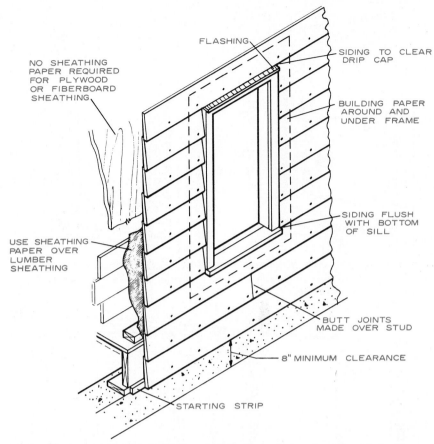

NO SHEATHING PAPER REQUIRED FOR PLYWOOD OR FIBERBOARD SHEATHING

FLASHING

SIDING TO CLEAR DRIP CAP

BUILDING PAPER AROUND AND UNDER FRAME

SIDING FLUSH WITH BOTTOM OF SILL

USE SHEATHING PAPER OVER LUMBER SHEATHING

BUTT JOINTS MADE OVER STUD

8" MINIMUM CLEARANCE

STARTING STRIP

26-31. Window frame installation details for a framed wall with horizontal lap siding.

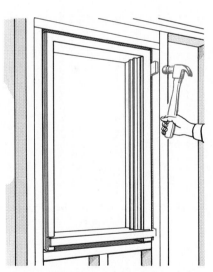

26-32c. When the diagonal measurements are equal, the unit is square.

26-32a. Install the window frame in the rough opening.

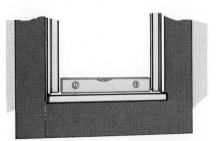

26-32b. Level the sill, wedge the frame with shingles, and tack it in place.

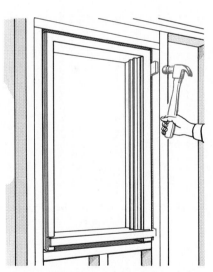

through the outside casings; or, if blind stops are provided, use 8d common nails. Fig. 26-32f-i. (When window frames are furnished with blind stops for installation, the sheathing should be installed 1½" back from the window rough openings. Fig. 26-33 and 26-34.) The nails should be spaced about 12" apart. They should penetrate the sheathing and/or the

26-32d. Shim the side jambs and recheck the diagonals.

trimmer studs and the header over the window. While nailing, open and close the sash to see that it works freely. The side and head casings are fastened in the same manner.

When a panel siding is used

26-32b. Check the sill and jamb with the level and square. Fig. 26-32c-e. When everything is in order, use 16d galvanized nails

26-32e. *Measure the distance between the side jambs to be sure they are equidistant at all points. Shim as necessary to correct.*

26-32h. *Install the sash. If the unit is out of square, the meeting rails will not be parallel.*

26-32i. *Pack insulation between the jambs and the trimmer studs.*

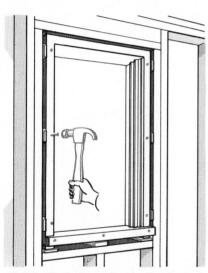

26-32f. *Nail the side jambs through the shims.*

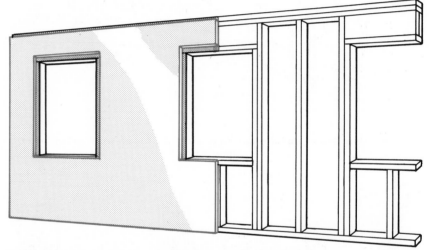

26-33a. *To accomodate the installation of a window unit with wood blind stops, the sheathing is nailed 1½" back from the rough opening.*

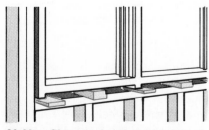

26-32g. *Shim under the raised jamb legs and at the center of long sills and mullions.*

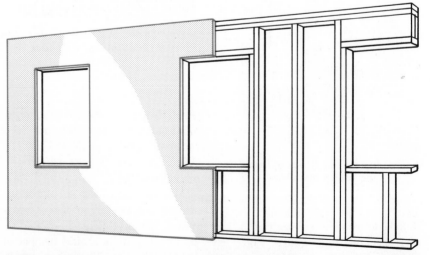

26-33b. *For the installation of windows that are nailed in place through the casing, the sheathing is nailed even with the inside of the rough opening.*

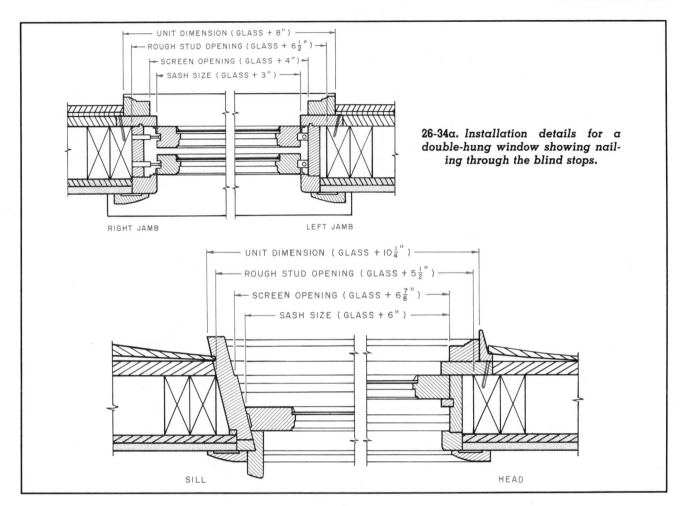

26-34a. *Installation details for a double-hung window showing nailing through the blind stops.*

26-34b. **Some window units are available with a vinyl covering to eliminate painting. A vinyl anchorage flange and windbreak then serves as a blind stop (see arrows).**

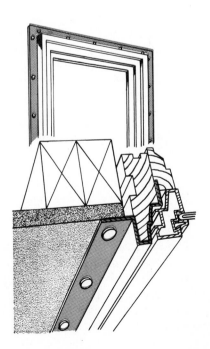

without sheathing, the windows are usually installed after the siding is in place. Before installing the window, place a ribbon of calking sealant over the siding at the location of side and head casings. Fig. 26-35. Install the win-

26-34c. *The window is installed with 1¾" galvanized nails through the vinyl anchoring flange. The outside wall covering is applied over this flange. With this method, there are no exposed nails on the exterior of the window.*

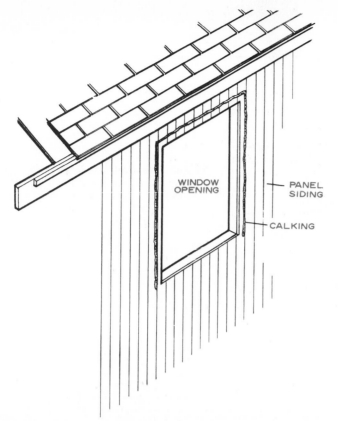

WINDOW OPENING

PANEL SIDING

CALKING

26-35. *Calk around the window opening at the location of the head and side casings before installing the window frame over panel siding.*

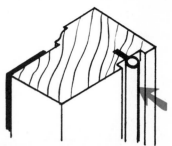

26-37. *A built-in calking on the back of window casings is sometimes provided by the manufacturer.*

framing is then constructed on the foundation wall with the window frames already set in place. The sills are usually installed later.

WINDOW SCREENS AND STORM PANELS

Window screens and storms may be an integral part of the window frame, or they may be separate units of wood, metal, or

dow as described earlier. Then place some calking at the junction of the siding and the sill and install a small molding such as a quarter-round. Fig. 26-36. If required, place metal flashing over each window.

On the back of the casing of some windows the manufacturer provides a built-in calking. This is a bulb-type vinyl that remains pliable and provides a weathertight fit. Fig. 26-37.

Basement Windows

Basement window units are made of wood, plastic, or metal. Figs. 26-38 & 26-39. In most cases the sash is removed from the frame. The frame is set into the concrete forms for a poured wall, and the wall is poured with the window frame in place. If the windows are to be set into a concrete block wall, special blocks are available to accommodate the various types of frames. The floor

26-36. *Installing a double-hung window frame over panel siding. This frame is installed by nailing through the casing into the studs. Note the molding under the sill.*

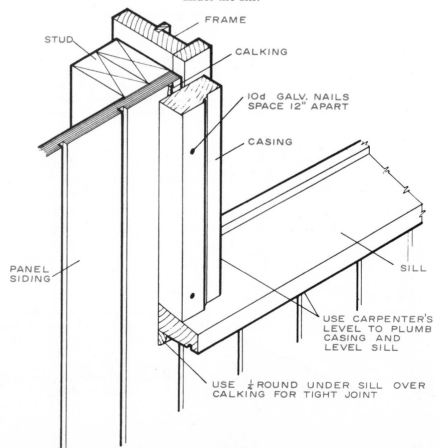

FRAME

STUD

CALKING

10d GALV. NAILS SPACE 12" APART

CASING

PANEL SIDING

SILL

USE CARPENTER'S LEVEL TO PLUMB CASING AND LEVEL SILL

USE ¼ ROUND UNDER SILL OVER CALKING FOR TIGHT JOINT

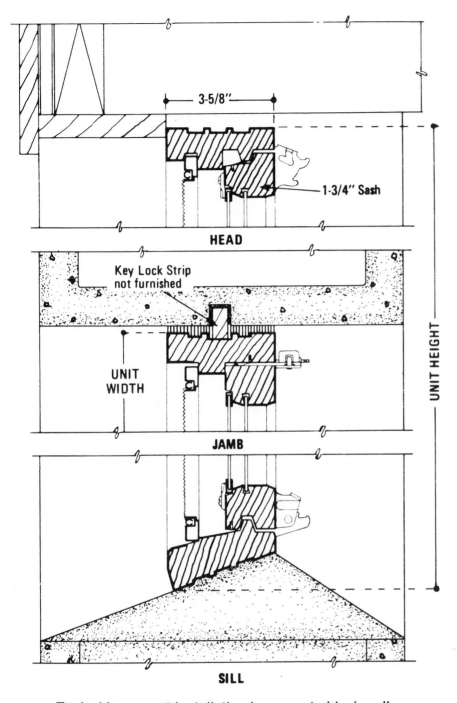

HEAD

3-5/8"

1-3/4" Sash

Key Lock Strip
not furnished

UNIT
WIDTH

JAMB

UNIT HEIGHT

SILL

Typical basement installation in concrete block wall.

26-38a. *Installation details for a wood basement window unit.*

SIZES

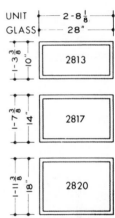

	UNIT	2-8 1/8
	GLASS	28"

1-3 3/8 / 10"		2813
1-7 3/8 / 14"		2817
1-11 3/8 / 18"		2820

26-38b. *Some wood basement window units are dual hinged so that they may be opened from either the top or the bottom. The insert shows typical sizes of wood basement window units.*

plastic. Fig. 26-40. For double-hung windows, separate units are designed so that the screens and storm panels may be stored within the unit. In this type the lower sash is usually the only one that is screened. The unit is pro-vided with three tracks; the up-per window remains in the upper position, the lower window may be slid up out of the way and the screen brought down for warm weather. For cold weather the screen is stored in the upper posi-tion, and the storm panel is low-ered.

Casement, awning, hopper, and sliding window units have the storm panel set into a rabbet in the outside of the window sash. The screen is a separate insert which is installed for the summer months. Screens used with dou-ble-hung, sliding, and hopper win-dows are installed on the outside. For casement and awning type windows that open out, the screen is installed on the inside

PUTTYLESS GLAZING DETAILS

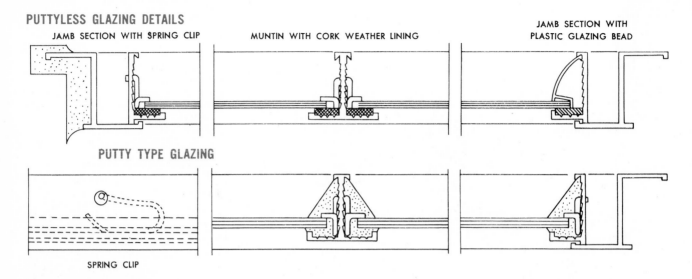

JAMB SECTION WITH SPRING CLIP   MUNTIN WITH CORK WEATHER LINING   JAMB SECTION WITH PLASTIC GLAZING BEAD

PUTTY TYPE GLAZING

SPRING CLIP

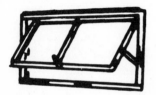

26-39. *Installation details for a metal basement window unit.*

WINDOW WITH SCREEN SHOWING INSTALLATION DETAILS

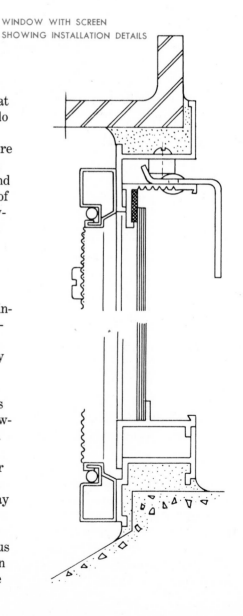

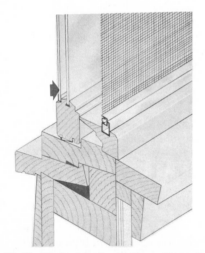

26-40. *If double-insulating glass is not used, a storm panel (arrow) can be installed as part of the window frame on most window styles.*

with a projecting handle for opening and closing the window sash.

WINDOW CONDENSATION

Condensation (formation of moisture) on windows is the result of improved heating systems, better insulation, and houses that are tightly built. Homes today do not breathe through the many small openings that existed before the extensive use of storm sash, weather stripping, insulation, and automatic heating systems. All of these improvements are fuel savers and add to people's comfort, but they do promote annoying and damaging condensation and steps should be taken to prevent it.

Condensation on the inside window surfaces results from differences in outside and inside temperatures and from the humidity conditions inside the home. Warm, humid air in the interior of the home, when temperatures are cold outside, reaches the dew-point necessary to condensation. Keeping the relative humidity within the home at a point lower than necessary for condensation to occur is the most effective way of preventing condensation on windows. The recommended indoor relative humidity for various outside temperatures is shown in Fig. 26-41. These maximum safe

With Relative Humidity Conditions as shown in this Table there will be no Condensation

Outside temperature as shown			Inside temperature 70°F	
			Relative Humidity	
Below		− 20°F	Not over	15%
− 20	to	− 10°F	" "	20%
− 10	"	− 0°F	" "	25%
− 0	"	10°F	" "	30%
10	"	20°F	" "	35%
Above		20°F	" "	40%
It is important to prevent excess humidity				

26-41. *Recommended indoor relative humidity for various outside temperatures.*

humidities for the home are not only better for the windows, but they will also improve paint performance and insulation and will eliminate problems with structural members.

There are three ways to reduce humidity:
• Controlling sources of humidity. For instance, venting all gas burners and clothes dryers to the outdoors and using kitchen and bathroom exhaust fans helps remove excess moisture from the air.
• Winter ventilation of homes. Because outside air usually contains less water vapor, it will "dilute" the humidity of inside air. This takes place automatically in older houses through constant infiltration of outside air.

• Proper heating. Dry heat will reduce the relative humidity. It will counterbalance most or all of the moisture produced by modern living.

Fog on the lower corners of windows now and then is not serious. However, *excessive* condensation, condensation that blocks entire windows with fog or frost and produces water droplets, can stain woodwork and in some cases, even damage the wallpaper or plaster. Condensation on windows is easily seen and can be removed. More serious is excessive moisture in the walls and insulation, where it cannot be seen. High humidity resulting in condensation can contribute greatly to the deterioration of a house and to the discomfort of its occupants.

QUESTIONS

1. What are the principal types of windows?
2. How does a casement window differ from an awning window?
3. What is the main purpose of a stationary window?
4. What is a jalousie window?
5. Where is a window schedule usually found?
6. What information is contained in a window schedule?
7. What type of window installation requires that the sheathing be installed 1½" back from the window rough opening?
8. Why are basement windows the first windows to be installed in a house?

Exterior Doors and Frames 27

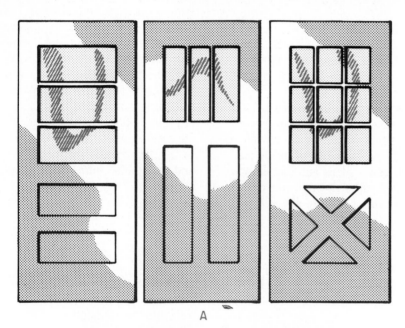

A

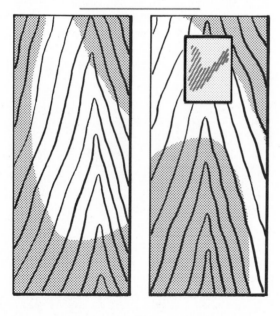

B

27-1. Wood doors are available in a large number of styles: A. Three types of panel doors. B. Two types of flush doors.

Exterior doors are made from wood or metal and are available in several styles. Figs. 27-1 & 27-2. Care should be taken to select a door that is correct for the architectural style of the house. The exterior trim around the main entrance door can vary in architectural design from a simple casing to a molded or plain pilaster with a decorative head casing. Decorative designs should always be in keeping with the architecture of the house. Fig. 27-3. Many combinations of door and entry designs for every kind of house are available along with millwork items which are adaptable to many styles. Figs. 27-4 & 27-5. For a house with an entry hall having no windows, it is usually desirable to have glass in the main exterior door.

TYPES OF EXTERIOR DOORS

Flush Doors

Flush doors are made with plywood or other suitable facing applied over light framework onto a core of suitable thickness. Figs. 27-6 & 27-7 (page 286). There are two types of cores: hollow and solid. Fig. 27-8. Solid core construction is woodblock or particleboard and is generally preferred for exterior doors. Solid core construction minimizes warping, particularly in cold climates where differences in humidity occur on opposite sides of a door.

27-2. One manufacturer of prehung steel doors offers a flush and three panel door styles which can be adopted to various architectural styles.

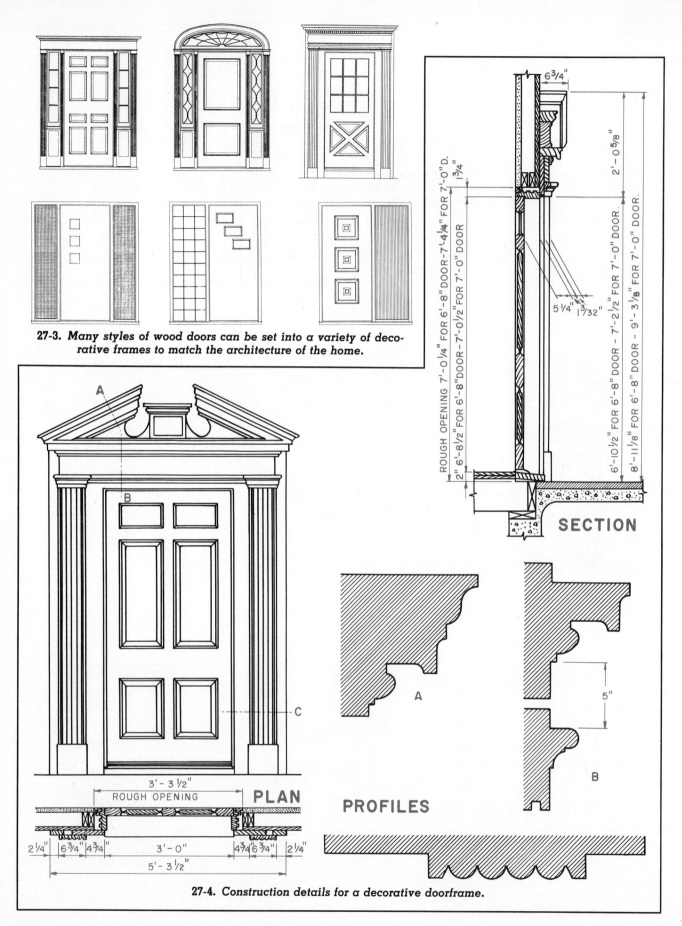

27-3. Many styles of wood doors can be set into a variety of decorative frames to match the architecture of the home.

SECTION

ROUGH OPENING 7'-0 1/4" FOR 6'-8" DOOR-7'-4 1/4" FOR 7'-0" D.

1 3/4"

6 3/4"

2'-0 5/8"

2'-6-8 1/2" FOR 6'-8"DOOR-7'-0 1/2" FOR 7'-0" DOOR

5 1/4" 1 3/32"

6'-10 1/2" FOR 6'-8" DOOR - 7'-2 1/2" FOR 7'-0" DOOR

8'-11 1/8" FOR 6'-8" DOOR - 9'-3 1/8" FOR 7'-0" DOOR.

PLAN

3'-3 1/2"
ROUGH OPENING

2 1/4" | 6 3/4" 4 3/4" | 3'-0" | 4 3/4" 6 3/4" | 2 1/4"

5'-3 1/2"

A

B

C

PROFILES

A

B

5"

27-4. Construction details for a decorative doorframe.

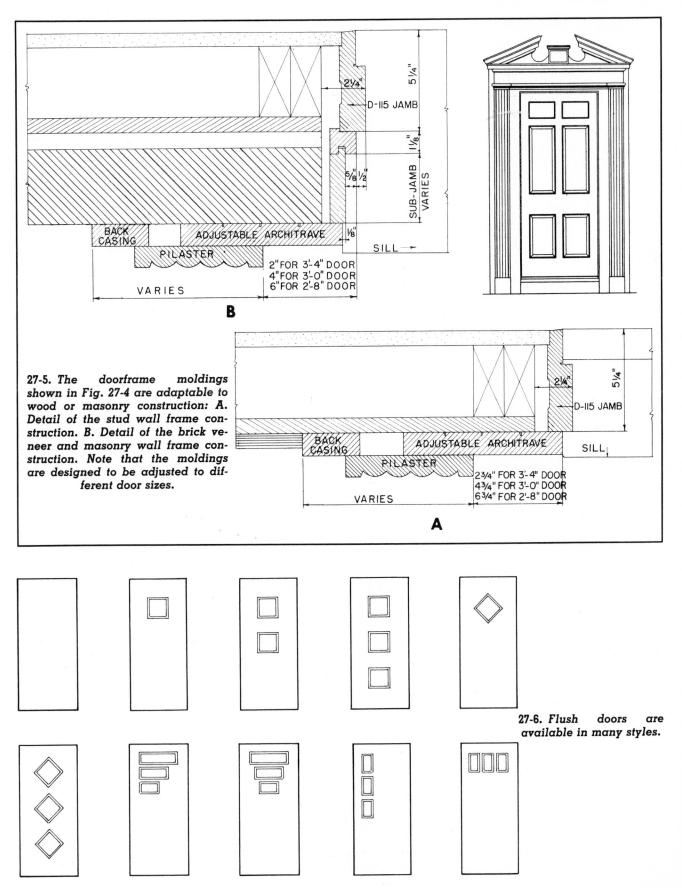

27-5. The doorframe moldings shown in Fig. 27-4 are adaptable to wood or masonry construction: A. Detail of the stud wall frame construction. B. Detail of the brick veneer and masonry wall frame construction. Note that the moldings are designed to be adjusted to different door sizes.

B

2¼"
D-115 JAMB
5¼"
1⅛"
SUB-JAMB VARIES
⅝" ½"
⅛"
BACK CASING
ADJUSTABLE ARCHITRAVE
PILASTER
SILL
2"FOR 3'-4" DOOR
4"FOR 3'-0" DOOR
6"FOR 2'-8" DOOR
VARIES

A

2¼"
D-115 JAMB
5¼"
BACK CASING
ADJUSTABLE ARCHITRAVE
PILASTER
SILL
2¾" FOR 3'-4" DOOR
4¾" FOR 3'-0" DOOR
6¾" FOR 2'-8" DOOR
VARIES

27-6. Flush doors are available in many styles.

285

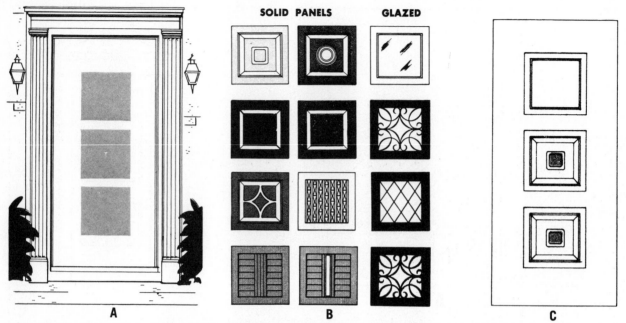

27-7. Flush doors can be individually styled by selecting from a variety of insert panels. The panels in B can be mounted on the shaded areas of the door at A. C shows one possibility.

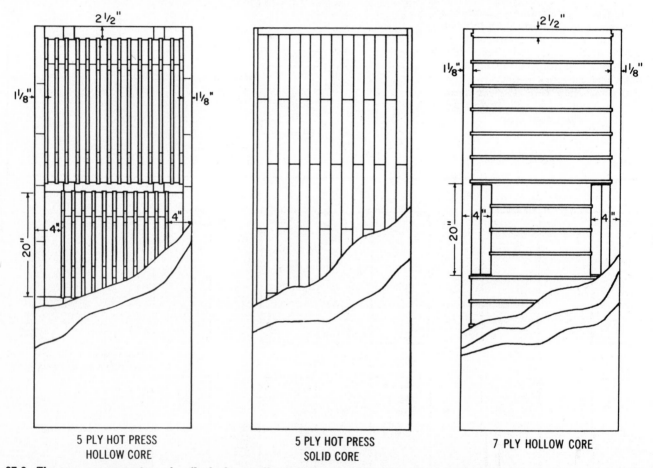

5 PLY HOT PRESS
HOLLOW CORE

5 PLY HOT PRESS
SOLID CORE

7 PLY HOLLOW CORE

27-8. The core construction of a flush door will vary considerably with the manufacturer. The construction details shown here are an example of the techniques used by one manufacturer. Note the built-up areas at the edges near the center for installation of the lock set.

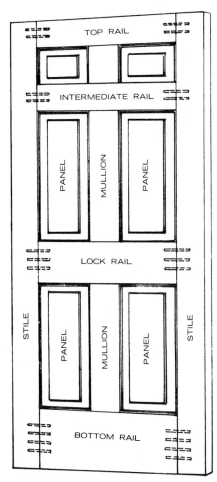

27-9. Parts of a six-panel door.

27-10. Panel doors are available in many styles.

Panel Doors

Panel doors consist of stiles (solid vertical members), rails (solid cross-members), and panels (thinner parts filling spaces between the stiles and the rails). Fig. 27-9. Many types with various wood or glass panels are available. Fig. 27-10.

Glazed Doors

French, or glazed, doors consist of stiles and rails with a space divided into lights by bars called muntins. Such doors are often hung in openings leading to porches or terraces. French doors may be hung singly or in pairs with a half-round molding stop between them. Fig. 27-11.

Sliding Glass Doors

Sliding glass doors are available with either wood or metal frames. Fig. 27-12. The glass may be 1″, ⅝″, or ¼″ thick insulating plate glass, depending on the local climate. These units are available in various combinations of stationary or operating doors in widths from

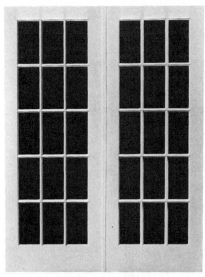

27-11. French doors.

27-12. A wood siding glass door with regular muntins.

30" to 120". The door operation may be specified as right- or left-hand sliding (as viewed from the outside). Fig. 27-13. Snap-in muntins can be added to create a traditional appearance. Fig. 27-14.

Combination Doors

Combination storm and screen doors of wood or metal are available in several styles. Fig. 27-15. Panels which include screen and storm inserts are normally located in the upper portion of the door. Some types have self-storing features similar to window combination units. Heat loss through metal combination doors is greater than through similar wood doors. Weather-stripping an exterior door will reduce both air infiltration and frosting of the glass on the storm door during cold weather.

DOORFRAMES

A doorframe surrounds a door to conceal or beautify structural building parts. The doorframe consists of the doorjamb, the sill, interior trim, exterior trim, and other molding, depending on the architectural design of the building. Figs. 27-16 & 27-17.

The doorjamb is the part of the frame which fits inside the masonry opening or rough frame opening. Jambs may be wood or metal. Wood has been the traditional material, but steel and aluminum have gained in popularity and are not uncommon in residential building. The jamb has three parts: the two side jambs and the head jamb across the top. Exterior doorjambs have a stop as part of the jamb. The stop is the portion of the jamb which the face of the door closes against. The jamb is 1⅛" thick with a ½" rabbet serving as a stop.

Wood jambs are manufactured in two standard widths: 5¼" for lath and plaster and 4½" for dry

WOOD DOOR SIZES

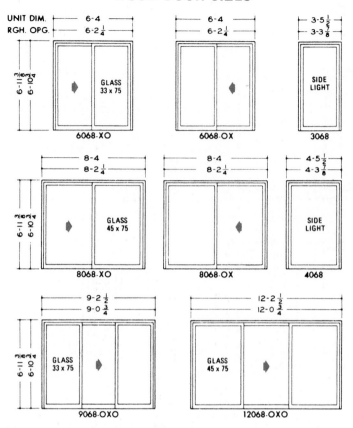

NUMBERING SYSTEM

No. 30, 40, 60, 80, 90 & 120—Unit Width
No. 68—Unit Height
X—Operating Door

O—Stationary Door
(Numbering figured as viewed from outside)

27-13a. A table from a manufacturer's catalog illustrating wood sliding glass doors.

wall. Jambs may easily be cut to fit walls of less thickness. If the jamb is not wide enough, strips of wood are nailed on the edges to form an extension. Jambs may also be custom-made to any size to accommodate various wall thicknesses.

Standard metal jambs are available in the following widths for lath and plaster, concrete block, brick veneer, etc.: 4¾", 5¾", 6¾", and 8¾". For dry-wall construction the common widths available are 5½" and 5⅝".

The sill is the bottom member in the doorframe. It is usually made of oak for wear resistance. When softer wood is used for the sill, a metal nosing and wear strips are included.

The brick mold or outside casings are designed and installed to serve as stops for the screen or combination door, which is 1⅛" thick. The stops are provided for by the edge of the jamb and the exterior casing thickness. Fig. 27-16.

Doorframes may be purchased knocked down (K.D.) or preassembled with just the exterior casing or brick mold applied. In some cases, they come preassem-

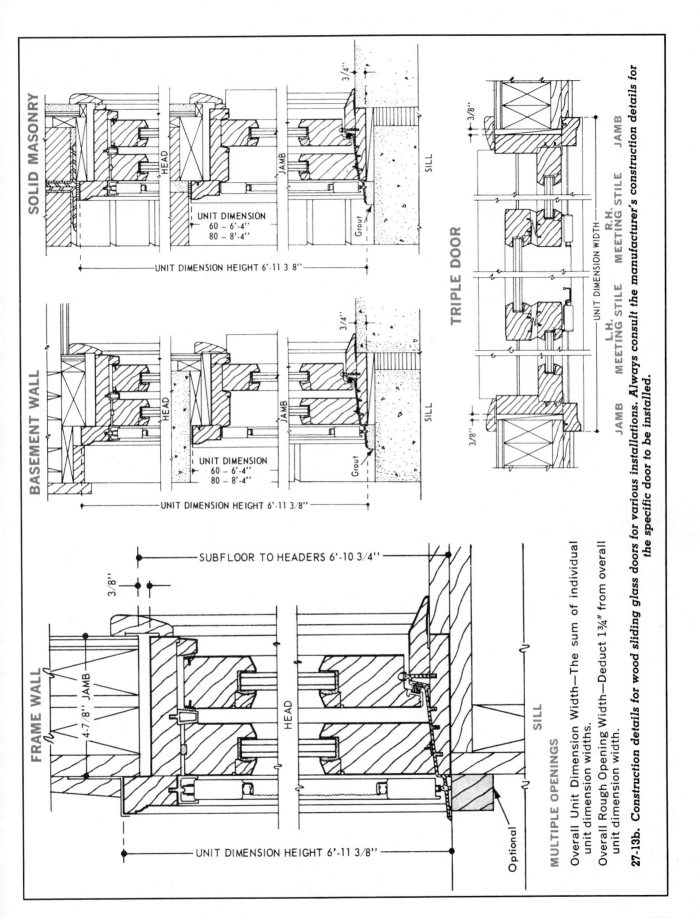

SOLID MASONRY

HEAD

UNIT DIMENSION
60 – 6'-4''
80 – 8'-4''

JAMB

3/4''

SILL

Grout

UNIT DIMENSION HEIGHT 6'-11 3 8''

BASEMENT WALL

HEAD

UNIT DIMENSION
60 – 6'-4''
80 – 8'-4''

JAMB

3/4''

SILL

Grout

UNIT DIMENSION HEIGHT 6'-11 3/8''

TRIPLE DOOR

3/8''

3/8''

UNIT DIMENSION WIDTH

JAMB   L.H.   R.H.   JAMB
MEETING STILE   MEETING STILE

FRAME WALL

SUBFLOOR TO HEADERS 6'-10 3/4''

3/8''

4-7/8'' JAMB

HEAD

SILL

Optional

UNIT DIMENSION HEIGHT 6'-11 3/8''

Construction details for wood sliding glass doors for various installations. Always consult the manufacturer's construction details for the specific door to be installed.

MULTIPLE OPENINGS

Overall Unit Dimension Width—The sum of individual unit dimension widths.

Overall Rough Opening Width—Deduct 1¾'' from overall unit dimension width.

27-13b. Construction details for wood sliding glass doors.

REGULAR MUNTIN ARRANGEMENTS

33" glass   45" glass   57" glass

DIAMOND MUNTIN ARRANGEMENTS

33" glass   45" glass   57" glass

27-14. A variety of muntins are available for different glass sizes.

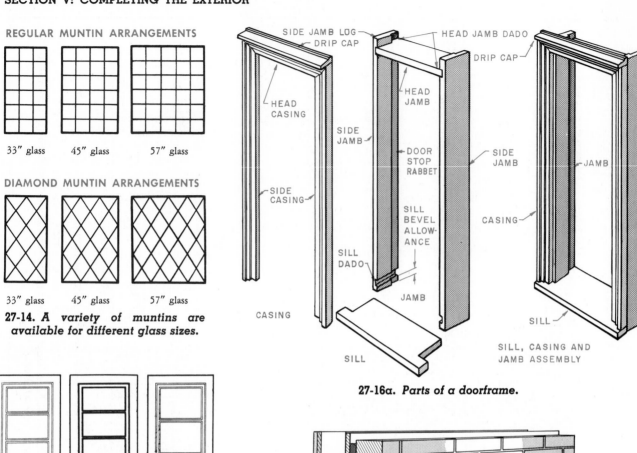

27-16a. Parts of a doorframe.

27-15. Combination doors are made in many styles.

bled with the door hung in the opening. Fig. 27-18. When the doorframe is assembled on the job, nail the side jambs to the head jamb and sill with 10d casing nails. Then nail the casings to the

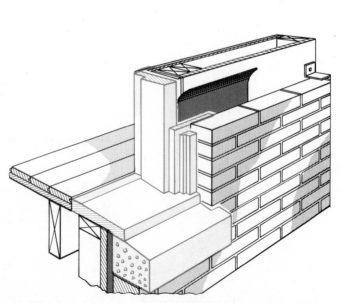

27-16b. Doorframe installed in brick-veneer construction.

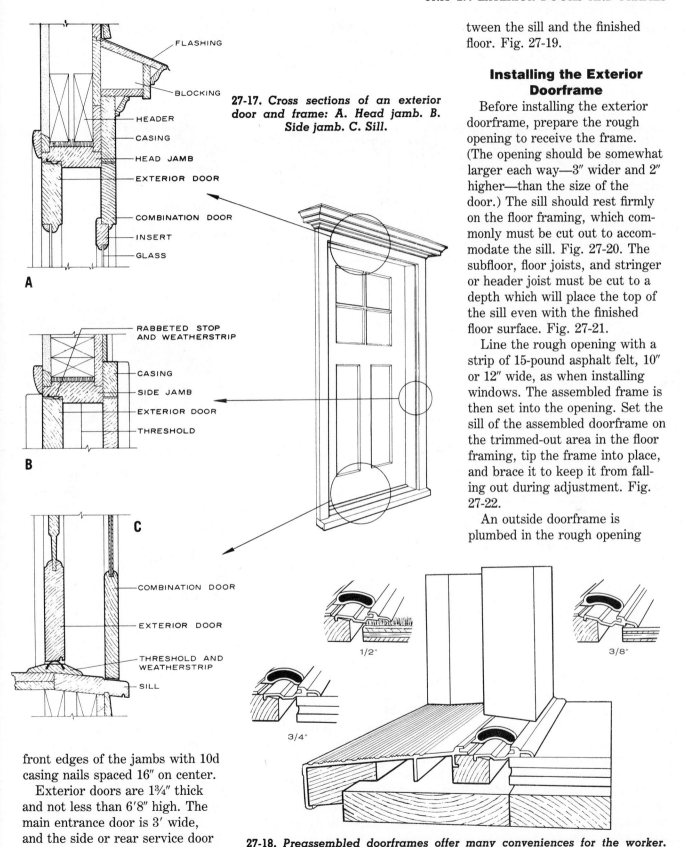

27-17. *Cross sections of an exterior door and frame: A. Head jamb. B. Side jamb. C. Sill.*

A — FLASHING, BLOCKING, HEADER, CASING, HEAD JAMB, EXTERIOR DOOR, COMBINATION DOOR, INSERT, GLASS

B — RABBETED STOP AND WEATHERSTRIP, CASING, SIDE JAMB, EXTERIOR DOOR, THRESHOLD

C — COMBINATION DOOR, EXTERIOR DOOR, THRESHOLD AND WEATHERSTRIP, SILL

1/2″   3/8″   3/4″

27-18. *Preassembled doorframes offer many conveniences for the worker. This one features a sill which is adjustable to eliminate trimming the floor joists.*

tween the sill and the finished floor. Fig. 27-19.

Installing the Exterior Doorframe

Before installing the exterior doorframe, prepare the rough opening to receive the frame. (The opening should be somewhat larger each way—3″ wider and 2″ higher—than the size of the door.) The sill should rest firmly on the floor framing, which commonly must be cut out to accommodate the sill. Fig. 27-20. The subfloor, floor joists, and stringer or header joist must be cut to a depth which will place the top of the sill even with the finished floor surface. Fig. 27-21.

Line the rough opening with a strip of 15-pound asphalt felt, 10″ or 12″ wide, as when installing windows. The assembled frame is then set into the opening. Set the sill of the assembled doorframe on the trimmed-out area in the floor framing, tip the frame into place, and brace it to keep it from falling out during adjustment. Fig. 27-22.

An outside doorframe is plumbed in the rough opening

front edges of the jambs with 10d casing nails spaced 16″ on center.

Exterior doors are 1¾″ thick and not less than 6′8″ high. The main entrance door is 3′ wide, and the side or rear service door is 2′8″ wide. A hardwood or metal threshold covers the joint be-

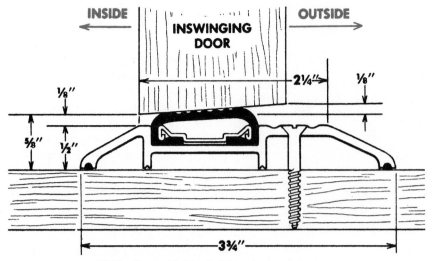

27-19. *A metal threshold with a vinyl insert.*

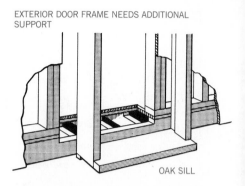

EXTERIOR DOOR FRAME NEEDS ADDITIONAL SUPPORT

OAK SILL

FLOOR MUST BE NOTCHED OUT FOR SILL

27-20. *Close-up view of the floor framing trimmed out to receive the doorframe.*

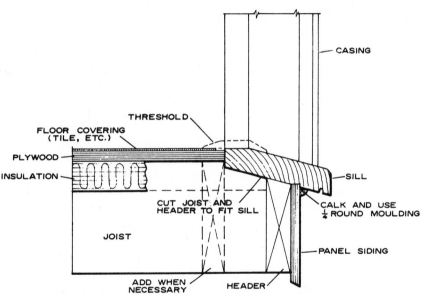

27-21. *The top of the doorframe sill should be set even with the surface of the finish floor.*

27-22. *Installing an exterior door-frame.*

with wood shingles used as wedges. These are inserted at intervals up the side jambs, between the jambs and the trimmer studs. Check the sill with a level, and wedge it up as necessary. Insert the side jamb wedges. Drive the lower wedges on each side alternately until the space between the side jamb and the trimmer stud is exactly the same on both sides. Then drive a 16d casing nail through the side casing and into the trimmer studs on each side, near the bottom of the casing, to hold the sill in position. Drive the nails in only partway. Do not drive any nails all the way in until all the nails have been placed and a final check has been made for level and plumb.

Next place the level against one of the side jambs and adjust the remaining wedges on that side until the jamb is perfectly true and plumb. Repeat the same pro-cedure on the other side. Make a final check for level and plumb. Fasten the frame in place with 16d casing nails driven through the casings into the trimmer studs and the door header. Nails are placed ¾" from the outer edges of the casings and spaced about 16" on center. Set all nails with a nail set.

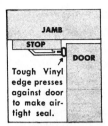

JAMB
STOP
DOOR
Tough Vinyl edge presses against door to make air-tight seal.

Side of Door **Head of Door** **Door Bottom**

27-23. *There are many kinds of weather stripping available to reduce air infiltration. Shown here are two types: one for the head and side of the door and a second for installation on the door bottom.*

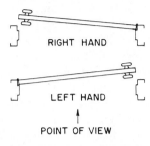

SINGLE DOOR

HAND OF DOOR MAY BE DETERMINED BY REFERRING TO SKETCHES BELOW. DOOR MUST ALWAYS SWING AWAY FROM POINT VIEWED.

RIGHT HAND

LEFT HAND

POINT OF VIEW

PAIRS OF DOORS

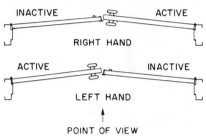

HAND OF DOORS IS DETERMINED BY LOCATION OF ACTIVE LEAF WHEN DOORS SWING AWAY FROM POINT VIEWED.

INACTIVE   ACTIVE

RIGHT HAND

ACTIVE   INACTIVE

LEFT HAND

POINT OF VIEW

27-24. *Determining the hand of a door.*

After the finish flooring is in place, a hardwood or metal threshold with a plastic weather strip covers the joint between the floor and the sill. Fig. 27-21. Thresholds are installed under exterior doors to close the space allowed for clearance. Weather stripping should be installed around exterior door openings to reduce drafts. Fig. 27-23.

HANDLING THE DOOR AT THE JOB SITE

A door is an important part of the building and has many functions. It guards the building and its possessions, insures privacy, protects against the elements, and lends beauty, refinement, and character to the building. A door is a high-grade precision-made item of cabinetwork and should be treated as such. Proper care and finishing of a door will insure maximum service and satisfaction.

Doors should not be delivered to the building site until after the plaster or concrete is dry, and then the doors should be:
• Stored under cover in a clean, dry, well-ventilated building, not in damp, moist, or freshly plastered areas.
• Stored on edge on a level surface.
• Sealed immediately on the top and bottom edges if they are to be stored at the job site for more than one week.
• Handled with clean gloves; bare hands leave finger marks and soil stains.
• Handled carefully. When moving doors, carry them. Do not drag a door except on the bottom end, and then only if it is protected by scuff strip or skid shoes. Do not drag one door across another.
• Conditioned to the average moisture content of the locality before hanging.
• Finished as soon as the doors are hung in the opening.
• Kept away from abnormal heat, dryness, or humidity. Sudden changes, such as forced heat to dry out a building, should be avoided.
• Straight. Before hanging, warp or bow can usually be eliminated by laying (or piling) the door (or doors) flat under weight. Bow or warp is due to stress forces in the door, usually caused by unequal moisture conditions on the two sides of the door. Improper installation of hinges can also be the cause. When moisture differential is the cause, the door will usually straighten when the moisture equalizes. When improper installation is the cause, hinges should be adjusted.

DETERMINING THE HAND OF A DOOR

A door is designated as having right-hand or left-hand swing. The hand of a door is determined by the location of the hinges when the door is viewed from the outside. For example, if the hinges are on the right when the door is viewed from the outside, the door is considered a right-hand door. Fig. 27-24. In general, the outside of a door is the side from which the hinges are not visible when the door is closed. However, the outside of a closet door is the room side.

FITTING A DOOR

The first step in fitting a door is to determine from the floor plan which edge of the door is the hinge edge and which is the lock edge. Mark both door edges and the corresponding jambs accordingly.

Carefully measure the height of the finished opening on both side jambs and the width of the opening at top and bottom. The finished opening should be perfectly rectangular, but it may not be. Regardless of the shape of the opening, the job is to fit the door accurately to the opening. A well-fitted door, when hung, should conform to the shape of the finished opening, less a clearance allowance of 1/16" at the sides and on top. For an exterior door with a sill and no threshold, the bottom clearance should be 1/16" above the sill. For a door with a threshold, the bottom clearance should be 1/8" above the threshold. The sill and threshold, if any, should be set in place before the door is hung. Lay out the measured dimensions

of the finished opening, less allowances, on the door.

Check the doorjambs for trueness and transfer any irregularities to the door lines. Plane the door edges to the lines, setting the door in the opening frequently to check the fit. The lock edge of a door must be beveled so that the inside edge will clear the jamb (at point A in Fig. 27-25) when the door is opened. The bevel required for this clearance is laid out by drawing a line from the point where the hinge pin will be located (B in Fig. 27-25) to the door's other side at the point where it intersects with the door stop (point C). Then place a T

bevel on the face of the door and set the blade so that it is parallel to line AC. Fig. 27-25. As shown in the illustration, this can be easily done by placing the blade against the inside edge of the framing square. Plane the edge as necessary, checking frequently with the T bevel to determine the correct angle. When all the planing has been completed, use a piece of sandpaper to form a slight radius on all edges to remove the sharpness.

As an aid in fitting the door, a door jack similar to the one shown in Fig. 27-26 should be constructed. The jack will hold the doors upright for planing edges and for the installation of hardware. Commercially made holders are also available. Fig. 27-27.

HANGING A DOOR

The hinge most frequently used for hanging doors on a residential building is the loose-pin butt mortise hinge. Fig. 27-28. This has two rectangular leaves pivoted on a pin which is called a loose pin because it can be removed. The

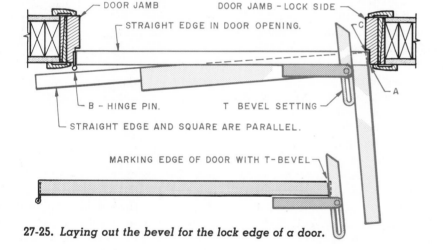

27-25. *Laying out the bevel for the lock edge of a door.*

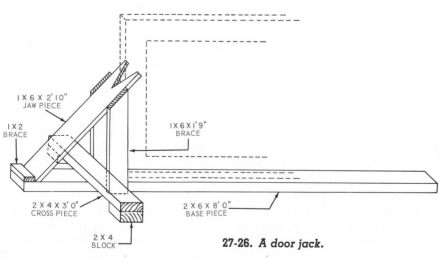

27-26. *A door jack.*

27-27. *A commercially made sash and door holder.*

hinge is called a mortise hinge because the leaves are mortised into gains cut in the edge of the door and in the hinge jamb of the doorframe.

After the door has been properly fitted, the first step in hanging it is to lay out the locations of the hinges on the edge of the door and the hinge jamb. Exterior doors usually have three hinges. The following distances may be specified: the vertical distance between the top of the door and the top of the top hinge, and the vertical distance between the top of the finish floor and the bottom of the bottom hinge. If these distances are not specified, the distances customarily used are those shown in Fig. 27-29. The middle hinge is located midway between the other two. The size of a loose-pin butt mortise hinge is designated by the length of a leaf in inches. For an exterior door a 3½″ or 4″ hinge is recommended.

Set the door in the frame and force the hinge edge of the door against the hinge jamb with the wedge marked A in Fig. 27-29. Then insert a 4d finish nail between the top of the door and the head jamb and force the top of the door up against the nail with the wedge marked B in Fig. 27-29. Since a 4d finish nail has a diameter of ¹⁄₁₆″ (which is the standard top clearance for a door), the door is now at the correct height.

Measure out the distance from the top of the door to the top of the top hinge and from the floor up to the bottom of the bottom hinge. Mark these locations with a ½″ chisel or a knife. If a chisel is used, hold it so that the bevel of the chisel is toward the location of the hinge. For example, when marking the bottom hinge, the bevel on the chisel should be held up, and when marking the top hinge, the bevel on the chisel should be held down. Hold the

chisel with the cutting edge in a level position so that it is in contact with both the jamb and the edge of the door. Apply pressure and make a small cut into both surfaces to mark the position of the hinge.

When marking for the center hinge, remember that the location line is to the center of the hinge; if a 4″ hinge is used, measure 2″ on one side of the location line and mark this point with a chisel. To help avoid mistakes it is best to pencil a small X on the side of the chisel mark where the gain for the hinge will be cut.

Remove the door from the opening. Place the door in a door jack and lay out the outlines of the gains on the edge of the door

using a hinge leaf or a hinge butt gauge as a marker. Fig. 27-30. The door-edge hinge setback, shown in Fig. 27-28, should not be less than ⅛″. It is usually made about ¼″. Fig. 27-31. Lay out gains of exactly the same size on the hinge jamb. Chisel out the gains to a depth equal to the thickness of the hinge leaf.

Separate the leaves on the hinges by removing the loose pins. Screw the leaves into the gains on the door and the jamb. Make sure that the leaf in which the pin will be inserted is in the up position when the door is hung in place. Hang the door in place, insert the loose pins, and check the clearances at the side jambs. If the clearance along the hinge

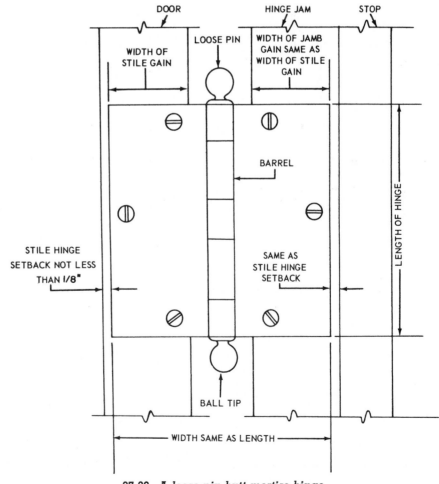

27-28. *A loose-pin butt mortise hinge.*

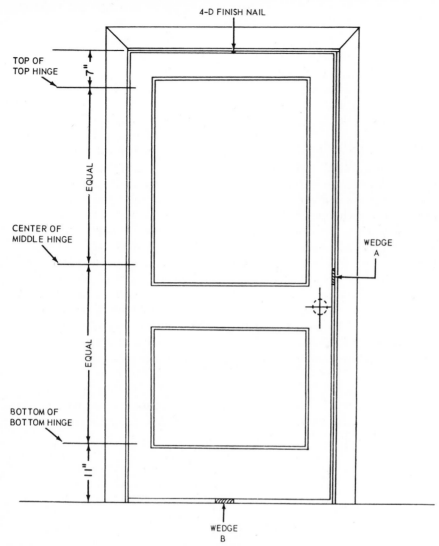

27-29. **Distances commonly used in laying out hinge locations on the door and the doorjamb.**

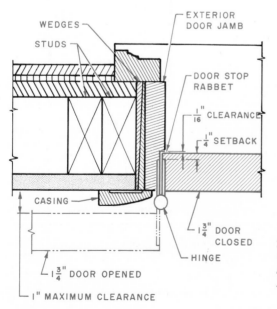

27-30. *A hinge butt gauge.*

jamb is too large (more than 1/16″) and that along the lock jamb is to small (less than 1/16″) extract the pins from the hinges and remove the door. Then remove the hinge leaves from the gains and slightly deepen the gains. If the clearance along the hinge jamb is too small and that along the lock jamb is too large, the gains are too deep. This can be corrected by shimming up the leaves with strips of cardboard placed in the gains under the hinges.

Hinge Butt Routing

A special template is available for hinge butt routing. Fig. 27-32a. The metal template may be adjusted for most common hinge spacings, and it is easily mounted on the door by driving six nails to hold the templates securely on the door. This template guides the router so that the hinge mortises are cut quickly and accurately to size and location. After the gains or mortises are cut on the door, the template guide can be transferred to the doorjamb for cutting the hinge mortises to match those on the door. Because the bits leave a radius at the corner of the cut, it is necessary to chisel the corners square for the

27-31a. *The door hinge should be set back sufficiently to allow the door to clear the casing when the door is swung wide open. With a 1¾″ exterior door and 4″ butt hinges, the maximum clearance is 1″, as specified on the chart in Fig. 27-31b.*

The following table gives the clearances for trim of regular stock size butt hinges for wood or hollow metal doors. The clearance is estimated on butt hinges set back ¼" for doors up to 2¼" and ⅜" for doors 2½" to 3" in thickness. Where trim presents a specific problem in determining the proper width of the butt hinges for a door, take twice the thickness of the door, plus the thickness of the trim and deduct ½" for doors up to 2¼" in thickness, and ¾" for doors 2½" to 3" in thickness.

Thickness of Door (Inches)	Size of Butt Hinge (Inches)	Maximum Clearance (Inches)
1⅜	3 x 3	¾
	3½ x 3½	1¼
	4 x 4	1¾
1⁹⁄₁₆	4 x 4	1⅜
	4½ x 4½	1⅞
	5 x 5	2⅜
	6 x 6	3⅜
1¾	4 x 4	1
	4½ x 4½	1½
	5 x 5	2
	6 x 6	3
1⅞	4½ x 4½	1¼
	5 x 5	1¾
	6 x 6	2¾
2	4½ x 4½	1
	5 x 5	1½
	6 x 6	2½
2¼	5 x 5	1
	6 x 6	2
	6 x 8	4
2½	5 x 5	¾
	6 x 6	1¾
	6 x 8	3¾

27-31b. *Trim clearances for wood and metal doors.*

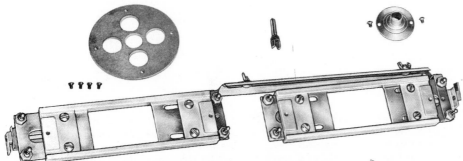

27-32a. *A door- and jamb-butt template and router accessories.*

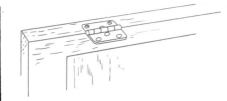

27-32b. *Round corner butt hinges save time when installed with the hinge butt router and door and jamb template.*

hinges. It is also possible to purchase hinges designed with round corners. Fig. 27-32b.

INSTALLING A LOCK SET

Lock sets come in many styles, from very simple to ornate. Fig. 27-33. Some are mounted in the center of the door with large decorative plates (escutcheons) behind the knob. Fig. 27-34. The installation instructions for lock sets, particularly the number and size of the holes to be bored in the door, will vary with the manufacturer. Always refer to the instructions which accompany the specific lock set to be installed. The general procedure for installing a lock set is as follows:

1. Open the door to a convenient working position and place wedges under the bottom near the outer edge to hold the door steady.

2. Measure up 36" from the floor to locate the height of the lock set.

3. Fold and apply the template (which comes with the lock set) to the edge of the door bevel. Fig. 27-35. Mark the center of the door edge and the center of the hole on the door face through the guides on the template. Fig. 27-36. If a boring jig is used, no template is needed. Fig. 27-37.

4. Bore a hole of the recommended diameter in the face of the door. It is recommended that the holes be bored from both sides to prevent splitting. Bore the hole on one side until the point of the bit breaks through

27-33. Entrance door locks used on residences and on smaller commercial buildings.

27-34. A decorative touch is added to the front entrance door by choosing from the wide selection of ornamental escutcheons which are available. The escutcheons not only add beauty to the entrance but also provide protection against finger marks and scratches. When replacing old or damaged locks, the decorative escutcheons may be used to hide unsightly scars and holes.

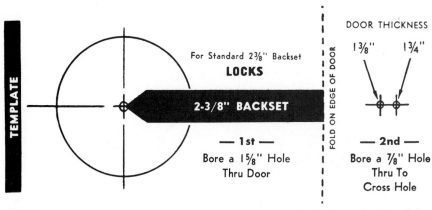

TEMPLATE

For Standard 2⅜" Backset
LOCKS

2-3/8" BACKSET

FOLD ON EDGE OF DOOR

DOOR THICKNESS

1⅜"   1¾"

— 1st —
Bore a 1⅝" Hole
Thru Door

— 2nd —
Bore a ⅞" Hole
Thru To
Cross Hole

27-35. *A typical template for locating the center of the holes to be bored for a lock set.*

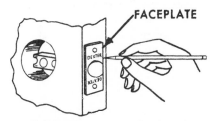

FACEPLATE

27-38. *Marking around the faceplate.*

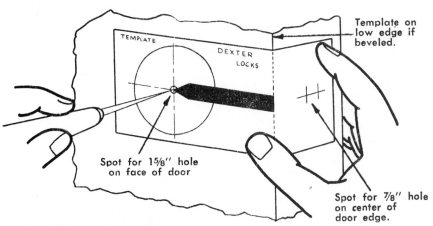

TEMPLATE

DEXTER LOCKS

Template on low edge if beveled.

Spot for 1⅝" hole on face of door

Spot for ⅞" hole on center of door edge.

27-36. *Hold the template on the dotted line and place on the door edge. Mark the door through the template with an awl or nail.*

27-39. *When a marking tool is available, it is not necessary to mark around the faceplate as shown in Fig. 27-38. The marking tool is inserted in the hole bored in the edge of the door. It is aligned parallel with the edge of the door and then given a sharp blow with a hammer to outline the area to be chiseled out for the latch faceplate.*

and then complete from the other side.

5. Bore a hole of the recommended diameter in the center of the door edge for the latch.

27-37. *When a boring jig is available, it is not necessary to use the template for marking the door previous to boring the holes for the lock set. The jig, when properly adjusted and clamped in position on the door, insures an accurate and rapid boring of the door.*

6. Insert the latch in the hole in the door edge. Keep the faceplate parallel to the edge of the door and mark with a sharp pencil around the faceplate. Fig. 27-38.

A marking tool may also be used to mark the position of the faceplate. Fig. 27-39.

7. Remove the latch. Chisel out the marked area so that the latch faceplate will be mounted flush with the edge of the door. Fig. 27-40.

8. Install the latch with the curved surface of the latch facing in the direction of the door closing. Insert and tighten the screws. Fig. 27-41.

9. Install the exterior knob by inserting the knob with the spindle into the latch. Make certain that the stems are positioned correctly through the latch holes and pressed flush against the door. Fig. 27-42.

10. Install the interior knob by placing it over the stem and align-

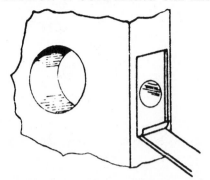

27-40. *Chisel out the marked area. The latch faceplate should mount flush with the edge of the door.*

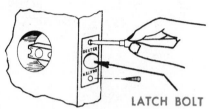

LATCH BOLT

27-41. *Installing the latch.*

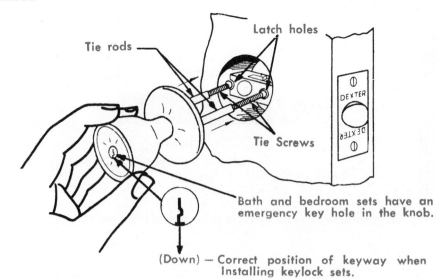

Tie rods

Latch holes

Tie Screws

Bath and bedroom sets have an emergency key hole in the knob.

(Down) — Correct position of keyway when Installing keylock sets.

27-42. *Installing the exterior trim assembly. Tie rods and tie screws must go through the holes in the latch.*

ing the screw guides with the stems. Push the assembly flush with the door, insert the screws, and tighten until the lock set is firm. Fig. 27-43.

11. Locate the strike on the doorjamb opposite the faceplate of the latch. To locate the strike, place it over the latch in the door. Then carefully close the door against the stops. The strike plate will hang on the latch in the clearance area between the door edge and the jamb. Push the strike plate in against the latch and, with a pencil, mark the top edge of the strike plate on the jamb. Then hold the pencil against the door edge and draw a line down the face of the strike plate.

12. Open the door and hold the strike against the doorjamb just under the line previously marked. Make sure that the line marked on the face of the strike is aligned with the edge of the jamb. Mark around the strike and chisel out the marked area so that the strike will mount flush with the surface of the jamb.

13. Make a clearance hole for the latch bolt by drilling a $15/16''$ hole $1/2''$ deep in the doorjamb on the center line of the screws from top to bottom. Install the strike and tighten the screws. Fig. 27-44.

GARAGE DOORS

There are many types and sizes of garage doors. Fig. 27-45. The standard single door is 9' wide and 6½' or 7' high. Double doors are usually 16' × 6½' or 7'. There are many architectural styles available to match the style of the home. Fig. 27-46. To give the door a distinctive custom look, it can be trimmed at any time with easily mounted mold-

SCREW GUIDES

SCREW HOLES

27-43. *Installing the interior knob.*

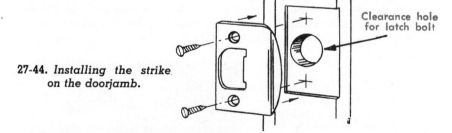

Clearance hole for latch bolt

27-44. *Installing the strike on the doorjamb.*

Stock Sizes	Standard Sizes	
8' x 7'	8' x 6'6"	15' x 6'6"
9' x 7'	9' x 6'6"	15' x 7'
10' x 7'	10' x 6'6"	16' x 6'6"
16' x 7'	18' x 7'	

27-45. *Garage doors are available in many sizes.*

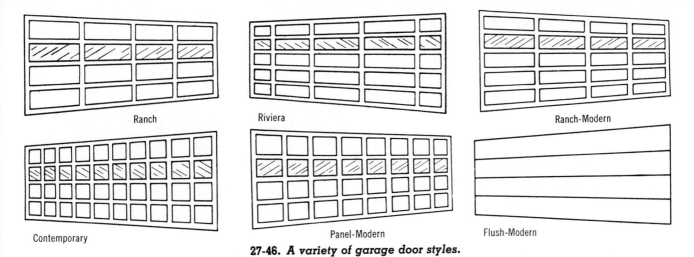

Ranch

Riviera

Ranch-Modern

Contemporary

Panel-Modern

Flush-Modern

27-46. A variety of garage door styles.

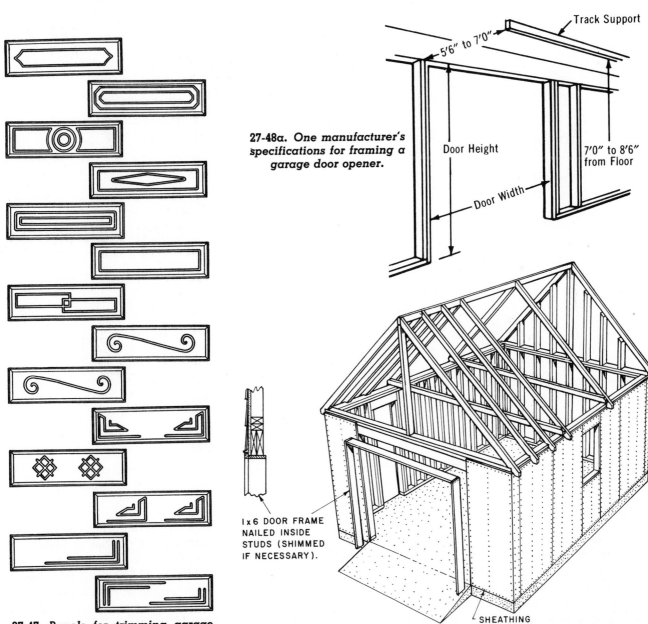

27-48a. One manufacturer's specifications for framing a garage door opener.

Track Support

5'6" to 7'0"

Door Height

7'0" to 8'6" from Floor

Door Width

1 x 6 DOOR FRAME NAILED INSIDE STUDS (SHIMMED IF NECESSARY).

SHEATHING

27-47. Panels for trimming garage doors come in many styles.

27-48b. Finished doorjamb assembled and ready for installation.

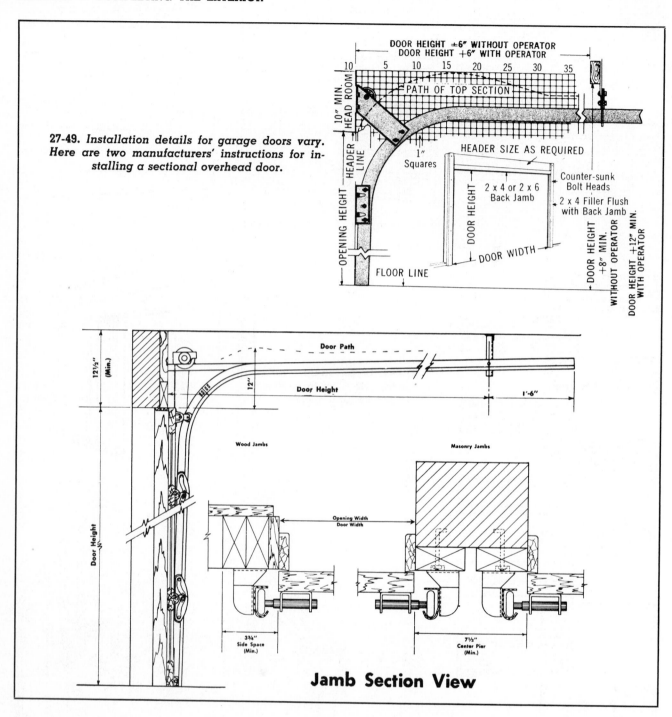

27-49. *Installation details for garage doors vary. Here are two manufacturers' instructions for installing a sectional overhead door.*

Jamb Section View

ings, rosettes, or monogram plates. Fig. 24-47.

The various types of doors and hardware, with complete instructions for their installation, can be obtained from local suppliers. Figs. 27-48 & 27-49. The three most commonly used doors are

the hinged, the overhead swing, and the overhead sectional. Fig. 27-50. Occasionally, folding sliding doors are used.

Hinged doors open outward and are held in position with door holders. These doors are the least expensive and the easiest to in-

stall. However, when the door is standing open, it has no protection from rain and snow.

Sliding folding doors are hung from a track above the door. If the track is hung on the outside, the doors are subject to weathering. If the track is hung on the in-

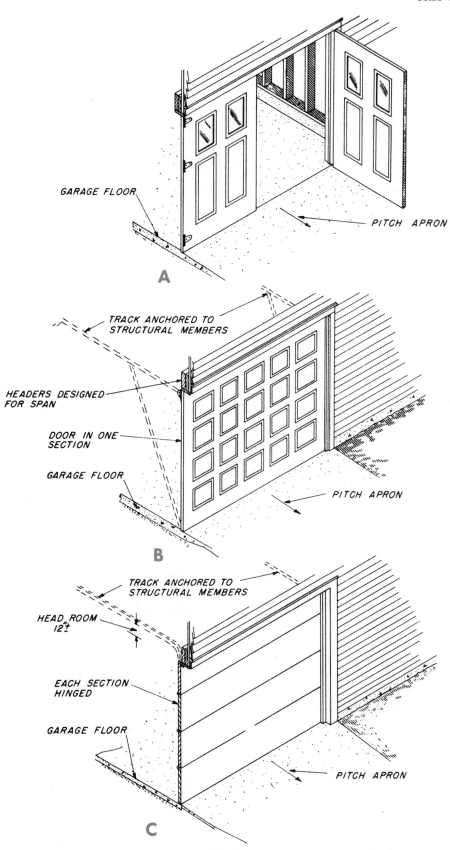

GARAGE FLOOR

PITCH APRON

A

TRACK ANCHORED TO
STRUCTURAL MEMBERS

HEADERS DESIGNED
FOR SPAN

DOOR IN ONE
SECTION

GARAGE FLOOR

PITCH APRON

B

TRACK ANCHORED TO
STRUCTURAL MEMBERS

HEAD ROOM
12±

EACH SECTION
HINGED

GARAGE FLOOR

PITCH APRON

C

27-50. *Types of garage doors: A. Hinged. B. Overhead swing. C. Overhead sectional.*

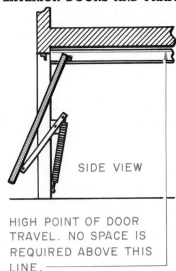

SIDE VIEW

HIGH POINT OF DOOR TRAVEL. NO SPACE IS REQUIRED ABOVE THIS LINE.

27-51. *The swing door must be moved outward slightly at the bottom as it is opened.*

side, the doors can fold against one another in several thicknesses, or the track can be curved along the inside wall.

Overhead doors are made in two types: as a single-section (swing) door and as four or five sections hinged together. The swing-up door with the single section operates on a pivot principle with the track mounted on the ceiling and rollers located at the center and top of the door. Fig. 27-51. The sectional overhead door has rollers at each section fitted into a track at the side of the door and the ceiling. It requires more headroom above the opening than the swing door but is by far the most widely used for residential building. Clearance required above the top of the sectional overhead doors is usually about 12″. However, low-headroom brackets are available when such clearance is not possible.

Overhead doors are well protected from rain and wind, and snow and ice offer no particular problem. They are somewhat more difficult to install and more expensive than the hinged doors.

27-52. *This garage door is counterbalanced with a torsion spring.*

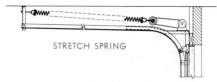

27-53. *This garage door is counterbalanced with a stretch spring.*

The overhead door has a pair of counterbalance springs mounted on it to help support the weight of the door so that it may be easily opened. These counterbalance springs are of two types: torsion and stretch. Figs. 27-52 and 27-53.

Power door operators are also available. These are electronically controlled by a wall-mounted button within the home or garage or by a portable battery-powered transmitter inside the car. Electric door operators can be installed during construction, or they can be added by the homeowner later.

The bottom edge of a garage door should be scribed and cut to conform to the garage floor. An application of weather stripping is recommended for the bottom rail. It seals any minor irregularities in the floor and acts as a cushion in closing.

The header beam over garage doors should be designed for the snow load which might be imposed on the roof. In wide openings, this may be a steel I-beam or a built-up wood section. For spans of 8' or 9', two doubled 2 × 10s of high-grade Douglas fir or similar species are commonly used when only snow loads must be considered. If floor loads are also imposed on the header, a steel I-beam or wideflange beam is usually selected.

QUESTIONS

1. List five kinds of exterior doors.
2. Name the parts of the doorframe.
3. What is the standard thickness of an exterior door?
4. What is the minimum width door recommended for a main entrance?
5. When the exterior doorframe is set into the opening, how far above the subfloor must the sill project?
6. When a door is fitted to the jamb, why is the lock edge of the door beveled?
7. What are the three types of garage doors most commonly used?

Exterior Wall Coverings 28

Today's home builder can select from a wide variety of easy-care materials for the exterior walls of the home. Fig. 28-1. Various materials, shapes, and surface treatments are used to produce over five hundred different wall coverings. An entirely different design effect can be achieved by changing the type of exterior covering on a house. Because it has such a great effect on a home's overall appearance and ease of maintenance, the exterior wall covering should be selected with great care.

A wide variety of exterior coverings is possible because changing any one of the following factors can produce a new kind of covering:

• Material used. Wood products used include solid wood and such man-made wood materials as plywood, hardboard, and particle board. Masonry, either solid or veneer, may be of brick, stone, or

28-1. *In this home, brick siding was used on the first level and bevel siding on the second level.*

stucco. Asphalt materials are also used for siding in two forms: as rolled products and as shingles. Common metals used are aluminum and steel, both of which are usually prefinished. Vinyl plastic is also a popular, easy-care exterior wall material.

• Shape and form. Some of the common shapes and forms in which exterior wall covering material is manufactured are bevel and drop siding, vertical tongued and grooved material, large panels, boards and battens, shingles and shakes, and rolled material.

• Surface treatment. Siding can be smooth or rough sawn, plywoods and hardboards can be textured, overlays of fiber and/or plastics can be added, and materials can be prefinished with paints, enamels, plastics, and other finishes. Most building supply dealers display samples of the wide variety of materials available.

With the exception of solid wood and man-made wood materials, most exterior wall coverings are applied either by bricklayers or masons or by specialty building construction workers. Therefore only those materials most often used in the construction of homes will be considered here. Specific emphasis will be placed on materials commonly installed by the carpenter.

WOOD SIDING

One of the materials most characteristic of the exteriors of North American houses is wood siding. The essential properties required for wood siding are good painting characteristics, easy working qualities, and freedom from warp. These properties are present to a high degree in the cedars, eastern white pine, sugar pine, western white pine, cypress, and redwood. They are present to a good degree in western hemlock, ponderosa pine, spruce, and yellow poplar and to a fair degree in Douglas fir, west-

ern larch, and southern yellow pine.

Exterior siding materials should be select grade and should be free from knots, pitch pockets, and waney edges. The moisture content at the time of application should be the same that it would attain in service. This is about 12%, except in the dry southwestern United States, where the moisture content should average about 9%.

Wood siding is made in many shapes and sizes and with various edge treatments. The common types are:
• Bevel.
• Drop.
• Board.
Some types, such as bevel siding, must be installed horizontally. Others, such as board siding, may be installed either horizontally (clapboard) or vertically (board and batten). Fig. 28-2.

Vertical siding is commonly applied to the gable ends of houses, over entrances, and sometimes on large wall areas. It may consist of plain-surfaced matched boards, patterned matched boards, or square-edge boards covered at the joint with a batten strip. Fig. 28-2.

Matched vertical siding should preferably be not more than 8″ wide. It should be fastened with two 8d nails not more than 4′ apart. Backer blocks placed between studs provide a good nailing base. The bottom of the boards should be undercut to form a water drip.

Bevel Siding. Plain bevel siding is made in nominal 4″, 5″, and 6″ widths with 7/16″ butts, in 6″, 8″, and 10″ widths with 9/16″ butts, and in 6″, 8″, 10″, 12″ widths with 11/16″ butts. Fig. 39-3a. The top edge is 3/16″ for all sizes. Bevel siding is generally furnished in random lengths varying from 4′ to 16′.

305

	Board	Channel Rustic	Drop	Bevel
	Board Board Clap- and on board Batten Board	(Board and Gap)	T&G Shiplap Patterns Patterns	Plain Rabbeted Edge
Patterns	Available surfaced or rough textured		Available in 13 different patterns. Some T&G (as shown), others ship-lapped.	Plain Bevel may be used with smooth face exposed or sawn face exposed for textured effect.
Application And Nailing	Recommended 1" minimum overlap. Use 10d Siding nails as shown.	May be applied horizontally or vertically. Has ½" lap and 1¼" channel when installed. Use 8d Siding nails as shown for 6" widths. Wider widths nail twice per bearing.	6d Finish nails for T&G, 8d Siding nails for shiplap.	Recommend 1" minimum overlap on plain bevel siding. Use 6d Siding nails as shown.
Available Grades *Most commonly used	No. 1 Common* No. 2 Common* No. 3 Common Or Select Merchantable* Construction* Standard	No. 1 Common* No. 2 Common* No. 3 Common Or C&Btr*, D*, E	No. 1 Common* No. 2 Common* No. 3 Common Or C&Btr*, D*, E	All species except WRC & IWP B&Btr*, C*, D **WRC** Clear-VG-All Heart* A*, B*, Rustic* **IWP** Supreme*, Choice*, Quality
Seasoning	Shipped 15% moisture content or less when specified.	Shipped 15% moisture content or less when specified.	Shipped 15% moisture content or less when specified.	Usually shipped at 12% or less moisture content.

	Bungalow	Dolly Varden	Log Cabin	Tongue & Groove
	Plain Rabbeted Edge	Rabbeted Edge		Plain
Patterns	Thicker and wider than Bevel Siding. Sometimes called "Colonial." Plain bungalow may be used with smooth face exposed for textured effect.	Thicker than Bevel Siding. Rabbeted edge.	1½" at thickest point.	Available in smooth surface or rough surface.
Application And Nailing	Same as for Bevel siding, but use 8d Siding nails.	Same as for Rabbeted Bevel Siding but use 8d Siding nails.	Nail 1½" up from lower edge of piece. Use 10d Casing nails.	Use 6d Finish nails as shown for 6" widths or less. Wider widths, face nail twice per bearing with 8d Siding nails.
Available Grades *Most commonly used	see Bevel Siding Grades	all species except IWP B&Btr*, C*, D **IWP** Supreme* Choice* Quality	No. 1 Common* No. 2 Common* No. 3 Common	No. 1 Common* No. 2 Common* No. 3 Common Or C&Btr*, D*, E
Seasoning	Usually shipped at 12% or less moisture content.	Usually shipped at 12% or less moisture content.	Shipped 15% moisture content or less when specified.	Shipped 15% moisture content or less when specified.

28-2. Siding use guide. IWP = Idaho white pine. WRC = western red cedar. The other letters stand for grades.

28-3. Anzac siding has a thick butt edge which accentuates the horizontal lines of the siding.

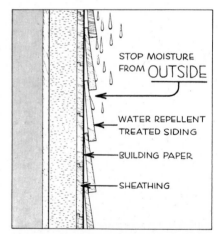

28-4. To keep moisture out, the siding should be treated with a water repellent. The sheathing should be covered with a water-repellent building paper.

One variation of bevel siding is the Anzac pattern. The Anzac siding pattern was derived from a New Zealand design. This pattern has a relatively thick butt edge which produces a heavy shadow line under each course. On the face of the pattern are two grooves. The deep upper groove acts as a water barrier, and the lower groove is a guideline which aids in aligning the siding properly as it is applied. The back of the siding is machined so that it will lie flat against the studs or sheathing. Fig. 28-3.

Drop Siding. Drop siding is generally ¾" thick, has a flat back, and is made in a variety of patterns with either matched or shiplap edges. Fig. 28-2. All patterns of drop siding may be applied horizontally. Some patterns may also be applied vertically; for example, at the gable ends of a house.

Drop siding is designed to be applied directly to the studs, and it thereby serves as both sheathing and exterior wall covering.

28-5. Boards that have been cut to length should have the fresh cut ends treated with a water repellent.

It is widely used in this manner in farm structures, sheds, and garages in all parts of North America, and for houses in mild climates. When drop siding is used over and in contact with other material such as sheathing or sheathing paper, water may work through the joints and be held between the sheathing and the siding. This condition can lead to paint failures and decay. Such conditions are not common when the sidewalls are protected by a good roof overhang. When drop siding is applied vertically, it should also be protected by a wide overhanging roof. Otherwise, water flowing down the face of the siding is led into the joint and held there.

Board Siding. Square-edge or clapboard siding made of $^{25}/_{32}"$ board is occasionally selected for architectural effect. In this case wide boards are generally used. Some of this siding is also beveled at the top of the back when used as clapboard siding. This allows the boards to lie rather close to the sheathing, thus providing solid nailing. Fig. 28-2.

In board and batten siding, when wide square-edged boards are used, they are subject to considerable expansion and contraction because of their width. The batten strips covering the joints should be nailed to only one siding board so that the adjacent board can swell and shrink without splitting the boards or the batten strip. Fig. 28-2.

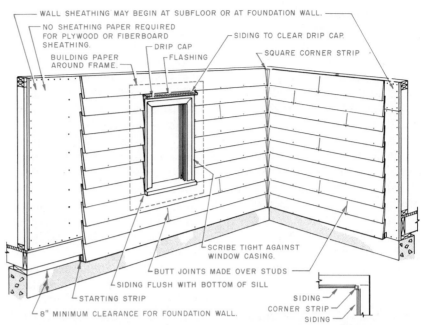

28-6. An exterior wall with wood sheathing and lap siding.

INSTALLATION OF WOOD SIDING

Before application, exterior wood siding should be treated with a water repellent. This will improve finish performance no matter what type of finish is used. Fig. 28-4.

Some siding is pretreated (or preprimed) by the lumber mill. If this has been done, mill instructions accompanying the siding should be followed carefully.

If the siding has not been pretreated, a water repellent is most easily and effectively applied before the siding is put in place. This may be done either by dipping or by brushing the water repellent on the face, back, ends, and edges of each piece. If the siding is stacked to dry after such treatment, stickers (strips of wood) placed between tiers also should be treated with a water repellent. Many boards will be cut to length as the siding is put in place. Freshly cut surfaces also should receive a liberal treatment with a water repellent. Fig. 28-5.

If wood sheathing has been used, the exterior of the building must first be covered with building paper with a 4" lap. The siding may be nailed into the sheathing at 24" intervals. Nails should penetrate at least 1½" into the studs. Fig. 28-6.

For matched horizontal siding, the weather exposure (that part of the siding which will not be overlapped by another piece of siding) is predetermined by its machined edge treatment. In the applications, each succeeding course is installed up tight against the preceding course. Since the spacing is predetermined, it is not necessary to lay out the spacing, as it is for plain or bungalow siding.

Laying out the Spacing of Vertical Siding

When laying out board and batten siding, measure the length of the wall on which the siding is to be installed and carefully lay out the spacing of the boards and battens. The spacing will have to be increased or decreased between the boards so that the widths of the underboards will appear to be the same. The underboards are then cut accurately to length and installed according to this layout. Care should be taken that the boards are plumb as they are installed.

When all the underboards are in place, the battens or the overboards (for board on board siding) are installed. Again, make sure that these members are also plumb. The installation of the overboards is much more critical because they will give the wall its finished appearance.

Laying out the Spacing of Plain Bevel or Bungalow Siding

The spacing for plain bevel or bungalow siding should be carefully laid out before the first board is applied. Siding starts with the bottom course of boards at the foundation. Fig. 28-7. Sometimes the siding is started on a water table, which is a projecting member at the top of the foundation to throw off water. Fig. 28-8. Each succeeding course of bevel siding overlaps the upper edge of the previous course.

Determine the number of courses by measuring the distance from at least 1" below the bottom plate to the underside of the soffit and dividing that height by the maximum weather exposure of the siding. Fig. 28-9. To determine the maximum exposure, de-

duct the minimum overlap, or head lap, from the overall width (dressed dimensions) of the siding. The minimum head lap is 1" for 4" and 6" widths and 1¼" for widths over 6". The dressed dimensions for various sidings can be found in Fig. 28-10.

For example, if a nominal 10" bevel siding is used, the actual or dressed width is 9¼". Fig. 28-10. On 10" plain bevel siding, a minimum overlap of 1¼" is required. Therefore the maximum exposed surface of a 10" piece of siding would be 8" (9¼" − 1¼" = 8"). With a pair of dividers set at 8", make a trial layout on the sidewall beginning at the bottom and "walking off" this exposure dimension. The bottom of the board that passes over the top of the first-floor windows should coincide with the top of the window cap. Fig. 28-11. If the bottom of this board does not line up, adjust the

spacing for each board to something slightly less than 8". (NOTE: Eight inches is the maximum exposure, so do not adjust it to a greater width). Continue to modify the spacing, if at all possible, until the bottom edge of this piece of siding is even with the tops of the windows and doors. The location on the foundation wall for the bottom edge of the first piece of siding may also be adjusted slightly as an aid in making this alignment.

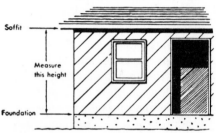

28-9. Measuring the vertical distance to be covered by the siding.

28-8. Bevel siding started on a water table (see arrow) at the foundation.

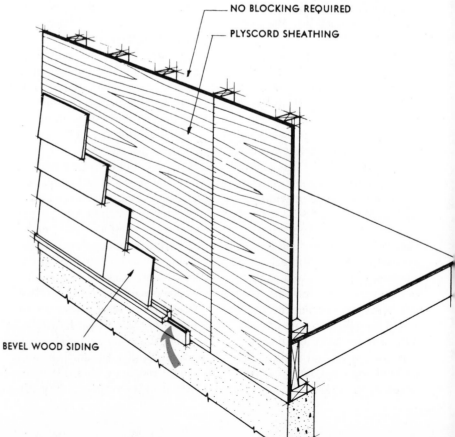

28-7. The beginning courses of siding at the foundation wall.

Product Description	Nominal Size		Dressed Dimensions		
	Thickness In.	Width In.	Thickness In.	Width In.	Lengths Ft.
Rustic And Drop Siding (D & M) If 3/8" or 1/2" T & G specified, same over-all widths apply.	5/8 1	4 5 6 8 10	9/16 23/32	3 1/8 4 1/8 5 1/8 6 7/8 8 7/8	Same
(Shiplapped, 3/8-in lap)	5/8 1	4 5 6	9/16 23/32	3 4 5	Same
(Shiplapped, 1/2-in. lap)	5/8 1	4 5 6 8 10 12	9/16 23/32	2 7/8 3 7/8 4 7/8 6 5/8 8 5/8 10 5/8	Same
Ceiling And Partition (S2S & CM)	3/8 1/2 5/8 3/4	3 4 5 6	5/16 7/16 9/16 11/16	2 1/8 3 1/8 4 1/8 5 1/8	Same
Bevel Siding Grades Bevel Siding / Western Red Cedar Bevel Siding available in 1/2", 5/8", 3/4" nominal thickness. Corresponding thick edge is 15/32", 9/16" and 3/4".	1/2 9/16 5/8 3/4 1	4 5 6 8 10 12	7/16 butt, 3/16 tip 15/32 butt, 3/16 tip 9/16 butt, 3/16 tip 11/16 butt, 3/16 tip 3/4 butt, 3/16 tip	3 1/2 4 1/2 5 1/2 7 1/4 9 1/4 11 1/4	Same
Wide Bevel Siding (Colonial or Bungalow)	3/4	8 10 12	11/16 butt, 3/16 tip	7 1/4 9 1/4 11 1/4	
Finish And Boards S-Dry S1S, S2S, S1S2E	3/8 1/2 5/8 3/4 1 1 1/4 1 1/2 1 3/4 2 2 1/2 3 3 1/2 4	2 3 4 5 6 7 8 and wider nominal	5/16 7/16 9/16 5/8 3/4 1 1 1/4 1 3/8 1 1/2 2 2 1/2 3 3 1/2	1 1/2 2 1/2 3 1/2 4 1/2 5 1/2 6 1/2 3/4 off	3' and longer. In Superior grade, 3% of 3' and 4' and 7% of 5' and 6' are permitted. In Prime grade, 20% of 3' to 6' is permitted.
Factory And Shop Lumber S2S*	1 (4/4) 1 1/4 (5/4) 1 1/2 (6/4) 1 3/4 (7/4) 2 (8/4) 2 1/2 (10/4) 3 (12/4) 4 (16/4)	5 and wider (4" and wider in 4/4 No. 1 Shop and 4/4 No. 2 Shop)	25/32 (4/4) 1 5/32 (5/4) 1 13/32 (6/4) 1 19/32 (7/4) 1 13/16 (8/4) 2 3/8 (10/4) 2 3/4 (12/4) 3 3/4 (16/4)	(See Rough Sizes Below)	6 ft. and longer in multiples of 1'

*These thicknesses also apply to Tongue & Groove (T&G).
See coverage estimator chart for T&G widths.

Minimum Rough Sizes   Thicknesses and Widths Dry or Unseasoned All Lumber (S1E, S2E, S1S, S2S)
80% of the pieces in a shipment shall be at least 1/8" thicker than the standard surfaced size, the remaining 20% at least 3/32" thicker than the surfaced size. Widths shall be at least 1/8" wider than standard surfaced widths.
When specified to be full sawn, lumber may not be manufactured to a size less than the size specified.

28-10. *Nominal and dressed dimensions for wood siding.*

The board spacing should be such that the maximum exposure will not be exceeded. This may mean that the boards will have less than the maximum exposure.

Application of Plain Bevel or Bungalow Siding

The application of the first course of siding determines the level and uniformity of all succeeding courses. To insure the proper application of the first course, proceed as follows:

1. Measure down with a tape from the top of the top plate if

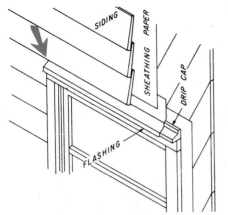

28-11. *Plan the courses of siding so that the bottom edge of the course running across the top of the window will be in alignment with the drip cap.*

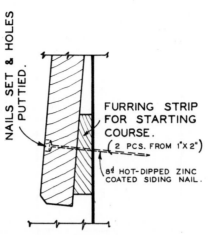

28-13. *The furring strip may be ripped at an angle to provide support along its full width. Notice that two furring strips were ripped from one piece of 1" × 2" stock.*

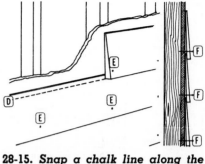

28-15. *Snap a chalk line along the top edge of the siding to locate the butt edge of the succeeding courses.*

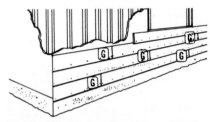

28-16. *Apply successive courses, making sure that all the vertical butt joints are staggered and that they fall on studs.*

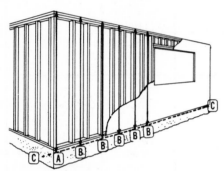

28-12. *Locating the first course of siding on an exterior wall.*

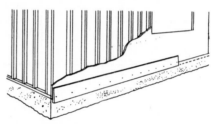

28-14. *Applying the first piece of siding with one nail at each stud location.*

possible. If not, measure from the underside of the soffit to a point at least 1" below the bottom plate. This point will usually be on the foundation. Make a mark on the foundation and record this measurement. Fig. 28-12, point A.

2. Repeat Step 1 along the sidewall at about 2' intervals using the same measured length each time. B, Fig. 28-12.

3. Snap a chalk line horizontally along the marks. With bevel, bungalow, or Anzac patterns, nail a furring strip about ⅜" above this line to provide support for the starting course. Fig. 28-13. The bottom edge of the first course of siding will be positioned

along the chalk line. Fig. 28-12, line C.

4. Begin the application by placing the butt edge of the siding along the chalk line. Start the first piece of siding at the end of the sidewall and work toward the other end. The first course is nailed to the bottom plate at points just below each stud. This will mark the nailing locations for the succeeding courses. Fig. 28-14.

5. The amount of overlap is measured along the top of the siding course to be lapped. Snap a chalk line along this mark to locate the butt edge of the second course. Fig. 28-15, line D.

6. The second course is also started at the end of the sidewall.

Align the siding horizontally along the chalk line and nail at each studbearing. E, Fig. 28-15. Use only one nail per bearing. Never nail through both courses of siding. F, Fig. 28-15. Tap the nail head flush with the siding surface. The siding courses should fit snug, not tight.

7. Repeat Steps 5 and 6 for the application of successive courses. Make certain that the vertical butt joints between boards are staggered along the sidewall and that they fall on studs, as shown at G in Fig. 28-16.

The siding should be carefully fitted and be in close contact with the adjacent piece. Some carpenters fit the boards so tightly that they have to spring the boards in place. Tight-fitting butt joints are obtained by cutting the closure board of each course approximately ¹⁄₁₆" too long. Bow the piece slightly to get the ends in

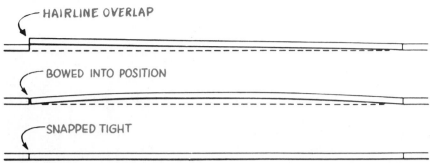

28-17. *To fit siding tightly, cut it about ¹⁄₁₆″ too long, bow it into position, and then snap it tight.*

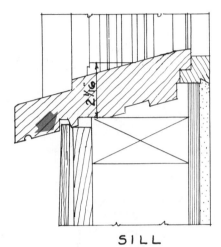

SILL

28-18a. *Bevel siding under window sills should fit in the groove provided.*

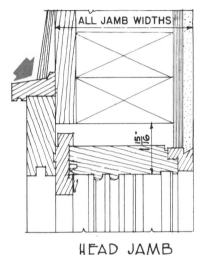

HEAD JAMB

28-18b. *Siding over doors and windows should rest on the drip cap.*

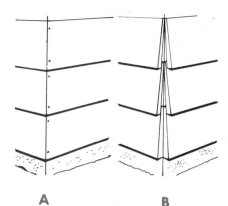

A   B

28-19. *Finishing an outside corner: A. Mitered corners. B. Siding installed and ready for metal corners.*

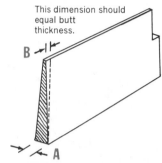

This dimension should equal butt thickness.

28-20. *Mitering of bevel siding corners must be done carefully to obtain a good joint. To lay out and cut the joint, measure the butt thickness at A. Measure back along the top edge a distance at B equal to the butt thickness shown at A. Then connect these two points as shown by the dotted line. With the saw blade set at about a 47° angle, make the cut beginning at the butt end.*

28-21. *Bevel siding at an inside corner. The bevel siding is butted against a square wood corner strip.*

position, and then snap it into place. Fig. 28-17. This assures a tight joint. Loose-fitting joints allow water to get behind the siding. The water can cause paint deterioration around the joints and also set up conditions conducive to decay at the ends of boards.

Siding that passes under a window sill should be cut to fit the groove provided in the bottom of the sill. Fig. 28-18a. Siding installed over doors and windows should be set on the drip cap. Fig. 28-18b.

8. Trim the last course of siding to fit under the eaves and apply a molding if required.

9. Outside corners may be mitered, covered with corner boards, or capped with metal cor-ners. Fig. 28-19. For mitered corners the siding is cut to length and mitered before application. Fig. 28-20. Corner boards are installed before siding application. The siding is then cut to length and butted against the corner boards. If metal corners are used, the siding is cut off even with the outside corner of the building. The metal corners are applied after the siding is in place.

Inside corners are cut before application and butted against a square wood corner strip approximately 1⅛″ × 1⅛″ in size. Fig. 28-21.

Using a Story Pole

You may prepare a story pole when installing horizontal siding to insure accuracy and increase efficiency. Select a straight piece of 1″ × 2″ stock for the siding story pole. Place it under the soffit against a dominant wall of the

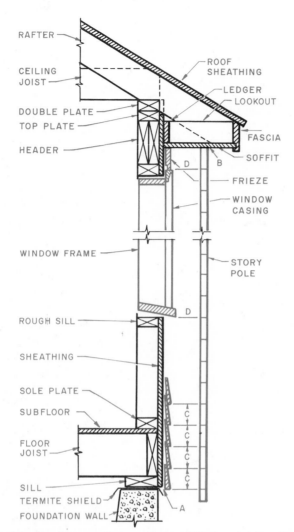

28-23. Transferring the marks of the story pole to the house.

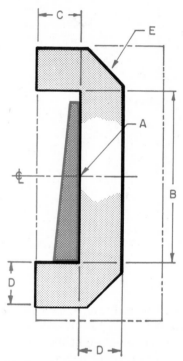

28-22. The story pole should extend from the underside of the soffit to the bottom edge of the first piece of siding as shown from A to B. Hold the story pole in position against the building and lay out the spacing on the story pole as shown at C. Check to be certain that the bottom edge of the piece of siding over the window is even with the top of the window as shown at D.

28-24. Laying out a siding gauge or "preacher".

house, usually the front, and mark the total height. Determine the number of courses and the spacing as described in the section on spacing of bevel siding earlier in this unit. Lay out the spacing on the story pole and check the layout against the building. Fig. 28-22.

Hold the story pole in position up against the soffit. Transfer the marks from the story pole to the house on all corners and on all window and door casings. Fig. 28-23. Make sure that the bottom

marks are clearly visible on the foundation. Snap a chalk line on the bottom marks around the perimeter of the house. Then install the siding as described previously, beginning with Step 4.

Using a Siding Gauge or "Preacher"

A siding gauge, or "preacher," is a small hardwood block used for accurately marking siding pieces to fit between two window casings or a window and a door casing. If corner boards are used

at the corners, it may also be used for marking siding between a corner board and a window or door casing. To make a siding gauge, select a piece of ⅜″ or ½″ hardwood long enough to accommodate the width siding used and proceed as follows:

1. Center the siding on the block of hardwood material. A, Fig. 28-24.

2. Lay out the width of the siding plus ¼″ for clearance. B, Fig. 28-24.

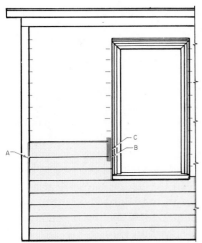

28-25. *Using the preacher to mark the length of a piece of siding.*

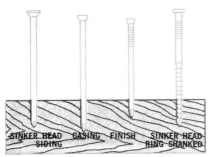

28-27. *Nails used for the application of siding.*

3. Lay out the thickness of the siding plus approximately ⅝". C, Fig. 28-24.

4. Allow about 1" around all of the inside cuts. D, Fig. 28-24.

5. Cut off the corners as shown at E in Fig. 28-24.

Figure 28-25 illustrates how to use the preacher for marking siding between a corner board and a window casing. Cut the end of the siding to fit against the corner board on the spacing mark. A, Fig. 28-25. Align the other end of the siding with the spacing mark on the window casing. B, Fig. 28-25. Place the preacher over the siding and hold it tight against the casing. Holding a pencil against the edge of the preacher, draw a line along the face of the siding. C, Fig. 28-25. Cut the sid-

ing to finish length, position it on the marks, and nail it in place.

Siding Nails and Nailing

Good nails and nailing practices are a must for the proper application of wood siding. Nails should be long enough to penetrate into studs (or studs and wood sheathing combined) at least 1½". When this much penetration is not possible, threaded nails are recommended for increased holding power. Do not nail siding only to composition or pressed fiber sheathing. The nails must penetrate the studs. Nail locations and recommended nail sizes are shown in Fig. 28-2 in the "Application and Nailing" section. However, the following data about nails will be very helpful in the selection and use of the right nail for the application.

NAIL REQUIREMENTS

The following characteristics are essential for nails used on wood siding. Such nails:

- Should be rust-resistant, preferably rust proof.
- Must not cause the siding to discolor or stain.
- Should not cause splitting, even when driven near the end or edge of siding.
- Should have adequate strength to avoid the need for predrilling.
- Should be able to be driven easily and rapidly.
- Should not emerge or "pop" at any time after being driven flush with the siding.
- Should not cause an unsightly visible pattern on the sidewall.

Two types of nails which have these characteristics are:
- High tensile strength aluminum nails.
- Galvanized nails.

High tensile strength aluminum nails are corrosion-resistant and will not discolor or deteriorate the wood siding. They are economical when the nail count per pound is considered, although they are somewhat more expensive than the common galvanized nail. Fig. 28-26.

There are two kinds of galvanized nails: the mechanically plated and the hot-dipped. Mechanical plating is an extremely successful process which provides a nail with a uniform coating, giving it outstanding corrosion resistance. With nails that are hot-dipped, the degree of coating protection varies.

NAIL DESIGN

The design of a nail influences the ease with which it can be driven and its holding power. Nail design includes the head, shank, and point. The basic types of nail heads are illustrated in Fig. 28-27. Nail shanks may be smooth or threaded. Nails that are smooth shanked will loosen under extreme temperature changes. Increased holding power may be obtained by using a ring-threaded or

Nail Size Specification

28-26. *Comparison of aluminum and hot-dipped galvanized nails.*

Size	Length (Inches)		Siding Nails (Count per lb.)		Approx. lbs. Per 1,000 B.F. of Siding	
	*	**	*	**	*	**
6d	1⅞"	2"	566	194	2	6
7d	2⅛"	2¼"	468	172	2½	6½
8d	2⅜"	2½"	319	123	4	9
10d	2⅞"	3"	215	103	5½	11

*Aluminum   **Hot-dipped Galv.

spiral-threaded nail shank. The commonly used nail points include:

• Blunt—reduces splitting.
• Diamond—most widely used.
• Needle—tops in holding but tendency to cause splitting.

For the best possible holding power with the least splitting, a blunt or medium diamond point and a blunt or medium needle point with a ring-threaded shank are recommended.

NAILING RECOMMENDATIONS

When nailing mitered corners or when nailing near the end of a piece, predrill the nail hole or blunt the nail point to avoid splitting the wood. With lapped siding, in order to prevent splitting and to allow expansion clearance, nail just above the lap and not through the tip of the undercourse. Fig. 28-28.

Specific nailing recommendations for standard siding patterns are shown in Figs. 28-2 & 28-29. All recommendations refer to nailing at every stud, if siding

courses are laid up horizontally. If siding is installed vertically, use 2 × 4 blocking between studs. Fig. 28-30. The blocking should be placed at top and bottom, and intermediately, at not more than 24″ on center.

Bevel and Bungalow Siding. Face-nail with one nail per bearing only. Use 8d siding nails for ¾″ thicknesses; 6d nails for thinner pieces. When applying plain bevel and bungalow siding, drive the nails so that the shank just clears the tip of the preceding course. For rabbeted bevel and bungalow siding, set each course to allow an expansion clearance of ⅛″. Drive the nails about 1″ from the lower edge of the course.

Shiplap and Rustic Siding. Face-nail with two siding nails per bearing for patterns wider than 6″. Space each nail about halfway between the center and the edge of a piece. For narrower courses, one nail per bearing is enough. Drive the nail 1″ from the overlapping edge. Use 8d siding nails for 1″ thicknesses, 6d for thinner pieces.

Tongue-and-Groove Siding. Siding 4″ or 6″ wide should be blind-nailed through the tongue with 6d finish nails. Use one nail per bearing. For wider patterns,

face-nail with two 8d siding nails per bearing.

Board and Batten Siding. Space the underboards about ½″ apart and fasten with one 8d siding nail per bearing, driven through the center of the piece. Fasten batten strips with one 10d siding nail per bearing, driven through the center of each piece so that the nail shank passes between the underboards. Variations of the board and batten with recommended nailing procedures are shown in Fig. 28-31.

Exterior Corner Treatment

Wood siding is commonly joined at the exterior corners by corner boards, mitered corners, metal corners, or alternately lapped corners. Fig. 28-32. The method of finishing the wood siding at exterior corners is influenced somewhat by overall house design.

Corner boards are used with bevel or drop siding and are generally made of nominal 1″ or 1¼″ material, depending upon the thickness of the siding. The boards may be plain or molded, depending on the architectural treatment of the house.

The corner boards and the window and door trim may be applied to the sheathing, with the siding

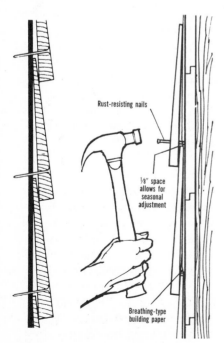

28-28. Correct nailing procedure for lapped siding.

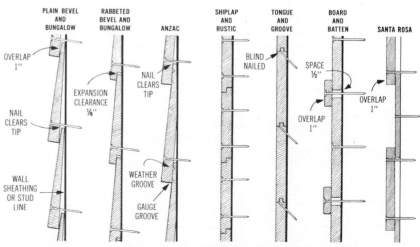

28-29. Suggested nailing methods for typical siding patterns.

28-30. *To provide backing when applying vertical siding, put 2″ × 4″ blocks between the studs at no more than 24″ on center.*

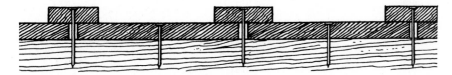

STANDARD BOARD AND BATTEN: One 8d siding nail is driven midway between edges of the underboard, at each bearing. Then apply batten strips and nail with one 10d siding nail at each bearing so that shank passes through space between underboards.

SPECIAL BATTENS: A T-shaped batten or standard batten nailed over a vertical nailing strip, is nailed exactly the same as the standard method; however, in this case an exceptionally good bearing is provided while driving nail through the batten.

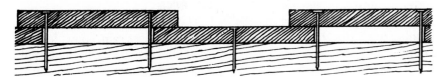

BOARD ON BOARD: Apply underboards first, spacing them to allow 1½-inch overlap by outer boards at both edges. Use standard nailing for underboards, one 8d siding nail per bearing. Outer boards must be nailed twice per bearing to insure proper fastening. Nails, having some free length, do not hold outer boards so rigidly as to cause splitting if there is "movement" from humidity changes. Drive 10d siding nails so that shanks clear edges of underboard approximately ¼-inch. This provides sufficient bearing for nailing, while allowing clearance to enable underboard to expand slightly.

REVERSE BATTEN: Nailing is similar to board on board. Drive one 8d nail per bearing through center of under strip, and two 10d siding nails per bearing through outer boards.

28-31. *Nailing details for various board and batten applications.*

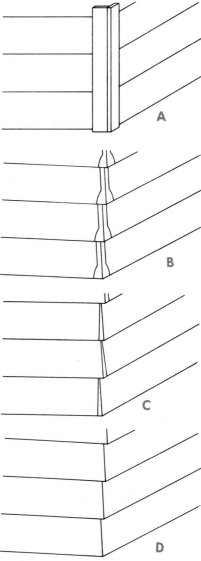

28-32. *Methods of treating bevel siding on an outside corner: A. Corner boards. B. Metal corners. C. Alternately lapped corners. D. Mitered corners.*

fitted tightly against the narrow edge of the corner boards and against the trim. When this method is used, the joints between the siding and the corner boards or trim should be calked or treated with a water repellent. Sometimes corner boards and trim around windows and doors are applied over the siding, a method that minimizes the entrance of water into the ends of

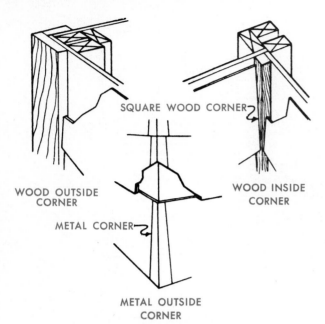

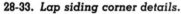

28-33. *Lap siding corner details.*

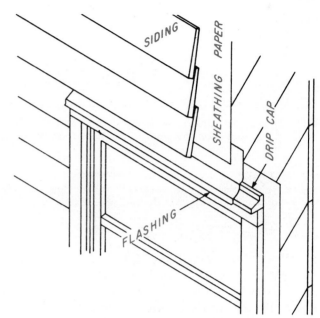

28-34. *Flashing should be used above windows and doors.*

the siding. This method works better for panel siding than for bevel siding. Fig. 28-33.

Mitered corners, sometimes used with the thicker patterns, should be cut in a miter box and must fit tightly and smoothly for the full depth of the miter. To maintain a tight fit at the miter, it is important that the siding be properly seasoned before delivery and protected from rain when stored at the site. The ends should be set in white lead when the siding is applied, and the exposed faces should be primed immediately after it is applied. Nail mitered ends to the corner posts, not to each other.

Metal corners are made of light-gauge metals, such as aluminum or galvanized iron, and are used with bevel siding as a substitute for mitered corners. Fig. 28-33. They can be purchased at most lumberyards. The application of metal corners takes less experience than is required to make good mitered corners or to fit siding to corner boards. Metal corners should be set in white lead paint.

Alternately lapped corners are fitted so that every other piece of siding has the end exposed. Fig. 28-32.

Interior Corner Treatment

Interior corners of siding are butted against a corner strip of nominal 1" or 1¼" material, depending upon the thickness of the siding. Fig. 28-33.

Preventing Outside Moisture Problems

Poor construction detailing may enable water to seep into the siding, eventually causing paint or finish deterioration. Poor construction may also result in inadequate insulation, causing discomfort and high heating bills. To avoid these problems take the following precautions at the points where trouble may occur:

CAREFUL FITTING

With any siding pattern, good joints are essential. Accurate cutting of pieces is the only way to insure the proper fit.

Bevel courses should have sufficient lap to prevent wind-driven

rain from working up between courses. FHA Minimum Property Standards require a 1" lap.

Gutter joints and downspouts are other areas which must be carefully fitted.

CALKING

The sealing of all joints helps protect against rain, snow, fog, and wind. It is particularly important at the butt joints of short length siding laid vertically. Use a nonhardening calking compound.

FLASHING

Flashing is necessary to drain away water at places where horizontal surfaces meet the siding. These places include the areas over door and window frames and around dormers. Figs. 28-34 & 28-35. Where siding returns against a roof, the siding should not be fitted tight against the shingles, but should have a clearance of 2". Windblown water working into the back of the siding is a potential cause of paint failure. Siding cannot dry out quickly where there is a tight fit. Fig. 28-35.

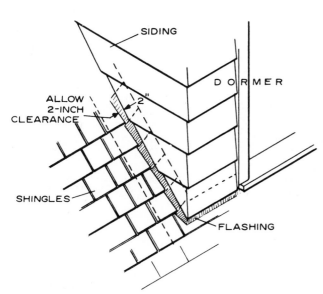

28-35. *Flashing should be used around dormers at the intersection of the siding and the roof.*

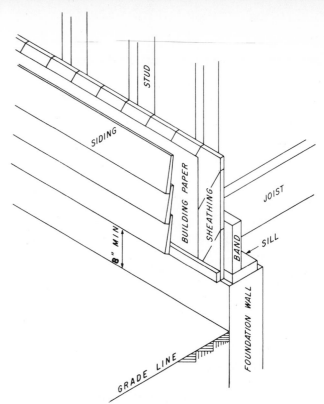

28-37. *Exterior siding should begin at least 8″ above the grade line.*

Flashing should be anchored tightly. It should extend well under the siding and sufficiently over edges and ends of well-sloped water tables to prevent water from running in behind siding or jambs. A bead of calking should be laid under the flashing to help seal out moisture. Fig. 28-36.

FOUNDATION LINES

The lowest edge of the siding should be at least 8″ above ground level. Fig. 28-37. Water (often present at the base of a foundation due to landscaping) and high humidity can cause finish difficulties and structural problems as well.

It is particularly important that the end grain at the bottom of vertical siding be given a water repellent treatment. The use of a drip cap at the lower edge of the siding will help direct water away from the foundation.

PLYWOOD AND OTHER MANUFACTURED MATERIALS

Plywood Siding

Plywood sheets are often used in gable ends, sometimes around windows and porches, and occasionally as overall exterior wall covering. Figs. 28-38 & 28-39. Plywood siding comes in many grades and surface textures, providing almost unlimited freedom of design for all types of construction. Fig. 28-40. For best results with painted surfaces, specify "medium-density overlaid" ply-

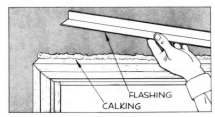

28-36. *Calking should be used under the flashing.*

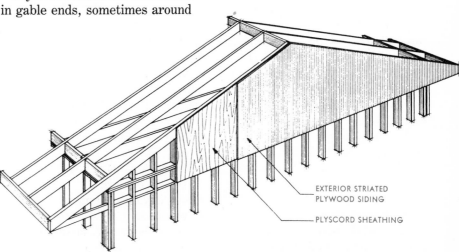

NOTE: PROVIDE ADEQUATE ATTIC VENTILATION THROUGH GABLE END OR SOFFIT OF ROOF OVERHANG

28-38. *Plywood used as an exterior covering for a gable.*

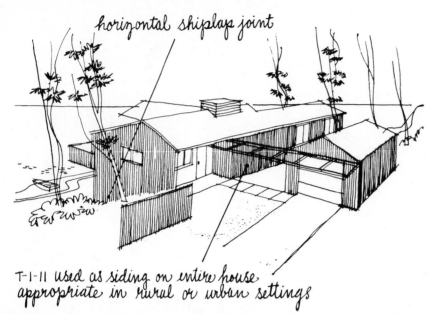

horizontal shiplap joint

T-1-11 used as siding on entire house appropriate in rural or urban settings

28-39. *Plywood panel siding used for the complete exterior of a home. Note the use of the shiplap joint to make a continuous surface when greater lengths are required.*

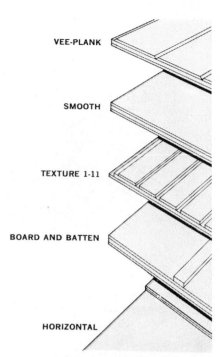

VEE-PLANK

SMOOTH

TEXTURE 1-11

BOARD AND BATTEN

HORIZONTAL

28-40. *Five of many plywood siding styles available.*

wood. This grade has a resin-impregnated fiber surface that is heat-fused to the panel faces. It takes paint well and holds it longer. Sheet siding can be applied directly to studs, thus eliminating the need for sheathing. Fig. 28-41.

Plywood siding is strong. Tests conducted by United States Forests Products Laboratory at Madison, Wisconsin, proved that plywood as thin as ¼″, when nailed directly to studs, provides more than twice the relative rigidity and more than three times the relative strength of 1″ × 8″ lumber sheathing nailed horizontally to studs.

Plywood siding can be applied horizontally or vertically. The joints can be battens, V-grooves, or flush joints. Sometimes plywood is installed as lap siding. Figs. 28-42 & 28-43.

Plywood panel (sheet) siding and plywood lap siding come in 8′ standard and 12′, 14′, and 16′ special lengths. Plywood panel siding also comes in 9′ lengths.

Fig. 28-44. Lap siding may be 12″, 16″, or 24″ wide. Plywood siding will cover large areas and cut installation time. The wide exposure of the lap siding or the use of panel siding is in keeping with the style of the modern ranch home.

APPLICATION OF PLYWOOD PANEL SIDING

Plywood panel siding is normally installed vertically but may be installed horizontally. All edges of panel siding should be backed with framing members or blocking. Fig. 28-45. To prevent staining of the siding, galvanized, aluminum, or other noncorrosive nails are recommended. Fig. 28-46.

To apply panel siding follow these suggestions:
● Single wall construction with plywood paneling less than ½″ thick will permit a maximum stud spacing of 16″ on center when no building paper, corner bracing, or sheathing is used. When sheathing is used, the maximum stud

28-41. *Applying a texture 1-11 plywood siding panel.*

spacing may be increased to 24″ on center. No building paper or corner bracing is required.
● Nail 6″ on center on panel edges and 12″ on center at intermediate supports. Fig. 28-42.

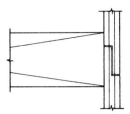

Shiplap—Horizontal or Vertical Joint

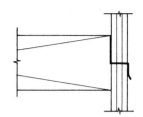

Flashed—Horizontal Joint galv. or alum. flashing

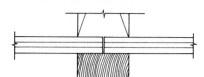

Batten (panel only)—Vertical Joint

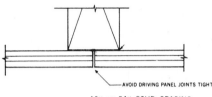

Butt —Vertical Joint

AVOID DRIVING PANEL JOINTS TIGHT

Note:

When finish is paint, prime all panel edges prior to application. Otherwise treat edges with water repellants or caulk joints. Building paper may be omitted under panel sidings; also under lapped and bevelled siding when plywood sheathing is used. Joints may occur away from studs when plywood or board sheathing is used.

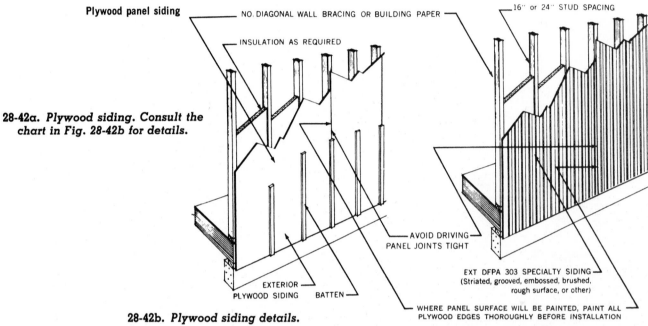

Plywood panel siding

NO. DIAGONAL WALL BRACING OR BUILDING PAPER

INSULATION AS REQUIRED

16" or 24" STUD SPACING

28-42a. *Plywood siding. Consult the chart in Fig. 28-42b for details.*

AVOID DRIVING PANEL JOINTS TIGHT

EXTERIOR PLYWOOD SIDING   BATTEN

EXT DFPA 303 SPECIALTY SIDING
(Striated, grooved, embossed, brushed, rough surface, or other)

WHERE PANEL SURFACE WILL BE PAINTED, PAINT ALL PLYWOOD EDGES THOROUGHLY BEFORE INSTALLATION

28-42b. *Plywood siding details.*

Plywood siding direct to studs (Plywood continuous over two or more spans) / Recommendations apply to all species groups.

Panel Siding	Maximum Stud Spacing c. to c. (inches)	Nail Size & Type[1]	Nail Spacing (inches)	
Minimum Plywood Thickness (inch)			Panel Edges	Intermediate
⅜	16	6d ⎧ non-corrosive	6	12
½	24	6d ⎪ siding or casing	6	12
⅝	24	8d ⎨ (galv. or alum.)	6	12
T 1-11 (⅝)	16	8d ⎩	6[2]	12

Lap or Bevel Siding[3]					
Typical Width (inches)	Min. Lap Siding Thickness (inch)	Min. Bevel Butt Thickness (inch)			
12, 16 or 24	⅜	9/16	16	6d ⎧ non-corrosive	One nail per	4" at vertical
	½		20	8d ⎨ siding or casing	stud along	joint; 8" at studs if
	⅝		24	8d ⎩ (galv. or alum.)	bottom edge	siding wider than 12"

Notes: (1) Nails through battens must penetrate studs at least 1".
(2) Use single nail on shiplap edges slant-driven to catch both edges. Can nail to ⅜" from panel edge, but do not set nails. Nails may be set if placed on both sides of joint instead of slant-driven.
(3) Minimum head-lap 1½".

Look for these typical APA grade-trademarks.

M.D. OVERLAY

GROUP 1 EXTERIOR PS 1-74 000 (APA)

Sanded Grades

A-C

GROUP 2 EXTERIOR PS 1-74 000 (APA)

Specialty Panels

303 SIDING 16 oc

GROUP 3 EXTERIOR PS 1-74 000 (APA)

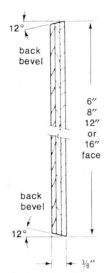

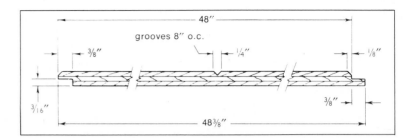

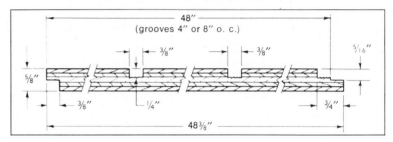

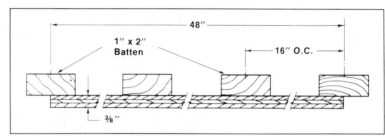

28-43. *Details for plywood panel siding.*

28-44. *Applying plywood lap siding in 16' lengths.*

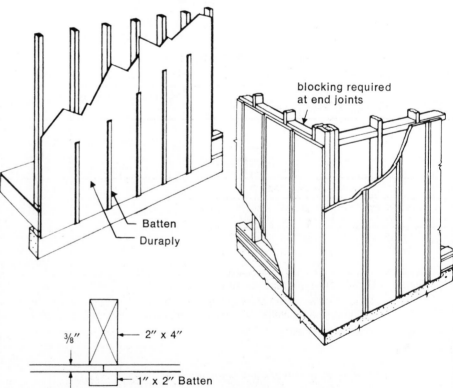

28-45. *Siding details for board and batten and for panel siding.*

320

Framing and Nailing Schedule						
Panel Siding Thickness	5/16″	3/8″	1/2″ grooved	1/2″ flat	5/8″ grooved	5/8″ flat
Single Wall Construction						
Maximum Stud Spacing	—	16″ o.c.	16″ o.c.	24″ o.c.	16″ o.c.	24″ o.c.
Nail Size	—	6d	8d	8d	8d	8d
Over 3/8″ Sheathing						
Maximum Stud Spacing	24″	24″	24″	24″	24″	24″
Nail Size	6d	6d	6d	6d	8d	8d
Approximate Nail Spacing*						
Edges	6″	6″	6″	6″	6″	6″
Intermediate Members	12″	12″	12″	12″	12″	12″

*Use non-corrosive casing, siding, or box nails.

28-46. Framing and nailing schedule for plywood panel siding.

Plywood lap or bevel siding

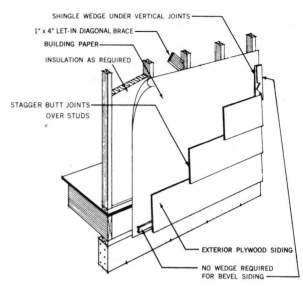

28-47. Plywood lap siding applied to an exterior wall without sheathing.

APPLICATION OF PLYWOOD LAP SIDING

To apply lap plywood siding, follow these suggestions:

• In single wall construction the maximum stud spacing is 16″ on center, and "let-in" corner bracing must be installed to meet FHA requirements. Install building paper between the siding and the studs. Fig. 28-47. The minimum head lap is 2″. Wedges are installed at butt ends and at corners. When sheathing is used, corner bracing is not required and the maximum stud spacing is 24″ on center. The minimum head lap for 12″ widths or less is 1″. For widths over 12″ use a 1½″ head lap.

• Use a 3/8″ thick starter strip for the first course. Fig. 28-48.

• Coat the edges of the siding with a primer before application.

• Vertical joints should be staggered. These joints must be centered over studs with a tapered wedge at least 1⅝″ wide behind the joint. Fig. 28-48.

• Use 8d noncorrosive casing or

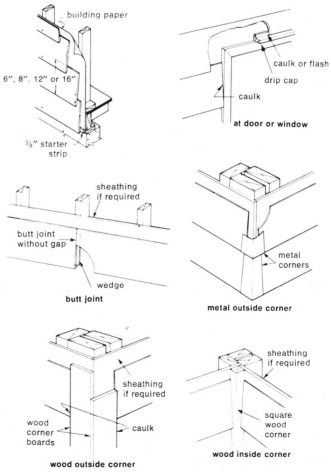

28-48. Typical lap siding details.

321

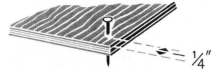

28-49. *Nails for lap siding should be placed ¼" back from the edge of the plywood.*

Panel thickness	Single wall construction			Over sheathing		Approximate nail spacing		
	Maximum stud spacing	Nail Size		Maximum stud spacing	Nail Size	Edges	Intermediate members	Distance from edge
7/16"	16" o.c.	6d		24"	8d	4"	8"	½"

28-50. *Hardboard framing and nailing schedule.*

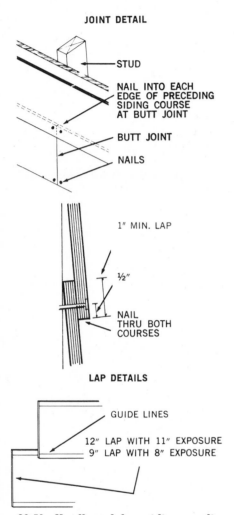

JOINT DETAIL

STUD

NAIL INTO EACH EDGE OF PRECEDING SIDING COURSE AT BUTT JOINT

BUTT JOINT

NAILS

1" MIN. LAP

½"

NAIL THRU BOTH COURSES

LAP DETAILS

GUIDE LINES

12" LAP WITH 11" EXPOSURE
9" LAP WITH 8" EXPOSURE

28-51. *Hardboard lap siding application details.*

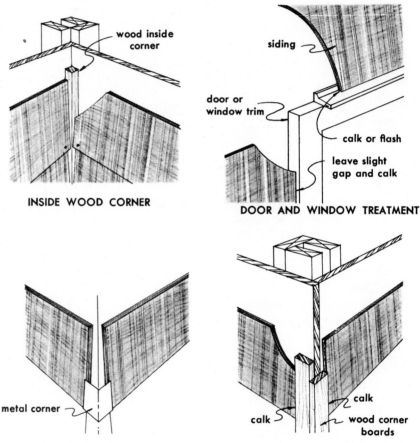

wood inside corner

INSIDE WOOD CORNER

siding

door or window trim

calk or flash

leave slight gap and calk

DOOR AND WINDOW TREATMENT

metal corner

OUTSIDE METAL CORNER

calk

calk

calk

wood corner boards

OUTSIDE WOOD CORNER

28-52. *Corner treatment for hardboard lap siding.*

box nails. Insert one nail at each stud on the bottom edge of the siding. At all vertical joints nail 4" on center for siding 12" wide or less. Nail 8" on center for siding 16" wide or more. All nails should be placed ¼" back from the edge of the plywood. Fig. 28-49. Set

and putty all casing nails. Box nails are driven flush.

Hardboard and Particleboard Siding

The hardboard exterior sidings are fabricated for use as lap siding or as large panels up to 16' in length by 4' in width. They are impregnated with a baked-on tempering compound. This process produces tough, dense, grainless sidings that will not split or splinter and are highly re-

sistant to denting. Panel surfaces are completely free of imperfections.

Hardboard sidings reinforce wall construction, go up quickly, and can be easily worked with both power and regular woodworking tools. Hardboards take paint, enamel, and stain and hold the finish longer than lumber. The sidings are available unprimed or factory primed. Factory-primed siding can be exposed to the weather for as long as 60 days prior to application of the finishing coats.

Particleboard can also be used as siding material. It is available in many of the same shapes and surfaces as hardboard or plywood and is applied in the same manner.

APPLICATION OF HARDBOARD SIDING

Hardboard siding may be applied over sheathed walls with studs spaced not more than 24″ on center or over unsheathed walls with studs spaced not more than 16″ on center. The lowest edge of the siding should be at least 8″ above the finished grade level.

When hardboard siding is applied directly to studs or over wood sheathing, moisture-resistant building paper or felt (non-vapor-barrier) should be laid directly under the siding.

As in all frame construction, a vapor barrier must be used in all insulated buildings and in uninsulated buildings located in areas where the average January temperature is below 40° F. The vapor barrier is installed next to the heated wall. The installed vapor barrier must be continuous, with tight joints and with any breaks or tears repaired.

When applying hardboard siding, use rustproof siding nails. Nail only at stud locations and on special members around doors and windows. Fig. 28-50. Nails must be kept back ½″ from the ends and edges of the siding pieces. Fig. 28-51.

At inside corners, siding should be butted (with approximately ¹⁄₁₆″ space) against a 1⅛″ × 1⅛″ wood corner member. Outside corners may be 1⅛″ wood corner boards, or metal corners may be used. Fig. 28-52. Calking should be applied wherever the siding butts against corner boards, windows, and door casings. Fig. 28-52.

Installing Flat Panels. Flat panels are installed vertically. All joints and panel edges should fall on the center of framing members. If it is necessary to make a joint with a panel that has been field cut and the shiplap joint removed, use a butt joint. Butter the edges with calking and bring to light contact. Do not force or spring panels into place. Leave a slight space where siding butts against window or door trim and calk. Fig. 28-53.

Installing Lap Siding. Start the application by fastening a wood starter strip (⅜″ × 1⅜″) along the bottom edge of the sill. Fig. 28-54. Level and install the first course of siding with the bottom edge at least ⅛″ below the starter strip. Fasten the first course by nailing 1½″ from drip edge of siding and ½″ from butt end.

Install subsequent siding courses using a minimum overlap of 1″. Fig. 28-51. Butt joints should occur only at stud locations. Factory-primed ends should be used for all vertical butt joints which will not be covered. Adjacent siding pieces should just touch at butt joints, or a ¹⁄₁₆″ space may be left and filled with a butyl calk. Never force or spring siding into place.

WOOD SHINGLES AND SHAKES

Wood shingles and shakes are widely used for wall coverings, and a large selection is available. Wall shingles come in lengths of 16″, 18″, and 24″. They may be prefinished or finished after installation. Handsplit shakes come in lengths of 18″, 24″, and 32″.

Shingles

Shingles are usually separated into four grades. The first grade is composed of clear shingles, all heart, all edge grain. The second grade consists of shingles with

clear butts and allows defects in that part of the shingle that will normally be covered in use. The third grade includes shingles that have defects other than permitted in the second grade. The fourth grade is a utility grade for undercoursing on double-coursed sidewall applications or for interior accent walls.

Shingles are made in random widths. In the No. 1 grade, they vary from 3″ to 14″, with only a small proportion of the narrow width permitted. Shingles cut uniformly to widths of 4″, 5″, or 6″ are also obtainable. They are known as dimension or rebutted-and-rejointed shingles. These are shingles with edges machine-trimmed so as to be exactly parallel, and with butts retrimmed at precise 90° angles. Dimension shingles are applied with tight-fitting joints to give a strong horizontal line. They are available with the natural "sawed" face or with one face sanded smooth. These shingles may be applied either single- or double-coursed.

APPLICATION OF SHINGLES

There are two basic ways to apply shingles: single-course and double-course. In single-coursing, shingles are applied much as in roof construction, but greater weather exposures are permitted. Shingle walls have two layers of shingles at every point, whereas shingle roofs have three-ply construction. Fig. 28-55 (A). To obtain architectural effect with deep bold shadow lines, shingles are frequently laid in double courses. Fig. 28-55 (B and C). Double-coursing allows for the application of shingles at extended weather exposures over undercoursing-grade shingles which are less expensive. When double-coursed, a shingle wall should be tripled at the foundation line (by using a double underlay). Fig. 28-56.

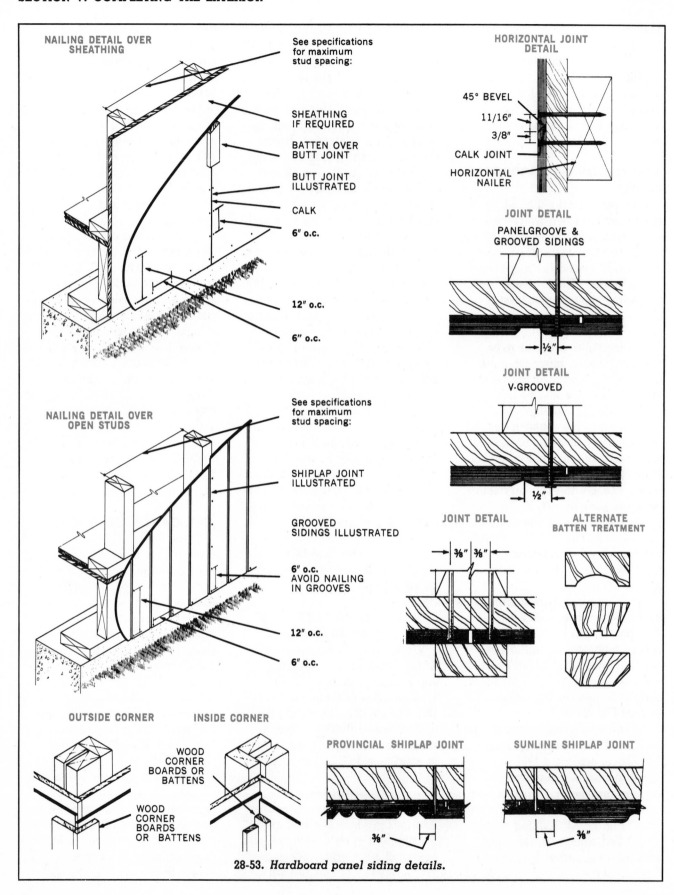

NAILING DETAIL OVER SHEATHING

See specifications for maximum stud spacing:

SHEATHING IF REQUIRED

BATTEN OVER BUTT JOINT

BUTT JOINT ILLUSTRATED

CALK

6″ o.c.

12″ o.c.

6″ o.c.

NAILING DETAIL OVER OPEN STUDS

See specifications for maximum stud spacing:

SHIPLAP JOINT ILLUSTRATED

GROOVED SIDINGS ILLUSTRATED

6″ o.c. AVOID NAILING IN GROOVES

12″ o.c.

6″ o.c.

OUTSIDE CORNER

INSIDE CORNER

WOOD CORNER BOARDS OR BATTENS

WOOD CORNER BOARDS OR BATTENS

HORIZONTAL JOINT DETAIL

45° BEVEL

11/16″

3/8″

CALK JOINT

HORIZONTAL NAILER

JOINT DETAIL
PANELGROOVE & GROOVED SIDINGS

½″

JOINT DETAIL
V-GROOVED

½″

JOINT DETAIL

⅜″ ⅜″

ALTERNATE BATTEN TREATMENT

PROVINCIAL SHIPLAP JOINT

⅜″

SUNLINE SHIPLAP JOINT

⅜″

28-53. Hardboard panel siding details.

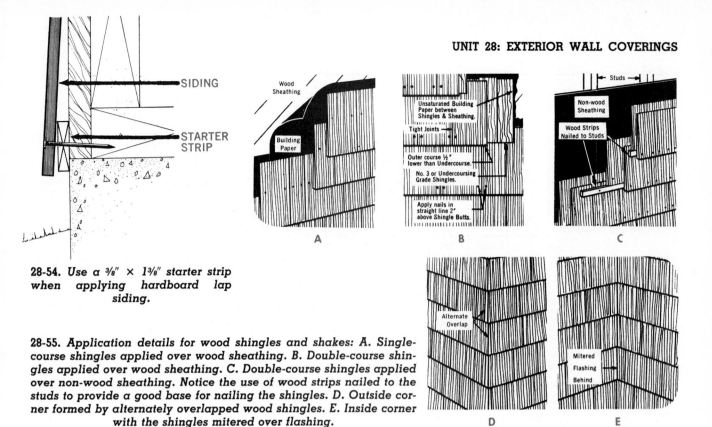

28-54. *Use a ⅜″ × 1⅜″ starter strip when applying hardboard lap siding.*

28-55. *Application details for wood shingles and shakes: A. Single-course shingles applied over wood sheathing. B. Double-course shingles applied over wood sheathing. C. Double-course shingles applied over non-wood sheathing. Notice the use of wood strips nailed to the studs to provide a good base for nailing the shingles. D. Outside corner formed by alternately overlapped wood shingles. E. Inside corner with the shingles mitered over flashing.*

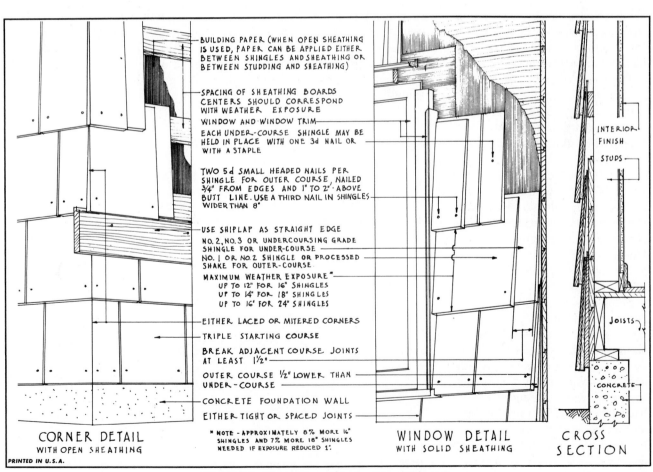

28-56. *Double-course shingle sidewall detail.*

	Approximate coverage of one square (4 bundles) of shingles based on following weather exposures												
LENGTH AND THICKNESS 16″ x 5/2″	3½″	4″	4½″	5″	5½″	6″	6½″	7″	7½″	8″	8½″	9″	9½″
	70	80	90	100*	110	120	130	140	150I	160	170	180	190
LENGTH AND THICKNESS 16″ x 5/2″	10″	10½″	11″	11½″	12″	12½″	13″	13½″	14″	14½″	15″	15½″	16″
	200	210	220	230	240†	....	....	....	....	....	....	....	....
LENGTH AND THICKNESS 18″ x 5/2¼″	3½″	4″	4½″	5″	5½″	6″	6½″	7″	7½″	8″	8½″	9″	9½″
	....	72½	81½	90½	100*	109	118	127	136	145½	154½I	163½	172½
LENGTH AND THICKNESS 18″ x 5/2¼″	10″	10½″	11″	11½″	12″	12½″	13″	13½″	14″	14½″	15″	15½″	16″
	181½	191	200	209	218	227	236	245½	254½	....	....	....	....
LENGTH AND THICKNESS 24″ x 4/2″	3½″	4″	4½″	5″	5½″	6″	6½″	7″	7½″	8″	8½″	9″	9½″
	....	....	....	....	....	80	86½	93	100*	106½	113	120	126½
LENGTH AND THICKNESS 24″ x 4/2″	10″	10½″	11″	11½″	12″	12½″	13″	13½″	14″	14½″	15″	15½″	16″
	133	140	146½	153I	160	166½	173	180	186½	193	200	206½	213†

Notes: * Maximum exposure recommended for roofs.
I Maximum exposure recommended for single-coursing on sidewalls.
† Maximum exposure recommended for double-coursing on sidewalls.

28-57. Wood shingle exposure and coverage chart. The thickness dimension represents the total thickness of a number of shingles. For example, 5/2″ means that 5 shingles, measured across the thickest portion, when green, measure 2 full inches.

When the wall is single-coursed, the shingles should be doubled at the foundation line. For recommended exposures see Fig. 28-57.

The spacing for the shingle courses is determined the same way as described for bevel siding. When shingles are applied over fiberboard or gypsum sheathing, horizontal 1″ × 4″ nailing strips should first be nailed to the studs. Fig. 28-58. The on-center spacing of these strips should be the same distance as the weather exposure chosen for the shingles, to provide a good base for nailing. Shingles may be staggered for rustic effect. Fig. 28-59.

Shingles should be applied with rust-resistant nails. At least ¼″ space should be allowed between shingles of the same course. It is frequently recommended that no shingle should be laid that is more than 8″ in width. Shingles wider than this should be sawed or split and nailed as two shingles.

For double-coursing, each outer course shingle should be secured with two 5d (1¾″) small head, rust-resistant nails driven about two inches above the butts, ¾″ in from each side. Additional nails should be driven about four inches apart across the face of the shingle. Fig. 28-55 (B and C). Single-coursing involves the same number of nails, but they can be shorter (3d, 1¼″) and should be blind-nailed not more than 1″ above the butt line of the next course. Fig. 28-55 (A). Never

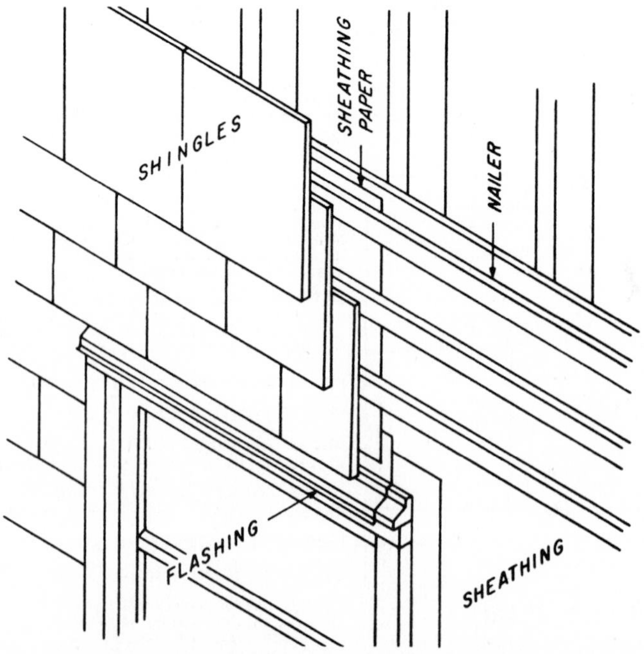

28-58. The application of wood shingles over non-wood sheathing. Note the use of the nailing strip.

28-59. *Staggered shingles give a rustic appearance.*

drive the nail so hard that its head crushes the wood.

Outside corners should be constructed with an alternate overlap of shingles between successive courses. Fig. 28-55 (D). Inside corners may be mitered over a metal flashing. Fig. 28-55 (E). They may also be made by nailing an S4S strip, 1½" or 2" square, in the corner, after which the shingles of each course are jointed to the strip.

Shakes

There are three kinds of shakes:
• Handsplit-and-resawn.
• Tapersplit.
• Straightsplit.

The handsplit-and-resawn shakes have split faces and sawn backs. Cedar logs are first cut into desired lengths. Blanks or boards of proper thickness are split and then run diagonally through a bandsaw to produce two tapered shakes from each blank.

Tapersplit shakes are produced largely by hand, using a sharp-bladed steel froe and a wooden mallet. The natural shingle-like taper is achieved by reversing the block end-for-end with each split.

Straightsplit shakes are produced in the same manner as tapersplit shakes except that by splitting from the same end of the block the shakes acquire the same thickness throughout. Fig. 28-60.

APPLICATION OF SHAKES

Maximum recommended weather exposure with single-course application is 8½" for 18" shakes, 11½" for 24" shakes, and 15" for 32" shakes. Fig. 28-61. The nailing normally is concealed in single-course applications; that is,

28-60. *Using a froe to split shakes in the 1800's. Today, tapersplit and straightsplit shakes are still produced largely by hand.*

the nailing is done at points slightly above (about 1") the butt line of the course to follow.

Double-course application requires an underlay of shakes or regular cedar shingles. With handsplit-resawn or with tapersplit shakes, the maximum weather exposure is 14" for 18" shakes and 20" for 24" shakes. If

	Approximate sq. ft. coverage of one square of handsplit shakes based on these weather exposures											
	5½"	6½"	7"	7½"	8"	8½"	10"	11½"	13"	14"	15"	16"
18" x ½" to ¾" Handsplit-and-Resawn	55*	65	70	75**	80	85†	. . .	. . .	. . .	140‡	. . .	. . .
18" x ¾" 1¼" Handsplit-and-Resawn	55*	65	70	75**	80	85†	. . .	. . .	. . .	140‡	. . .	. . .
24" x ⅜" Handsplit	. . .	65	70	75***	80	85	100††	115†	. . .	. . .	. . .	. . .
24" x ½" to ¾" Handsplit-and-Resawn	. . .	65	70	75*	80	85	100**	115†	. . .	. . .	. . .	. . .
24" x ¾" to 1¼" Handsplit-and-Resawn	. . .	65	70	75*	80	85	100**	115†	. . .	. . .	. . .	. . .
32" x ¾" to 1¼" Handsplit-and-Resawn	. . .	. . .	. . .	. . .	. . .	. . .	100*	115	130**	140	150†	. . .
24" x ½" to ⅝" Tapersplit	. . .	65	70	75*	80	85	100**	115†	. . .	. . .	. . .	. . .
18" x ⅜" True-Edge Straightsplit	. . .	. . .	. . .	. . .	. . .	. . .	. . .	. . .	. . .	100	106	112‡
18" x ⅜" Straightsplit	65*	75	80	90	95	100†	. . .	. . .	. . .	. . .	. . .	. . .
24" x ⅜" Straightsplit	. . .	65	70	75*	80	85	100	115†	. . .	. . .	. . .	. . .
15" Starter-Finish Course	Use supplementary with shakes applied not over 10" weather exposure.											

Notes:
    * Recommended maximum weather exposure for 3-ply roof construction.
  ** Recommended maximum weather exposure for 2-ply roof construction.
 *** Recommended maximum weather exposure for roof pitches of ⁴/₁₂ to ⁸/₁₂.
    † Recommended maximum weather exposure for single-coursed wall construction.
  †† Recommended maximum weather exposure for roof pitches of ⁸/₁₂ or steeper.
   ‡ Recommended maximum weather exposure for double-coursed wall construction.

28-61. *Wood shake exposure and coverage chart.*

straightsplit shakes are used, the exposure may be 16″ for 18″ shakes and 22″ for 24″ shakes. Butt-nailing of shakes is required with double-course application.

Use rust-resistant nails, preferably hot-dipped zinc-coated. The 6d size normally is adequate, but longer nails may be required, depending on the thickness of the shakes and the weather exposure. Do not drive nailheads into the shake surface. The methods for constructing the corners when applying shakes are shown in Fig. 28-62.

Finishing Wood Shingles and Shakes

Red cedar shingles and shakes are well equipped by nature to endure without any protective finish or stain. In this state, the wood will eventually weather to a silver or dark gray. The speed of change and final shade depend mainly on atmosphere and climate.

Bleaching agents may be applied, in which case the wood will turn an antique silver gray. So-called natural finishes, which are lightly pigmented and maintain the original appearance of the wood, are available commercially. Stains, whether heavy or semi-transparent, are readily "ab-sorbed" by cedar, and it also takes paint well. Quality finishes are strongly recommended because they will prove most economical on a long-term basis.

ASBESTOS-CEMENT SIDING AND SHINGLES

Asbestos-cement siding and shingles come in various sizes and colors. They should be applied in accordance with the manufacturer's directions.

Wood sheathing should be used under asbestos-cement shingles and siding. The siding or shingles are laid over a waterproof paper applied over the sheathing. The courses are laid out in the same manner as for horizontal bevel siding, or a story pole may be used. Noncorroding nails should be used, and care should be taken when driving nails not to crack the shingles. Vertical joints should be flashed with 4″ wide strips of saturated felt laid under each joint. Fig. 28-63.

The treatment of corners may be similar to those used for wood siding. However, in most cases, the manufacturers will suggest the type of corner treatment best suited for their products. The corners should be flashed with a wide strip of asphalt paper underlay applied vertically.

STUCCO SIDEWALL FINISH

Stucco when properly used makes a good wall finish. Fig. 28-64. It may be a natural cement color or colored as desired. If stucco is to be applied on houses more than one story high, balloon framing should be used in the outside walls. With platform framing, shrinkage of the joists and sills of the platform may cause an unsightly bulge or break in the stucco at that point.

Stucco is applied over lath. Three acceptable types of lath are:
● Zinc-coated or galvanized metal, with large openings (1.8 pounds per square yard) or small openings (3.4 pounds per square yard).
● Galvanized woven-wire fabric.

28-62. Wood shake corner treatment.

28-63. Asbestos-cement shingles over plywood sheathing.

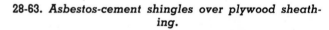

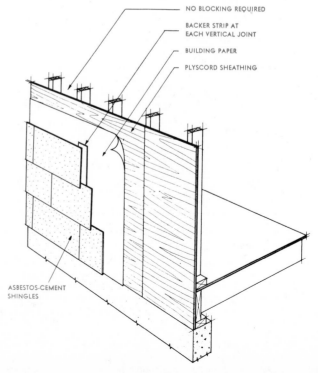

NO BLOCKING REQUIRED

BACKER STRIP AT EACH VERTICAL JOINT

BUILDING PAPER

PLYSCORD SHEATHING

ASBESTOS-CEMENT SHINGLES

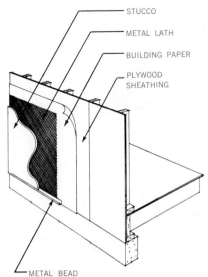

28-64. *Stucco applied as an exterior covering over plywood sheathing.*

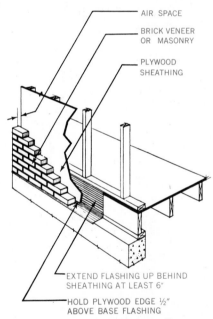

28-65. *Brick veneer as a wall covering over wood-frame construction.*

This material may be 18-gauge wire with 1″ maximum mesh, 17-gauge wire with 1½″ maximum mesh, or 16-gauge wire with a 2″ maximum mesh.

• Galvanized welded-wire fabric. This may be made of 16-gauge wire with 2″ × 2″ mesh and waterproof paper backing, or it may be 18-gauge wire with 1″ × 1″ mesh but no paper backing.

The lath should be kept at least ¼″ away from the sheathing so that the stucco can be forced through the lath and embedded completely. Galvanized furring nails, metal furring strips, or self-furring lath are available for this spacing. Nails should penetrate the wood at least ¾″. Where fiberboard or gypsum sheathing is used, the length of the nail should be such that at least ¾″ penetrates into the wood stud.

Stucco Plaster

The plaster should be one part portland cement, three parts sand, and hydrated lime equal to 10% of the cement by volume. It should be applied in three coats to a total thickness of 1″. The first coat should be forced through the lath and worked so as to embed the lath at all points. Keep fresh stucco shaded and wet for three days. Do not apply stucco when the temperature is below 40° F. It sets very slowly, and there may be a freezing hazard before it has set.

Portland-cement stucco that has been commercially prepared may also be used. It should be applied according to the manufacturer's instructions.

MASONRY VENEER

Brick or stone veneer is often used for part or all of the wall covering over wood frame walls. Fig. 28-65. Although high in initial cost, it is frequently used as a wall covering in residential construction because of the low maintenance. Brick or stone veneer is applied by a skilled mason.

METAL AND PLASTIC SIDING

There are many kinds of metal and plastic sidings that can be applied over sheathing. Metal sid-

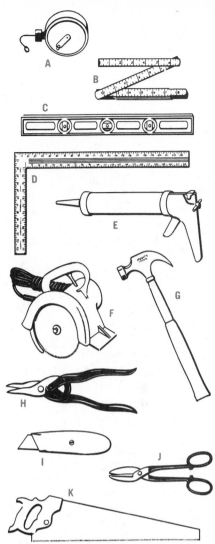

28-66. *Basic tools necessary for the application of metal or plastic siding: A. Chalk line. B. Rule. C. Level. D. Carpenter's square. E. Calking gun. F. Power saw (with a plywood cutting blade). G. Hammer. H. Double-action aviation snips. I. Utility knife. J. Tin snips. K. Crosscut saw.*

ing, either aluminum or steel, has a baked-on finish that requires little maintenance or care. Solid vinyl siding in various colors is also available. Standard hand tools such as those shown in Fig. 28-66 may be used to apply these siding materials.

Application of Metal and Plastic Siding

The interlocking joints of the pieces of siding and the accessory

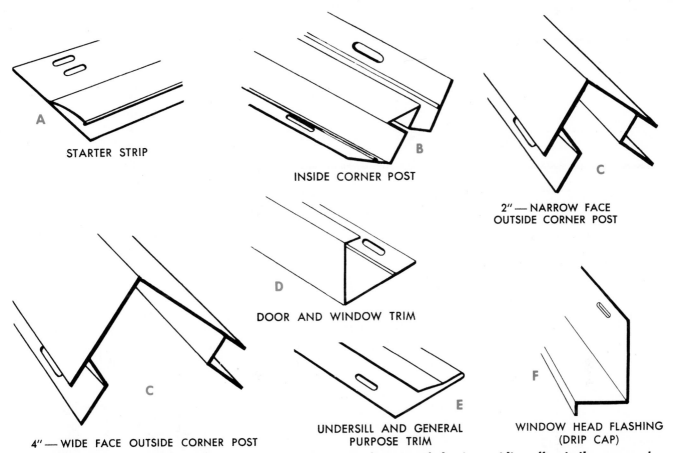

A STARTER STRIP

B INSIDE CORNER POST

C 2" — NARROW FACE OUTSIDE CORNER POST

D DOOR AND WINDOW TRIM

C 4" — WIDE FACE OUTSIDE CORNER POST

E UNDERSILL AND GENERAL PURPOSE TRIM

F WINDOW HEAD FLASHING (DRIP CAP)

28-67. *Typical siding accessories. These are for vinyl, but manufacturers of aluminum siding offer similar accessories for the application of their particular product. The picture on page 330 shows how these accessories are used.*

items vary with the manufacturer. Before starting, inspect and plan the job in accordance with the manufacturer's instructions for the material to be applied. Be sure to use the nail recommended to avoid corrosion and stains. Drive the nails firm and snug, but never so tight as to cause waves in the siding.

The general procedure for the application of metal or plastic siding is as follows:

1. Lay out the courses on the sidewalls of the building the same way as for wood bevel siding, or use a story pole. Fig. 28-22.

2. Run a chalk line for the starter strip around the house. Install the starter strip. A, Fig. 28-67. All nails should be driven so that the head is only slightly flush to the material. Do not drive the nails hard enough to bind the material tightly.

3. Nail the inside corner posts 12" on center before the rest of the siding is applied. The bottom of the corner posts should be aligned with the chalk line. The siding later fits into channels on both sides. B, Fig. 28-67.

4. Nail the outside corner posts 12" on center before the siding is applied. C, Fig. 28-67. Here also, siding fits into channels on the sides. Fill the cavity behind a wide corner post with a backer board or wood strips. To close in the open lower end of a wide corner post, cut a piece of "J" channel to the proper length (about 6⅛"). Miter it by cutting as shown in Fig. 28-68. Nail it to the wall so that the bottom end of the corner post will fit over it.

5. Attach the door and window trim along the sides of the doors and windows. The gables are also trimmed with this accessory. The same trim is used at the base of a wall intersecting a sloping roof, as on a breezeway. D, Fig. 28-67.

6. Install undersill and general purpose trim under windows and at the tops of walls against soffit moldings, furring where necessary to preserve alignment with the adjacent panels. This may also be used inverted to lock lower cut edges of siding-courses which are located above the level of the starter strip, such as at porch floors and cellar bulkheads. E, Fig. 28-67.

7. Place the first siding panel in the starter strip and lock it securely. Fig. 28-69. Backerboards, if used, are dropped into place.

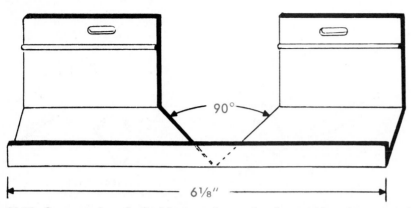

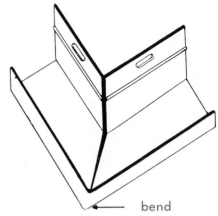

bend

28-68. *Suggested method of bottom closure for the outside corner post shown at C in Fig. 28-67. This piece must be cut and installed before the corner post is nailed in place.*

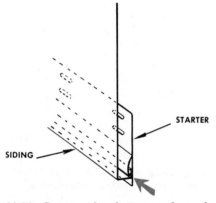

SIDING

STARTER

28-69. *Secure the bottom edge of the first piece of siding in the starter strip before nailing the top edge.*

Nail the panel and install succeeding courses similarly. Nails must never be driven so tightly as to cause distortion of the siding.

Allowance should be made for expansion and contraction by leaving a ¼″ space at joints, channels, and corner posts. Field-cut end joints may be made with a sharp knife, hack saw blade, or tin snips. On the end to be overlapped, cuts should be made according to the manufacturer's instructions. For best appearance, overlap the siding away from areas of greatest traffic. Stagger end laps a minimum of 24″ and in such a way that one is not di-

**REVERSIBLE SHUTTER BLIND
BLIND SIDE**

28-70. *This house has two-panel louvered shutters which add to the architectural effect. They also provide the homeowner with an opportunity to brighten the home with additional color when decorating.*

**REVERSIBLE SHUTTER BLIND
SHUTTER SIDE**

28-71. *Shutters mounted adjacent to a double-hung window. The shutters from this manufacturer are reversible; they might be used as a blind with louvers or as a shutter with a paneled effect.*

332

Sizes

1⅛" thick			
Widths (Pair)			
2'0"			2'10"
2'4"			3'0"
2'6"			3'4"
2'7"			3'8"
2'8"			4'0"
Heights			
1'3"	2'5"	3'7"	5'3"
1'7"	2'7"	3'11"	5'7"
1'9"	2'11"	4'3"	5'11"
1'11"	3'1"	4'7"	6'3"
2'1"	3'3"	4'11"	6'7"
2'3"			6'11"

28-72. This chart shows a variety of shutter sizes available from one manufacturer. The sizes will vary with the manufacturer and the material from which the shutter or blind is made.

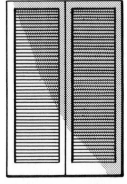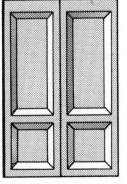

28-73a. Examples of a few of the many styles and patterns of shutters and blinds which are available.

rectly above another unless separated by three courses. The end of the uncut panel should lap over the end of the cut panel.

8. Cut and fit the siding as necessary around windows and doors.

9. Attach window head flashing or window and door trim above the windows and doors as application reaches these levels, furring out where necessary for alignment with adjacent panels. See F, Fig. 28-67.

10. Trim the last course of siding to fit under the eaves. Install undersill and general purpose trim, furring where necessary to maintain proper panel angle. Engage the top of the panel with undersill and general purpose trim and lock at lower edge of panel as usual. Nail to secure when necessary.

11. To complete the gables, install windows and door trim above the windows at the gable ends. Cut the siding to the proper angle and install.

12. Calk where required.

13. Finish by washing down the siding to remove fingerprints and soil. Clean up all scrap material around the house.

SHUTTERS AND BLINDS

Shutters, or blinds, are used on today's homes more for architectural effect than for any functional purpose. Fig. 28-70. Historically they were intended as a means of protection from enemy attack or the weather. Shutters are particularly popular with the New England style Cape Cod home, on which they are usually installed adjacent to double-hung windows. Fig. 28-71.

Shutters are made from wood, nylon, vinyl, or aluminum. They are available in a variety of sizes and styles to enhance the architectural effect desired. Figs. 28-72 & 28-73. When shutters are in-

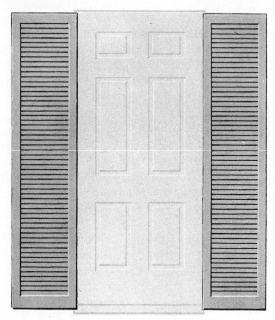

1 Panel

28-73b. *Blinds mounted adjacent to an entrance door.*

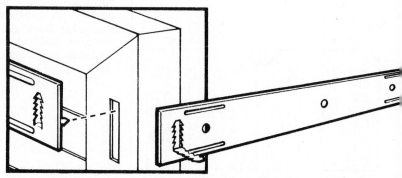

28-74. *Shutter mounting bracket. This mounting bracket is fastened to the wall. The projection on the mounting bracket snaps into a plastic insert at the rear of the shutter. With this mounting system, the shutter can be easily removed for cleaning or painting.*

stalled for a functional purpose, they are hinged to the window trim or doorjamb and will close and lock over the opening. Shutters used only as decoration are woodscrewed or held with special clips after the siding is applied and finished. Fig. 28-74.

QUESTIONS

1. What factors are combined to produce the wide variety of exterior wall coverings?

2. What type of wood siding serves as both sheathing and exterior wall coverings?

3. What is the maximum exposure for a piece of 8″ bevel siding?

4. What is the advantage of using a story pole when installing bevel siding?

5. What is the purpose of a "preacher"?

6. List several requirements for nails used on wood siding.

7. List several precautions that should be taken to prevent outside moisture problems.

8. When installing plywood siding, what type of surface should be specified?

9. List several advantages of hardboard siding.

10. What are the two methods of shingle sidewall application?

11. When using shingles as a sidewall covering, under what conditions must nailing strips be used?

12. What are the three kinds of shakes available?

13. When using asbestos-cement siding and shingles, what type of sheathing must be used?

14. When stucco is used as a sidewall finish, what holds it in place?

15. When installing metal and plastic siding, why aren't the nails driven in tightly?

16. When making a lap joint on metal or plastic siding, in which direction should the overlap be made?

17. What is the purpose of shutters and blinds on today's homes?

SECTION VI

Completing the Interior

Thermal Insulation and Vapor Barriers

29

Insulation is the property of material which slows down the transmission of energy in the form of heat, sound, or electricity. In construction, insulation is usually thought of in relation to heat transmission, although sound is an equally important item to consider. (Sound insulation is discussed in the next unit.) Most building materials—and even the air space between studs—have some heat insulation properties. However, by themselves they are not sufficient for modern homes that must have heating, air conditioning, and other climate control equipment. Insulation materials are products that are used in addition to ordinary building materials for the specific purpose of retarding the passage of heat. Insulation keeps homes warmer in the winter and cooler in the summer.

Better and more efficiently insulated homes are being built by contractors to please quality-conscious buyers. Upgrading of insulation beyond the minimum FHA standards pays good dividends in the form of increased comfort, reduced heating and air conditioning costs, and smaller, less expensive furnaces, cooling equipment, and duct work. Maximum insulation also makes electric heating feasible. In warm climates, the use of insulation with air conditioning is justified because operating costs are reduced and units of smaller capacity are required.

INSULATING MATERIALS

Insulation is manufactured in a variety of forms and types, each with advantages for specific uses. Materials commonly used for insulation may be grouped into the following general classes:

- Flexible (blanket and batt).
- Loose fill.
- Reflective.
- Rigid (structural and nonstructural).
- Miscellaneous types.

Flexible Insulation

Flexible insulation is manufactured in two forms: *blanket* and *batt*. Blanket insulation is furnished in rolls or packages. It comes in widths suited to 16″ and 24″ stud and joist spacing. A, Fig. 29-1. These pieces can be easily cut to length to fit various size openings. Fig. 29-2. Usual thicknesses are 1½, 2, and 3 inches. The body of the blanket is made of felted mats of mineral or vegetable fibers, such as rock wool, glass wool, wood fiber, or cotton. Organic insulations are treated to

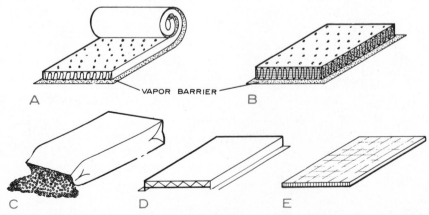

29-1. Types of insulation: A. Blanket. B. Batt. C. Loose fill. D. One type of reflective. E. Rigid.

VAPOR BARRIER

29-2. To cut insulation, place it on a piece of scrap plywood or 2 × 4, compress the material with one hand, and cut it with a sharp knife. When cutting faced insulation, keep the facing up.

make them resistant to fire, decay, insects, and vermin. Most blanket insulation is covered with paper or other sheet material with tabs on the sides for fastening to studs or joists. One covering sheet serves as a vapor barrier to resist movement of water vapor and should always face the warm side of the wall. Aluminum foil, asphalt, and plastic-laminated paper are common barrier materials.

Batt insulation is also made of fibrous material preformed to thicknesses of 4″ and 6″ for 16″ and 24″ joist spacing. B, Fig. 29-1. It is supplied with or without a vapor barrier. Fig. 29-3. One friction-type fiberglass batt is supplied without a covering and is designed to remain in place without the normal fastening methods.

Loose Fill Insulation

Loose fill insulation is usually supplied in bags or bales and placed by pouring, blowing, or packing by hand. C, Fig. 29-1. Materials used include rock and glass wool, wood fibers, shredded

redwood bark, cork, wood pulp products, vermiculite, sawdust, and shavings.

Fill insulation is best used between first-floor ceiling joists in unheated attics. It is also used in sidewalls of existing houses that were not insulated during construction. Where no vapor barrier was installed during construction, suitable paint coatings, as described later in this unit, should be used for vapor barriers.

Reflective Insulation

Most materials reflect some radiant heat. Radiant heat is heat that flows through air in a direct line from a warm surface to a cooler one. Materials high in reflective properties include aluminum foil, sheet metal with tin coating, and paper products coated with a reflective oxide composition. Such materials can be used in enclosed stud spaces, in attics, and in similar locations to retard heat transfer by radiation.

Reflective insulations are equally effective whether the re-

flective surface faces the warm or cold side. However, the reflective surface must face an air space at least ¾″ deep. Where a reflective surface contacts another material, the reflective properties are lost and the material has little or no insulating value.

Reflective insulation is more effective in preventing summer heat flow through ceilings and walls. It should be considered more for use in the warmer climates than in the North.

Sometimes, reflective insulation of foil is applied to blanket insulation and to the stud-surface side of gypsum lath. The type of reflective insulation shown in D, Fig. 29-1 has air spaces between the reflective surfaces. Metal foil suitably mounted on some supporting base also makes an excellent vapor barrier.

Rigid Insulation

Rigid insulation is manufactured in sheets and other forms. E, Fig. 29-1. The most common types of fiberboard are made from processed wood, sugar cane, or other vegetable products. The insulation may also be made from such inorganic materials as glass fiber.

Rigid insulation may be structural or nonstructural. Structural insulating boards, in densities ranging from 15 to 31 pounds per cubic foot, are used as building boards, roof decking, sheathing, and wallboard. While they have moderately good insulating properties, their primary purpose is structural.

Roof insulation is nonstructural and serves mainly to provide thermal resistance to heat flow. It is called slab or block insulation and is manufactured in rigid units ½″ to 3″ thick and usually 2′ × 4′ in size.

In house construction, the most common forms of rigid insulation

29-3. *Installing flexible insulation. Note that the vapor barrier is placed so that it faces the warm side of the wall. Sometimes an additional thin sheet of plastic is added to completely seal the wall. This should always be done when insulation without a vapor barrier is used.*

29-4
*Thermal Properties Of
Various
Building Materials
Per Inch Of Thickness*

Material	Thermal Conductivity K	Thermal Resistance R	Efficiency as as insulator Percent
Wood	0.80	1.25	100.0
Air Space[1]	1.03	0.97	77.6
Cinder Block	3.6	0.28	22.4
Common Brick	5.0	0.20	16.0
Face Brick	9.0	0.11	8.9
Concrete (Sand and Gravel)	12.0	0.08	6.4
Stone (Lime or Sand)	12.5	0.08	6.4
Steel	312.0	0.0032	0.25
Aluminum	1416.0	0.00070	0.06

[1]Thermal properties are for air in a space and apply for air spaces ranging from ¾ to 4 inches in thickness.

are sheathing and decorative coverings in sheets or in tile squares. Sheathing board is made in thicknesses of ½″ and ²⁵⁄₃₂″. It is coated or impregnated with an asphalt compound to provide water resistance. Sheets are made 2′ × 8′ for horizontal application and 4′ × 8′ or longer for vertical application.

Miscellaneous Types

There are several other kinds of insulation. Blanket insulation may be made up of multiple layers of corrugated paper. Sometimes lightweight vermiculite and perlite aggregates are used in plaster to increase its thermal resistance.

Foamed-in-place insulation includes sprayed and foam types. Sprayed insulation is usually inorganic fibrous material blown against a clean surface which has been primed with an adhesive coating. It is often left exposed for acoustical purposes.

Expanded polystyrene and urethane plastic foams may be molded or foamed in place. Urethane insulation may also be applied by spraying. Polystyrene and urethane in board form can be obtained in thicknesses from ½″ to 2″.

INSULATING VALUES

The thermal properties of most building materials are known, and the rate of heat flow, or coefficient of transmission, for most combinations of construction can be calculated. This coefficient, or *U*-value, is a measure of heat transmission between air on the warm side and air on the cold side of the construction unit. It is defined as the amount of heat (in Btu's) transmitted in 1 hour, through 1 square foot of surface, for each degree Fahrenheit difference in temperature between the inside and outside air. (One Btu, or British thermal unit, is the amount of heat which will raise the temperature of 1 pound of water 1 degree Fahrenheit.)

The insulating value of a wall will vary with different types of construction, with materials used in construction, and with different types and thicknesses of insulation. Comparisons of *U*-values may be used to evaluate different combinations of materials and insulation based on overall heat loss, potential fuel savings, influence on comfort, and installation costs. The amount of insulation required to obtain a desired *U*-value can be determined for any type of construction.

Information regarding the calculated *U*-values for typical constructions with various combinations of insulation may be found in "Thermal Insulation from Wood for Buildings: Effects of Moisture and Its Control," published by Forest Products Laboratory, Madison, Wisconsin.

The table in Fig. 29-4 provides some comparison of the individual insulating values of various building materials. These are expressed as *k* values, or heat conductivity. Heat conductivity is defined as the amount of heat (in Btu's) that will pass in 1 hour through 1 square foot of material 1 inch thick per 1° F temperature difference between faces of the material. Simply expressed, *k* represents heat loss. The lower this numerical value, the better the insulating qualities.

Building materials are also rated on their resistance or *R* value, which is merely another expression of insulating value. Fig. 29-4. The *R* value is usually expressed as the total resistance of the wall or of a thick insulating blanket or batt, whereas *k* is the rating per inch of thickness. $R = 1/k$. Thus, if the *k* value of 1 inch of insulation is 0.25, the resistance, *R*, is 1/0.25, or 4.0. If there are 3 inches of this insulation, the *R* value is 3 × 4.0, or 12.0.

Climate must also be taken into consideration when choosing building materials, insulation, and heating and cooling equipment. The map in Fig. 29-5 shows the

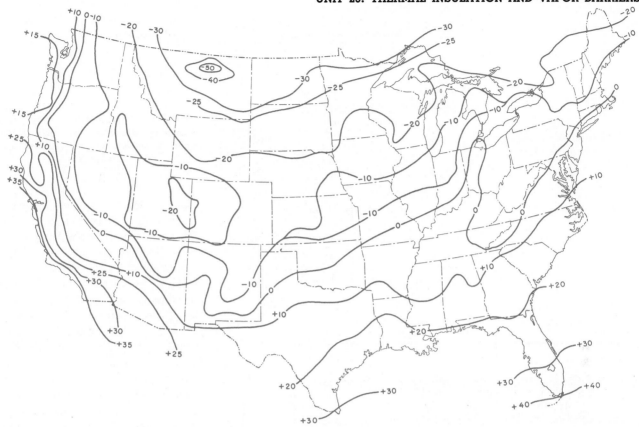

29-5. *This map of the United States indicates the lowest temperatures occuring in each zone during an average winter.*

average winter low temperatures found in different areas of the United States. Such data is used in determining the size of the heating plant required after calculating heat loss. This information is also useful in figuring the amount of insulation needed for walls, ceilings, and floors.

ACCEPTABLE COMFORT LEVELS

The amount of insulation necessary to provide indoor climate comfort can be determined accurately. The thermal properties of all common building materials are known, and the U-value for any combination of construction and insulation can be calculated.

Studies by heating engineers of home heating and air conditioning requirements have resulted in a number of U-value design standards recommended for new construction.

These standards are based on the geographical location of the structure and on the cost and type of fuel used.

The most widely used recommendations are in the *All-Weather Comfort Standard*, developed cooperatively by electric power suppliers, equipment makers, and material manufacturers, and in the supplementary performance standards created by the National Mineral Wool Insulation Association. Basically, these standards establish three degrees of comfort which may be attained by varying the amounts of insulation installed in ceilings, walls, and floors of homes.

● The *Maximum Comfort Standards* specify required amounts of insulation for houses in the coldest sections of the country and for energy-saving houses in any location.

● The *Moderate Comfort Standards* are for houses in the midsection of the country.

● The *Minimum Comfort Standards* apply to all homes as the minimum recommendations of many building codes.

In recent years, the need to conserve energy has caused insulation standards to be raised. The table in Fig. 29-6 shows the currently recommended U-values. Also listed on the table are the installed insulation requirements in terms of resistance units, or R numbers, which will provide the desired U-value in the building. This system of specifying insulation by total thermal resistance, instead of thickness, stems from

29-6a.

U-Values And Insulation Requirements

Maximum Comfort Standard		
	U-Value	Insulation "R" Number
Ceilings	0.02	R38-42
Walls	0.05	R19
Floors over unheated spaces	0.05	R22
Moderate Comfort Standard		
	U-Value	Insulation "R" Number
Ceilings	0.03	R30-33
Walls	0.05	R19
Floors over unheated spaces	0.05	R19-22
Minimum Comfort Standard		
	U-Value	Insulation "R" Number
Ceilings	0.05	R19
Walls	0.07	R11
Floors over unheated spaces	0.07	R13

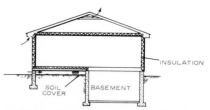

29-7. A conventional one-story house is insulated in walls, floors over unheated crawl spaces, and ceilings.

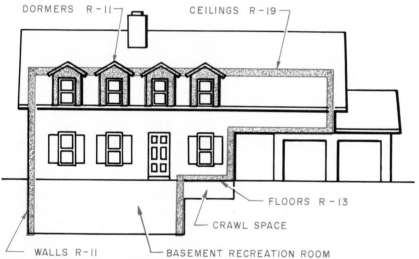

DORMERS R-11 — CEILINGS R-19

FLOORS R-13

CRAWL SPACE

WALLS R-11 — BASEMENT RECREATION ROOM

29-6b. Minimum insulation standards included in many local codes and recommended by many utilities. These recommendations are the maximum standards for the deep south and the western coastal regions.

the fact that insulating materials vary in density, type of surface, and heat conductivity.

In choosing the best, good, or minimum insulation standards in any climate zone, it should be remembered that comfort and operating economy are dual benefits of insulation. Insulating for maximum comfort automatically provides maximum economy of operation, and reduces initial costs of heating and cooling equipment to a minimum.

WHERE TO INSULATE

To reduce heat loss from the house during cold weather in most climates, all walls, ceilings, roofs, and floors that separate heated from unheated spaces should be insulated. Fig. 29-7.

In houses with unheated crawl spaces, insulation should be placed between the floor joists or around the wall perimeter. If flexible insulation is used, it should be well supported between joists by slats and a galvanized wire mesh or by a rigid board. The vapor barrier should be installed toward the subflooring. Fig. 29-8. Press-fit or friction insulation fits tightly between joists and requires only a small amount of support to hold it in place. Reflective insulation is often used for crawl spaces, but only one dead-air space (between the insulation and the subflooring) should be assumed in calculating heat loss when the crawl space is ventilated. A ground cover of roll roofing or plastic film such as polyethylene should be placed on the soil of crawl spaces to decrease the moisture content of the space as well as of the wood members.

In 1½-story houses, insulation should be placed along all walls, floors, and ceilings that are adjacent to unheated areas. Fig. 29-9. These include stairways, dwarf (knee) walls, and dormers. Provisions should be made for ventilation of the unheated areas.

Where attic space is unheated and a stairway is included, insulation should be installed around the stairway as well as in the first-floor ceiling. Fig. 29-10. The door leading to the attic should be weather-stripped to prevent heat loss. Walls adjoining an unheated garage or porch should also be insulated.

In houses with flat or low-

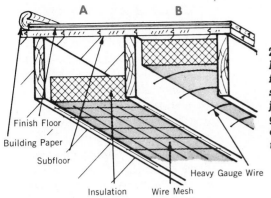

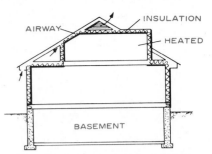

29-8. *Methods of installing insulation between floor joists: A. Wire mesh is stapled to the edges of the joists. B. Pieces of heavy-gauge wire pointed at each end are wedged between the joists to support the insulation.*

Finish Floor
Building Paper
Subfloor
Heavy Gauge Wire
Insulation
Wire Mesh

29-9. *Insulating a 1½-story house.*

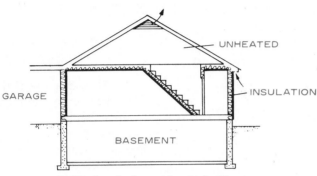

29-10. *Insulating unheated attic space.*

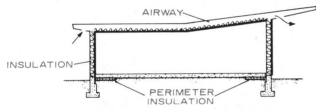

29-11a. *When insulating a flat roof, leave an air space for ventilation. For houses built on slabs, use perimeter insulation under the slab.*

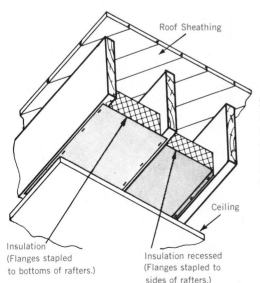

Roof Sheathing

29-11b. *Methods of installing flexible insulation between rafters. Regardless of method, it is important that a space be provided between the insulation and the roof sheathing to permit air circulation.*

Insulation
(Flanges stapled to bottoms of rafters.)

Insulation recessed
(Flanges stapled to sides of rafters.)

Ceiling

pitched roofs, insulation should be used in the ceiling area with sufficient space allowed above for unobstructed ventilation between the joists. Insulation should be used along the perimeter of houses built on slabs. A vapor barrier should be included under the slab. Fig. 29-11.

In the summer, outside surfaces exposed to the direct rays of the sun may attain temperatures of 50° F or more above shade temperatures and, of course, tend to transfer this heat toward the inside of the house. Insulation in the walls and in attic areas retards the flow of heat, improving summer comfort conditions.

Where air conditioning systems are used, insulation should be placed in all exposed ceilings and walls in the same manner as when insulating against cold-weather heat loss. Shading of glass against direct rays of the sun and the use of insulated glass will aid in reducing the air conditioning load.

Ventilation of attic and roof spaces is an important addition to insulation. Without ventilation, an attic space may become very hot and hold the heat for many hours. Obviously, more heat will be transmitted through the ceiling when the attic temperature is 150° F than if it is 100° to 120° F. Ventilation methods suggested for protection against cold-weather condensation apply equally well to protection against excessive hot-weather roof temperatures.

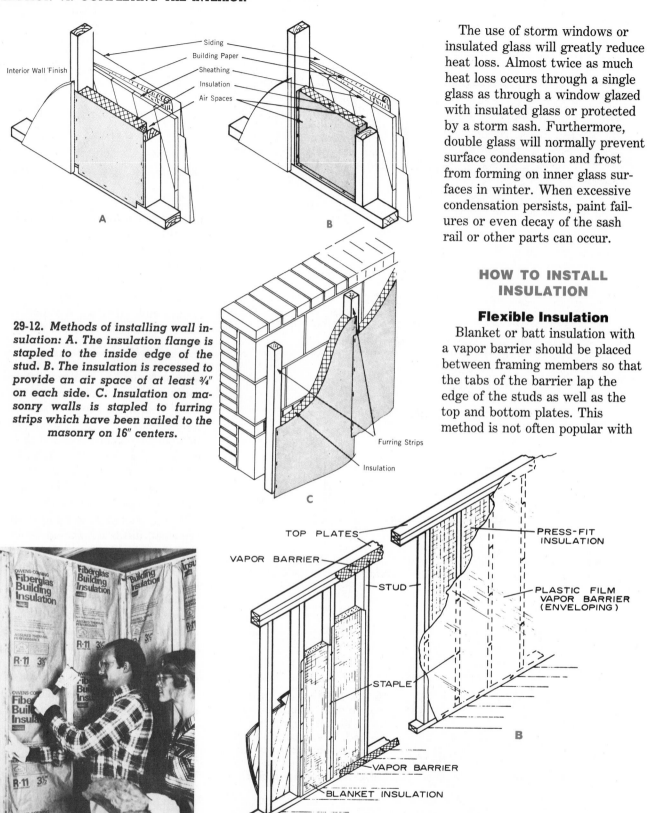

Siding
Building Paper
Sheathing
Insulation
Air Spaces
Interior Wall Finish

A

B

29-12. *Methods of installing wall insulation: A. The insulation flange is stapled to the inside edge of the stud. B. The insulation is recessed to provide an air space of at least ¾" on each side. C. Insulation on masonry walls is stapled to furring strips which have been nailed to the masonry on 16" centers.*

Furring Strips

Insulation

C

TOP PLATES
VAPOR BARRIER
STUD
STAPLE
PRESS-FIT INSULATION
PLASTIC FILM VAPOR BARRIER (ENVELOPING)
VAPOR BARRIER
BLANKET INSULATION
A
B

The use of storm windows or insulated glass will greatly reduce heat loss. Almost twice as much heat loss occurs through a single glass as through a window glazed with insulated glass or protected by a storm sash. Furthermore, double glass will normally prevent surface condensation and frost from forming on inner glass surfaces in winter. When excessive condensation persists, paint failures or even decay of the sash rail or other parts can occur.

HOW TO INSTALL INSULATION

Flexible Insulation

Blanket or batt insulation with a vapor barrier should be placed between framing members so that the tabs of the barrier lap the edge of the studs as well as the top and bottom plates. This method is not often popular with

29-13. *Installing batt insulation with a hand stapler following the method shown in B, Fig. 29-12.*

29-14. *Insulating a frame wall: A. Blanket insulation installed with a vapor barrier at the plates. B. Press-fit insulation installed with a plastic film vapor barrier.*

the contractor because it is more difficult to apply the dry wall or rock lath (plaster base). However, it assures a minimum amount of vapor loss compared to the loss when tabs are stapled to the sides of the studs. Fig. 29-12. A hand stapler is commonly used to fasten the insulation and the barriers in place. Fig. 29-13.

To protect the head and sole plate as well as the headers over openings, it is good practice to use narrow strips of vapor barrier material along the top and bottom of the wall. A, Fig. 29-14. Ordinarily, these areas are not covered too well by the barrier on the blanket or batt.

For insulation without a barrier (press-fit or friction type), a plastic film vapor barrier such as 4-mil polyethylene is commonly used to envelop the entire exposed wall and ceiling. B, Fig. 29-14. It covers the openings as well as window and door headers and edge studs. This system is one of the best from the standpoint of resistance to vapor movement. Furthermore, it does not have the installation inconveniences encountered when tabs of the insulation are stapled over the edges of the studs. After the dry wall is installed or plastering is completed, the film is trimmed around the window and door openings.

Reflective Insulation

Reflective insulation, in single-sheet form with two reflective surfaces, should be placed to divide the space formed by the framing members into two approximately equal spaces. For example, insulation between studs should be placed so as to leave an equal amount of air space on each side of the insulation. Some reflective insulations include air spaces and are furnished with nailing tabs. This type is fastened to the studs in such a way as to

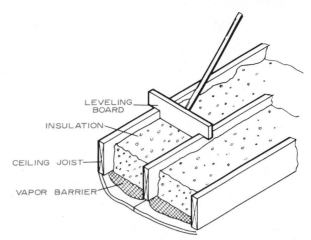

29-15. *Installing loose fill insulation in a ceiling. Note the use of the leveling board.*

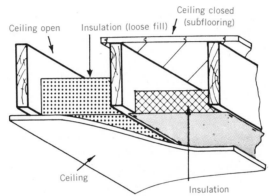

29-16. *Loose fill or batt insulation may be installed in a ceiling under an unheated attic.*

provide at least a ¾" space on each side of the reflective surfaces.

Loose Fill Insulation

Loose fill insulation is commonly used in ceiling areas and is poured or blown into place. Fig. 29-15. A vapor barrier should be used on the warm side (the bottom, in case of ceiling joists) before insulation is placed. A leveling board will give a constant insulation thickness. Fig. 29-15.

Thick batt insulation is also used in ceiling areas. Fig. 29-16. Batt and fill insulation can be combined to obtain the desired thickness. The vapor barrier is placed against the back of the ceiling finish. Ceiling insulation 6" or more thick greatly reduces heat loss in the winter and also provides summertime protection.

Rigid Insulation

Rigid insulation is nailed to sloping rafters through 1" strips of wood. Sheathing is then nailed to the strips, and the shingles are applied. Fig. 29-17. On a roof with wood decking, the insulation is fastened with nails or adhesive and covered with built-up roofing. Fig. 29-18.

PRECAUTIONS IN INSULATING

Areas over door and window frames and along side and head jambs also require insulation. Because these areas are filled with small sections of insulation, a vapor barrier must be used around the opening as well as over the header above the openings. Fig. 29-19. Enveloping the entire wall eliminates the need for this type of vapor barrier installation.

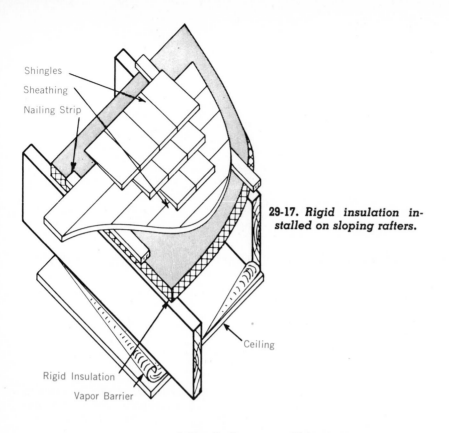

Shingles
Sheathing
Nailing Strip

29-17. *Rigid insulation installed on sloping rafters.*

Ceiling

Rigid Insulation
Vapor Barrier

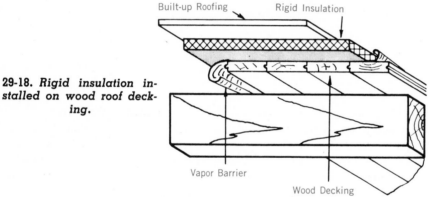

Built-up Roofing   Rigid Insulation

29-18. *Rigid insulation installed on wood roof decking.*

Vapor Barrier

Wood Decking

In 1½- and 2-story houses and in basements, the area at the joist header at outside walls should be insulated and protected with a vapor barrier. Fig. 29-20.

Insulation should be placed behind electrical outlet boxes and other utility connections in exposed walls to minimize condensation.

VAPOR BARRIERS

Some discussion of vapor barriers has been included previously because vapor barriers are usually a part of flexible insulation.

However, further information is given in the following paragraphs.

Most building materials are permeable to water vapor. This presents problems because considerable water vapor is generated in a house from cooking, dishwashing, laundering, bathing, humidifiers, and other sources. During cold weather, this vapor may pass through wall and ceiling materials and condense in the wall or attic space. Subsequently, in severe cases, it may damage the exterior paint and interior finish, or even promote decay in structural

members. For protection, a material highly resistant to vapor transmission, called a *vapor barrier*, should be used on the warm side of a wall or below the insulation in an attic space.

Among the effective vapor barrier materials are asphalt laminated papers, aluminum foil, and plastic films. Most blanket and batt insulations are provided with a vapor barrier on one side, some of them with paper-backed aluminum foil. Foil-backed gypsum lath or gypsum boards are also available and serve as excellent vapor barriers.

The effectiveness of vapor barriers is rated by their perm values. (1 perm = 1 grain of vapor transmission per square foot, per hour, for each inch of mercury vapor pressure difference.) Low perm values indicate high resistance to vapor transmission. The perm values of vapor barriers vary, but ordinarily it is good practice to use barriers which have values less than ¼ (0.25) perm. Although a value of ½ perm is considered adequate, aging reduces the effectiveness of some materials.

To obtain a positive seal against vapor transmission, wall-height rolls of plastic film vapor barriers should be applied over studs, plates, and window and door headers. Application of the plastic film is discussed on pages 342-343. This system, called "enveloping," is used over insulation having no vapor barrier or to insure excellent protection when used over any insulation. The plastic should be fitted tightly around outlet boxes and sealed if necessary.

A ribbon of sealing compound around an outlet or switch box will minimize vapor loss at this area. Cold-air returns in outside walls should consist of metal ducts to prevent vapor loss and subsequent paint problems.

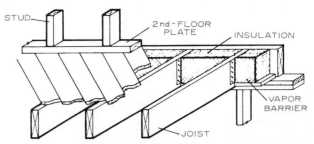

29-20. *A vapor barrier and insulation should be installed in the joist space at outside walls.*

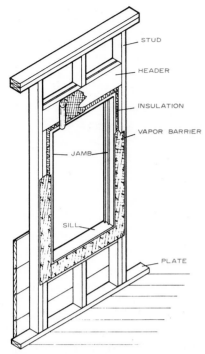

29-19. *A vapor barrier should be applied over the insulation around openings.*

Paint coatings on plaster may be very effective as vapor barriers if materials are properly chosen and applied. They do not, however, offer protection during construction, and moisture may cause paint blisters on exterior paint before the interior paint can be applied. This is most likely to happen in buildings that are constructed during periods when outdoor temperatures are 25° F or more below inside temperatures. Paint coatings cannot be considered a substitute for the membrane types of vapor barriers. However, they do provide some protection for houses where other types of vapor barriers were not installed during construction.

Of the various types of paint, one coat of aluminum primer followed by two decorative coats of flat wall or lead and oil paint is quite effective. For rough plaster or for buildings in very cold climates, two coats of the aluminum primer may be necessary. A primer and sealer of the pigmented type, followed by decorative finish coats or two coats of rubber-base paint, are also effective in retarding vapor transmission.

No type of vapor barrier can be considered 100% resistive, and some vapor leakage into the wall may be expected. Therefore the flow of vapor to the outside should not be impeded by materials of relatively high vapor resistance on the cold side of the vapor barrier. For example, sheathing paper should be of a type that is waterproof but not highly vapor resistant. This also applies to "permanent" outer coverings or siding. The vapor barrier itself should have a low perm value to prevent the passage of moisture to the cold side of the barrier. This will reduce the danger of condensation on cold surfaces within the wall.

QUESTIONS

1. What is insulation?
2. List several types of commercial insulating materials.
3. List some of the materials used for loose fill insulation.
4. What is a Btu?

5. To reduce heat loss, what areas in a structure should be insulated?
6. When installing flexible insulation, to what should the tabs be fastened?
7. When discussing vapor barriers, what is meant by "enveloping"?

Sound Insulation

30

Development of the "quiet" home is becoming more and more important. In the past, sound insulation was more important in apartments, motels, and hotels than in private homes. However, the use of household appliances, television, radio, and stereo systems has increased the noise levels in homes. House designs now often include a family room or "active" living room as well as a "quiet" living room. These rooms should be isolated from the remainder of the house. Sound insulation between the bedroom area and the living area is usually needed, as is isolation of the bathrooms and lavatories. Insulation against outdoor sounds is also desirable. Thus sound control has become a vital part of house design and construction and will be even more important in the coming years.

HOW SOUND TRAVELS

Sound is transmitted by waves. Noises inside a house, such as loud conversation or a barking dog, create sound waves which radiate outward from the source through the air until they strike a wall, floor, or ceiling. These surfaces vibrate as a result of the fluctuating pressure of the sound waves. Because the surface vibrates, it conducts sound to the other side in varying degrees, depending on the construction. Fig. 30-1.

The resistance of a building element, such as a wall, to the passage of airborne sound is rated by its Sound Transmission Class (STC). The higher the number, the better the sound barrier. The approximate effectiveness of walls with varying STC numbers is shown in Fig. 30-2. Most authorities agree that a floor or wall in a multi-occupancy residence should have an STC rating of at least 45, while 50 is considered premium construction. Below 40, privacy and comfort may be impaired because loud speech can be heard as a murmur.

Sound travels readily through the air and also through some materials. When airborne sound strikes a conventional wall, the studs act as sound conductors unless they are separated in some way from the covering material. Electrical switches or convenience outlets placed back to back in a wall readily pass sound. Faulty construction, such as poorly fitted doors, often allows sound to travel through. Thus good con-

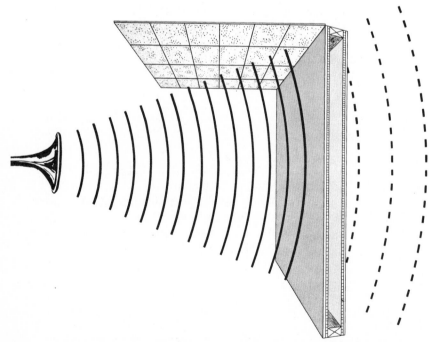

30-1. A wall will conduct sound to the other side in varying degrees, depending on its construction.

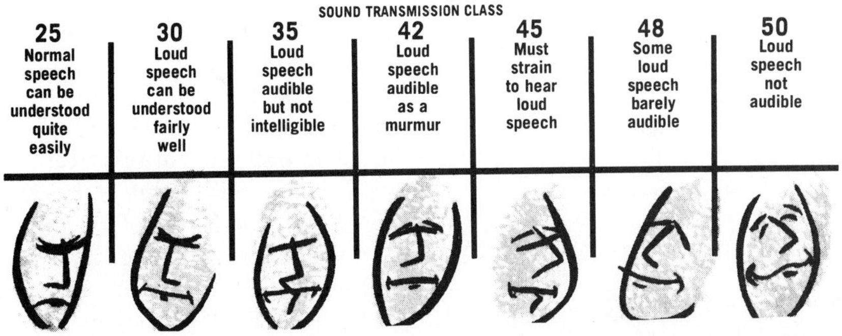

SOUND TRANSMISSION CLASS

25	30	35	42	45	48	50
Normal speech can be understood quite easily	Loud speech can be understood fairly well	Loud speech audible but not intelligible	Loud speech audible as a murmur	Must strain to hear loud speech	Some loud speech barely audible	Loud speech not audible

30-2a. *STC numbers have been adopted by acoustical engineers as a measure of the ability of structural assemblies to reduce airborne noise. The higher the number, the more effective the sound barrier.*

Location	Degree of Privacy	Stc Requirement	Example of Wall Construction Needed
Rural (20 db. background noise level)	High	60 or over	Solid dense masonry min. 12″ thick.
	Moderate	50 to 60	3⅝″ metal studs, 2 layers of ⅝″ fire-stop each side, 1½″ fiber glass insulation in cavity.
	Low	45 to 50	Double solid partition.
Suburban (30 db. background noise level)	High	55 or over	Triple solid partition.
	Moderate	45 to 55	2″ x 4″ wood studs, resilient channel one side, ⅝″ firestop both sides, 1½″ fiber glass insulation in cavity.
	Low	40 to 45	2″ x 4″ wood studs, resilient channel one side, ⅝″ firestop both sides.
City (40 db. background noise level)	High	45 or over	2½″ x 3″ gypsum ribs, 2½″ steel track, ½″ sound deadening board both sides, ⅝″ firestop each side
	Moderate	40 to 45	2″ x 4″ wood studs, 2 layers ⅝″ firestop each side.
	Low	35 to 40	2″ x 4″ wood studs, 2 layers ⅝″ firestop one side, 1 layer ⅝″ firestop other side.

30-2b. *This chart should be used only as a general guide to the partition performance required to meet specific sound control needs. Note that it does not incorporate the effects of interconnecting ducts, wiring, and plumbing.*

struction practices are important in providing sound-resistant walls.

SOUND INSULATION IN WALL CONSTRUCTION

Thick walls of dense materials such as masonry can stop sound. In a wood-frame house, however, an interior masonry wall results in increased costs and in struc-tural problems created by heavy walls. To economically provide a satisfactory sound-resistant wall has been a problem. At one time, sound-resistant frame construc-tion for the home involved much higher costs because it usually meant double walls or suspended ceilings. However, a relatively simple system has been developed using sound-deadening insulating board along with a gypsum board outer covering. This provides good sound control at only slight additional cost. A number of com-binations, providing different STC ratings, are possible with this system.

As Fig. 30-2a showed, a wall should have an STC rating of 45 or more to provide sufficient re-sistance to airborne sound. Gyp-sum wallboard or lath and plaster are commonly used for partition walls. A and B, Fig. 30-3. How-ever, an STC rating of 45 cannot be obtained with this construc-tion. An 8″ concrete block wall has the minimum rating, but this construction is not always practi-cal in a wood-frame house. C, Fig. 30-3.

In construction of a partition wall, its cost as related to the STC rating should be considered. Good STC ratings can be obtained in a wood-frame wall by using the combination of materials shown in D and E of Fig. 30-3. One-half inch of sound-deadening board nailed to the studs, followed by a lamination of ½″ gypsum wall-board, will provide an STC value of 46 at a relatively low cost. Fig. 30-4. A slightly better rating can be obtained by using ⅝″ gypsum wallboard rather than ½″. A very satisfactory STC rating of 52 can be obtained by using resilient clips to fasten ⅜″ gypsum backer boards to the studs, followed by adhesive-laminated ½″ fiberboard. E, Fig. 30-3. This method further isolates the wall covering from the framing.

A similar isolation system con-sists of resilient channels nailed horizontally to 2″ × 4″ studs spaced 16″ on center. The chan-nels are spaced 24″ apart verti-cally, and ⅝″ gypsum wallboard is screwed to the channels. An STC rating of 44 is thus obtained at a moderately low cost. Fig. 30-5.

A double wall, which may con-

WALL DETAIL	DESCRIPTION	STC RATING
A — 16", 2 x 4	½" GYPSUM WALLBOARD	32
	⅝" GYPSUM WALLBOARD	37
B — 16", 2 x 4	⅜" GYPSUM LATH (NAILED) PLUS ½" GYPSUM PLASTER WITH WHITECOAT FINISH (EACH SIDE)	39
C	8" CONCRETE BLOCK	45
D — 16", 2 x 4	½" SOUND DEADENING BOARD (NAILED) ½" GYPSUM WALLBOARD (LAMINATED) (EACH SIDE)	46
E — 16", 2 x 4	RESILIENT CLIPS TO ⅜" GYPSUM BACKER BOARD ½" FIBERBOARD (LAMINATED) (EACH SIDE)	52

30-3. Sound insulation of single walls.

sist of 2″ × 6″ or wider plate and staggered 2″ × 4″ studs, is sometimes constructed for sound control. One-half-inch gypsum wallboard on each side of this wall results in an STC value of 45. A, Fig. 30-6. Two layers of ⅝″ gypsum wallboard add little, if any, additional sound-transfer resistance. B, Fig. 30-6. However, when 1½″ blanket insulation is added to double wall construction, the STC rating increases to 49. C, Fig. 30-6. This insulation may be installed as shown, or placed between studs on a single wall. A single wall with 3½″ of insulation is low in cost and will resist sound transfer much better than an open stud space.

The use of ½″ sound-deadening board and a lamination of gypsum wallboard in the double wall will result in an STC rating of 50. D, Fig. 30-6. The addition of blanket insulation to this combination will likely provide an even higher value, perhaps 53 or 54. This system, with single-wall construction, might also be used to insulate exterior walls against street noises.

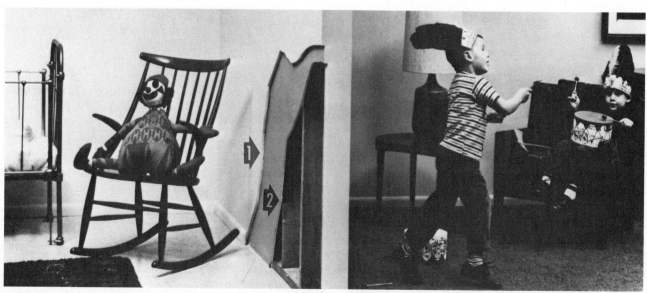

30-4. Gypsum wallboard (arrow # 1) installed over a sound-deadening board (arrow # 2) will help assure quiet living.

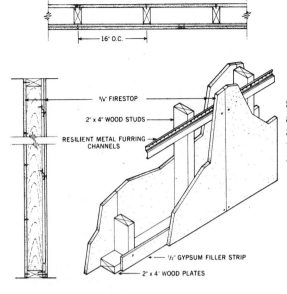

30-5. Sound transmission can be reduced by fastening the interior wall covering to resilient metal furring channels.

⅝" FIRESTOP

2" x 4" WOOD STUDS

RESILIENT METAL FURRING CHANNELS

½" GYPSUM FILLER STRIP

2" x 4" WOOD PLATES

WALL DETAIL	DESCRIPTION	STC RATING
A 16" 2x4	½" GYPSUM WALLBOARD	45
B 2x4	⅝" GYPSUM WALLBOARD (DOUBLE LAYER EACH SIDE)	45
C 2x4 BETWEEN OR "WOVEN"	½" GYPSUM WALLBOARD 1½" FIBROUS INSULATION	49
D 2x4	½" SOUND DEADENING BOARD (NAILED) ½" GYPSUM WALLBOARD (LAMINATED)	50

30-6. Sound insulation of double walls.

SOUND INSULATION IN FLOOR-CEILING CONSTRUCTION

Sound insulation between an upper floor and the ceiling of a lower story involves not only resistance to airborne sounds but also to impact noises. Impact noise results when an object strikes or slides along a wall or floor surface. Footsteps, dropped objects, and furniture being moved all cause impact noise. It may also be caused by the vibration of a dishwasher, food-disposal apparatus, or other equipment. In all instances, the floor is set into vibration by the impact or contact, and sound is radiated from both sides of the floor.

A method of measuring impact noise has been developed and is commonly expressed as the Impact Noise Rating (INR). [INR ratings in some publications are being abandoned in favor of IIC (Impact Insulation Class) ratings.]

The higher the INR, the better the impact sound reduction. For example, an INR of -5 is better than -10, and $+5$ is better than 0.

INR performance standards for floors are based on criteria established by the FHA. Those criteria range from -8 to $+5$, depending on location. Figure 30-7 shows STC and approximate INR (decibel) values for several types of floor construction. A minimum floor assembly with tongued and grooved floor and ⅜" gypsum board ceiling has an STC value of 30 and an approximate INR value of -18. A, Fig. 30-7. This is improved somewhat by the construction shown in part B, and still further by the combination of materials in part C.

The value of isolating the ceiling joists from a gypsum lath and plaster ceiling by means of spring clips is illustrated in Fig. 30-8, part A. An STC rating of 52 and an approximate INR value of -2 result.

Foam rubber padding and carpeting improve both the STC and the INR values. The STC rating increases from 31 to 45 and the approximate INR from -17 to $+5$. B and C, Fig. 30-8. The ratings can probably be further improved by using an isolated ceiling finish with spring clips. The use of sound-deadening board and

DETAIL	DESCRIPTION	ESTIMATED VALUES	
		STC RATING	APPROX. INR
A — 16" — 2 x 8	FLOOR ⅞" T. & G. FLOORING CEILING ⅜" GYPSUM BOARD	30	-18
B — 2 x 8	FLOOR ¾" SUBFLOOR ¾" FINISH FLOOR CEILING ¾" FIBERBOARD	42	-12
C — 2 x 8	FLOOR ¾" SUBFLOOR ¾" FINISH FLOOR CEILING ½" FIBERBOARD LATH ½" GYPSUM PLASTER ¾" FIBERBOARD	45	-4

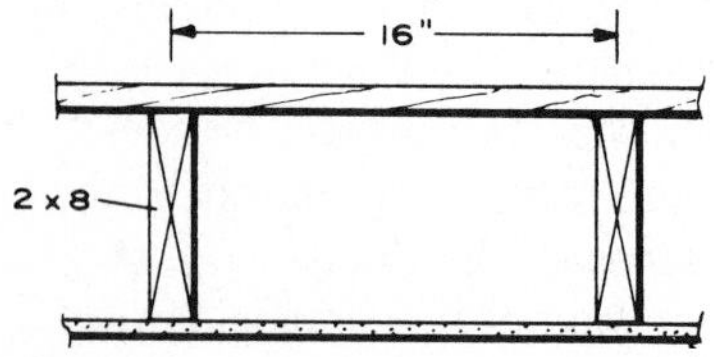

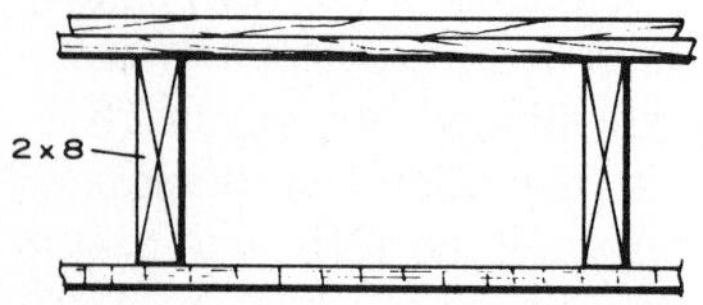

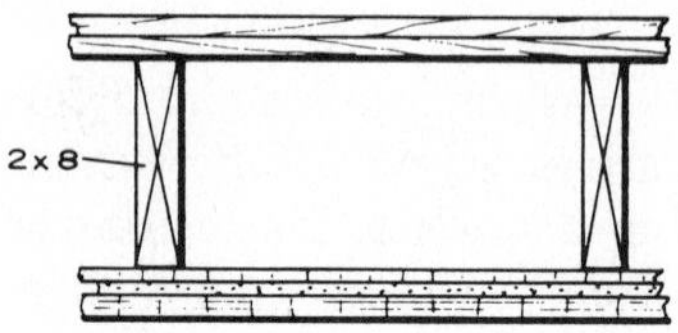

30-7. *STC and INR values in floor-ceiling combinations using 2″ × 8″ joists.*

tion costs. Separate joists with insulation between and a sound-deadening board between subfloor and finish provide an STC rating of 53 and an approximate INR value of −3. Other combinations are illustrated in Fig. 30-9.

SOUND ABSORPTION

Design of the "quiet" house can include another system of sound insulation, namely, sound absorption. Sound-absorbing materials do not necessarily have resistance to airborne sounds, but they can minimize the amount of noise by stopping the reflection of sound back into a room. Perhaps the

a lamination of gypsum board for the ceiling would also improve resistance to sound transfer.

An economical construction similar to (but an improvement over) the one shown in part C of Fig. 30-8 has an STC value of 48 and an approximate INR of +18. It consists of the following: (1) a pad and carpet over ⅝" tongued and grooved plywood underlayment, (2) 3″ fiberglass insulating batts between joists, (3) resilient channels spaced 24″ apart across the bottom of the joists, and (4) ⅝" gypsum board screwed to the bottom of the channels and finished with taped joints.

The use of separate floor joists with staggered ceiling joists below provides reasonable values but adds a good deal to construc-

30-8. *STC and INR values in floor-ceiling combinations using 2″ × 10″ joists.*

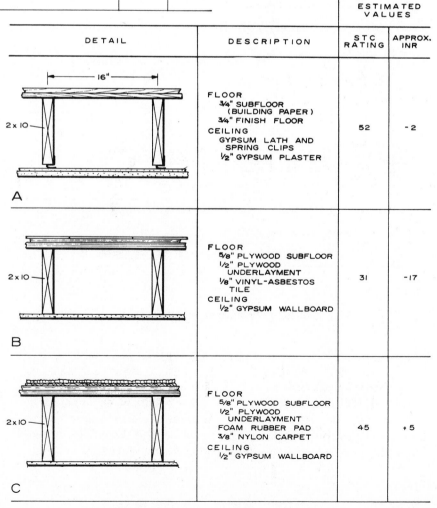

DETAIL	DESCRIPTION	ESTIMATED VALUES	
		STC RATING	APPROX. INR
A — 16" — 2 x 10	FLOOR ¾" SUBFLOOR (BUILDING PAPER) ¾" FINISH FLOOR CEILING GYPSUM LATH AND SPRING CLIPS ½" GYPSUM PLASTER	52	-2
B — 2 x 10	FLOOR ⅝" PLYWOOD SUBFLOOR ½" PLYWOOD UNDERLAYMENT ⅛" VINYL-ASBESTOS TILE CEILING ½" GYPSUM WALLBOARD	31	-17
C — 2 x 10	FLOOR ⅝" PLYWOOD SUBFLOOR ½" PLYWOOD UNDERLAYMENT FOAM RUBBER PAD ⅜" NYLON CARPET CEILING ½" GYPSUM WALLBOARD	45	+5

Floor on sleepers

Weight lbs./sq. ft.	Floor	STC	INR	Test No.
10.9	.075" sheet vinyl on ⅝" T&G plywood underlayment	52	−2	L-224-5&6
11.7	44 oz. Carpet on 40 oz. hair pad on ⅝" T&G Plywood underlayment	52	+27	L-224-7&8
13.0	25/32" wood strip flooring nailed sleepers	53	0	L-224-9&10

Note: Flooring is fastened to sleepers on ½" insulation board over a ½" plywood subfloor on wood joists with insulation between. ⅝" gypsum board ceiling with taped joints is screwed to resilient channels.

Separate ceiling joists

Weight lbs./sq. ft.	Floor	STC	INR	Test No.
11.0	25/32" hardwood strip flooring on ½" plywood subfloor	53	−6	L-224-12&13
10.7	44 oz. Carpet on 40 oz. hair pad on 1⅛" plywood subfloor (2-4-1 T&G)	51	+29	L-224-14&15

Note: Conventional wood joist floor system with finish floor applied directly on plywood. Separate 2 × 4 ceiling joists, with insulation, and ⅝" gypsum board ceiling with taped joints nailed directly to joists.

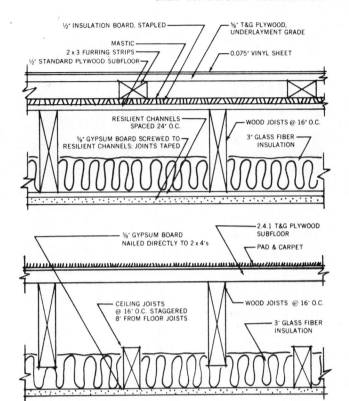

30-9. *Two of the sound resistant floor-ceiling combinations which can be obtained with plywood construction.*

most commonly used sound-absorbing material is acoustic tile or panels. Numerous tiny sound traps on the surfaces may consist of tiny drilled or punched holes, fissured surfaces, or a combination of both. Wood fiber or similar materials are used in the manufacture of the tile and panels, which are usually processed to provide some fire resistance.

Acoustic tile and panels are most often used in the ceiling and other areas, such as above a wall wainscoting, where they are not subject to excessive mechanical damage. Paint or other finishes which fill or cover the tiny holes or fissures for trapping sound will greatly reduce its efficiency.

QUESTIONS

1. How does sound travel?
2. What does STC stand for?
3. What do the higher numbers of STC indicate?
4. Why does an 8" concrete block wall have a lower STC rating compared to traditional methods of wood-frame wall construction?

5. What is impact noise?
6. What does INR stand for?
7. Which is considered better, an INR of −5 or an INR of +5?
8. What is the most commonly used sound absorbing material in house construction?

Interior Wall and Ceiling Finish

31

Before interior wall and ceiling finish is applied, insulation should be in place. Wiring, heating ducts, and other utilities should be roughed in.

Wood, gypsum, wallboard, plywood, and plaster make good finishes for covering the interior framed areas of walls and ceilings. They can serve as a base for paint and other finishes, including wallpaper, or be purchased already finished. The size and thickness of interior finish material should comply with local and national codes. Requirements for interior finish materials used in baths and kitchens normally will be more rigid because of the moisture conditions.

Although there are several types of interior finishes, lath and plaster has been about the most widely used. However, the use of dry-wall materials is increasing

steadily. Many builders select dry wall because of the time-saving factor. A plaster finish requires drying time before other interior work can be started, whereas dry wall does not. However, when gypsum dry wall is used as a wall finish, the framing lumber must have a low moisture content to prevent nail pops. These result when framing members dry out, causing the nail head to form small humps on the surface of the gypsum board. It is also very important when applying single-layer gypsum finish that the studs be in alignment. Otherwise, the wall may have a wavy, uneven appearance. Since there are advantages to both plaster and gypsum dry-wall finishes, all factors should be considered along with the initial cost and the future maintenance that may be required.

LATH AND PLASTER

A plaster finish requires a base upon which the plaster is applied. This base, or *lath*, is fastened to the framing members. It must have bonding qualities so that plaster adheres, or is keyed, to it. The most commonly used types of lath are the following:
- Gypsum.
- Insulating fiberboard.
- Metal.

Gypsum Lath

One of the most common types of plaster base used on sidewalls

and ceilings is gypsum lath. This lath has paper faces with a gypsum filler. It comes in 16″ × 48″ boards and is applied horizontally across the framing members. Fig. 31-1. For stud or joist spacing of 16″ on center, a ⅜″ thickness is used. For 24″ spacing, the thickness should be ½″.

This material can be obtained with a foil back that serves as a vapor barrier. If the foil faces an air space, it also has reflective insulating value. Gypsum lath may also be obtained with perforations, which improve the bond and increase the time the plaster remains intact when exposed to fire. The building codes in some cities require these perforations. A waterproof facing is provided on one type of gypsum board for use as a ceramic tile base when the tile is applied with an adhesive.

INSTALLING GYPSUM LATH

Vertical joints should be made over the center of studs or joists and nailed with 12- or 13-gauge gypsum-lathing nails 1½″ long with ⅜″ flat heads. Fig. 31-2. The nails should be spaced 5″ on center, or four nails for the 16″ height, and used at each stud or

31-2. *A gypsum-lathing nail.*

31-1. *Installing gypsum board lath.*

joist crossing. Some manufacturers specify ring-shank nails with a slightly greater spacing. Joints over heads of openings should not occur at the jamb lines. Fig. 31-3. Gypsum lath may also be used in constructions where metal studs are used for framing. The lath is secured to the studs by the use of special clips or tapping screws. Fig. 31-4.

Insulating Fiberboard Lath

Insulating fiberboard lath measuring 16″ × 48″ and ½″ thick is also used as a plaster base. It has greater insulating value than gypsum lath, but horizontal joints must usually be reinforced with metal clips.

INSTALLING INSULATING LATH

Insulating lath is installed much the same as gypsum lath, except that slightly longer blued nails should be used.

Metal Lath

Another type of plaster base is made of sheet metal. The metal is slit and expanded in various forms, such as diamond mesh, flat ribbed, and wire lath, to create innumerable openings for the keying of plaster. Fig. 31-5. Metal lath is usually 27″ × 96″ in size and galvanized or painted to resist rusting. Metal lath is usually installed on studs or joists spaced 16″ on center. The minimum weights to be installed on studs

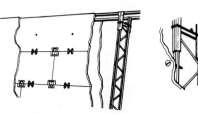

31-4. Gypsum lath installed on metal studs with special clips and tapping screws.

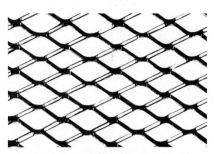

31-5. Diamond mesh metal lath.

or joists spaced 16″ on center are as follows:

• For walls—2.5 lbs. per sq. yd.
• For ceilings—3.4 lbs. per sq. yd. (if rib metal lath is used—2.75 lbs. per sq. yd.)

Metal lath is often used as a plaster base around tub recesses and other bath and kitchen areas. Fig. 31-6. It is also used when ceramic tile is applied over a plaster base. For these uses, the metal lath must be backed with water-resistant sheathing paper placed over the framing.

INSTALLING METAL LATH

Metal lath is applied horizontally over the waterproof backing with side and end joints lapped. It is nailed with No. 11 or No. 12 roofing nails long enough to provide about 1½″ penetration into the framing member or blocking. Fig. 31-7.

Plaster Reinforcing

Because some drying usually takes place in wood framing members after a house is completed, some shrinkage can be expected. This in turn may cause plaster

31-3. Gypsum board lath is nailed horizontally. Note that the joints are broken and that there is no joint at the jamb line in the doorway.

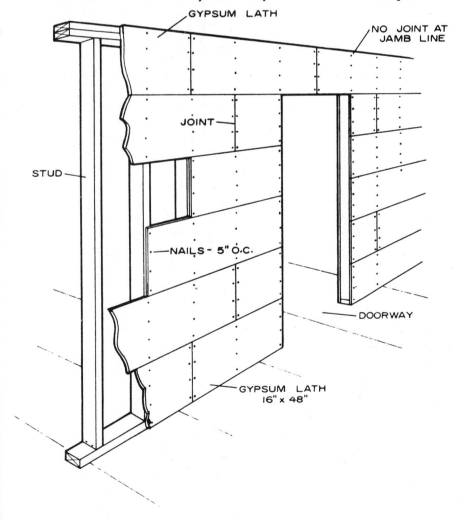

GYPSUM LATH

NO JOINT AT JAMB LINE

JOINT

STUD

NAILS - 5" O.C.

DOORWAY

GYPSUM LATH 16" x 48"

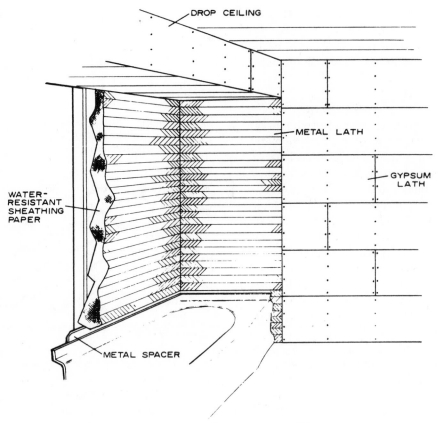

31-6. *Metal lath used as a plaster base around a tub recess.*

31-7. *A No. 11 or No. 12 roofing nail is used to apply metal lath.*

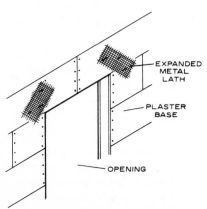

31-8. *Expanded metal lath is used to help minimize plaster cracks.*

cracks to develop around openings and in corners. To minimize this cracking, expanded metal lath is used in certain key positions over the plaster base as reinforcement. Strips of expanded metal lath about 10″ × 20″ should be placed diagonally across the upper corners of all window and door openings and tacked in place. Fig. 31-8.

Metal lath should also be installed under flush ceiling beams to prevent plaster cracks. Fig. 31-

9. On wood drop beams extending below the ceiling line, the metal lath is applied with self-furring nails to provide space for keying of the plaster.

Corner beads of expanded metal lath or perforated metal should be installed on all exterior corners. They should be applied plumb and level. The bead acts as a leveling edge when walls are plastered and reinforces the corner against damage, such as from moving furniture. Fig. 31-10.

Inside corners at the intersection of walls and ceilings should also be reinforced. A cornerite of metal lath or wire fabric is tacked lightly in place in these areas. Cornerites provide a key width of 2″ to 2½″ at each side for plaster. Fig. 31-11.

Plaster Grounds

Plaster grounds are strips of wood used as guides or strike-off edges when plastering. They are located around window and door openings and at the base of the walls. Grounds around interior door openings are often full-width pieces nailed to the door sides over the studs and to the underside of the header. They are 5″ in width, which coincides with standard jamb widths for interior walls with a plaster finish. Fig. 31-12. Narrow strip grounds might also be used around these interior openings. Fig. 31-13. These grounds are removed after plaster has dried.

The frames for window and ex-

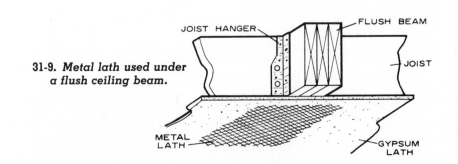

31-9. *Metal lath used under a flush ceiling beam.*

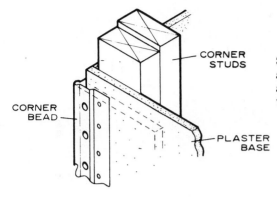

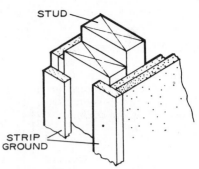

31-10. *A corner bead is installed at outside corners to serve as a leveling edge when the plaster is applied.*

31-13. *Narrow strip grounds are sometimes used around interior openings.*

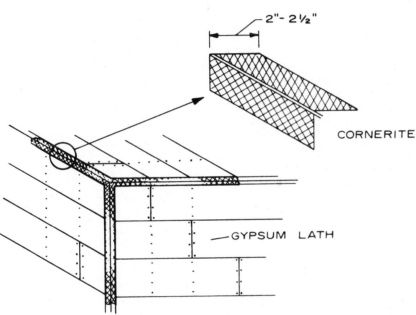

31-11. *A cornerite is installed at inside corners for reinforcement and to minimize plaster cracks.*

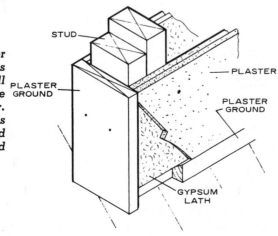

31-12. *A one-piece plaster ground whose width is equal to the finished wall thickness is applied to the trimmer studs and header. Do not drive the nails home. Nails will be pulled and the grounds removed after the plaster is dry.*

terior door openings are normally in place before plaster is applied. Thus the inside edges of the side and head jambs serve as grounds. The edge of the window sill may also be used as a ground, or a narrow ground strip ⅞" thick and 1" wide may be nailed to the edge of the 2" × 4" sill. The ⅞" × 1" grounds might also be used around window and door openings. Fig. 31-14. These are normally left in place and are covered by the casing.

A similar narrow ground, or screed, is used at the bottom of a wall for controlling the thickness of the plaster and providing an even surface for the baseboard and molding. Fig. 31-12. These strips are also left in place after the plaster has been applied.

Plaster Materials and Methods of Application

Plaster for interior finishing is made from combinations of sand, lime or prepared plaster, and water. Waterproof finishes for walls are available and should be used in bathrooms, especially in shower and tub recesses when tile is not used, and sometimes on the kitchen wainscot.

Plaster should be applied in three-coat or in two-coat double-up work. The minimum thickness over ⅜" gypsum or insulating lath should be about ⅜".

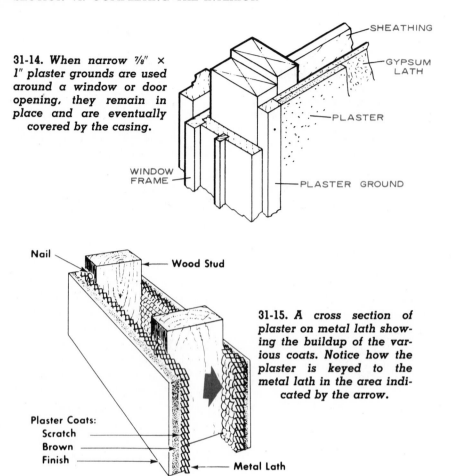

31-14. *When narrow ⅞″ × 1″ plaster grounds are used around a window or door opening, they remain in place and are eventually covered by the casing.*

SHEATHING

GYPSUM LATH

PLASTER

WINDOW FRAME

PLASTER GROUND

Nail

Wood Stud

31-15. *A cross section of plaster on metal lath showing the buildup of the various coats. Notice how the plaster is keyed to the metal lath in the area indicated by the arrow.*

Plaster Coats:
Scratch
Brown
Finish

Metal Lath

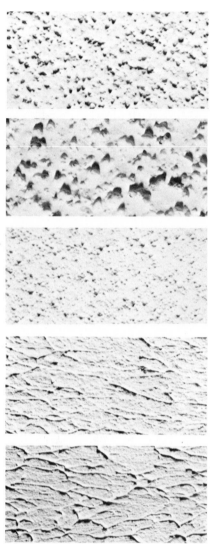

31-16. *Products offering a variety of textures are available for spray or roller application over gypsum wallboard.*

Three-coat work is used on metal lath and is usually at least ¾″ thick. The first plaster coat over metal lath is called the scratch coat. It is scratched, after a slight set has occurred, to insure a good bond for the second coat. The second coat is called the brown or leveling coat, and leveling is done during its application. The third coat is the finish coat. Fig. 31-15.

Double-up work, combining the scratch and brown coat, is used on gypsum or insulating lath. Leveling and plumbing of the walls and ceilings are done during application.

The final or finish coats are of two general types: the sand-float and the putty finish. In the sand-float finish, lime is mixed with sand, which results in a textured finish. The texture depends on the coarseness of the sand. Putty finish, used without sand, is smooth. This type is common in kitchens and bathrooms where a gloss paint or enamel finish is used, and in other rooms where a smooth finish is desired. Because of its durability, keene's cement is often used as a finish plaster in bathrooms.

Plastering should not be done in freezing weather without a source of constant heat. In normal construction, the heating unit is in place before plastering is started.

Insulating plaster, consisting of a vermiculite, perlite, or other aggregate with the plaster mix, may also be used for the finish coat.

DRY WALL

Dry wall is so called because it requires little if any water for application. Gypsum board, plywood, fiberboard, and similar sheet materials, as well as different forms and thicknesses of wood paneling, are classified as dry-wall finishes. Dry-wall materials are versatile, easily applied, decorative, and utilitarian. Fig. 31-16. Most of the several types of paneling can be used wherever excessive moisture is not a problem.

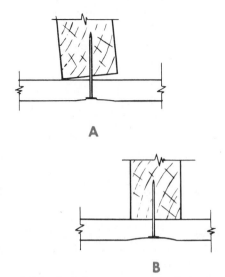

A

B

31-17. A. *A framing member has not been properly squared with the plate. This increases the possibility of puncturing the gypsum board paper with the nailhead. There is also the danger of a reverse twisting of the stud as it dries out, in which case the board will be loosely nailed and a "pop" will occur.* **B.** *The twisted stud has been squared before the application of the wallboard.*

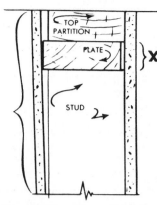

31-18a. *Improperly aligned studs, joists, or headers will result in the nailheads puncturing the paper or cracking the board.*

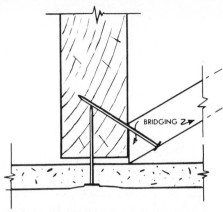

31-18b. *Protrusions, such as the bridging shown here, will puncture the face paper of wallboard. The bridging, which projects beyond the edge of the joists, also prevents the back of the board from being brought into contact with the nailing surface.*

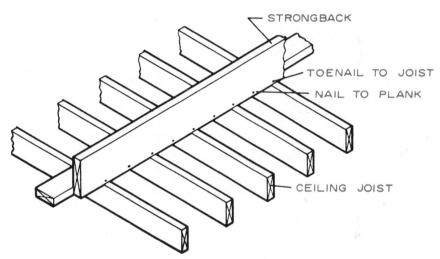

31-19. *A strongback used to align the joists.*

Some are even suitable for moist places like the bathroom and shower stall.

Paneling can be purchased unfinished or prefinished. The unfinished may be stained any color or given a clear finish to preserve its natural color. Some paneling comes with a factory-applied, baked-on plastic finish or a vinyl covering.

When thin sheet material such as gypsum board or plywood is used, the studs and ceiling joists must be in alignment to provide a smooth, even surface. Figs. 31-17 & 31-18. If ceiling joists are uneven, a "strongback" may be used to align the joists. Fig. 31-19.

The minimum thicknesses for plywood, fiberboard, and paneling and the required spacing of framing members are shown in Fig. 31-20.

31-20. *Minimum thicknesses for paneling applied to different framing spaces.*

Framing spaced (inches)	Thickness		
	Plywood	Fiberboard	Paneling
16	1/4"	1/2"	3/8"
20	3/8"	3/4"	1/2"
24	3/8"	3/4"	5/6"

31-21. *Maximum framing spacing recommended for various thicknesses of gypsum board.*

Installed long direction of sheet	Minimum thickness	Maximum spacing of supports (on center)	
		Walls	Ceilings
Parallel to framing members	3/8"	16"	
	1/2"	24"	16"
	5/8"	24"	16"
Right angles to framing members	3/8"	16"	16"
	1/2"	24"	24"
	5/8"	24"	24"

FASTENER	SPACING	QUANTITY REQUIRED
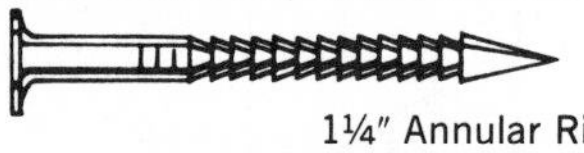 1¼″ Annular Ring Nail—12½ gauge; ¼″ dia. head with a slight taper to a small fillet at shank; bright finish; medium diamond point.	7″ c. to c. on ceilings—8″ c. to c. on walls	5¼ lbs./1000 sq. ft. approx. 325 nails/lb.
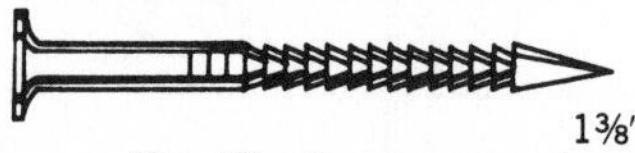 1⅜″ Annular Ring Nail (Specification same as above except for length)	7″ c. to c. on ceilings—8″ c. to c. on walls	5¼ lbs./sq. ft., approx. 321 nails/lb.
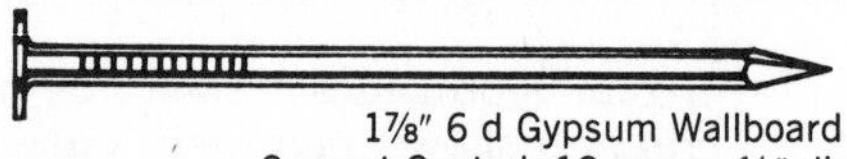 1⅞″ 6 d Gypsum Wallboard Nail— Cement Coated, 13 gauge, ¼″ dia. head	7″ c. to c. on ceilings—8″ c. to c. on walls	6¼ lbs./1000 sq. ft., approx. 275 nails/lb.
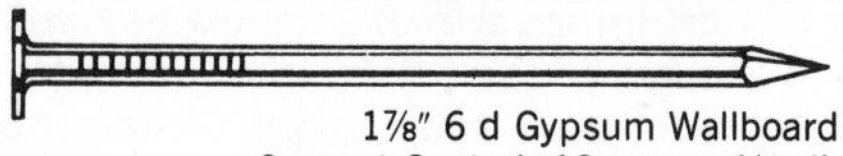 1⅞″ 6 d Gypsum Wallboard Nail— Cement Coated, 13 gauge, ¼″ dia. head	6″ c. to c. on ceilings—7″ c. to c. on walls	6¾ lbs./1000 sq. ft. approx. 278 nails/lb.
 1⅝″ 5 d Gypsum Wallboard Nail— Cement Coated, 13½ gauge, 15/64″ dia. head	6″ c. to c. on ceilings—7″ c. to c. on walls	5¼ lbs./1000 sq. ft. approx. 366 nails/lb.
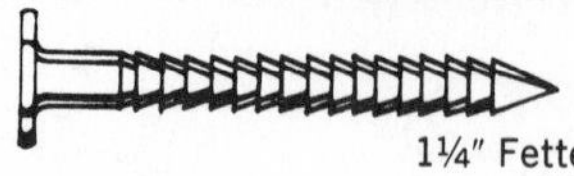 1¼″ Fetter Annular Ring Nail— 11 gauge; 5/16″ dia. head	6″ c. to c. on ceilings	6 lbs./1000 sq. ft. approx. 315 nails/lb.
 1⅛″ Matching Color Nail (Steel)	8″ c. to c. on walls	1½ lbs./1000 sq. ft. approx. 1,008 nails/lb.
 1⅞″ Matching Color Nail (Steel)	8″ c. to c. on walls	4½ lbs./1000 sq. ft. approx. 349 nails/lb.
 1⅛″ Matching Color Nail (Brass)	8″ c. to c. on walls	1¾ lbs./1000 sq. ft. approx. 901 nails/lb.
 1¼″ Drywall Screw— Type W	Framing spaced 16″ c. to c. 12″ c. to c. on ceilings—16″ c. to c. on walls Framing spaced 24″ c. to c. 12″ c. to c. on ceilings—12″ c. to c. on walls	Approx. 1000 screws/1000 sq. ft.
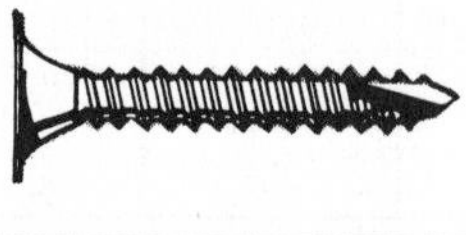 1″ Drywall Screw—Type S	12″ c. to c. on walls and ceilings	Approx. 875 screws/1000 sq. ft.
 1⅝″ Drywall Screw—Type S	16″ c. to c. on walls and ceilings when installed permanently without laminating adhesive; or as required for temporary mechanical attachment while laminating adhesive dries	Varies depending on 2 layer system used.
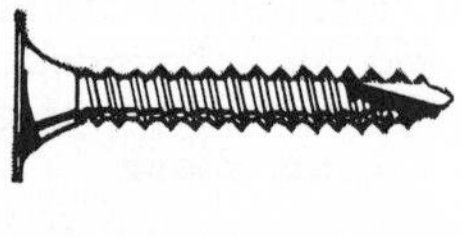 1″ Drywall Screw—Type S	12″ c. to c. in field of board and 8″ c. to c. staggered at vertical joints on walls— 12″ c. to c. on ceilings	Approx. 1100 screws/1000 sq. ft.

31-22. *Recommended gypsum wallboard fasteners for various applications.*

31-23. *The gypsum wallboard nail has a thin head and is ring shanked for greater holding power.*

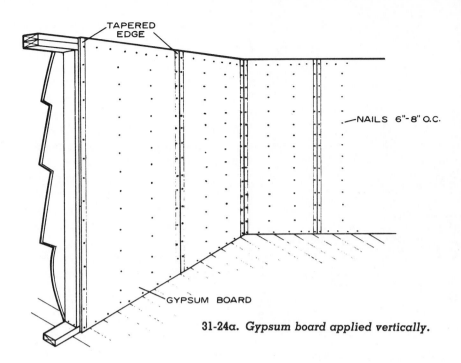

31-24a. *Gypsum board applied vertically.*

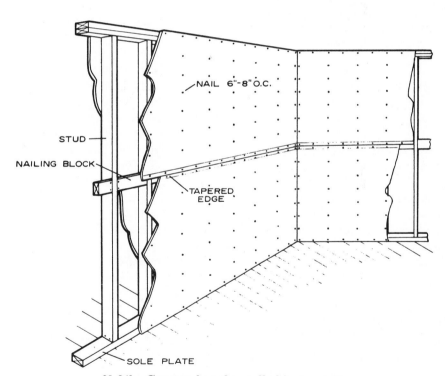

31-24b. *Gypsum board installed horizontally.*

Gypsum Board

Gypsum board is a sheet material made up of gypsum filler faced with paper. These sheets are normally 4' wide and 8' long, but may be obtained in lengths up to 16'. The edges along the length are tapered and, on some types, the ends are tapered also. Tapering allows for a filled and taped joint.

Some gypsum board has a foil back which serves as a vapor barrier on exterior walls. Prefinished gypsum board is also available for single-layer application in new construction. A thickness of ½" is recommended. For two-ply laminated applications, two ⅜" thick sheets are used. The maximum spacing of framing members for various thicknesses of gypsum board is shown in Fig. 31-21.

Installing Gypsum Board on Walls

Gypsum board may be applied with nails, screws, or adhesive. Fig. 31-22. The nails must have flat thin heads for flush driving without damage to the surface of the board. Fig. 31-23. Gypsum board ½" thick should be applied with a 5d nail (1⅝" long). For ⅜" thick material, use a 4d nail (1⅜" long). When ring-shank nails are used, a nail about ⅛" shorter will provide adequate holding power. Special screws will help prevent a bulging surface, sometimes referred to as a "nail pop," caused by the drying out of the framing members.

Nail pops are greatly reduced if the moisture content of the fram-ing members is less than 15% when the gypsum board is applied. When framing members have a high moisture content, it is good practice to let them approach moisture equilibrium be-fore application of the gypsum board.

Nails should be spaced 6" to 8" on the sidewalls and 5" to 7" on the ceiling, with a minimum edge distance of ⅜". Nail spacing is the

359

31-25. **Applying a sound-deadening board to the studs as an undercourse for gypsum board. This special-density fiberboard helps to reduce sound transmission between rooms.**

same for horizontal and vertical application. Fig. 31-24.

For studs or joists 16″ on center, screws should be spaced not more than 12″ apart on ceilings and 16″ apart on sidewalls. For studs 24″ on center, screws must not be spaced further than 12″ apart.

Horizontal application is best adapted to rooms in which full-length sheets can be applied because this reduces the number of vertical joints. Any joints necessary should be made at windows or doors. When this is not possible, the end joints should be staggered and centered on the framing members.

Horizontal nailing blocks between studs are not normally required when the studs are 16″ on center and the gypsum board is ⅜″ or thicker. However, if the spacing is greater or additional support at the joint is required, nailing blocks may be used. Fig. 31-24b.

In single-layer application, the 4′ wide gypsum sheets are installed vertically or horizontally on the walls after the ceiling has been covered. When the sheets are applied vertically, they cover three stud spaces if the studs are spaced 16″ on center and two if the studs are spaced 24″ on center. The edges of the gypsum board should be centered on studs and should make a very light contact with each other.

The laminated, two-ply method of gypsum application is begun by applying an undercourse of ⅜″ material vertically. To reduce sound transmission between rooms, sound-deadening panels are sometimes used as an undercourse. Fig. 31-25. The finish ⅜″ sheet is usually in room-size lengths. It is applied horizontally with an adhesive. Be certain to follow the manufacturer's recommendations when applying the adhesive. Nails used in the application of the finish gypsum wallboard should be driven with the head below the surface. The domed head of the hammer will form a small dimple in the wallboard. Fig. 31-26. Do not use a nail set. Care should be taken to avoid breaking the paper face of the gypsum board when nailing.

DOUBLE NAILING SYSTEM

A nail pop is caused by a movement of either the gypsum wallboard or the nail head in relation to the other. These pops may be prevented if the board is held tightly against the framing by the head of the nail at all times. The double nailing system incorporates a second nail in close proximity (2″) to the first to insure that the board is nailed tight. The wallboard is first nailed to each framing member with nails spaced approximately 12″ on center. This places five nails per 4′ width into each framing member. The board is then nailed around its perimeter. The top and bottom are fastened with one nail at each framing member. At the ends of the board, nails are spaced 7″ on center. Additional nails are then spaced approximately 2″ from each nail on the inner area of the board. Fig. 31-27.

As the second nail in each group of two nails is driven home, the worker can watch for any movement between the board and the head of the first nail driven. Movement indicates that the first nail is not holding the board tight and that it should be given additional blows with the hammer. Always begin nailing at the center of the board and work toward the ends, making sure to hold the gypsum board tight against the framing member.

CUTTING GYPSUM BOARD

Gypsum board may be cut to size by sawing. Another method is to score the finished side with an awl or knife. Fig. 31-28. Snap the board over a straight edge. Fig. 31-29. To complete the separation, score the back of the

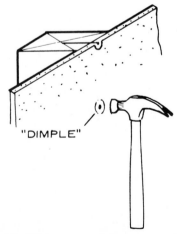

31-26. **When applying gypsum board, the final hammer blows should make a slight dimple on the face of the board around the nailhead. Avoid a heavy blow that would break the face paper or crush the gypsum core.**

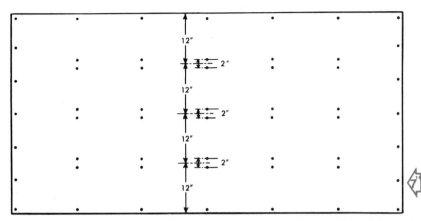

31-27. *The double nailing system for installing gypsum wallboard. Note that the nails are spaced 7″ on center at the panel ends.*

31-31. *Making a cutout for a door or window. Make two saw cuts to the correct depth, then score the snap.*

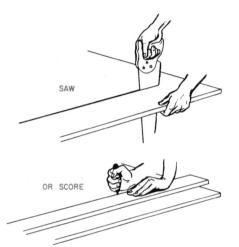

31-28. *Gypsum wallboard may be sawed, or it may be scored and snapped.*

31-29. *Break the gypsum core by placing the scored line of the sheet over the edge of the table or bench and snapping down. Hold the sheet firmly on the table and support the cutoff with your other hand.*

31-32. *Sanding the rough edge of gypsum board.*

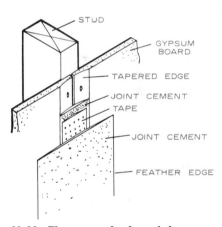

31-33. *The tapered edge of the gypsum board is filled with joint cement and taped. Additional joint cement is then applied and feathered out to provide a smooth surface.*

31-30. *Complete the separation by scoring the back paper of the board with a sharp knife. Then snap the board forward toward the face side.*

board with a sharp knife and snap the board forward toward the face side for a clean, straight joint break. Fig. 31-30.

Small cutouts for electrical outlets and other openings are made with a keyhole saw or a saber saw. To mark the location of the electrical outlet cutouts, hold the panel in place against the wall. With a wood block to protect the board, tap around the outlet with a hammer. An indentation of the outlet box will be made on the back of the board to use as a guide for cutting.

When notching gypsum board for door or window openings,

make two saw cuts to the correct depth. The final cut is made by scoring and snapping, the same as when cutting the sheet to size. Fig. 31-31. The cut edges of gypsum board may be smoothed with

#2 sandpaper wrapped around a wood block. Fig. 31-32.

JOINT TREATMENT

The joints between the panels are made smooth by applying joint cement, perforated tape, and

31-34. Applying joint cement with a wide spackling knife.

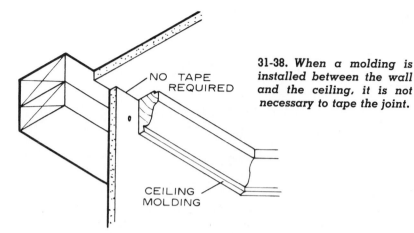

31-38. When a molding is installed between the wall and the ceiling, it is not necessary to tape the joint.

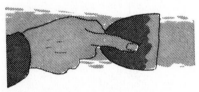

31-35. Press the perforated tape into the cement, forcing the excess cement from under the tape.

31-39. Filling the nail dimples with joint cement. Repeat with second and third coats of cement, if necessary, and sand smooth.

31-36. Spread a thin coat of cement over the tape. If necessary, follow with a second and third coat after each preceding coat has dried.

additional coats of joint cement, and then sanding the surface level with the wall surface. Fig. 31-33. Joint cement can be purchased in either premixed or powder form. The power is mixed with water to a soft putty consistency that can be easily applied.

Use a 5" wide spackling knife or a mechanical applicator to fill the joints with the cement. Fig. 31-34. Press the tape into the recess with a wide, flat knife until the joint cement is forced through the perforations in the tape. Fig. 31-35. Next cover the tape with additional cement, feathering the outer edges. Fig. 31-36. After the cement has dried, sand the joint lightly and then apply a second coat, again feathering the edges. Sometimes a steel trowel is used to apply the second coat. For best results, a third coat is applied and the edges are feathered beyond the second coat. After the joint cement is completely dry, sand the joint smooth and even with the wall surface.

To tape interior corners, fold the tape down the center to form a right angle. Fig. 31-37. Apply the cement in the corner and press the tape in place. Then finish the corner with joint cement and sand smooth when dry. Apply a second coat if necessary and

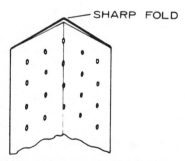

31-37. Fold the perforated tape down the center to form a right angle when taping interior corners.

31-40. Sanding the joints and nail dimples after the joint cement is completely dried. This is the last step in preparing the gypsum board for a decorative treatment.

sand it smooth, flat, and even with the wall surface. The same procedure is followed for interior corners between a wall and ceiling, or a molding of some type is installed. Fig. 31-38. To hide

Relative Humidity	0	20%	40%	50%	60%	70%	80%	90%	98%
Temp. ° F. 40	28 H	34 H	44 H	2 D	2½ D	3½ D	4½ D	9 D	37 D
60	13 H	16 H	20 H	24 H	29 H	38 H	2½ D	4½ D	18 D
80	6 H	8 H	10 H	12 H	13½ H	19½ H	27 H	49 H	9 D
100	3 H	4 H	5 H	6 H	8 H	10 H	14 H	26 H	5 D

H = hours   D = days   (24 hours)

31-41. Drying time for joint cements.

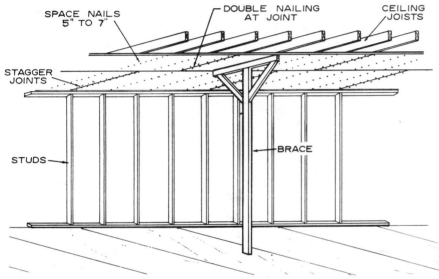

31-42. Gypsum board ceiling installation details. A brace is used to hold the material in position for nailing.

31-43. Installing gypsum board on the ceiling. Note the staggering of the end joints.

hammer indentations, fill them with joint cement and sand them smooth when they are dry. Usually a second coat is necessary. Figs. 31-39 & 31-40.

Temperature and humidity have a direct effect on the drying time of joint treatment products. Very little can be done to alter temperature and humidity under job conditions. However, care should be taken to note the average differences in drying time under different atmospheric conditions so that problems may be minimized. Joint treatment products must be thoroughly dry before successive coats and/or final decorations are applied. In all cases, a well-ventilated area assists in proper drying of these materials. The chart in Fig. 31-41 indicates the average drying periods for joint treatment products under different temperature and humidity conditions.

Installing Gypsum Board on the Ceiling

Gypsum board applied to the ceiling is nailed to ceiling joists or to the bottom chord of a truss. Nails are spaced 5″ to 7″ apart and dimpled in the same manner as when applying gypsum board to sidewalls. During installation, the gypsum board can be held in place with one or two braces about 1″ longer than the height of the ceiling. Fig. 31-42. Joints should be staggered and centered over framing members. Fig. 31-43. Joint treatment is the same for the ceiling as for the walls.

Plywood Paneling

Plywood paneling usually comes in sheets that are 4′ wide by 8′ long. However, other lengths are available. Some types are narrower than 4′ for greater ease of handling or to create interesting patterns. Plywood is available in a number of species. Because of

31-44. *The indentations of age can be felt as well as seen in this antiqued plywood paneling used as an accent wall.*

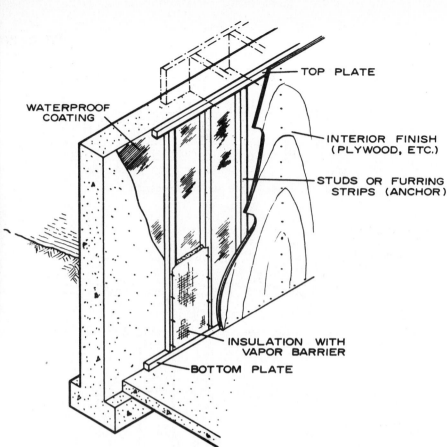

31-45a. *Plywood paneling applied to furring strips on a basement wall. A waterproof coating has been applied to the cement. Insulation with a vapor barrier has been installed between the furring strips.*

31-45b. *Basement paneling applied to a foamed plastic insulation with wallboard adhesive.*

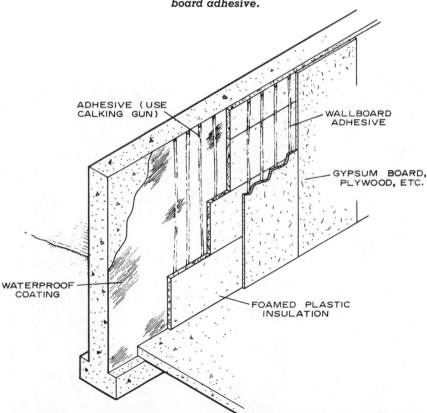

its beauty, warmth, and ease of maintenance, plywood paneling may be used for accent walls or to cover entire room wall areas. Fig. 31-44. Plywood panels can be applied either vertically or horizontally, as long as solid backing is provided at all edges. For framing spaced 16″ on center, plywood ¼″ thick is considered minimum. For framing 20″ or 24″ on center, ⅜″ plywood is the minimum thickness. Sometimes the wall is first covered with ⅜″ or ½″ gypsum board, and the plywood is then applied to the surface of the gypsum board.

Installing Plywood Paneling

On exterior masonry walls above or below grade, be sure the wall is properly waterproofed before the studding or furring is applied. Where extreme humidity may cause condensation on the inside of the exterior masonry wall, apply a vapor barrier, paper, or film to prevent moisture penetration to the panel. Fig. 31-45.

Furring strips must be used on masonry and plaster walls. When paneling over an existing wall

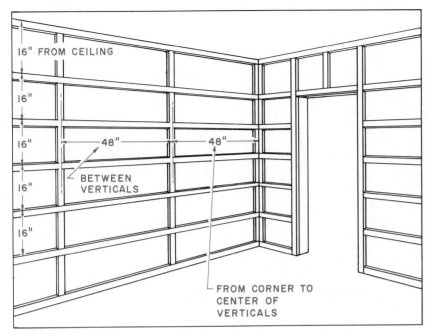

31-46. *Correct placement of furring strips in preparation for paneling.*

anchors. Fig. 31-49. Uneven furring strips can be leveled by placing shims in the low spots and driving a nail through the furring strip and the shims to hold them in place. Fig. 31-50.

APPLYING FURRING STRIPS TO PLASTER WALLS

If the paneling is to be applied on a plaster wall, the furring strips should be nailed horizontally to the studs, starting at the floor line and continuing up the wall every 16″. Nail vertical trips every 48″ to support the panel edges. The furring strips are shimmed as necessary with wood shingles to obtain a flush surface.

that is not masonry or plaster, sand any uneven or rough spots to obtain a flush surface. If sanding will not remove severe unevenness, furring strips should be used. Paneling may be applied directly to furring strips or studs. However, for additional strength, fire resistance, and sound deadening, ⅜″ or ½″ gypsum wall board is recommended as a backing behind plywood paneling. Plywood sheathing ⁵⁄₁₆″ thick is also ideal for application to the studs as a backing under finish paneling.

APPLYING FURRING STRIPS TO MASONRY WALLS

On masonry walls, apply furring strips horizontally every 16″. Fig. 31-46. Allow a clearance of at least ¼″ between the top furring strip and the ceiling and between the bottom furring strip and the floor. The top furring strip is nailed to the bottom edge of the ceiling joists or to a nailing block. Fig. 31-47. Insert vertical strips every 48″ to support the panel edges. Figs. 31-46 & 31-48. The

furring strips are attached to the masonry walls with masonry nails, screws, nails driven into shields or wood dowels, nail anchors, adhesive anchors, or bolt

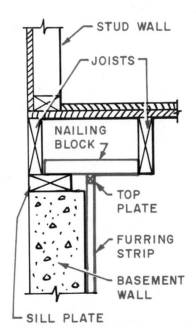

31-47a. *When the wall to be furred runs parallel to the joists, install a nailing block to which the top plate can be nailed.*

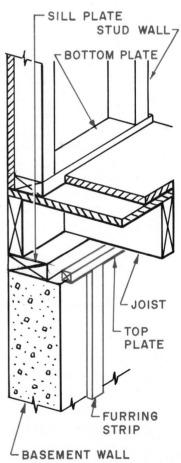

31-47b. *When the wall to be furred runs at right angles to the joists, nail the top plate (top furring strip) to the underside of the joists.*

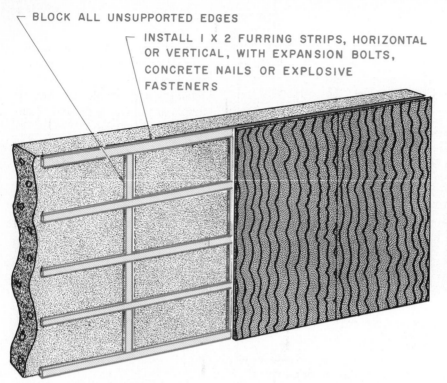

BLOCK ALL UNSUPPORTED EDGES

INSTALL I X 2 FURRING STRIPS, HORIZONTAL OR VERTICAL, WITH EXPANSION BOLTS, CONCRETE NAILS OR EXPLOSIVE FASTENERS

31-48. *Furring strips must be installed so that all panel edges are supported. Attach the furring strips securely to the masonry wall.*

31-50. *Wood shingles are used as shims behind furring.*

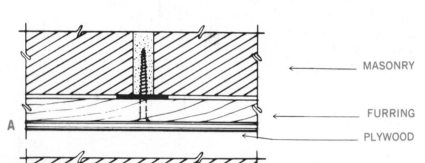

MASONRY

FURRING

PLYWOOD

MASONRY

FURRING

PLYWOOD

MASONRY

SUBFURRING

PLYWOOD

31-49. *Methods of attaching furring strips to a masonry wall: A. Insert wood dowels in the masonry to which the furring strips can be wood screwed or nailed. B. Use nail anchors or adhesive anchors. C. Bolt anchors may be used for attaching a 1″ × 3″ subfurring to the wall, then attaching the 1″ × 2″ furring to the 1″ × 3″ as if to studs.*

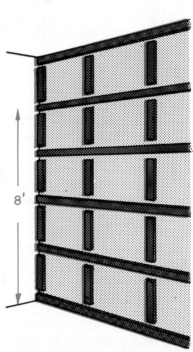

8′

31-51. *A wall over 8′ high must have a horizontal furring strip positioned so that the top of an 8′ panel will be aligned with the center of the furring strip at that point.*

FURRING STRIPS FOR SPECIAL APPLICATIONS

For walls over 8′ high, additional furring strips are nailed horizontally with the center of one of the strips 8′ high, addi-

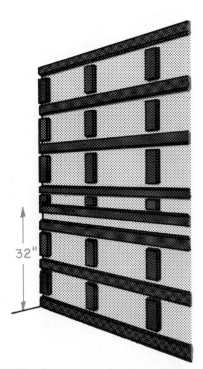

31-52. *An extra furring strip is nailed horizontally at the wainscot height.*

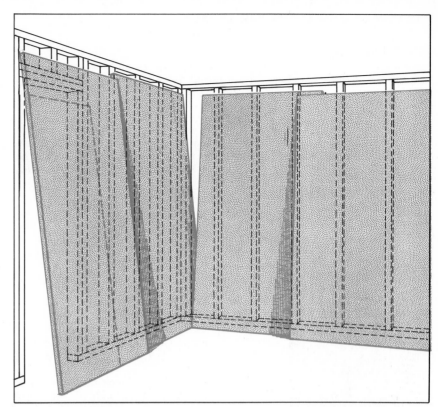

31-53. *Arrange the panels around the room to show the best pattern of color variations in both daylight and artificial light.*

tional furring strips are nailed horizontally with the center of one of the strips 8′ from the floor and another at the ceiling. Fig. 31-51.

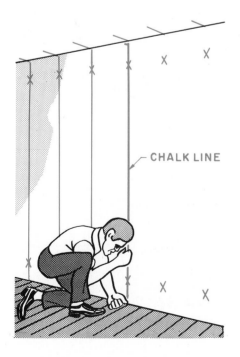

31-54. *Snapping a chalk line on the wall to indicate the center of each wall stud.*

When wainscoting is installed, nail an extra strip horizontally at the wainscot height. This is usually about 32″, since three 32″ pieces can be cut from a full 8′ panel (96″). Fig. 31-52.

LAYING OUT THE JOB

Set up the panels around the room to plan their sequence. Arrange them so that the natural color variations form a pleasing pattern. Fig. 31-53. For most interiors, it is practical to start paneling from one corner and work around the room. After deciding in which corner to start the paneling, stack the panels in the correct sequence so that as the job proceeds, the panels may be removed from the stack in the proper order.

INSTALLING PANELS

Accurately measure the height of the wall in several places. The panels should be cut so that they have a ¼″ clearance at top and bottom. If the location of the studs is not visible, locate them and mark the center of each stud with a chalk line. Also lightly mark the stud center locations on the floor and ceiling to serve as a guide when nailing each panel in position. Fig. 31-54. If the panel has grooves, these will usually be spaced to line up with the studs. Fig. 31-55.

Place the first panel in position and butt it to the adjacent wall in the corner. Make sure the panel is perfectly plumb and the outer edge is directly over the chalk line which marks the center of the stud. Fig. 31-56. If this edge does not fall directly on the stud, cut the other edge of the panel so that it will. In most cases, the corner where the paneling begins

GROOVE SPACING

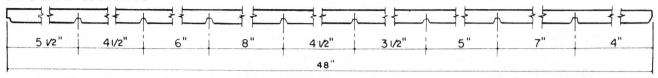

31-55. *A typical groove spacing for plywood paneling. The groove locations appear to be randomly spaced. However, when the dimensions are added together, the grooves fall on 16" and 24" centers. In this way, the panel can be nailed through the grooves to the studs.*

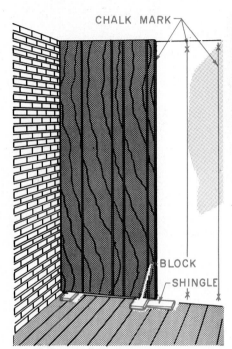

31-56. *Set the first panel in position, making certain that the edge is plumb.*

31-57. *Use a block and a shingle to position the panel at the correct height.*

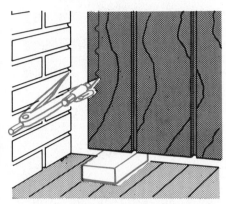

31-58. *Scribing the panel edge to fit the adjacent wall.*

31-59. *The panel edges must meet on the center of a stud or furring strip.*

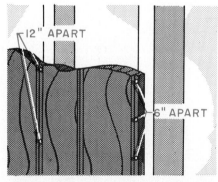

31-60a. *Nail spacing for plywood panels applied directly to the studs.*

is irregular. There may be a fireplace, concrete blocks, or uneven plaster. Scribe the panel with a small compass to insure a perfect fit. Position the panel at the proper height by setting it on a block and shimming it with a shingle to allow for ¼" clearance at the top and bottom. Fig. 31-57. When the panel is set perfectly plumb and at the correct height, set the compass for the amount to be cut off and scribe the line as shown in Fig. 31-58. Cut the panel along this line to fit the irregular wall.

Set the panel back in place against the wall to which it has been scribed and block the panel up as previously to the correct height. Fasten the panel to the wall. It may be applied with nails or adhesive; this will be discussed later. Butt the second panel to the first. With the first panel properly positioned on the studs, the edges of the remaining 4' panels will also land on stud centers, assuming that the stud spacing is uniform across the wall surface. Fig. 31-59. When paneling is being applied over a backer board with adhesive, the panel edges do not need to meet on a stud.

FASTENING PANELS WITH NAILS

For paneling applied directly to studs, use 4d finish nails spaced 6" along the panel edges and 12" elsewhere in the grooves. Fig. 31-60. If the paneling is applied over furring, use 2d nails spaced 8" apart along the panel edges and 16" apart elsewhere. Fig. 31-61.

Interior plywood paneling/Recommendations apply to all species groups.

Plywood Thickness (inch)	Max. Support Spacing (inches)	Nail Size & Type	Nail Spacing (inches)	
			Panel Edges	Intermediate
¼	16[1]	4d casing or finish	6	12
⅜	24	6d casing or finish	6	12

Notes: (1) Can be 20″ if face grain of paneling is across supports.

31-60b. *Nailing recommendations for interior plywood paneling.*

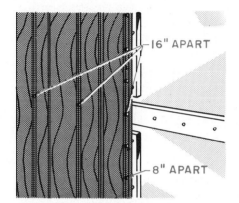

31-61. *Nail spacing for plywood paneling applied over furring.*

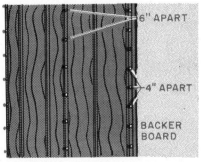

31-62. *Nail spacing for plywood paneling applied over backer board.*

For paneling applied over backer board, use 6d finish nails spaced 4″ along the panel edges and 6″ elsewhere. Fig. 31-62. Countersink the nails 1/32″ below the surface. These holes can be filled later with a putty stick to match the color of the panel. Colored nails which blend with the wood finish eliminate the need for countersinking and puttying.

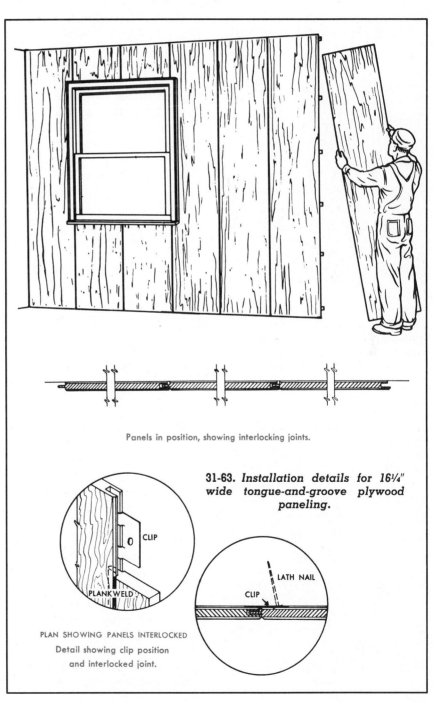

Panels in position, showing interlocking joints.

31-63. *Installation details for 16¼″ wide tongue-and-groove plywood paneling.*

PLAN SHOWING PANELS INTERLOCKED
Detail showing clip position and interlocked joint.

31-64. Applying adhesive to the edges of the studs or furring strips.

31-65. After the panel has been nailed at the top for hinge action, pull the panel out from the wall and block it in this position to allow the adhesive to partially set.

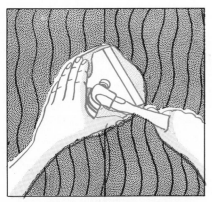

31-66. Tapping the panel with a soft wood block to press it firmly into place. You may put a cloth under the block to further protect the paneling.

31-67. Cutting an opening for an electrical outlet box.

Plywood panels 16¼″ wide are available with a groove at one edge and a tongue on the other. The panel is attached to the wall with a metal clip which slips into the groove and is nailed to the stud or furring strip. The tongue on the next panel is then placed in the groove of the first, which covers up the nailing clip. This method of applying plywood paneling provides secure and completely invisible nailing. Fig. 31-63.

FASTENING PANELS WITH ADHESIVE

Panel adhesive or other adhesives may be used instead of nails. Be sure to follow the manufacturer's instructions. After the panels are properly cut and fitted, the adhesive is applied to the studs, furring strips, or backing. Fig. 31-64. Apply the adhesive in continuous ⅛″ wide beads or in intermittent beads 3″ apart to all stud or furring strip surfaces. Apply a continuous ⅛″ wide bead at the corners and around cutouts. Position the panel and press it firmly against the adhesive. Place three or four finishing nails across the top of the panel to hold it in place.

Pull the bottom of the panel 8″ to 10″ away from the wall, allowing the nails at the top to serve as a hinge. Hold the panel out with a spacer block for 8 to 10 minutes to allow the adhesive to dry. Fig. 31-65. Remove the spacer block and reposition the panel to fit perfectly. Place a

clean block of soft wood against the panel and tap the block with a hammer or rubber mallet to obtain a full surface contact. Fig. 31-66.

If a panel is not flush with the stud or furring strip surfaces, small finishing nails may be needed to hold the panel in position until the adhesive acquires full strengh.

CUTTING PLYWOOD PANELS

When cutting panels with a crosscut saw or table saw, cut with the face side up. If an electric hand saw or saber saw is used, cut with the face side down. Never use a rip saw, since this will tear the veneer on the edge of the panel.

When cutting the panel for outlets, locate the opening by chalking the edges of the outlet box and carefully fitting the panel loosely over the chalked box. Strike the face of the panel sharply several times with the heel of your hand to transfer the box outline to the back of the panel. Drill pilot holes in the corners from the back side; then cut out the outlet hole from the front side with a keyhole saw. Fig. 31-67. If a saber saw is used, a plunge cut can be made to eliminate the drilling of the pilot holes at the corners.

MOLDINGS

There are several styles of wood and metal moldings for wood paneling. Pine moldings are sometimes used. They can be

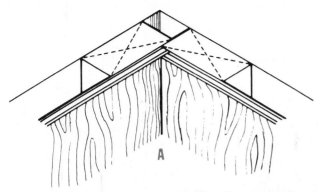

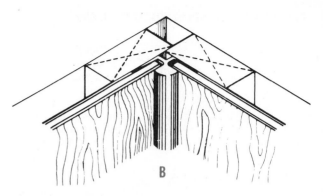

31-68a. *Corner details: A. An inside corner with the first panel butted into the corner and the second panel scribed to the face of the first. B. An inside corner trimmed with a veneer-faced aluminum molding. C. An outside corner mitered. D. An outside corner trimmed with a veneer-faced aluminum molding.*

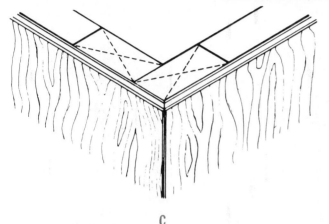

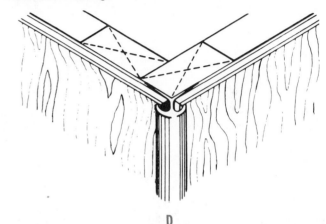

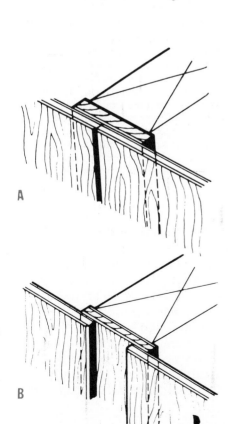

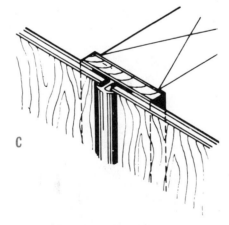

31-68b. *Joint details between panels: A. Shallow V-joint. B. Wide joint using ¼″ × 2½″ furring strips of matching or contrasting paneling. If a prefinished furring strip is not used, the strip should be finished before the panels are installed. C. Veneer-faced aluminum molding installed as a divider strip between panels.*

stained to harmonize with the prefinished paneling. Prefinished moldings to match the panel finish are also available.

Accurate measurements are essential for a good, professional-looking molding job. Measure along the ceiling line for the cove or crown molding. Measure along the floor for the exact length of the base and shoe molding. Do not assume that the ceiling and floor are the same length. Wood moldings should be scribed, mitered at 45°, or coped as described in Unit 35, "Interior Trim." Construction details showing the use of various metal and wood moldings at corners, doors, windows, floors, and ceilings are illustrated in Fig. 31-68.

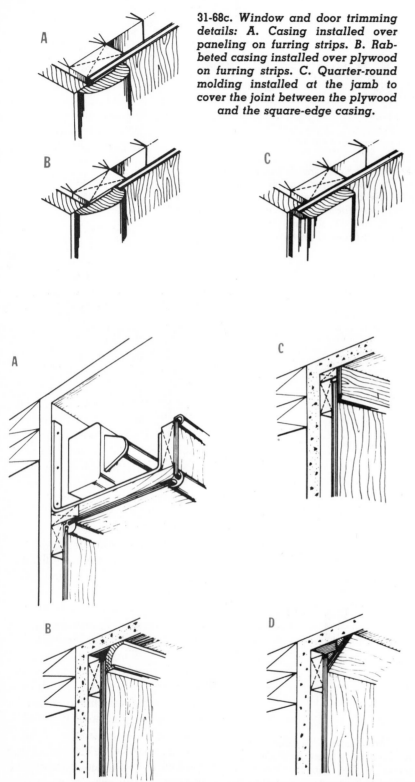

31-68c. *Window and door trimming details: A. Casing installed over paneling on furring strips. B. Rabbeted casing installed over plywood on furring strips. C. Quarter-round molding installed at the jamb to cover the joint between the plywood and the square-edge casing.*

31-68e. *Plywood installation details at the ceiling: A. Cove lighting framed and covered with plywood paneling. Note the use of the veneer-faced aluminum cap and inside and outside corner moldings. B. Crown molding. C. A strip of prefinished paneling cut from leftover pieces and scribed to the ceiling with a quarter-round attached at the bottom edge. D. A strip of prefinished paneling ripped at 45° and installed at the ceiling line.*

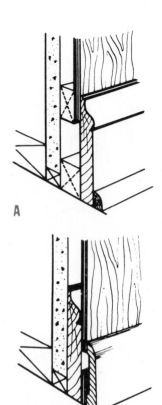

31-68d. *Base installation details: A. This method is used for installing wainscoting when it is desirable to gain a few inches of wall height. The panel is held up off the floor, and a piece of thicker furring is attached to the wall at the floor. The base is then nailed at the top and bottom to the two furring strips. B. This method is frequently used in remodeling. The walls are furred out, and the thickness of the old base is used as the bottom furring strip. The new base is then nailed to the face of the plywood paneling. This same method may be employed for new construction by using a furring strip at the floor line.*

Hardboard and Fiberboard

Hardboard and fiberboard are applied in the same manner as plywood. Hardboard should be at least ¼″ thick when applied over open framing spaced 16″ on center. It should be at least ⁷⁄₁₆″ thick for framing spaced 24″ on center. When ⅛″ hardboard is used, a rigid backing of some type is required. Fiberboard in tongue-and-

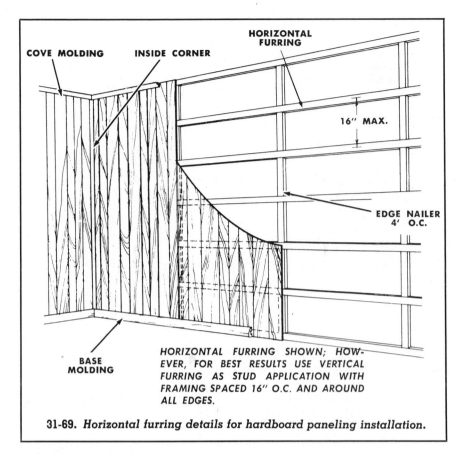

31-69. *Horizontal furring details for hardboard paneling installation.*

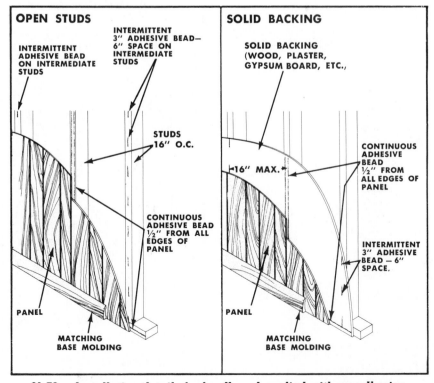

31-70a. *Installation details for hardboard applied with an adhesive.*

groove planks or sheet form must be ½″ thick when framing members are spaced 16″ on center and ¾″ thick for 24″ spacing. For best results, vertical furring should be used with hardboard paneling. However, horizontal furring can be applied 16″ on center over studs spaced 48″ on center. Fig. 31-69.

The paneling may be nailed or applied with an adhesive. Fig. 31-70. Nails should penetrate into the studs at least ¾″ and should be spaced 4″ on center at all joints and along the edges. At all intermediate supports, nail 8″ on center. Nails around the perimeter of the panel should be ¼″ from the edge. Fig. 31-71. Wood or metal moldings are used as trim and are applied in the same manner as over plywood. Fig. 31-72.

Wood Paneling

Many kinds of wood are made into paneling. For example, a rustic or informal look can be obtained with knotty pine, white pocket Douglas fir, sound wormy chestnut, and pecky cypress. The panels can be cut plain or with a tongue and groove. Fig. 31-73. These may be finished natural or stained and varnished. Wood paneling may be used to cover one or more walls or partial walls of a room.

Only thoroughly seasoned wood paneling should be used. The moisture content should be near the average it reaches in service, about 8% in most areas. However, in the dry southwestern United States, it should be about 6%, and in the southern and coastal areas of the country, about 11%. Allow the material to reach correct moisture content by storing it in the area in which it will be installed in such a way that air may circulate around all surfaces of the boards. Wood paneling on the inside of an exterior

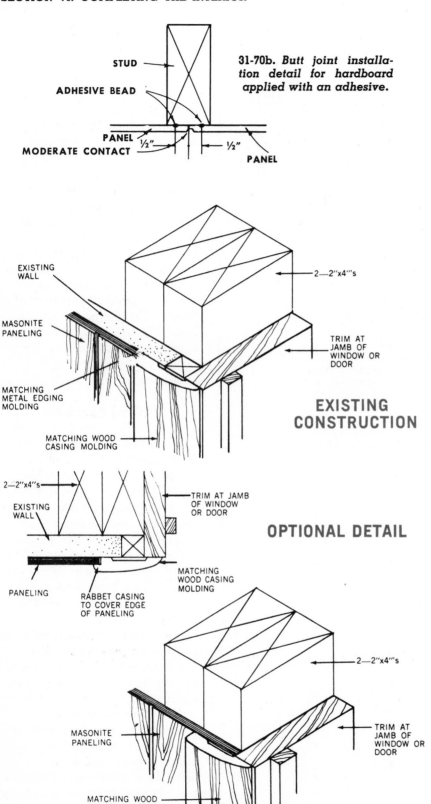

STUD

ADHESIVE BEAD

31-70b. *Butt joint installation detail for hardboard applied with an adhesive.*

PANEL
MODERATE CONTACT

½″   ½″

PANEL

EXISTING WALL

MASONITE PANELING

MATCHING METAL EDGING MOLDING

MATCHING WOOD CASING MOLDING

2—2″x4″'s

TRIM AT JAMB OF WINDOW OR DOOR

EXISTING CONSTRUCTION

2—2″x4″'s

EXISTING WALL

TRIM AT JAMB OF WINDOW OR DOOR

OPTIONAL DETAIL

PANELING

RABBET CASING TO COVER EDGE OF PANELING

MATCHING WOOD CASING MOLDING

2—2″x4″'s

TRIM AT JAMB OF WINDOW OR DOOR

MASONITE PANELING

MATCHING WOOD CASING MOLDING

NEW OR UNFINISHED CONSTRUCTION (OPEN STUDS)

31-72a. Window and door trim details.

31-71. *Hardboard nailing details.*

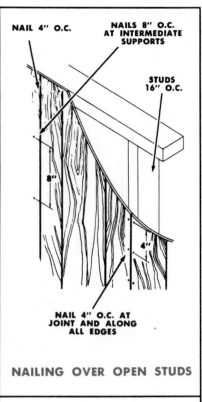

NAIL 4″ O.C.

NAILS 8″ O.C. AT INTERMEDIATE SUPPORTS

STUDS 16″ O.C.

8″

4″

NAIL 4″ O.C. AT JOINT AND ALONG ALL EDGES

NAILING OVER OPEN STUDS

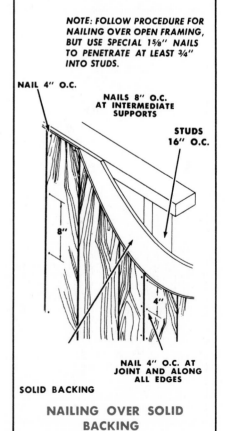

NOTE: FOLLOW PROCEDURE FOR NAILING OVER OPEN FRAMING, BUT USE SPECIAL 1⅝″ NAILS TO PENETRATE AT LEAST ¾″ INTO STUDS.

NAIL 4″ O.C.

NAILS 8″ O.C. AT INTERMEDIATE SUPPORTS

STUDS 16″ O.C.

8″

4″

NAIL 4″ O.C. AT JOINT AND ALONG ALL EDGES

SOLID BACKING

NAILING OVER SOLID BACKING

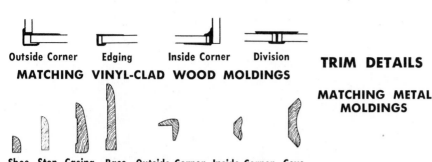

Outside Corner   **Edging**   **Inside Corner**   **Division**

MATCHING VINYL-CLAD WOOD MOLDINGS

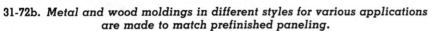

Shoe   Stop   Casing   Base   Outside Corner   Inside Corner   Cove

31-72b. *Metal and wood moldings in different styles for various applications are made to match prefinished paneling.*

TRIM DETAILS

MATCHING METAL MOLDINGS

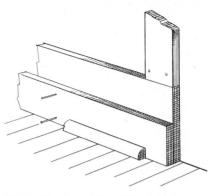

31-75. *The baseboard is nailed to a 1″ × 8″ furring strip at the floorline.*

31-73a. *Six popular tongue-and-groove paneling patterns. Most retail lumberyards carry two or three of these patterns in stock.*

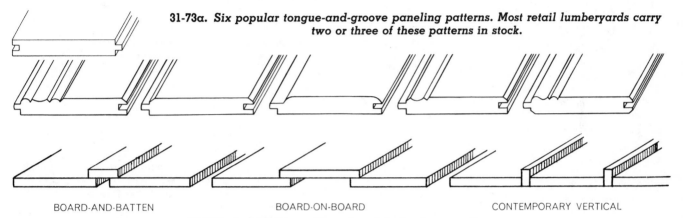

BOARD-AND-BATTEN   BOARD-ON-BOARD   CONTEMPORARY VERTICAL

31-73b. *Installation patterns for plain lumber paneling.*

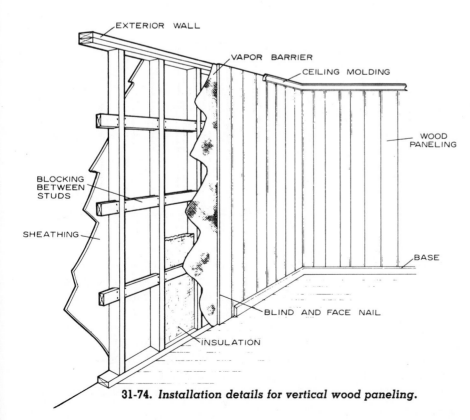

31-74. *Installation details for vertical wood paneling.*

wall should be installed over a vapor barrier and insulation.

Wood paneling should not be too wide; a nominal 8″ is recommended for most parts of the country. Boards wider than 8″ should not be used except when they have a long tongue or matched edge. The boards may be applied horizontally or vertically.

INSTALLATION OF VERTICAL WOOD PANELING

For paneling that is to be installed vertically, adequate blocking should be installed between the studs to provide nailing support. The blocking should not be more than 24″ O.C. Fig. 31-74.

A common practice when installing wood paneling is to nail a 1″ × 8″ board at the floor line. The 1″ × 4″ baseboard is then face-nailed to the 1″ × 8″ board.

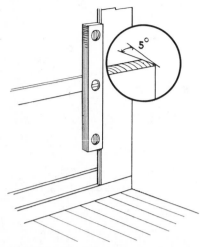

31-76. *The first piece of paneling to be installed is scribed in a plumb position to the adjacent wall and undercut about 5° to provide a tight joint in the corner.*

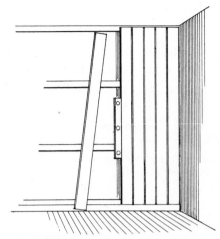

31-78. *Periodically check the paneling for plumb as the installation progresses.*

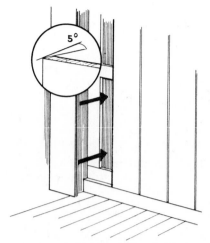

31-79. *Scribe and undercut the last piece of paneling to insure a tight fit against the adjacent wall.*

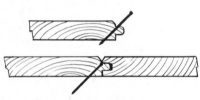

31-77. *Blind-nailing details for lumber paneling.*

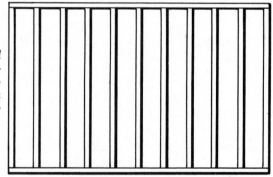

31-80. *Apply cove or crown molding at the intersection of wall and ceiling and a base shoe at the floor.*

Fig. 31-75. The ends of the vertical paneling will rest on the top edge of the 1″ × 4″ base. This is a much cleaner application than resting the paneling ends on the floor and applying the base to the face of the paneling.

Plumb and scribe the first piece of paneling to the wall. Undercut the edge about 5° to insure a snug fit against the wall. Fig. 31-76. Blind-nail all paneling in place using 5d or 6d casing or finishing nails. Fig. 31-77. Continue to install the pieces of paneling, checking for plumb periodically. If necessary, adjust slightly on each added panel until the pieces are again in plumb. The tongue and groove is used for the adjustment. Fig. 31-78.

On the last piece of paneling to

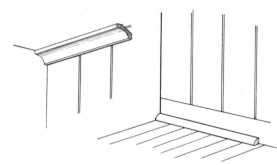

31-81. *When applying wood paneling horizontally, apply the nailing strips vertically 18″ on center. Begin the measurement at the wall on the end strips.*

be installed on a wall, the edge that is to fit into the corner should be scribed and undercut at about 5° angle. The groove of the panel can then be slipped over the tongue of the preceding piece and the panel snapped into place. Fig. 31-79. Apply a cove or a crown molding at the ceiling and wall in-

tersection and a base shoe at the floor. Fig. 31-80. If necessary, install quarter-round trim in the corners.

INSTALLATION OF HORIZONTAL WOOD PANELING

Horizontal paneling, while not

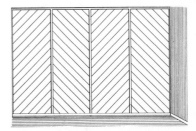

31-84. Wood paneling applied in a herringbone style.

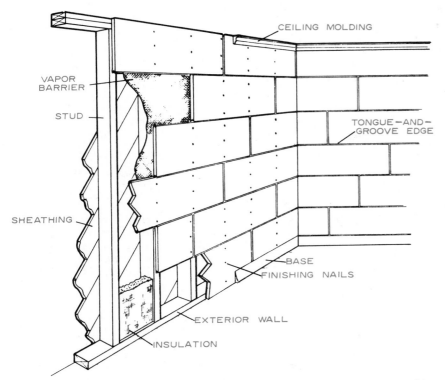

31-82. Wood paneling applied horizontally to studs. Face-nailing is used because of the wide paneling stock.

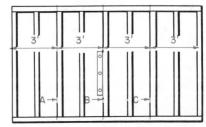

31-85. Locating the furring strips on a 12' wall for a herringbone application.

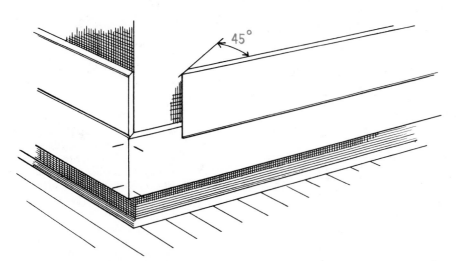

31-83. Wood paneling applied horizontally is mitered at outside corners. In this installation, instead of a baseboard, a "reveal" is shown.

as common as vertical paneling, has some advantages. This method of application requires fewer pieces to shape and install, and it is therefore much faster to apply. It also gives the room the appearance of being longer or larger with a lower ceiling.

Apply vertical nailing strips 18″ on center as shown in Fig. 31-81. Horizontal paneling may also be nailed directly to the studs. Fig. 31-82. Begin the paneling at the floor line, making certain that the first piece is installed level. Undercut the ends slightly to pro-

vide a tight joint at the inside corners. Outside corners should be mitered. Fig. 31-83. Blind-nail all the paneling as described for the vertical application, checking periodically to make certain the paneling remains level.

If no molding is to be used at the ceiling, undercut the last panel edge at a 5° angle to insure a snug fit against the ceiling line. When desired, apply cove or crown moldings at the ceiling and wall joints and apply base shoe at the floor. If necessary, quarter-round trim may be installed in the corners.

INSTALLING WOOD PANELING IN A HERRINGBONE PATTERN

The herringbone style of application is very interesting, but it is also the most demanding in craftsmanship. Fig. 31-84. Apply vertical nailing strips so that the space between them is evenly divided. For example, if the wall is 12' long, space the strips 18″ on center as shown in Fig. 31-85. When laying out, begin the actual

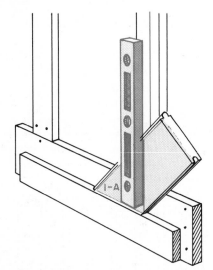

31-86. *Laying out the first piece of paneling (A). Baseboard installation is necessary to provide backing for nailing the paneling in place.*

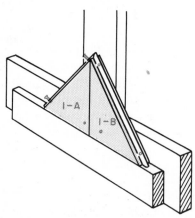

31-87. *Lay out and cut the second piece (1B) before nailing the first piece (1A) in place.*

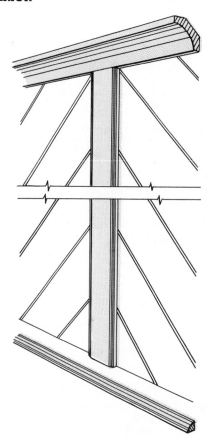

31-88. *Apply a molding strip at the vertical joint of the paneling. The molding should extend from the top of the baseboard to the underside of the ceiling molding.*

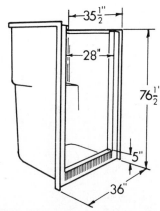

31-89a. *A shower cove bathing module made from plastic reinforced with fiberglass. Bathing modules are available in white and in several colors.*

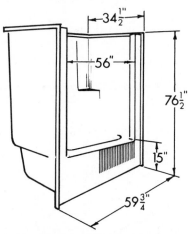

31-89b. *A bath and shower module. This module includes a bathtub and walled alcove. The tub has a flat rim for seating or to accommodate a shower door. The unit is plastic reinforced with fiberglass.*

measurement from the adjacent wall next to the end strip and measure to the center of the other strips. Make sure that each strip is plumb.

Install the baseboard as shown in Fig. 31-86. Then, using a long level, draw a plumb line at the center of every other nailing strip as shown at a, b, and c in Fig. 31-85. For a 12′ wall, these lines should be as close as possible to 36″ apart. Saw two pieces of paneling in the shape of a triangle

with the tongue on the long edge, as shown in Fig. 31-87. Install them with the vertical joint on one of the plumb lines. Extreme care should be taken to make the paneling butt even with the baseboard. A molding strip will be applied later at the vertical joint.

In the same manner, measure and cut to length the second pieces to be installed. Install each paneling piece all the way across the wall, building toward the top. Use the play in the tongue and groove to keep the panels aligned at the "V" (along the vertical joint). After all the pieces have been applied, install the cove or crown molding at the ceiling. Next apply the molding at the butt joint of the paneling. This

molding should stop at the baseboard and butt into the cove or crown molding at the ceiling. Fig. 31-88. A cove or quarter-round may be used at the corners and a base shoe at the floor if necessary.

Bathroom Wall Covering

When a complete, prefabricated shower or combination shower and tub stall is installed in the bathroom instead of a tub, special

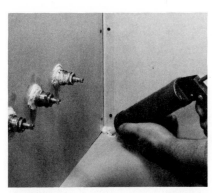

31-90. *Moisture-resistant gypsum wallboard installed in a tub alcove. Corners and fittings must be calked.*

END CAP   INSIDE CORNER

31-91. *Inside corner molding and end cap.*

wall finishes are not required. Fig. 31-89. When tubs are used, however, some type of waterproof wall covering is normally required around it to protect the wall.

There are several types of finish, including coated hardboard paneling and various ceramic, plastic, and similar tiles. Fig. 31-90.

Plastic-surfaced hardboard materials are applied in sheet form and are fastened with an adhesive or nails as described under "Installing Gypsum Board" or "Hard-

board" earlier in this unit. The method of application depends on the nature of the material. Moldings are placed on inside corners, on tub edges at the joints, and as end caps. Fig. 31-91. Several types of calking sealants are also available which will provide excellent results.

Ceramic, plastic, and metal tile is installed over water-resistant gypsum board. The adhesive is spread with a serrated trowel, and the tiles are pressed into place. A grout cement is inserted in the joints of the tile after the adhesive has set. The plastic, metal, or ceramic type of wall covering around the tub area is usually installed by subcontractors specializing in this craft.

QUESTIONS

1. List several materials that are commonly used for interior wall and ceiling finish.

2. What materials are most frequently used as a plaster base?

3. Why is plaster reinforced at certain key positions?

4. What material is used for plaster reinforcing?

5. What are plaster grounds?

6. What are the ingredients of plaster?

7. What two plaster coats are combined when they are applied on gypsum or insulating lath?

8. When plastering, what is the difference between a sand-float and a putty finish?

9. What materials are classified as dry-wall finish?

10. When applying gypsum board dry wall, what is meant by double nailing?

11. Describe how gypsum board is held in place on a ceiling during application.

12. When plywood is used as a wainscot, why is it usually about 32″ high?

13. Describe how an outlet box cutout is located on a plywood panel.

14. What is the recommended nominal width for wood paneling used in most parts of the country?

15. Which method of wood paneling installation will give a room the appearance of being longer or larger and having a lower ceiling?

16. List some waterproof materials that are commonly used for a bathroom wall covering.

Ceiling Tile and Suspended Ceilings

32

In addition to lath and plaster and the sheet materials discussed in Unit 31, insulating board or ceiling tile makes a good ceiling. Fig. 32-1. In most cases, this type of ceiling finish has excellent acoustic and insulating qualities. It is also fire retardant. Ceiling tile may be installed in several ways, depending on the type of ceiling or roof construction. When a flat surface is present, such as between beams of a beamed ceiling in a low sloped roof, the tiles are fastened with adhesive. When the tile is edgematched, it can be stapled in place. Another method is to apply a suspended ceiling with small metal or wood hangers which form supports for drop-in panels. The most common method of installing ceiling tile is with 1" × 3" wood furring strips nailed across the ceiling joists or roof trusses. Fig. 32-2a.

CEILING TILE

Standard ceiling tiles and acoustical tiles are fiberboard products made from natural wood or cane fibers. They are designed for decoration and sound insulation and are used in new construction as well as remodeling. The tiles are factory predecorated, requiring no painting or other finishing. Many designs, colors, and patterns are available in either the standard or acoustical type. Surface characteristics vary from smooth to various textured and sculptured effects. In acousti-

cal tiles, surface openings provide for sound absorption. These openings may be holes drilled or punched in various patterns, or they may be slots, striations, or fissures.

The most popular sizes of ceiling tile are 12" × 12" and 12" × 24". The 12" × 24" size is available with or without center scoring to represent two tiles. Ceiling tiles come in a nominal ½" thickness with an interlocking tongue-and-groove joint which provides for self-leveling and concealed attachment. Some tiles are available with beveled or curved edges. Fig. 32-2b. They also come in 16" × 16" and 16" × 32" sizes in nominal thicknesses from ½" to ¾".

Ceiling tile may be cemented to gypsum wallboard if the wallboard is at least ⅜" thick and nailed on not more than 16" centers. Ceiling tile may also be applied directly to a plaster ceiling, provided the plaster is solid and level. If the existing surface in either case does not meet these minimum requirements, furring strips should be installed to keep the new ceiling level and prevent future trouble.

Determining Room Layout

The following steps are taken for figuring the ceiling tile layout of a rectangular room. If an L-shaped room is to be tiled, divide the room into two rectangles or squares and figure accordingly.

A ceiling has a better appear-

ance if border tiles (those adjacent to the walls) are the same width on opposite sides of the room. To find the border tile width for the long walls of a room, follow these steps:

1. Measure one of the short walls in the room.

2. If this measurement is not an exact number of feet, add 12" to the inches left over.

32-1. *A suspended ceiling system with 2' × 4' drop-in panels. The embossed swirl design gives the effect of a notch-troweled plaster ceiling.*

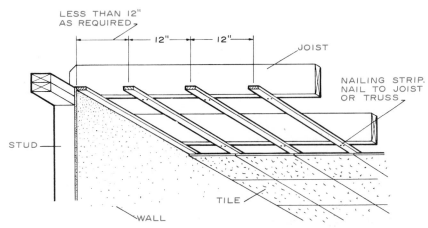

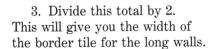

32-2a. *Ceiling tile installed on wood furring strips.*

BUTT (KERF)

T & G (KERF)

BUTT K & R

T & G FLANGE

32-2b. *Ceiling tile joint details.*

3. Divide this total by 2. This will give you the width of the border tile for the long walls.

Example:

Short wall = 10'8"	
Extra inches:	8"
Add:	12"
Divide:	2)20"
Border tile for long wall:	10"

To figure border tile for the short walls, follow the same procedure:

1. Measure one of the long walls in the room.

2. If this measurement is not an exact number of feet, add 12" to the inches left over.

3. Divide this total by 2. This will give you the width of the border tile for the short walls.

Example:

Long wall = 12'4"	
Extra inches:	4"
Add:	12"
Divide:	2)16"
Border tile for short wall:	8"

When using 16" × 16" tile, convert the wall measurement into inches. Divide this measurement by 16. Treat the extra inches the same way as with the 12" tile in the example above, but add 16" (instead of 12") and divide by 2.

Installing Furring Strips

Furring strips can be nailed directly to wood joists or through an existing ceiling into the joists to provide a solid base for stapling the tiles. The first two furring strips must be carefully placed so that the border tiles are properly aligned. Place the first furring strip flush against the wall at right angles to the ceiling joists. Nail the strip into place with two 8d nails at each joist location. Fig. 32-3.

If the joists are concealed by an existing ceiling, locate and mark the position of each joist before nailing the furring strips in place. Generally, joists are spaced 16" on center and are perpendicular to the long wall. If you are unable to determine the direction of the joist by tapping on the existing ceiling, check the floor above the joists. If the finished floor is wood, it will be nailed across the joists.

The second furring strip should be placed parallel to the first at the border width distance from the wall. In the example, the border width measurement is 10". Add ½" for the stapling flange, which overlaps the finished bevel edge of the tile. Thus the second furring strip should be placed so

32-3. *The first furring strip is nailed flush against a wall at right angles to the joists. Use two 8d nails at each joist crossing.*

32-4. *The position of the second furring strip depends upon the width of the border tile to be used. All strips thereafter are installed 12" on center.*

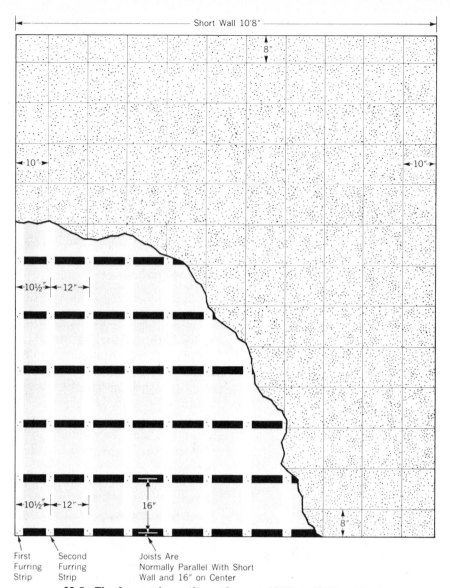

32-5. *The layout for ceiling tile on a 10'8" × 12'4" ceiling.*

First Furring Strip

Second Furring Strip

Joists Are Normally Parallel With Short Wall and 16" on Center

that the center is 10½" from the wall.

After installing the second strip, work across the ceiling, nailing the rest of the furring strips in place 12" on center from each other. Fig. 32-4. Because the border tile size was determined earlier, the next to last strip will be positioned automatically—in this case 10½" from the wall. The last strip should be nailed flush against the wall in the same manner as the first strip. Fig. 32-5.

If using 16" × 16" tile, nail the first strip flush against the wall. Nail the second strip at the appropriate border width distance from the wall and then nail remaining strips 16" on center.

Make certain that the furring strips are level by checking them with a straightedge. Correct unevenness by driving thin wood shims between the strips and the joists. Fig. 32-6.

Installing Special Furring Strips

If pipes or wire cables are located below ceiling joists or project less than 1½" below the existing ceiling structure, install double furring strips over the entire ceiling to avoid the projections. The first course should be spaced 24" to 32" on center and fastened directly to the joists. The second course is then applied perpendicular to the first, appro-

priately spaced to receive the ceiling tile. Fig. 32-7.

Pipes or ducts that project more than 1½" below the ceiling joists should be boxed in with furring strips before the ceiling is installed. After the ceiling tile has been installed, wood molding can be used to provide a finish detail for inside and outside edges.

Ceiling Layout

Construct two reference lines before beginning the tile application. These will align the first rows of tile installed in both di-

32-6. *Level the furring strips with pieces of wood shingle. Check the strips with a carpenter's level or pull a line taut from end to end.*

32-7. When cables or pipes are mounted below the joists, install a double layer of furring strips over the entire ceiling to eliminate the projection. The first layer of strips is nailed on about 24″ to 32″ centers parallel to the pipes.

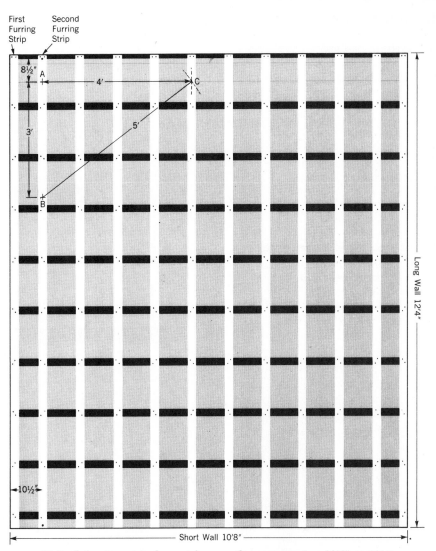

32-8. A furring strip layout for a ceiling measuring 10′8″ × 12′4″.

rections. To make certain these lines are accurate, follow these steps:

1. Partially drive a nail into both ends of the second furring strip, using the appropriate border tile measurement. In the example, the border tile measurement from the long wall is 10″. Thus the nails should be positioned at each end of the second furring strip, 10½″ from the long wall. The stapling flange of the tile, when installed, will line up with the exact center of the furring strip. Fig. 32-8.

2. Stretch a chalk line tightly between the two nails and snap it.

To be assured of proper tile alignment, the second reference line must be constructed exactly at a right angle to the first. This is done by constructing a 3′ × 4′ × 5′ right triangle, using the first reference line as a base.

1. Locate point A on the first reference line using the short wall border tile measurement. Fig. 32-8.

2. From point A measure in exactly 3′ along the reference line to locate point B.

3. Drive small nails into the furring strips at points A and B.

4. Starting with point A measure off exactly 4′ across the furring strips. Mark a small arc on the furring strip at this point.

5. From point B measure exactly 5′ toward the first arc, marking the point of intersection as point C.

6. Snap a chalk line through points A and C across all furring strips. This second reference line is perpendicular to the first. Fig. 32-9.

32-9. Snap a chalk line across the furring strips to indicate the location of the short wall border tile.

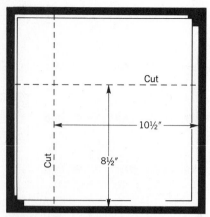

32-10. *Laying out the first tile so that it fits into the corner. Be sure to leave a flange (tongue). The tile will be nailed or stapled in place through the flange.*

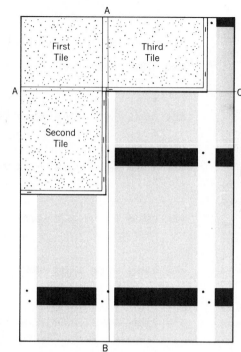

32-11. *Installation details for the first border tile along each wall.*

32-12. *Stapling ceiling tile. When finishing the border on the opposite side of the room, remove the stapling flanges from the tiles so that the tiles will fit against the walls. These tiles are face-nailed into the furring along the wall.*

Cutting Tile

Measure and cut each of the border tiles individually. Remove the tongue edge and leave the wide flanges for stapling. *Include the face and the flange of each tile in your measurement.* Cut the tile face up with a coping saw or a very sharp fiberboard knife.

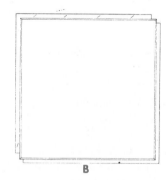

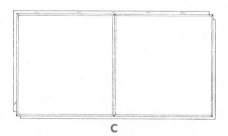

Cut the first tile so that it fits into the corner. In the example, the first tile would be cut 8½″ × 10½″. Fig. 32-10. After aligning the flanges with reference lines AB and AD, staple this tile into position. Fig. 32-11.

Cut a second tile so that one of the tongue edges fits into the corner tile and the stapling flange falls directly on line AB. Fig. 32-11.

Cut a third tile so that the tongue edge will fit into the corner tile and the stapling flange will fall directly on line AC. Fig. 32-11.

Work across the ceiling, installing about two tiles at a time along the borders and filling in between with full-sized tile. When you

32-13. *Stapling details: A. Use four staples to fasten each 12″ × 12″ tile in place; three staples in one flange, one in the other. B. Use five staples for each 16″ × 16″ tile; put four staples in one flange and one in the other. C. For 12″ × 24″ tile use six staples for each tile; one in the 12″ flange, five in the 24″ flange.*

reach the opposite wall, measure each border tile individually. Fasten the tile by stapling into remaining flange and face-nailing into furring along the wall. Finish off the ceiling with a crown or cove molding.

Fastening Tile

STAPLING

Using 9/16" staples, fasten each tile in place with three staples in the flange parallel to the reference line AB and one staple in the flange on the furring strip closest to the wall. Figs. 32-11 & 32-12. If you are using 16" × 16" tile, put four staples in the flange parallel to the reference line. Fig. 32-13.

When sliding the tiles into position, be sure they join snugly but *do not force them tightly together*. Fig. 32-14.

CEMENTING

Brush-on ceiling cement can be used only when installing tiles that have a tongue-and-groove joint detail. Acoustic cement should be used for butt-edge tile. Before starting to cement, plan the room layout. Then lay out the ceiling and cut the tile as described earlier. The reference lines can be snapped directly on the existing ceiling surface.

For 12" × 12" tile, put five daubs of the cement on the back of each tile with a paintbrush. Place one daub on each corner of the tile and one in the center. Each spot of cement should be about 1½" across and at least the thickness of a nickel. Keep the daubs of cement away from the edges of the tile to allow them room to spread when the tile is installed. Fig. 32-15.

Slip the first border tile into position and press it tightly against the ceiling. Fasten it in place with two ⅜" or 5/16" staples in each

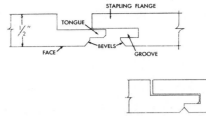

32-14. The tongue-and-groove joint is designed to give each tile room to expand and contract as the indoor climate changes. When sliding tiles into position, be sure to join them snugly, but do not force them tightly together.

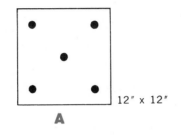

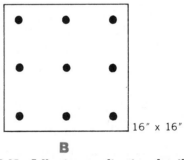

32-15. Adhesive application details: A. Put five daubs of cement on a 12" × 12" tile. B. Put nine daubs of cement on a 16" × 16" tile.

flange. The staples are needed to hold the tile in place long enough for the cement to set up.

Follow a similar procedure for 16" × 16" tile, but use nine daubs of cement. Place these daubs evenly over the back of the title. Put three staples in each flange of the 16" × 16" tile. Fig. 32-15.

The Ashlar Pattern

The ashlar pattern of installation creates an interesting design effect by staggering the title in "brickwork" fashion. Fig. 32-16. To install a ceiling in the ashlar pattern, proceed as follows:

1. Make chalk lines for the long and short walls as described earlier. After making a chalk line for the short wall (this is the AC line), snap a third chalk line (DE) 6" closer to the wall. Fig. 32-8.

2. Cut the first tile to fit into the corner where the first and second chalk lines cross. One of the stapling flanges should fall directly along the AB line and the other along the AC line. Staple into position.

3. Remove one of the tongue edges from the second tile so that the cut side will fit flush against the long wall and one of the stapling flanges falls on the AB line. The tile's remaining tongue edge

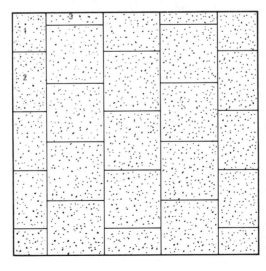

32-16. Ceiling tile installed using the ashlar pattern.

385

32-17. *A fluorescent lighting fixture installed directly under the ceiling tile.*

32-18. *Installing a recessed fixture in place of a 12" × 12" tile.*

32-19. *The recessed fixture installed, complete with lens frame.*

Slide Lock

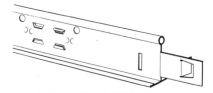

A. Main runner with splicer attached

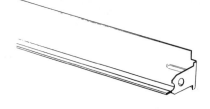

B. Cross tees—2-foot—4-foot

C. Wall molding

Custom Grid

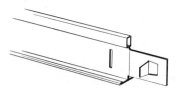

A. Main runner with splicer attached

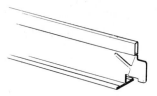

B. Cross tees—2-foot—4-foot

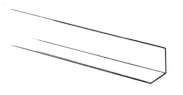

C. Wall molding

32-20. *Two of the many types of suspended ceiling grid components.*

should fit into the groove of the corner tile. Fig. 32-16.

4. Cut a third tile so that it will fit flush against the short wall. One of the stapling flanges should fall directly on line DE, and its remaining tongue edge should fit into the corner tile. Fig. 32-16.

5. Work across the ceiling, installing about two tiles at a time along the borders as indicated in Fig. 32-16. Fill in between them with full-sized tiles.

6. When you reach the opposite wall, measure each border tile individually. Fasten the tile in place by stapling into the remaining flange.

7. Finish off the ceiling with a crown or cove molding.

Ceiling Light Fixtures for Use with Ceiling Tile

Several styles of ceiling lights are made to go with ceiling tile. Fixtures can be standard, chandelier, or fluorescent. Fig. 32-17. Also available is a lighting fixture measuring 12" × 12", the same size as a standard ceiling tile. It can be easily inserted in place of a tile during the ceiling installation. An adapter plate is attached to the furring strips with four wood screws. Fig. 32-18. This adapter plate fits into the tongue and groove of the tiles that surround it and serves as a base for the fixture's other parts. After the junction box is installed and wired, the reflector dome is snapped into place. The lens frame is then attached to the adapter plate and pushed into place. Fig. 32-19.

SUSPENDED CEILINGS

Suspended ceilings consist of panels held in place by a grid system at a desired distance from the existing ceiling structure. The panels are made of fiberglass or plastic and are 24" × 48" or smaller. The grid system which supports these panels includes main runners, cross tees, and wall molding. Main runners are usually 12′ long and are spaced 2′ or 4′ on center. Cross tees are installed at right angles to the main runners. There are many types of suspended ceiling grid components. Two kinds are shown in Fig. 32-20.

The suspended ceiling reduces noise in two ways. It absorbs a large amount of the noise striking its surface. Also, because of its suspension, it does not transmit the sound vibrations into the framing above as readily as materials applied directly to framing.

A suspended ceiling is easily and quickly installed. It conveniently covers up bare joists, exposed pipes, and wiring, and it may be used to lower a high ceiling. Accessibility to unsightly valves, switches, and controls hidden by the suspended ceiling is no problem because the panel in question can merely be slid to one side.

The first step when installing a suspended ceiling is to determine the ceiling height. Sometimes, such as for a basement ceiling, it is desirable that the ceiling be as high as possible to provide maximum headroom. Care should be taken, however, to keep the top edges of the grid system at least 2″ to 2½″ below the bottom of the framing. This space is necessary for the insertion of the ceiling panels after the grid system is in place.

Mark or snap a level chalk line on each wall of the room, ¾″ (the width of the wall molding) above the intended height. This permits the wall molding to be installed below the chalk line and eliminates any undesirable marks on the wall below the ceiling level after installation is completed. To insure a level ceiling, check these lines carefully with a carpenter's level.

Determining Room Layout

First determine the direction of the main runners. In most cases, the main runners should be installed perpendicular to the ceiling joists and parallel to the long wall.

Determine where the first main runner will be placed. To locate the distance of the first main runner from the long wall, proceed as follows:

1. Measure the length of one of the short walls and convert to inches.
2. Divide this figure by 48″.
3. Take the number of inches left over (if any), and add 48″ to it.
4. Divide this figure by 2. You now have the distance that the first main runner should be placed from the long wall. This also will be the size of the long wall border panel.

Example:

Short wall = 10′4″ = 124″
Divide: 48)124 = 2, remainder 28
Add: 28″ + 48″ = 76″
Divide: 2)76 = 38″, The distance of main runner and long wall border panel size

NOTE: When using 24″ × 24″ ceiling panels, divide by 24″ and add 24″ instead of 48″.

Measure out this exact distance (in the example, 38″) on both short walls from the long wall. Fasten reference string A at the ceiling height line and stretch it between the two points. The first main runner will eventually be placed along this string. Fig. 32-21.

To determine the short wall border panel size and the location of cross tees, follow these steps:

1. Measure the length of one of the long walls and convert to inches.
2. Divide this figure by 24″.
3. Take the number of inches left over (if any), and add 24″ to it.
4. Divide this figure by 2.

Example:

Long wall = 18′8″ = 224″
Divide: 24)224 = 9, remainder 8
Add: 8″ + 24″ = 32″
Divide: 2)32 = 16″, Distance of first cross tee and short wall border panel size

Measure out this exact distance (in the example, 16″) on both long walls from the short walls. Fasten a second reference string, B, at the ceiling height line between these two points. Fig. 32-21. The first row of cross tees will be installed in line with this string.

Since walls are seldom perfectly straight, it is imperative that the second reference string be perpendicular to the first. When stretching this string, make certain that it is exactly at a 90° angle.

Installing Wall Molding

Fasten the metal molding to the walls, making certain that the top of the molding is in line with the level chalk mark. If the molding cannot be nailed directly to the wall, hang a suspended main runner in place of the regular wall molding. For inside corners, lap one piece of molding over the other. Outside corners are formed by mitering the two wall moldings or by overlapping. Fig. 32-22. If the molding is to be nailed to a cinderblock wall, use concrete stud nails. Drive the nail between

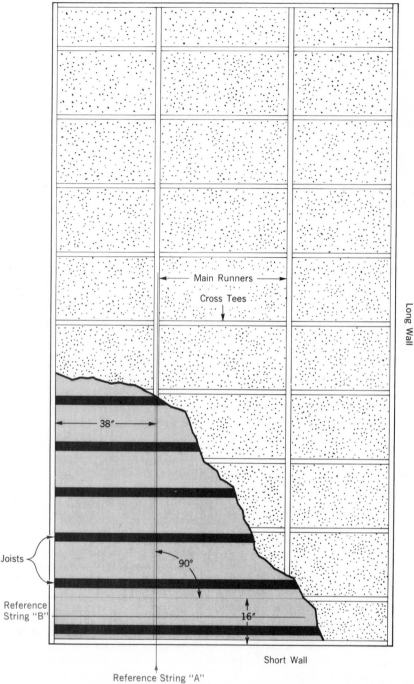

32-21. *Layout details for a 10'4" × 18'8" suspended ceiling using 24" × 48" panels.*

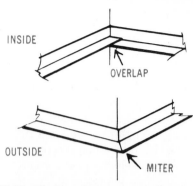

32-22. *Wall molding corner details.*

der measurement from 24". (In the example this measurement was 16"; 16" from 24" equals 8".)

2. Add 6" to remaining inches (6" + 8" = 14").

3. Cut the main runner by this amount. (In the example, the main runner is cut 14" from the end.) A, Fig. 32-23. NOTE: Be sure that a cross tee tab falls directly above reference string B. Figs. 32-21 & 32-23.

The main runners are suspended from the joists by hanger wires. To find the location of the first hanger wire, rest the cut end of the runner on the wall molding and directly above reference string A. See B, Fig. 32-23. Directly above any hole near the uncut end of the main runner fasten the first hanger wire to the existing ceiling structure. Run the wire through the hole in the main runner, but do not attach it permanently. Install the other hanger wires for this runner at 4' intervals.

Attach wires to the ceiling structure at each main runner location. Place them in line with the wires attached to the main runner. Figs. 32-23 & 32-24.

To insure that all main runners will be perfectly level, follow these steps:

1. Stretch a string across all hanger wires in the direction of the main runner. Attach each ref-

the mortar joint and the edge of the cinder block.

Installing Main Runners

To insure that the short wall border panels are of equal size, the main runners must be accurately cut. To cut a main runner, follow these steps:

1. Subtract the short wall bor-

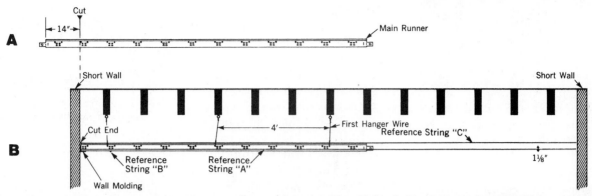

32-23. *Laying out the main runner for the room layout shown in Fig. 32-21. A. Fourteen inches should be cut from the end of the main runner. This end will rest on the wall molding. B. Suspend the main runner from the joists with wire. Use a string pulled taut to align the main runner. Make certain that the cross tee tab connection is located directly above the reference string B. This is essential so that a cross tee will be properly located to provide support for the border tile.*

32-24. *The hanger wires are attached to the joists at 4' intervals.*

32-25. *Fastening the main runner of the metal framework to the hanger wires.*

on the runner should be directly above reference string B. Insert a hanger wire into one of these holes. Let one end of the main runner rest on the wall molding and support the other end with a hanger wire.

If the grid is uneven, fasten additional hanger wires as needed to the holes in the main runners. Bend the wire at the top of the hole, insert it through the hole, and wrap any excess wire around itself. Fig. 32-25. More than one runner may be needed to reach the opposite wall. For longer lengths, connect the main runners by inserting the end tabs into precut holes and bending them over.

Attaching Cross Tees to Main Runners

The first set of cross tees will be installed directly above reference string B. Figs. 32-21 & 32-23. Follow these steps:

1. Measure the distance from the long wall to reference string A. (In the example, this is 38".)

2. Cut the cross tee to this length and install it.

3. Complete the installation of cross tees above reference string

erence string so that it is in line with the holes in the runner to which the hanger wires will be attached. Fig. 32-23, string C. For the system shown in Fig. 32-23, these holes should be located 1⅛" above the bottom edge of the runners. Thus string C is 1⅛" above string A. This string must be kept very tight.

2. Make sharp 90° bends where the wires come across the

string. The best method is to clamp pliers horizontally to the wire so that the bottom of the pliers is at string level. With the other hand, firmly bend the wire tightly against the bottom of the pliers.

Install the remaining runners that go perpendicular to the joists. Measure each new main runner individually. After cutting a main runner, two of the holes

389

32-26. *Installing the cross tees between the main runners.*

32-27. *Lay the ceiling panels into the grid formed by the main runners and the cross tees.*

B. Working in the same direction, install the balance of the cross tees in rows across the room. Fig. 32-26. The tees may be 2′ or 4′ apart.

Depending on the spacing of the main runners, the cross tees may be 24″ or 48″ long. If the cross tees are 48″ long and 24″ apart, and 24″ × 24″ ceiling panels are to be installed, attach 24″ cross tees between the 48″ cross tees, parallel to the main runners.

Cutting the Panels

Measure and cut each of the border panels individually. A panel is cut face up, with a coping saw or a very sharp fiberboard knife.

Installing Ceiling Panels

Drop-in panels are installed by resting these units on the flanges of cross tees and main runners. Fig. 32-27. Exercise care when handling ceiling panels to avoid marring the surface. Handle the panels by the edges, keeping the fingers off the finished side of the board as much as possible.

Ceiling Light Fixtures for Use with Suspended Ceilings

Standard ceiling fixtures and chandeliers may be used with a suspended ceiling. Recessed lighting can also be conveniently installed at any point. Simply install any one of several styles of translucent panels in place of a ceiling tile. Fig. 32-28.

32-28. *Translucent panels used with recessed lighting fixtures.*

32-29. *The fluorescent fixture shown here is mounted on the suspended ceiling grid system.*

Fluorescent lighting fixtures can be suspended from the wood framing between the floor joists. Special mounting brackets can also be obtained for attaching the fixture directly to the suspended grid. Fig. 32-29. Some fluorescent fixtures are designed to fit flush against the ceiling surface. Vaulted ceiling lighting modules which fit the grid system are also available.

QUESTIONS

1. What are some of the advantages of ceiling tile and suspended ceilings?

2. When laying out a room for ceiling tile, why is it important that the border tile on opposite sides of the room be the same width?

3. Describe briefly the procedure for determining the width of the border tile for the ceiling.

4. Why is the distance between the wall and the first furring strip different from the distance between centers on all other furring strips?

5. What tools may be used for cutting the tile?

6. What size staples should be used when installing ceiling tile?

7. Describe the ashlar pattern.

8. Why is the chalk line snapped above the location of the wall moldings for suspended ceilings?

9. Calculate the width of the border tile for a ceiling using 24″ × 24″ ceiling panels. The room measures 14′6″ × 23′4″.

10. List some of the lighting systems that can be used with a suspended ceiling.

Finish Flooring

33

Finish flooring is the final wearing surface applied to a floor. Many materials are used as finish flooring, each one having properties suited to a particular usage. Durability and ease of cleaning are essential in all cases. Specific service requirements may call for special properties such as resistance to hard wear, comfort to users, and attractive appearance.

There is a wide selection of flooring materials. Hardwoods are available as strip flooring in a variety of widths and thicknesses and as random-width planks, par-

quetry, and block flooring. Other materials include plain and inlaid linoleum, cork, asphalt, vinyl, rubber, plastic vinyl, ceramic tile, and wall-to-wall carpeting. When a finish floor other than wood is to be installed on a subfloor of wood boards, an underlayment is required. This may be plywood, hardboard, or particleboard.

HARDWOOD STRIP FLOORING

Strip flooring is the most widely used type of hardwood flooring. Practically all species are

produced in this form. As the name implies, it consists of flooring pieces cut in narrow strips of varying thicknesses. The thinner strips are used chiefly in remodeling when new floor is laid over old and it is not desirable to reduce room height by using thick, more expensive flooring.

The most common hardwoods for strip flooring are oak, maple, beech, birch, and pecan. Oak, the most plentiful, is the most popular by far. It constitutes about 80% of the residential hardwood flooring in the United States.

Despite its extensive usage, strip flooring should not be considered common-place. Its popularity is due to its high quality. No two hardwood floors are exactly alike. Each has individuality of character and beauty of grain. Most floors of this type are composed of strips of uniform width. Interesting patterns are formed by use of stock selected for variations in color or other natural irregularities. Attractive designs also may be achieved with strips of random widths. Most hardwood strip flooring today is tongued and grooved at the factory so that each piece joins the next one snugly when laid. Fig. 33-1.

Sizes

Strip flooring of oak, maple, beech, birch, and pecan is manufactured in a variety of sizes, ranging in width from 1″ to 3½″ and in thickness from 5/16″ to 33/32″.

33-1. Strip flooring. A. Side- and end-matched. B. Side-matched. C. Square-edged.

391

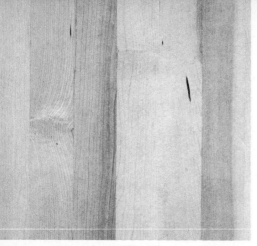

33-2. Typical first grade MFMA (Maple Flooring Manufacturers' Association) maple flooring.

The standard thicknesses for tongued and grooved strip flooring are ½″, ⅜″, and 25/32″. Since strips of these thicknesses are the most commonly chosen for homes, they are produced in greatest volume and are available at lower cost than special thicknesses. The 25/32″ strips, used the most, are manufactured in four widths: 1½″, 2″, 2¼″, and 3¼″. Most popular is the 2¼″.

Square-edge oak strip flooring is 5/16″ thick and comes in widths of 1″, 1⅛″, 1½″, and 2″. Square-edge maple, beech, birch, and pecan strip flooring comes in two thicknesses: 25/32″ and 33/32″. Widths of 2½″ and 3½″ are available in each thickness. The length of the strips in a bundle of flooring varies, but average lengths are specified for each grade. Some strips may be as long as 16′.

Grading

Through two major trade associations, the principal American producers of hardwood strip flooring have adopted uniform grading rules and regulations for commercial practice. Approved by the Bureau of Standards of the U.S. Department of Commerce, these rules and regulations are enforced rigidly, in part by the organizations themselves. As a result, dealers and consumers are assured of high quality flooring, and

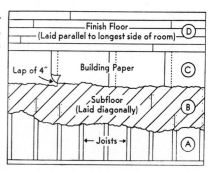

33-3. Cutaway view of floor, showing details of construction: A. Joist. B. Subfloor. C. Building paper. D. Finish floor.

the industry is protected against sharp practices. The two organizations are the National Oak Flooring Manaufacturers' Association, with headquarters at Memphis, Tennessee, and the Maple Floor Manufacturers' Association, Chicago, Illinois. Every bundle of flooring produced by a member of either association is identified as to grade and is guaranteed to meet all established specifications. Usually the manufacturer's name and a mill mark of identification are stated on each bundle.

The hardwood flooring grades in Canada are identical to those for the United States.

There are no official grading rules for plank, parquet, and block flooring. Generally the different grades correspond to those of strip flooring. Hardwood strip flooring grades are based principally on appearance. Since all regular grades have adequate strength, durability, and resistance to wear, these qualities are not factors. Chiefly considered are such characteristics as knots, streaks, pin wormholes and, in some cases, sapwood and variations in color. Slight imperfections in processing also are factors.

OAK

Oak is classified into two grades of quarter-sawed stock and four grades plain-sawed. In descending order the quarter-sawed grades are: Clear and Select. Plain-sawed grades are: Clear, Select, No. 1 Common, and No. 2 Common.

In Clear Grade Oak the amount of sapwood is limited. Otherwise, variations in color are disregarded in grading. Red oak and white oak ordinarily are separated, but

that does not affect their grading. In most cases the average length of strip flooring pieces is greater in the higher grades.

MAPLE, BEECH, BIRCH

Rules governing the grading of maple, beech, and birch are virtually identical for all three species. Neither sapwood nor varying natural color is considered a defect in standard grades. These standard grades are first, second, and third, with first being the highest. Fig. 33-2. Each of these grades also is available in a special grade selected for uniformity of color.

PECAN

Pecan is processed in six standard grades. Two of the grades specify all heartwood, and one specifies bright sapwood. Otherwise color variation is not considered.

STORAGE AND DELIVERY

Certified hardwood flooring is kiln-dried at the factory to a low moisture content. However, moisture content later equalizes itself to the moisture conditions in the area where the flooring is used. The flooring must be protected from the elements during storage and delivery to guard against excessive shrinkage or expansion, which may cause cracks or buckling after the floor has been laid. Manufacturers who ship hardwood flooring in closed boxcars

33-4. *Laying asphalt-coated building paper over the subfloor in preparation for the finish floor. NOTE: The building paper will cover up the nails in the subfloor which indicate the location of the floor joists. It is therefore advisable to mark the location of the floor joists on the walls before laying the building paper. Later chalk lines may be snapped on the building paper to indicate the location of the floor joists below for nailing the strip flooring.*

recommend the following precautions in handling:

• Do not unload, truck, or transfer hardwood flooring in rain or snow. Cover it with tarpaulin if the atmosphere is foggy or damp.

• Flooring should be stored in airy, well-ventilated buildings, preferably with weathertight windows that will admit sunlight.

• Do not pile flooring on storage floors that are less than 18″ from the ground and which do not have good air circulation underneath.

• Do not store or lay flooring in a cold or damp building. Wait until the plaster and concrete work have dried thoroughly and all but the final woodwork and trim have been installed.

• Especially in winter construction, the building in which the flooring is to be used should first be heated to 70° F. Then the flooring should be stored in the building at least four or five days before being laid. This permits the flooring to reach a moisture

content equivalent to that of the building.

INSTALLATION

Installation of the finish flooring should be the last construction operation in a house. Fig. 33-3. All plumbing, electrical wiring, and plastering should be completed before the application of the finish floor is begun. Only the final interior trim work should remain.

Before laying the flooring, the bundles should be opened and the flooring spread out and exposed to warm, dry air for at least 24 hours, preferably 48 hours. Moisture content of the flooring should be 6% for the dry Southwest, 10% for southern and coastal areas, and 7% for the remainder of North America.

Preparing the Subfloor

Just before installation of the finish flooring is to begin, the subfloors should be examined carefully and any defects corrected. Raised nails, for instance,

should be driven down and loose or warped boards replaced. The subfloors should be swept thoroughly and scraped if necessary to remove all plaster, mortar, or other foreign materials. These precautions must be taken if the finish flooring is to be laid properly.

The last operation before actual installation of the finish flooring should be the application of a good quality asphalt-coated building paper over the subflooring. Fig. 33-4. The paper will protect the finish floor and the interior of the home from dust, cold, and moisture which might seep through the floor seams. In the area directly over the heating plant, it is advisable to lay double weight building paper or standard insulating board. This will protect the finish floor from excessive heat which might shrink the boards.

Where to Start

Before the actual laying of strip

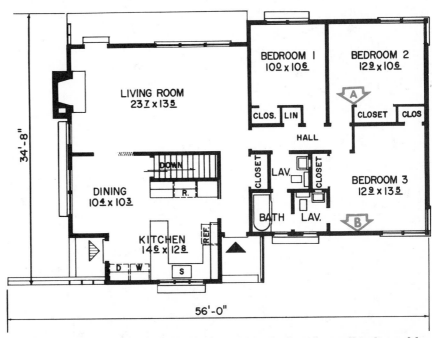

33-5. *The first piece of strip flooring should be laid at the wall indicated by arrow A. At arrow B a piece of strip flooring is reversed as shown in Fig. 33-6. This will permit blind-nailing into the closet area.*

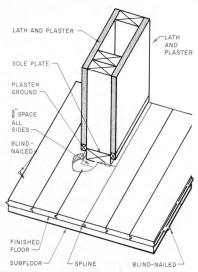

33-6. A spline is inserted between the two pieces of flooring at the wall line indicated by arrow B in Fig. 33-5.

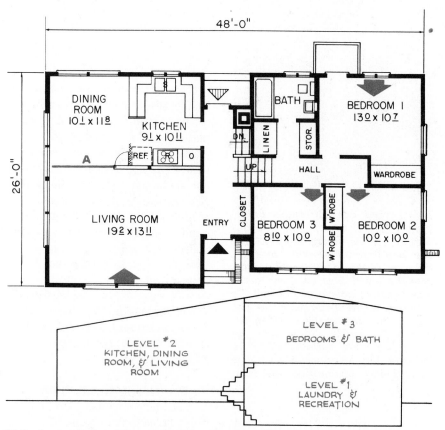

33-7. In a split-level home, laying strip flooring is simplified. Each level is done independently, which eliminates many of the problems of lining up courses in several rooms.

flooring begins, the floor plans of the house should be studied. Careful consideration must be given to the area of the house in which the installation will begin. Strip flooring should be laid so that there is an uninterrupted flow from one room through the hall to the other rooms. Plan the job to eliminate having to align in the hall the courses of strip flooring from two rooms.

For example, in Fig. 33-5 if the courses from bedrooms 1 and 2 are laid through the doorways and into the hall, it is very unlikely that they will line up. Therefore it is best to start the first strip of flooring against the wall indicated by arrow A in bedroom 3. Fig. 33-5. Work across the bedroom and through the door in to the hall. Continue along the hall wall into the living room.

Work the courses across the hall and the first few courses of the living room together. As the courses approach the rear wall of the hall, work through the doorways into bedrooms 1 and 2, and complete the laying of the bedroom floors. Then return to the living room and finish laying the

living room floor, working toward the rear of the house.

In working in this manner the closets in bedrooms 1 and 2 will require placing the grooves of two strips together, with a spline between the grooves. The tongues can then be blind-nailed on each piece. This will permit the courses of strip flooring in the closet to be blind-nailed as well as the pieces which will go across the bedroom floor toward the back wall. Fig. 33-6.

In a split-level, as shown in Fig. 33-7, the first strip of flooring is laid against the back wall of bedroom 1. The floor is laid through the door, across the hall, and into the two front bedrooms. The living room is done separately, and should be started at its front wall and laid in the direc-

tion of the arrow across the floor toward the dining room. If hardwood flooring is to be laid in the dining room, continue through the archway and to the rear wall of the house. If another type of flooring is to be used in the dining room, the strip flooring should terminate at the wall of the dining room. Fig. 33-7, line A.

When different floor coverings are laid in adjoining rooms and a doorway connects the two rooms, the hardwood flooring should terminate under the center of the door when the door is closed. The second type of flooring should begin at this point. In this way, when the door is closed, only the flooring of the room in which the individual is standing may be seen. When finish floors of different materials come together in an

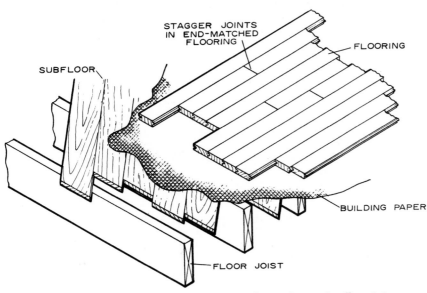

33-8. *Strip flooring should be laid at right angles to the floor joists.*

33-9. *Laying out the strip flooring in preparation for cutting the end pieces so that the joints are staggered.*

archway between two rooms, the hardwood floor is usually laid through the archway and even with the wall line of the adjacent room.

Arrangement of Flooring Pieces

In most houses strip flooring presents the most attractive appearance when laid lengthwise of the room's longest dimension. In some cases, however, it is considered acceptable to lay the flooring crosswise. This is true only if the rooms are sufficiently wide. Wood-strip flooring is usually laid at right angles to the direction of the joists under the largest room on that story. Fig. 33-8. Since interior thresholds are omitted today, the flooring should run continuously between adjoining rooms.

Proper placing of the strips of flooring calls for the use of the shorter lengths in closets and in the general floor area. The longer pieces should be used at entrances and doorways and for starting and finishing in a room. This arrangement results in maximum attractiveness of the floor as a whole. Care should be taken to stagger the end joints of the flooring pieces so that several of them are not grouped closely together. Fig. 33-9.

Importance of Adequate Nailing

Adequate nailing is absolutely essential in the finish floor as well as the subfloor. Insufficient nailing may result in loose or squeaky flooring, one of the most annoying deficiencies a house may develop. The building paper should be chalklined at the joists as a guide in nailing the strip flooring. Tongue-and-groove flooring should be blind-nailed. The nails should be driven at an angle of about 45° or 50° at the point where the tongue leaves the shoulder. Square-edge flooring should be face-nailed. It is recommended that nail heads be countersunk. Figure 33-10 includes information about the types of nails and the spacing commonly used.

Laying Strip Flooring over Subflooring

Cover the subfloor with building paper, preferably 15-pound, asphalt-saturated felt. Fig. 33-4. Stretch a string the length of the room between two nails placed about 8″ from a side wall. Many walls are not perfectly true. By lining up the first courses of flooring at a uniform distance from the string rather than from the wall

Nail Schedule

Tongued and Grooved Flooring Must Always Be Blind-Nailed, Square-Edge Flooring Face-Nailed		
Size Flooring	**Type and Size of Nails**	**Spacing**
(Tongued & Grooved) 25/32 x 3¼	7d or 8d screw type, cut steel nails or 2″ barbed fasteners*	10-12 in. apart
(Tongued & Grooved) 25/32 x 2¼	Same as above	Same as above
(Tongued & Grooved) 25/32 x 1½	Same as above	Same as above
(Tongued & Grooved) ½ x 2, ½ x 1½	5d screw, cut or wire nail. Or, 1½″ barbed fasteners*	8-10 in. apart
Following flooring must be laid on wood subfloor		
(Tongued & Grooved) ⅜ x 2, ⅜ x 1½	4d bright casing, wire, cut, screw nail or 1¼″ barbed fasteners*	6-8 in. apart
(Square-Edge) 5/16 x 2, 5/16 x 1½	1-in. 15-gauge fully barbed flooring brad, preferably cement coated.	2 nails every 7 in.

*If steel wire flooring nails are used they should be 8d, preferably cement coated. Newly developed machine-driven barbed fasteners, used as recommended by the manufacturer, are acceptable.

33-10a. *Nailing schedule for wood flooring.*

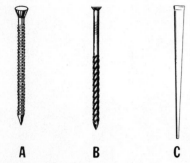

33-10b. Flooring nails: A. Barbed. B. Screw. C. Cut steel.

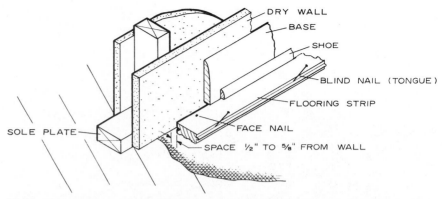

33-12. Cross section of a wall showing the first piece of strip flooring nailed in place. NOTE: With lath and plaster walls, sometimes the ground is kept up off the subfloor about ⅞", and the edge of the first piece of strip flooring is set about even with the wall line. The flooring is then allowed to expand under the ground.

33-11. To line up the first course of flooring, stretch a string or snap a chalk line about 8" from the wall. The first course is face-nailed.

itself, a straighter course is assured. Fig. 33-11.

Place a long piece of flooring with the groove edge about ½" to ⅝" from a side wall and the groove end nearest an end wall. The ½" to ⅝" space is for expansion. It will be hidden by shoe molding. Face-nail the flooring piece near the end. Fig. 33-12. Measure as you nail toward the other end, maintaining the same fixed distance from the guide string. Fig. 33-11. Do likewise in lining up succeeding pieces in the course. Drive one nail at each joist crossing, or every 10" to 12" if the joists run parallel to the flooring.

Observe nailing recommendations of the flooring manufacturer. For example, with flooring ²⁵⁄₃₂" thick and 1½" or more wide, 7d or

8d screw nails or cut steel nails are best. If steel wire flooring nails are used, they should be 8d, preferably cement coated. Fig. 33-10.

After face-nailing the first course, toenail the pieces through the tongue edge, following the same spacing. Fig. 33-13. Drive the nail at the point where the tongue of the flooring leaves the shoulder, at an angle of about 45° or 50°. Fig. 33-14. Each nail should be driven down to the point where another blow or two might cause the hammer to damage the edge of the strip. Use a nail set to drive the nail the rest of the way home. Fig. 33-15. The best nailing procedure is to stand on the strip, with toes in line with the outer edge, and strike the nail from a stooping position which will bring the hammer head square against the nail.

Fit the groove edges of pieces in each succeeding course with the tongues of those in the preceding course. Fit the groove end of each piece in a course with the tongue end of the previous pieces.

When a piece of flooring cannot be readily found to fit the remaining space in a course, cut one to size. Lay a piece down in reverse

33-13. Succeeding courses are toenailed into the flooring at the point where the tongue leaves the shoulder.

position from that in which it will be nailed. Draw a line at the point where it should be sawed. Fig. 33-16. Be sure the piece is reversed for this marking so that the tongue end will be cut off. The groove end is needed for joining with the tongue end of the previous piece.

Stagger the joints of neighboring pieces so that they will not be

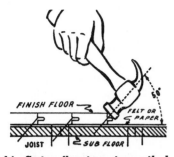

33-14. Strip flooring is nailed in place with the nail driven at about a 50° angle to the floor.

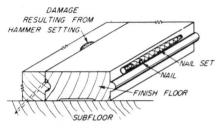

33-15. When setting nails in strip flooring, many workers place the nail set on top of the nail as shown here. The nail set is struck with a sharp blow against its side. This sets the nail and at the same time tends to drive the piece of flooring up tightly against the piece previously laid.

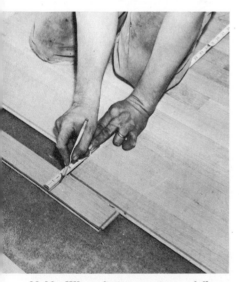

33-16. When fitting a piece of flooring to the remaining space in a course, place it in reverse position for marking so that the tongue end is cut off. The groove end is needed for joining with the tongue of the previous piece.

33-17. When installing strip flooring, it is customary for two people to work together in the room. One person nails the flooring in place, while the second works ahead laying out pieces, staggering joints, and cutting the end pieces to length. Notice in this picture that the man laying the flooring is using a nailing machine rather than nailing by hand as shown in Fig. 33-13.

grouped closely. Fig. 33-17. A joint should not be closer than 6" to another in a previous course. To provide for this, arrange several pieces in their approximate positions in succeeding courses before they are nailed. After nailing three or four courses, place a piece of scrap flooring at intervals against the tongue edges of pieces in the last course. Strike the scrap piece a couple of good hammer blows. This drives the nailed flooring pieces up tightly. Fig. 33-18.

To fit flooring around a jutting door frame, place it flush against the frame. Fig. 33-19. Measure the gap between the face of the previous piece and the groove edge of the piece to be nailed. Where the jutting begins, draw a straight line on the flooring to the same distance as the width of the gap. Fig. 33-19. Do the same on the other side of the door frame.

33-18. After nailing three or four courses, place a piece of scrap flooring along the tongue edge of the flooring already laid. Drive the flooring up tight with two or three sharp hammer blows.

33-19. The flooring should be fitted around a doorframe or other projection. Place the strip tight against the frame, measure the gap between the face edge of the previous piece and the groove edge of the piece to be nailed. Mark this distance, and the width of the frame, on the flooring and notch it out.

Draw a straight line connecting the ends of these lines. Cut the flooring along the lines.

On reaching the opposite side of the room, you will find there is no

33-20. *When laying the last few courses in the room, there will not be enough space to swing a hammer for toenailing. Therefore these courses are face-nailed. Place the last few courses in position, and pull the flooring up tightly when nailing. When prying the pieces with the crowbar, put something behind the crowbar to protect the plaster. A piece of paper is used here. However, a scrap piece of flooring or other piece of wood is better, particularly if it is a little longer and will span between the studs in the wall so that the plaster is not punctured.*

space between the wall and the flooring to permit toenailing of the last two courses. Therefore just fit the pieces in. Face-nail the last few courses, at the same time pulling the flooring up tightly by exerting pressure against it with a chisel or crowbar driven into the subflooring. Fig. 33-20. Protect molding with cardboard or a piece of scrap lumber. After the last course, if the remaining space is too large to be covered by the shoe molding, rip pieces of flooring to the proper width and insert them in the spaces. Face-nail these strips. If they are very narrow, drill holes for the nails to prevent splitting.

Laying Strip Flooring over Concrete

Wood subflooring ordinarily is omitted in homes built on concrete slabs. In such a dwelling the slab serves the major function of subflooring because it offers a strong, solid support for the finish floors and provides a working surface for the building mechanics. Omission of wood subflooring in a

slab home has a definite cost advantage, since it permits savings both in material and labor. An adequate nailing surface for the finish floors must be provided, of course. The method described here is economical and results in finish floors of ample strength and resiliency. Relatively new, it has gained widespread popularity in recent years. Fig. 33-21.

1. Sweep the slab clean and prime the surface. After the primer is dry, snap chalk lines 16" apart at right angles to the direction the flooring will run. Cover the lines with rivers of adhesive applied to a width of about 2". The adhesive can be an asphalt mastic designed for bonding wood to concrete or a suitable adhesive of another type designed for the same purpose. Fig. 33-22. (If heating is in the slab, the adhesive should be resistant to heat.)

2. The bottom sleepers should be 1" × 2" strips treated with wood preservative. Imbed the strips in the adhesive and also secure them to the slab with 1½" concrete nails, approximately 24"

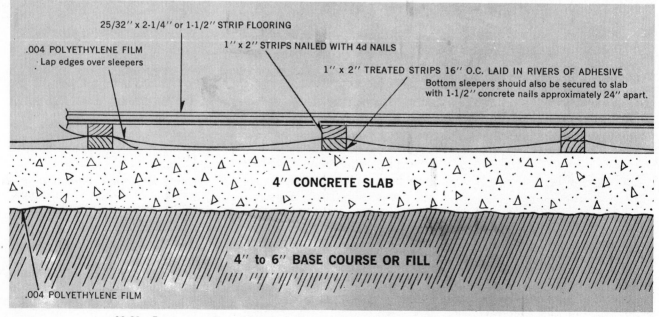

25/32" x 2-1/4" or 1-1/2" STRIP FLOORING

.004 POLYETHYLENE FILM
· Lap edges over sleepers

1" x 2" STRIPS NAILED WITH 4d NAILS

1" x 2" TREATED STRIPS 16" O.C. LAID IN RIVERS OF ADHESIVE
Bottom sleepers should also be secured to slab with 1-1/2" concrete nails approximately 24" apart.

4" CONCRETE SLAB

4" to 6" BASE COURSE OR FILL

.004 POLYETHYLENE FILM

33-21. *A cross section of hardwood strip flooring installed over a concrete slab.*

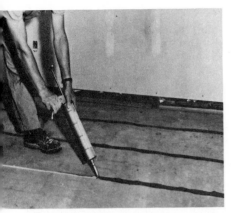

33-22. *Applying adhesive to the concrete slab for installation of the sleepers.*

33-23. *The bottom sleepers are placed over the ribbons of adhesive and nailed in place with concrete nails.*

33-24. *Polyethylene film is laid over the first course of sleepers.*

apart. Fig. 33-23. The sleepers should be of random lengths and laid end to end with slight spaces between the ends, not butted tightly together.

33-25. *The second course of 1" × 2" sleepers is nailed with 4d nails through the polyethylene film and into the bottom sleepers.*

3. After all bottom sleepers have been installed, 0.004 polyethylene film should be laid over the strips. Fig. 33-24. Join the polyethylene sheets by lapping edges over sleepers.

4. The second course of 1 × 2s (which do not have to be preservative treated) would be nailed with 4d nails, 16" to 24" apart. Nail through the top sleeper and the polyethylene into the bottom sleeper. Fig. 33-25.

5. Install the flooring at right

33-26. *The strip flooring is laid over the sleepers at right angles.*

angles to the sleepers by blind-nailing to each sleeper, driving at an angle of approximately 50°. Fig. 33-26. Nails should be the threaded or screw type, cut steel, or barbed. Two adjoining flooring courses should not have joints on the same sleeper. Each strip should bear on at least one sleeper. Provide a minimum of ½" clearance between flooring and wall to allow for expansion. Fig. 33-27.

33-27. *This drawing shows correct nailing procedure when tongue-and-groove strip flooring is installed over subflooring and a single sleeper is used over a concrete slab.*

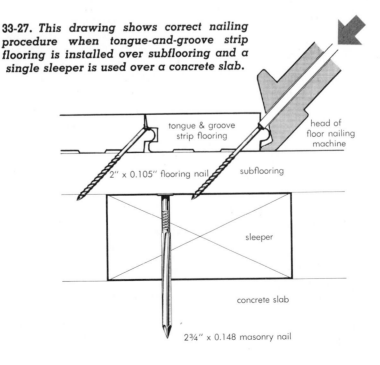

399

33-28. *Although flooring is sanded at the mill, additional sanding is required after the floor is laid for a good finishing job.*

33-29. *There are several finishing materials available for hardwood floors. However, most hardwood flooring producers recommend a floor seal. Floor seal is a tough wear-resistant finish that can be mopped on easily with a long-handled lamb's wool or sponge applicator or a clean string mop. The floor should first be swept and dusted.*

FINISHING

Sanding

Although certified unfinished hardwood flooring is smoothly surfaced at the factory, scratches and other slight marks caused by handling usually show after the

floor has been laid. These may be removed by sanding or scraping. The work should be done by a specialist in that line. An electrically operated sanding machine generally is used for this work, since sanding or scraping by hand is laborious and time-consuming. Fig. 33-28. For fine floors, most manufacturers advise at least four sandings, starting with No. 2 sandpaper and graduating down to No. ½, No. 0, and No. 00. A final buffing with No. 00 or No. 000 sandpaper assures an even smoother surface. Many authorities say that best results are obtained when the final traverse is made by hand.

Stain

It is important that the first coat of stain or other finish be applied the same day as the last sanding. Otherwise the wood grain will have risen, and the finish consequently will be slightly rough. Stain is not used if the finish is to retain the natural color of the wood. If stain is used, it is applied first, before wood filler or

33-30. *After a few square feet of floor seal has been applied and allowed to penetrate, the excess seal should be wiped off with clean cloths. With some brands of seal, however, the excess is rubbed into the wood. The room should be well ventilated during all finishing operations.*

other finishes. It should be put on evenly, preferably with a brush 3″ or 4″ wide.

Wood Filler

Paste wood filler customarily is used to fill the minute surface crevices in oak and other hardwoods with large pores. It gives the floor the perfectly smooth surface required for a lustrous appearance. Filler is applied after stains and sometimes after floor seals but always before other finishing materials. It should be allowed to dry 24 hours before the next operation is begun. Wood filler may be colorless, or it may contain pigment to bring out the grain of the wood. For residential oak flooring, wood filler is always recommended.

Types of Finish

A finish for hardwood floors ideally should have the following qualities:
- Attractive appearance.
- Durability.
- Ease of maintenance.
- Capacity for being retouched in worn spots without revealing a patched appearance.

A finish applied to high grade flooring of considerable natural beauty should also be transparent in order to accentuate that beauty. The three principal types of finishes are floor seal, varnish, and shellac. Lacquer also is used occasionally.

FLOOR SEAL

Floor seal, a relatively new material, is being used on an increasingly large scale for residential as well as heavy duty flooring. It differs from other finishes in this important respect: rather than forming a surface coating, it penetrates the wood fibers, sealing them together. In effect it becomes a part of the wood itself. It wears only as the wood wears, does not chip or scratch, and is practically immune to ordinary stains and spots. While it does not provide as shiny an appearance as other finishes, it has the advantage of being easily retouched. Worn spots may be refinished without presenting a patched appearance. Floor seals are available either with or without color.

It is difficult to give specific directions for applying floor seal because directions of different manufacturers vary widely. Generally it is applied across the grain first, then smoothed out in the direction of the grain. A wide brush, a squeegee, or a wool applicator may be used. Fig. 33-29. After a period of 15 minutes to 2 hours, depending on specific directions of the manufacturer, the excess seal should be wiped off with clean cloths or a rubber squeegee. Fig. 33-30.

For best results the floor then should be buffed with No. 2 steel wool. An electric buffer makes this task relatively simple. Fig. 33-31. If a power buffer is not available, a sanding machine equipped with steel wool pads may be used, or the buffing may be done by hand. Although one application of seal sometimes is sufficient, a second coat frequently is recommended for new floors or floors just sanded. Floor seal is a complete finish in itself. However, it may also be used as a base for a surface finish such as varnish.

VARNISH

Varnish presents a glossy appearance and is quite durable. It is fairly resistant to stains and spots, but shows scratches. It is difficult to patch worn spots without leaving lines of demarcation between old and new varnish. New types of varnish dry in eight hours or less. Like other types of

finish, varnish is satisfactory if properly waxed and otherwise maintained.

Precise directions for application of varnish usually are stated on containers. Varnish made especially for floors is much preferred. So-called all-purpose varnish ordinarily is not so durable when used on floors. As a rule three coats are required when varnish is applied to bare wood. Two coats usually are adequate when wood filler has been used or a coat of shellac has been applied. Cleanliness of both floor and applicator is essential for smooth finish. Drying action is hastened when room temperature is at least 70° F and plenty of ventilation is provided.

SHELLAC

One of the chief reasons shellac is so widely used is because it dries quickly. Workers, starting with floors in the front of a house and moving toward the rear, may begin applying the second coat by the time they have finished the first. Shellac spots rather readily if water or other liquids remain on it long. It is transparent and has a high gloss. It does not darken with age as quickly as varnish.

Shellac to be used on floors should be fresh or at least stored in a glass container. If it remains too long in a metal container, it may accumulate salts of iron which will discolor oak and other hardwoods containing tannin. Shellac should not be mixed with cheaper resins, but before use should be thinned with 188-proof No. 1 denatured alcohol. The recommended proportion is 1 quart of thinner to a gallon of 5-pound cut shellac. A wide brush that covers three boards of strip flooring is the most effective and convenient size. Strokes should be long and even with laps joined smoothly.

33-31. *After the floor seal has dried completely, buff the floor lightly with a pad of fine steel wool on the polishing machine. Wax may also be polished with the machine pictured here by detaching the steel wool pad and attaching a brush provided with the machine.*

The first coat on bare wood will dry in 15 to 20 minutes. After drying, the floor should be rubbed lightly with steel wood or sandpaper, then swept clean. A second coat should be applied and allowed to dry for 2 to 3 hours. The floor should be rubbed again with steel wool or sandpaper and swept. A third coat then is applied. If necessary, the floor may be walked on in about 3 hours, but preferably it should remain out of service overnight.

LACQUER

Lacquer is a glossy finish with about the same durability as varnish. Because it dries so rapidly, it requires considerable skill in application. Worn spots may be retouched with fairly good results, since a new coat of lacquer dissolves the original coat.

If possible, lacquer should be applied with a spray gun. The first coat or sealer should be sanded with a 150 or 180 grit aluminum oxide or garnet abrasive paper. Additional top coats are then applied. Unlike other materials, lacquer does not have to be sanded between coats unless some imperfection occurs and the spot has to be sanded smooth.

Wax

All hardwood floors should be waxed after the finish has dried thoroughly. In some cases two or three coats are recommended for best results. Wax not only imparts a lustrous sheen to a floor, but forms a protective film that prevents dirt from penetrating the wood pores. When wax becomes dirty, it is easily removed and new wax applied.

Hardwood floor wax is available in paste or liquid form. The liquid is known as rubbing wax. Considered about equal in performance, both forms are applied in much the same manner. Usually the wax is mopped on with a cloth, then polished after an interval of 15 to 30 minutes with a soft cloth, a weighted floor brush, or an electric polisher. The latter is preferred, for it eliminates a great deal of labor and does the job equally well, if not better. Some electric polishers apply the wax and polish it in the same operation. Power-driven polishing machines, as well as sanding and buffing machines helpful in the earlier finishing steps, can usually be rented.

Prefinished Hardwood Flooring

When hardwood flooring is purchased in unfinished form, sanding and other finishing operations are performed after the floor has been installed. Some manufacturers produce flooring which is completely prefinished at the factory. It is ready for use immediately after being laid. All species and types of hardwood flooring may not be readily available in this

33-32. A few examples of pattern or parquet floors.

form. However, a manufacturer equipped to make factory finished flooring usually can furnish it in any species and type ordinarily produced unfinished.

OTHER TYPES OF HARDWOOD FLOORING

Plank Flooring

One of the oldest types of hardwood floors, plank flooring, dates back to the handicraft era. In its crude early form it was widely used in medieval Europe, and it later became a popular flooring for American colonial homes. Its use has been increasing in homes, clubs, and other buildings where an atmosphere of rugged informality is desired. Colonial plank flooring derived much of its charm from its rough effects and interesting irregularities, which were unintentional. This charm is retained in modern plank flooring even though it is now a precision-made product.

Oak is preferred for plank floors. Production of planks ordinarily is confined to that species, but they can also be obtained in solid walnut and East Indian teak, as well as in various veneers. The planks usually come in random widths. Generally the pieces are tongued and grooved, with square or matched ends, but sometimes they are produced

with square edges and ends. Frequently the edges of planks are beveled to reproduce the effect of the large cracks which characterize early handhewn plank floors. The wood pegs by which old plank floors were fastened also are simulated. This is done by gluing wood plugs in holes on top of the countersunk screws fastening the planks to the subfloor.

Pattern or Parquet Flooring

Pattern floors, also known as parquet and design floors, appeared as early as the 14th century in Europe. While today, as then, they are the most elaborate and expensive type of hardwood flooring, they have been simplified to a great extent. Early pattern floors presented bizarre effects which would be incompatible with present architectural styles. Today's parquetry uses squares, rectangles, and herringbone patterns to achieve an almost infinite variety of effects. Fig. 33-32. Literally hundreds of designs are available, many featuring various species or different shades of the same species. Most parquet flooring is of oak, although it also is produced in maple, beech, birch, walnut, mahogany, East Indian teak, and ebonized wood. The latter consists of dyed white maple or holly. Parquetry is manufactured in short lengths of individual pieces. Each piece must be cut to exact dimensions so that it will match perfectly the dimensions of another piece, or multiples thereof. Customarily tongued and grooved and endmatched, the pieces are laid separately, either by nailing or setting in mastic.

Block Flooring

Block flooring is really a form of parquetry, since it constitutes a definite pattern. It differs from conventional parquetry in that the pieces are assembled into square

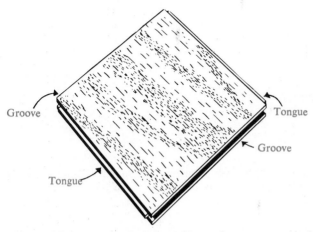

33-33a. *Wood block flooring. Note the tongue and groove locations.*

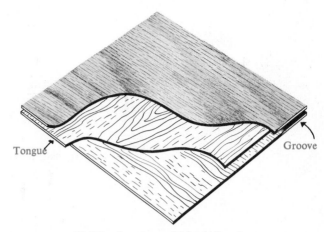

33-33b. *Laminated block flooring.*

or rectangular blocks at the factory. They are held together by various means on the back, sides, or ends with matching tongue-and-groove edges. Fig. 33-33. These prefabricated blocks are laid as units, either in mastic or by nailing.

Installing Plank, Block, and Parquet Flooring

Each piece in plank flooring, in addition to being blind-nailed, should be face-nailed or screwed about every 30″. The ends of each piece also should be fastened to the subfloor with countersunk screws. Wood plugs glued in the plank conceal screwheads and lend a decorative effect.

Parquet flooring, including prefabricated blocks, may be nailed over a wood subfloor or old finish floor, or it may be laid in mastic over concrete. Preferably the subfloor should cross the joists at right angles rather than on a diagonal. When the subfloor is installed before the type of finish flooring has been decided upon, it is best to lay it at a 60° angle. This "compromise" will make it adaptable to any type of flooring.

LINOLEUM AND TILE FLOORS

Use ¼″ underlayment for lino-leum, asphalt tile, or vinyl tile floors. It may be laid directly on $^{24}\!/_{32}$″ tongue-and-groove wood flooring strips, maximum width 3¼″, with a joist spacing of 16″ on center. Where these finish floors are used for the floor covering in one room and wood floors are used in adjacent rooms over a subfloor of common level, a suitable base floor is required for the nonwood finish flooring. This base floor may also be tongue-and-groove flooring or plywood. The thickness of the base floor plus the thickness of the nonwood finish floor should equal the thickness of the wood finish floors in adjacent rooms so that the floors will be at the same level.

Linoleum

Linoleum is manufactured in thicknesses ranging from ¹⁄₁₆″ to ¼″ and is generally 6′ wide. It is made in various grades, in plain colors, inlaid, or embossed. Linoleum may be laid on wood or plywood base floors, but not on concrete slabs on the ground. Since linoleum follows the contour of the base floor over which it is laid, it is essential that the base be uniform and level. When wood floors are used as a base, they should be sanded smooth and be level and dry. When plywood base floors are used, the sheets should be carefully joined together. After the base floor is correctly prepared, the adhesive is applied. The linoleum is then laid and thoroughly rolled to insure complete adhesion to the floor.

Asphalt and Vinyl Tile

Asphalt tile is widely used as a covering over concrete slabs and is sometimes used over an underlayment. It is the least costly of the commonly used floor-covering materials. This tile is about ⅛″ thick and 9″ × 9″ or 12″ × 12″. Most types of asphalt tile are damaged by grease and oil and for that reason are not recommended for use in kitchens.

It is important that the subfloor or base be suitably prepared. Otherwise, the finish floor will not give satisfactory performance. Most manufacturers provide directions on the preparation of the base and recommend the type of adhesive that is best for their product. The tile should be laid according to the manufacturer's directions. When a wax finish is recommended, the wax should have a water base. Vinyl tile can be used in the same manner as asphalt tile. It is impervious to grease and oil.

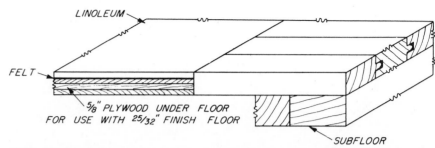

33-35. *The underlayment, in this case plywood, is usually thinner than the hardwood floor to compensate for the thickness of the tile or linoleum.*

Labels in figure: LINOLEUM, FELT, 5/8″ PLYWOOD UNDER FLOOR FOR USE WITH 25/32″ FINISH FLOOR, SUBFLOOR

33-34. *Particleboard is frequently used as an underlayment for a tile floor.*

Rubber Tile

Rubber tile flooring is resilient, noise-absorbing, waterproof, and highly wear resistant. It may be applied over wood subfloors or concrete floor slabs, except slabs on ground. The finish may be plain or marbleized in various designs, with the colors running throughout the body of the tile. Rubber tile is made in square shapes ranging in size from 4″ × 4″ to 18″ × 18″. It also comes in rectangular shapes ranging from 9″ × 18″ to 9″ × 36″. Thickness is from 1/8″ to 3/16″. The tile is generally laid in a waterproof rubber cement and thoroughly rolled.

Wood subfloors for rubber tile should be above grade. If the subfloor is plywood, rubber tile may be laid directly on the wood surface. Make sure that plywood joints do not coincide with rubber tile joints, but come midway between the tile joints. If tile is laid on a solid wood subfloor, the floor should first be sanded smooth and sealed. A layer of 15- to 30-pound saturated lining felt should then be bonded to the surface. Joints in the tile should not coincide with subfloor joints. Otherwise, expansion or contraction of the subfloor will also affect the

rubber tile. The tile should be installed following the manufacturer's recommendations for both method and materials.

Installing a Tile Floor

Tile can be laid directly over wood flooring or particleboard. Fig. 33-34. If a hardwood floor is to be used in one room and the adjacent room is to have a tile floor, a base is required for the tile so that the floors will be exactly the same height. In other words, the thickness of the base floor plus the thickness of the tile should equal the thickness of the finish floor in an adjacent room. Fig. 33-35. Some kinds of tile require an underfelt that must be applied over the floor before the tiles are installed. Figures 33-36 through 33-48 and the accompanying text describe how to install 9″ × 9″ vinyl-asbestos tile. The same general procedure applies to other types and sizes of tile.
• Prepare the surface on which the tile will be laid. The appearance of the finished floor will depend a great deal on the condition of the subfloor. Make sure the floor is smooth and completely free of wax, paint, varnish, grease, or oil. Holes or cracks in concrete subfloors should be filled with crack filler. Plane down high spots and renail loose boards of wood floors. Fig. 33-36.
• If a wood subfloor is only a single layer or if it is a double-layer

33-36. *Planing down high spots.*

33-37. *Covering the old floor with underlayment.*

floor and the boards are in bad condition, the old floor should be covered with an underlayment. Fig. 33-37.

33-38. *Allow a distance equal to the thickness of a matchbook between the panels of underlayment.*

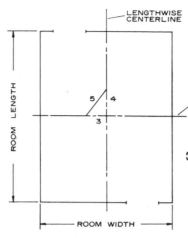

33-41. *The two lines must be exactly perpendicular*

33-39. *Using a nailing machine to install underlayment.*

33-40. *Using a framing square to draw a perpendicular from the centerline.*

• Allow a little less than ⅟₃₂″, or the thickness of a paper matchbook, between each panel of underlayment or plywood to allow for expansion. Stagger the joints as shown. Fig. 33-38.

• Nail the underlayment or plywood with coated or ring-grooved nails at least every four inches along all edges and over the entire face of the panels. Most manufacturers of underlayment provide a nailing guide in the form of lines or dots stenciled on the surface to indicate the nail locations.

• The underlayment or plywood may be nailed with a nailing machine to speed up the operation. Fig. 33-39.

• Find the center points of the two end walls of the room. Connect these points with a chalk line and snap a line down the middle of the room.

• Locate the midpoint of this line. Use a framing square to draw a perpendicular line from this point. Fig. 33-40. If a square is not available, a tile may be used with one edge on the center line to establish the perpendicular.

• The center lines should be at an exact 90° angle (right angle) to each other. This can be assured by constructing a 3′ × 4′ × 5′ right triangle from the two lines and a diagonal. Fig. 33-41. In a

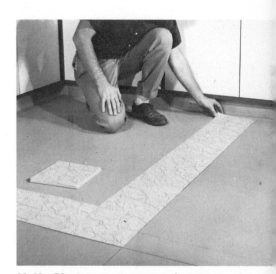

33-42. *Placing test rows of uncemented tile along the chalk lines.*

large room, a 6′ × 8′ × 10′ triangle can be used.

• Along the perpendicular line, strike a chalk line which extends to both of the side walls. Along the chalk lines, lay one test row of uncemented tile from the center point to one side wall and to one end wall. Fig. 33-42.

• The uncemented tiles are placed along the line with the edge of the tiles exactly on the line. The rows may be shifted by moving one tile at a time until the border space at the end of each of the rows is equal. Count the tiles in each row. If there is an even

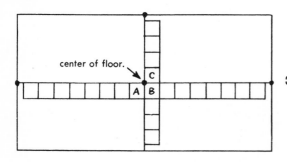

center of floor.

C
A B

33-43. *Layout for an even number of tiles.*

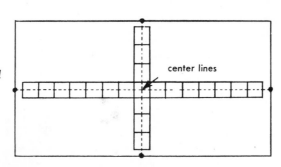

center lines

33-44. *Layout for an odd number of tiles.*

33-45. *Relocating the center line.*

number of tiles, those at the center of each row (tiles A, B, and C) should meet exactly on the center lines. Fig. 33-43.

• If there is an odd number of tiles, the center lines on the floor will bisect the tiles. Fig. 33-44.

• Border tiles should not be less than half a tile wide. Measure the distance between the wall and the last tile. If the distance is less than 2″ or more than 8″, move the center line parallel to that wall 4½″ closer to the wall.

• Locate the new center line with the chalk line and snap it. Fig.

33-45. Moving the center line closer to the wall creates wider borders. For example, if the borders are 1½″ wide, moving the center line 4½″ closer to one wall will take away the 1½″ border plus 3″ from the next tile. The border tile on that side will then be 6″ wide. On the opposite side, 4½″ will be gained. The border will thus become 6″ wide (4½″ + 1½″). Since the line is moved 4½″, half the size of one tile, the border tile remains uniform on both sides.

• Choose the correct adhesive for the tile. Spread a thin coat over one-fourth of the room. Do not cover the chalk lines. The adhesive may be trowled or brushed on, depending on the type of tile to be laid and the adhesive used.

• Allow the adhesive to dry about 15 minutes. Then test it for proper tackiness by touching lightly with the thumb. It should feel tacky but not stick to the thumb. If it sticks to the thumb, allow more drying time.

• For an even number of tiles, start at one inside corner and lay the first tile exactly in line with the marked center lines. For an odd number of tiles, center the first tile over the intersection of the center lines. The second tile can be laid adjacent to the first on one side. The third tile is laid adjacent to the first on the other side. Continue laying tiles along the center line and filling in between until the entire section is covered. Fig. 33-46. The remaining three sections can be covered in the same way. Do not slide tiles into position. Some kinds of tile require only pressing in place; others should be rolled after installation for better adherence.

• To fit border tiles, place a loose tile (A) exactly over the last tile in the row. Take another tile (B) and place it against the wall, overlapping tile A. Mark tile A with a pencil along the edge of B. Fig. 33-47.

• With household shears, tin-snips, or a knife, cut tile A along the pencil mark.

• The cut portion of tile A will fit exactly into the border space. Place the hand-cut edge against

33-46. *Diagram of installation sequence.*

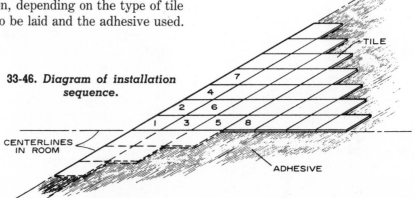

TILE

7
4
2 6
1 3 5 8

CENTERLINES IN ROOM

ADHESIVE

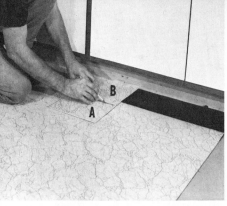

33-47. Marking a border tile to be cut.

the wall. Repeat this procedure until the border area is completely covered. A clearance of ⅛″ to ¼″ should be allowed at all sides for expansion. This space is covered with a cove base of the same resilient material as the tile or with a standard wood base. Wood base is usually lower in cost than the resilient cove base, but installation costs are somewhat

greater. When installing vinyl tile, each quarter of the room should be rolled with a linoleum roller or other smooth roller as it is completed.

● To cut around pipes or other obstructions in a room, first make a paper pattern to fit the space exactly. Then trace the outline onto the tile and cut along the outline. Some tile can be cut with scissors. Fig. 33-48.

● Vinyl-asbestos feature strips in solid colors can be used to create an unusual effect. Several colors can be combined to provide a room with individuality and an interesting complement to the room's decor. Spread the adhesive, allow it to dry, and lay the feature strips in place.

● Vinyl cove base can be installed

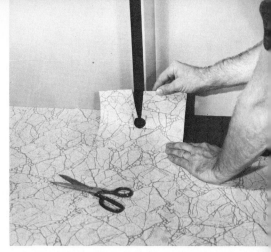

33-48. Fitting tile around a pipe.

along the walls of the room. After cutting the proper lengths to fit into place, apply the adhesive to the back of the vinyl cove base and press the material against the wall. This completes the floor installation.

QUESTIONS

1. What is finish flooring?
2. What are some of the finish flooring materials?
3. What are the most common hardwoods used for wood strip flooring?
4. What wood species is used most often for wood strip flooring?
5. What is the most popular thickness and width of strip flooring?
6. Who enforces the grading rules and regulations for hardwood flooring?
7. What are the grades of oak strip flooring?
8. What is the last operation before actual installation of finish flooring?
9. What factors must be taken into considera-

tion when deciding what area of the house to begin the installation?
10. When two different floor coverings are used in adjoining rooms, they should terminate under the center of the door when the door is closed. Why?
11. What kind and size of nails are recommended for laying strip flooring?
12. How is strip flooring installed over concrete?
13. What are some other types of hardwood flooring besides strip flooring?
14. What thickness of underlayment is used for linoleum floors?

Stairs

34

There are two general types of stairs: principal and service. The principal stairs are designed to provide ease and comfort and are often made a feature of house design. The service stairs lead to the basement or attic. They are usually somewhat steeper and constructed of less expensive materials.

Stairs may be built on the job or assembled from units built in a mill. All parts for a finish stairway can be purchased from a lumberyard as stock mill items. Stairways may have a straight, continuous run with or without an intermediate platform. They may also consist of two or more runs at angles to each other. Usually there is a platform at the angle. The turn may also be made by radiating treads called winders.

Figure 34-1 shows the stair patterns most often found in homes. Winders are not often used because they are not as safe as platforms. Fig. 34-2. The stairway for most homes is a straight, continuous run, although a stairway with a landing or platform is sometimes used to conserve space. Fig. 34-3. Details for stair building are shown in the stairwell section of most house plans. Fig. 34-4.

There are many different kinds of stairs, but all have two main parts in common: the treads people walk on and the stringers which support the treads. The simplest stairway has a pair of

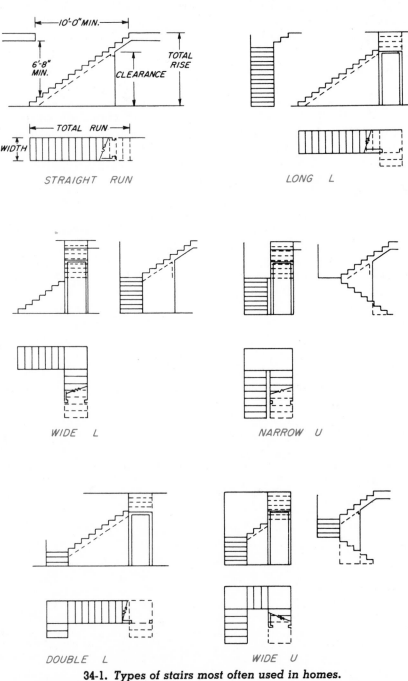

34-1. *Types of stairs most often used in homes.*

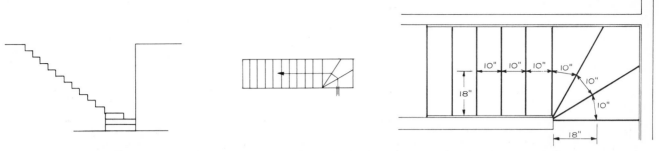

34-2. A stairway with winders. At the inner corner where all the winders meet, there is very little if any tread to support one's foot. This makes winders rather dangerous, and they are not often used.

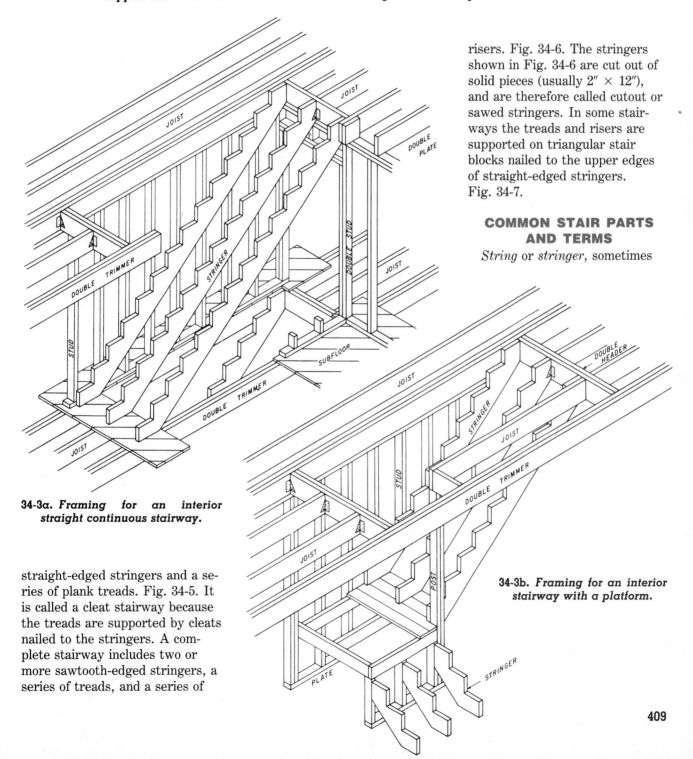

34-3a. Framing for an interior straight continuous stairway.

34-3b. Framing for an interior stairway with a platform.

risers. Fig. 34-6. The stringers shown in Fig. 34-6 are cut out of solid pieces (usually 2″ × 12″), and are therefore called cutout or sawed stringers. In some stairways the treads and risers are supported on triangular stair blocks nailed to the upper edges of straight-edged stringers. Fig. 34-7.

COMMON STAIR PARTS AND TERMS

String or *stringer*, sometimes

straight-edged stringers and a series of plank treads. Fig. 34-5. It is called a cleat stairway because the treads are supported by cleats nailed to the stringers. A complete stairway includes two or more sawtooth-edged stringers, a series of treads, and a series of

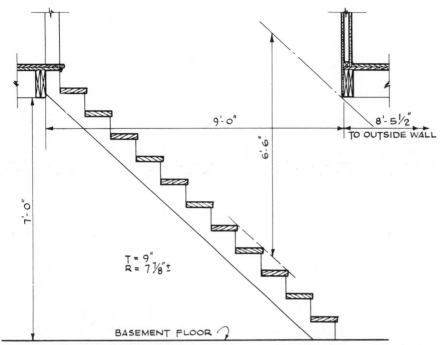

$T = 9"$
$R = 7\frac{1}{8}" \pm$

9'-0"

8'-5½"

TO OUTSIDE WALL

6'-6"

7'-0"

BASEMENT FLOOR

34-4. *A stairwell section drawing for a set of house plans. Note that the architect has indicated the riser to be 7⅞" and the tread to be 9".*

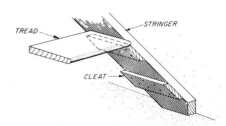

34-5. *The simplest type of stairs, a cleat stairway.*

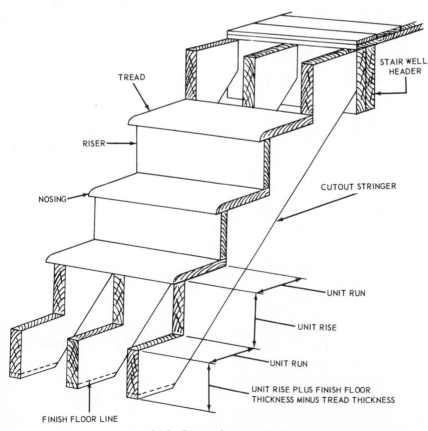

TREAD

STAIR WELL HEADER

RISER

NOSING

CUTOUT STRINGER

UNIT RUN

UNIT RISE

UNIT RUN

UNIT RISE PLUS FINISH FLOOR THICKNESS MINUS TREAD THICKNESS

FINISH FLOOR LINE

34-6. *Parts of a stairway.*

called a *carriage*, or *horse*. One of the inclined sides of a stair which support the treads and risers. Open (plain) stringers can be either rough or finish stock and are cut to follow the lines of the treads and risers. Closed stringers have parallel sides, with the risers and treads housed into them. The term also applies to any similar member, whether a support or not, such as finish stock placed outside the carriage on open stairs and next to the walls on closed stairs. Figures 34-5, 34-6, 34-7, 34-8, 34-9, & 34-10 show various kinds of stringers.

Riser. The vertical face of one step. Fig. 34-11.

Tread. The horizontal face of one step. Fig. 34-11.

Winders. Radiating or wedge-shaped treads at turns of stairs. Fig. 34-12.

Nosing. The projection of tread beyond the face of the riser. Fig. 34-11.

Railing. The protection on the open side of a run of stairs. Fig. 34-11.

Newel. The main post of the railing at the start of the stairs and the stiffening post at angles or platforms. Fig. 34-11.

Handrail. The top finishing piece on the railing to be grasped by the hand when going up or down the stairs. Fig. 34-11.

Balusters. The vertical members supporting the handrail on open stairs. Fig. 34-11. For closed

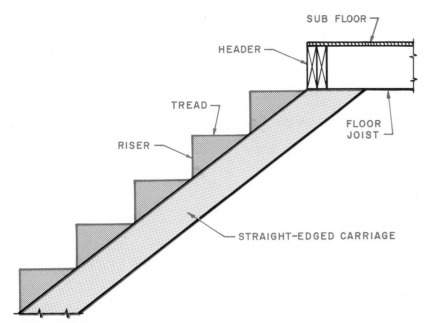

34-7a. *A built-up stringer made up of a straight-edged carriage with triangular blocks attached.*

TREAD

RISER

STRAIGHT CARRIAGE

STRAIGHT-EDGED CARRIAGE

34-7b. *If additional support is needed for a wide stairway, a third stringer is used in the center. The triangular blocks may be nailed to the edge or to the face of the straight stringer (carriage).*

stairs where there is no railing, the handrail is attached to the wall with brackets.

Platform. The intermediate area between two parts of a flight of stairs. Fig. 34-13, arrow 1.

Landing. The floor at the top or bottom of each story where the flight of stairs ends or begins. Fig. 34-13, arrow 2.

Total Rise. The total vertical distance from one floor to the next. Fig. 34-1.

Total Run. The total horizontal length of the stairs. Fig. 34-1.

FRAMING A STAIRWELL

When large openings are made in floors, such as for stairwells, one or more joists must be cut. The location of openings in the floor determines the method of framing. When the length of the stairway is parallel to the joists, the opening is framed as shown in Fig. 34-14. When the stairway is arranged so that the opening is perpendicular to the length of the joists, the framing should follow the details shown in Fig. 34-15. Figure 34-16, page 414, shows typical framing for a stair landing. Nailing and framing of stairwell openings should comply with the principles explained in Unit 13, "Floor Framing."

DESIGNING A STAIRWAY

Stairways are designed, arranged, and installed for safety, adequate headroom, and space for the passage of furniture. There are three important considerations when designing the stairway:

• The stair width.
• The headroom.
• The relationship between the height of the riser and the width of the tread.

Stair Width

Staircases must be wide enough to allow two people to pass com-

411

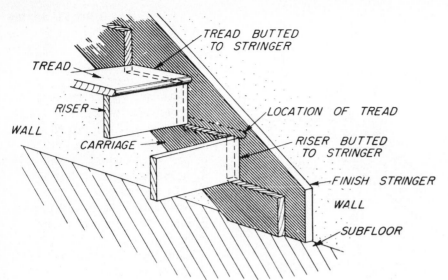

TREAD BUTTED TO STRINGER

TREAD

RISER

WALL

CARRIAGE

LOCATION OF TREAD

RISER BUTTED TO STRINGER

FINISH STRINGER

WALL

SUBFLOOR

34-8. *A finished wall stringer and carriage.*

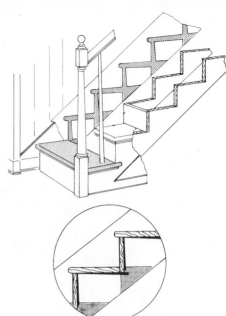

34-9. *A housed stringer. Note that the treads and risers are wedged from the underside with glue-covered wedges.*

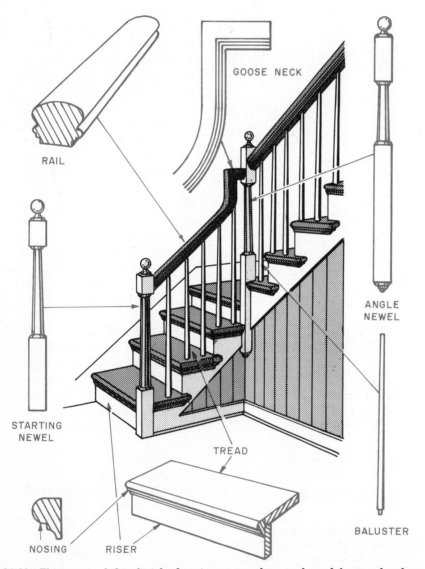

RAIL

GOOSE NECK

ANGLE NEWEL

STARTING NEWEL

TREAD

BALUSTER

NOSING

RISER

34-11. *The parts of this finished stairway can be purchased from a lumber-yard as stock mill items.*

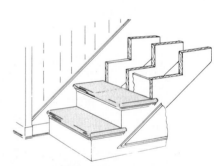

34-10. *A plain stringer.*

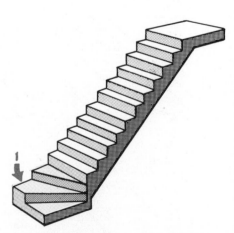

1

34-12. *A stairway with winders, shown at arrow 1.*

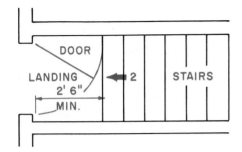

34-13. *The platform is shown at arrow 1, the landing at arrow 2.*

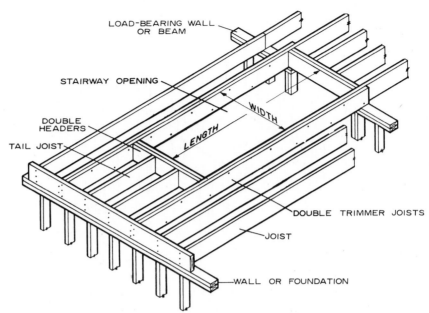

34-14. *A stairwell framed parallel to the joists.*

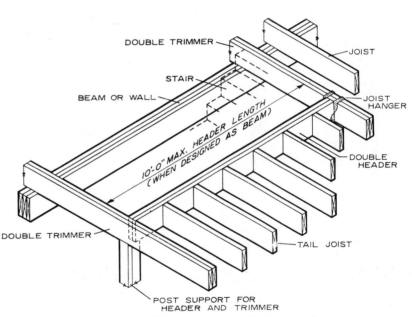

34-15. *A stairwell framed perpendicular to the joists.*

fortably on the stairs and to permit furniture to be carried up or down. The minimum width is 3'; however, 3½' is better. Fig. 34-17.

The width needed for the passage of furniture varies. Stairs which are open on one side, including open-well stairs, are best for moving large pieces of furniture. The furniture can usually be raised up over the handrails and newel posts.

Headroom

Headroom is the distance or clearance above a stair. It is measured from the outside corner of the tread and riser to the lowest point of the soffit or ceiling directly overhead. If two or more flights of stairs are arranged one above the other in the same stairwell (for example, cellar stairs under, or attic stairs over, the main staircase), getting enough headroom beneath the upper stair is a planning problem and must be carefully considered. Although the minimum required by the FHA is 6'8", this is usually not enough for the main stairway. It has been found from studies of

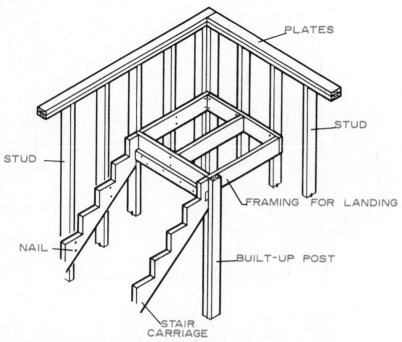

34-16. *Framing for a landing.*

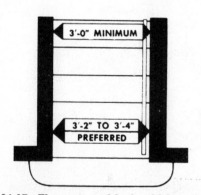

34-17. *The stair width should be adequate. Three feet is acceptable, but wider stairs are preferred.*

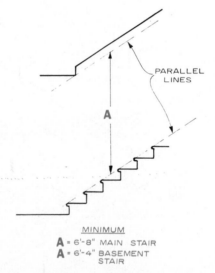

34-18. *Headroom should be a minimum of 6'8" for a main stair, as shown at A.*

34-19. *A stairway should be built with the proper rise and run: A. Stairway with a tread too wide and the riser too short. B. Stairway with a tread too narrow and the riser too high. C. Stairway with a correct tread width (10") and the correct riser height (7½").*

the dimensions of the average man or woman that headroom may vary with the steepness of the stairs, but should generally be between 7'4" and 7'7". This allows for the arm to be swung up over the head without hitting anything. Fig. 34-18.

Riser and Tread Relationship

It is very important that the stairway be built with the proper

rise and run. The relationship between the height of the riser and the width of the tread determines the ease with which the stairs may be ascended or descended. If the combination of run and rise is too great, the steps are tiring. There is a strain on the leg mus-

cles and the heart. If the combination is too short, the foot may kick the riser at each step. An attempt to shorten one's stride may be tiring. If one of the three rules

414

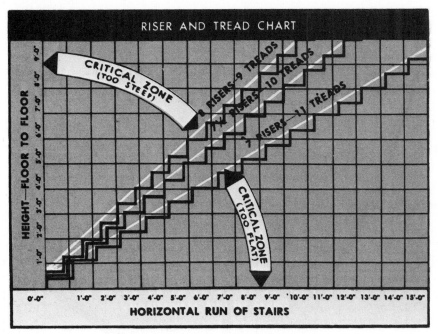

34-20. *Riser and tread relationships.*

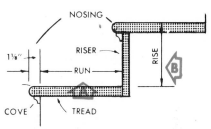

34-21. *The unit run (arrow A) is the distance from the face of one riser to the face of the next riser. It does not include the nosing. The unit rise (arrow B) is the distance from the top of one tread to the top of the next tread.*

described below is followed, the stairway will be easy and safe to ascend and descend. Figs. 34-19 & 34-20.

● The sum of two risers and one tread should be between 24″ and 25″. Acceptable therefore would be a riser 7″ to 7½″ and tread 10″ to 11″. (Example: 7½″ + 7½″ + 10″.)

● The sum of one riser and one tread should equal between 17″ and 18″. (Example: 7½″ + 10″.)

● The product obtained by multiplying the height of the riser by the width of the tread should be between 70″ and 75″. (Example: 7″ × 10″.) For the main staircase in a house, risers should not be higher than 7⅝″ or less than 7″, combined with a tread width of 10″ to 11″ (not including nosing).

LAYING OUT A STAIRWAY

The first step in stairway layout is to determine the *unit rise* and *unit run* per step. Fig. 34-21. The unit rise (height of one riser) is calculated on the basis of the total rise of the stairway, and the

fact that the unit rise for stairs should be about 7″.

The total rise is the vertical distance between the lower finish floor level and the upper finish floor level. It is given on the elevations and wall sections. Fig. 34-22. The actual distance, however, may vary slightly from the specified distance. Measure the actual distance. If both the lower and the upper floor are to be covered

with finish flooring of the same thickness, the vertical distance between the subfloor levels will be the same as the vertical distance between the finish floor levels. However, you may be measuring up from a finish floor (a concrete basement floor, for example) to a floor which will be covered with finish flooring. In this case, you must add the thickness of the upper floor finish flooring to the vertical distance between the lower finish floor and the upper subfloor.

Let's assume that the total rise, or vertical distance between finish floors, is 8′11″. The unit rise can be determined from the total rise as follows:

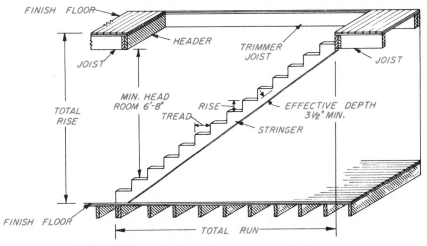

34-22. *Stairway information necessary for layout.*

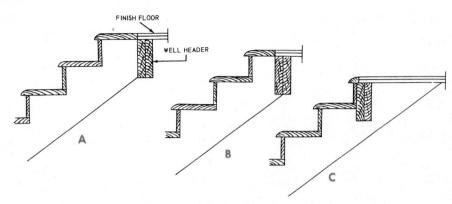

34-23. *Three methods of anchoring the upper end of a stringer.*

• The total rise is 107″ (8′11″ = 107″).

• A permissible unit rise is anything near 7″. Divide 107″ by 7″.

• The result (disregarding any fraction) is the total number of risers in the stairway. In this case, the total is 15 risers (107 ÷ 7 = 15).

• To get the unit rise, divide the total rise by the number of risers. The unit rise in this case is 7⅛″ (107 ÷ 15 = 7.13, or 7⅛).

Stair With Landing

This type is easier to climb, safer, and reduces the length of space required. The landing provides a resting point and a logical place to have a right angle turn. Landing near bottom with quarter-turn is basis of calling this type "dog-legged" or "platform" stairs.

Height Floor to Floor H	Number of Risers	Height of Risers R	Width of Tread T	Run		Run	
				Number of Risers	L	Number of Risers	L2
8′0″	13	7⅜″ +	10″	11	8′4″ + W	2	0′10″ + W
8′6″	14	7⁵⁄₁₆″ −	10″	12	9′2″ + W	2	0′10″ + W
9′0″	15	7³⁄₁₆″ +	10″	13	10′0″ + W	2	0′10″ + W
9′6″	16	7⅛″	10″	14	10′10″ + W	2	0′10″ + W

Straight Stairs

Simplest and least costly; requires a long hallway which may sometimes be a disadvantage. May have walls on both sides (closed string) or may have open balustrade on one side (open string).

Height Floor to Floor H	Number of Risers	Height of Risers R	Width of Treads T	Total Run L	Minimum Head Rm. Y	Well Opening U
	12	8″	9″	8′3″	6′6″	8′1″
8′0″	13	7⅜″ +	9½″	9′6″	6′6″	9′2½″
	13	7⅜″ +	10″	10′0″	6′6″	9′8½″
	13	7⅞″ −	9″	9′0″	6′6″	8′3″
8′6″	14	7⁵⁄₁₆″ −	9½″	10′3½″	6′6″	9′4″
	14	7⁵⁄₁₆″ −	10″	10′10″	6′6″	9′10″
	14	7¹¹⁄₁₆″ +	9″	9′9″	6′6″	8′5″
9′0″	15	7³⁄₁₆″ +	9½″	11′1″	6′6″	9′6½″
	15	7³⁄₁₆″ +	10″	11′8″	6′6″	9′11½″
	15	7⅝″ −	9″	10′6″	6′6″	8′6½″
9′6″	16	7⅛″	9½″	11′10½″	6′6″	9′7″
	16	7⅛″	10″	12′6″	6′6″	10′1″

Note: Dimensions shown under well opening "U" are based on 6′6″ minimum headroom. If headroom is increased well opening also increases.

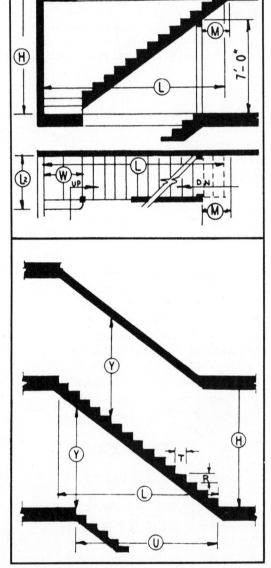

34-24. *Layout dimensions for some standard stairways.*

The unit run is equal to the width of a tread, less the width of the nosing. Fig. 34-21. It is calculated on the basis of the unit rise and a rule, discussed earlier, that the sum of one riser and one tread should equal between 17″ and 18″. Subtract the unit rise from this sum. Let's assume that the sum of one riser and one tread should be 17½″. If the unit rise is 7⅛″, the unit run will be 10⅜″ (17½″ − 7⅛″ = 10⅜″).

You now have all the information you need to lay out, cut, and install a cutout stringer except the total run of the stairway. The total run is equal to the unit run times the number of treads in the stairway. The total number of treads depends on the manner in which the upper end of the stairway is anchored to the upper landing. Three common types of anchorage are shown in Fig. 34-23.

In A, Fig. 34-23, there is a complete tread at the top of the stairway. This means that the number of treads in the stairway is the same as the number of risers. If there are 15 risers and the unit run is 10⅜″, the total run of the stairway is 12′11⅝″ (15 × 10⅜″ = 155⅝″, or 12′11⅝″).

In B, Fig. 34-23, there is only part of a tread at the top of the stairway. In this case, the number of complete treads is one less than the number of risers, or 14. The total run of the stairway is 14 × 10⅜″, plus the run of the partial tread at the top. This run may be shown in detail. If not, you will have to estimate it as closely as possible. Let's assume it's about 7″. The total run, then, is 12′8¼″ ÷ (14 × 10⅜″ = 145¼″; 145¼″ + 7″ = 152¼″, or 12′8¼″).

In C, Fig. 34-23, there is no tread at the top of the stairway. The upper finish flooring serves as a top thread. In this case the

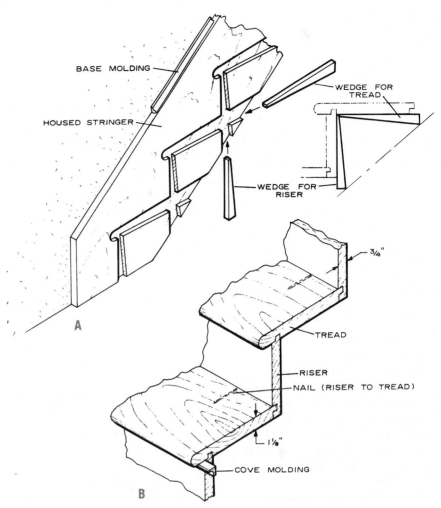

34-25. *Parts of stairs: A. A housed stringer. B. Risers and treads rabbeted and grooved for installation in a housed stringer.*

number of treads is one less than the number of risers, or 14. The total run is 12′1¼″ (14 × 10⅜″ = 145¼″, or 12′1¼″).

After you have calculated the total run of the stairway, drop a plumb bob from the stairwell header to the floor below. Measure off along the floor the total run, starting at the plumb bob. You have now located the anchoring point for the lower end of the stairway. Some standard stair layouts can be found in the chart in Fig. 34-24.

STRINGERS

The treads and risers are sup-ported by stringers, or carriages, that are solidly fixed in place, level and true, upon the frame-work of the building. The string-ers may be cut or routed to fit the outline of the treads and risers. A third stringer should be installed in the middle of the stairs when the treads are less than 1⅛″ thick or the stairs are more than 2′6″ wide.

In some cases rough stringers with rough treads nailed across them are used during the con-struction period. These are in-stalled for the convenience of the workers until wall finish is ap-plied. When the wall finish is

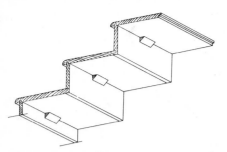

34-26. *A method of joining treads to risers.*

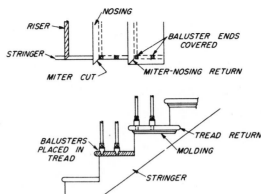

34-27. *An open stringer with the risers mitered to the stringer. The balusters are set into the treads and trimmed with a nosing return and a molding.*

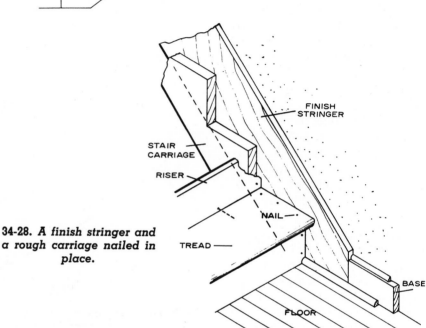

34-28. *A finish stringer and a rough carriage nailed in place.*

completed, the temporary stairway is removed and finished stairs, which usually have been made in a mill, are erected or built in place.

For a housed stringer, the wall stringer is routed out to the exact profile of the tread, riser, and nosing, with sufficient space at the back to take the wedges. Fig. 34-25. The top of the riser is rabbeted to fit into a groove in the bottom front of the tread. The back of the tread is rabbeted into a groove in the bottom of the next riser. The wall stringer is spiked to the inside of the wall. The treads and risers are fitted together and forced into the wall stringer housing, where they are set tight by driving and gluing wood wedges behind them. Fig. 34-26. The wall stringer thus shows above the profiles of the treads and risers as a finish against the wall. It is often made continuous with the baseboard of the upper and lower landing.

If the outside stringer (stair carriage) is an open stringer, it is cut out to fit risers and treads and nailed against the finish stringer. The edges of the risers are mitered with the corresponding edges of the stringer. The nosing of the tread is returned upon its outside edge along the face of the stringer. Fig. 34-27. Another method would be to butt the stringer to the riser and cover

the joint with an inexpensive stair bracket.

Figure 34-28 shows a finish stringer nailed in position on the wall and a rough carriage nailed in place against the stringer. If there are walls on both sides of the staircase, the other stringer and carriage are nailed in the same way. The risers are nailed to the riser cuts of the carriage on each side and butted against the stringers. The treads are nailed to the tread cuts of the carriage and butted against the stringers. This is the least expensive of the types described and perhaps the best type of construc-

tion to use when the treads and risers are to be nailed to the carriages.

Another method of fitting the treads and risers to wall stringers is shown in Fig. 34-29. The finish stringers are laid out with the same rise and run as the stair carriages, but they are cut out in reverse. The risers are butted and nailed from the back to the riser cuts of the wall stringers, and the assembled stringers and risers are laid over the carriage. The treads are butted to the stringers and nailed to the risers. Sometimes the treads are allowed to run underneath the tread cut

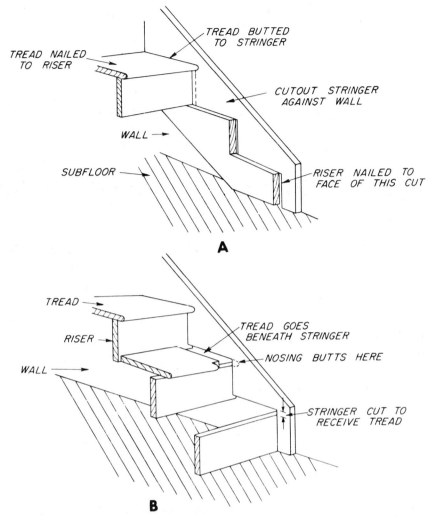

34-29. *In this method the stringer is cut out in reverse of the carriage. In A, the riser fits between the stringer and the carriage. The treads butt against the stringer. In B, both the tread and riser are fitted between the stringer and carriage, with the tread nosing notched and butted to the stringer.*

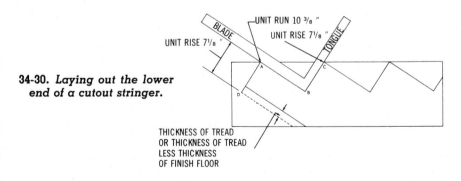

34-30. *Laying out the lower end of a cutout stringer.*

of the stringer. This makes it necessary to notch the tread at the nosing to fit around the finish stringer. Fig. 34-29.

Laying Out a Cutout Stringer

Cutout stringers for main stairways are usually made from 2″ ×

12″ stock. To lay out a cutout stringer, you must first determine how long a piece of stock you will need. Using the example from "Laying Out a Stairway," let's assume that the method of upper-end anchorage is the one shown in C, Fig. 34-23. In that case the total run of the stairway is 12′1¼″. The total rise in the example is 8′ 11″. On the framing square twelfth scale, measure the distance between a little over 12¹⁄₁₂″ on the blade and 8¹¹⁄₁₂″ on the tongue. You'll find that it comes to just about 15″. Therefore you'll need a piece of stock at least 15″ long. It is better to allow about 4″ more for waste and for the part that extends beyond the header at the upper end. Select or cut a piece about 19″ long. Proceed to lay out the stringer from the lower end as follows:

Set the framing square to the unit run and unit rise as shown in Fig. 34-30. Draw the line AB along the blade and the line BC along the tongue. AB indicates the first tread, BC the second riser. Reverse the square and draw line AD from A, perpendicular to AB, and equal in length to the unit rise.

Line AD indicates the first riser. The first riser has to be shortened, a process which is called dropping the stringer. Fig. 34-31. As you can see, in the completed stairway the unit rise is measured from the top of one tread to the top of the next. Let's assume that the bottom of the stairway is to be anchored on a finished floor, such as a concrete basement floor. If AD were cut to the unit rise, when the first tread was put on, the height of the first step would be the unit rise plus the thickness of the tread. To make the height of the first step equal to the unit rise, you must shorten AD by the thickness of a tread.

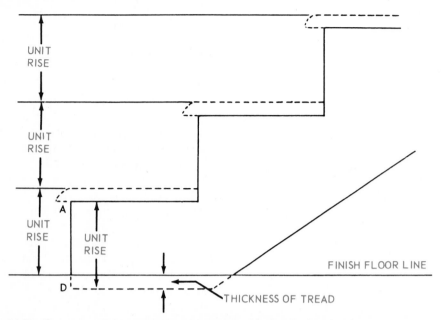

UNIT RISE

UNIT RISE

UNIT RISE

UNIT RISE

A

D

FINISH FLOOR LINE

THICKNESS OF TREAD

34-31. *Dropping the stringer to compensate for the thickness of the first tread keeps the unit rise uniform.*

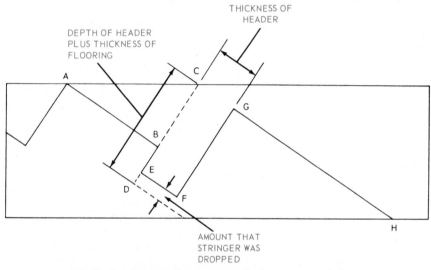

THICKNESS OF HEADER

DEPTH OF HEADER PLUS THICKNESS OF FLOORING

A

C

G

B

E

D

F

H

AMOUNT THAT STRINGER WAS DROPPED

34-32. *Laying out the upper end of a cutout stringer.*

If the bottom of the stringer is to be anchored on a subfloor to which finish flooring will be applied, shorten AD by the thickness of a tread less the thickness of the finish flooring. When you have shortened AD as required, proceed to step off the unit run and unit rise as many times as the stairway has treads—in this case, 14.

Finish the layout at the upper end as shown in Fig. 34-32. You are going to anchor the upper end by the method shown in C, Fig. 34-23, in which the stringer fits around the well header and extends beyond it to end level with the upper edges of the floor joists. Lay out the line AB, which indicates the last of the treads. Lay out the dotted line BC, which

indicates the face of the well header. Extend BC down to D, so that BC plus BD will equal the depth of the header plus the thickness of the flooring.

To make the stringer fit close up under the lower edge of the header, you must shorten BD by the amount the stringer was dropped, as was shown in Fig. 34-31. Draw EF equal in length to the thickness of the header. From F draw FG equal in length to the depth of the header, and from G draw GH. When the stringer is set in place, the edge indicated by GH will lie close up under the subflooring, level with the upper edges of the joists.

Carefully cut out the first stringer, set it in position, and check it. Then use this as a pattern for cutting one or two more.

Installing Stringers

Methods used in framing stairways and securing stringers vary in different areas of the country. Regardless of method, the object of the stair builder should be the installation of a structurally strong, safe stairway. A few suggested ways of securing stringers are shown in Fig. 34-33. In A, the upper rough stringer is notched to fit the stairwell header. The stringers are hung by means of a metal supporting strap. B, Fig. 34-33, shows the stringer notched out for the stairwell header and supported by a ledger strip. This method will slightly reduce the headroom underneath.

The method shown in C, Fig. 34-33, requires a larger well "opening" and is not used too often, yet it offers the full bearing of the rough stringer against the stairwell header. The support is a ledger strip. In D, Fig. 34-33, a piece of plywood sheathing is used as a bearing surface and ledger. The stringers are secured by nailing from the back. This

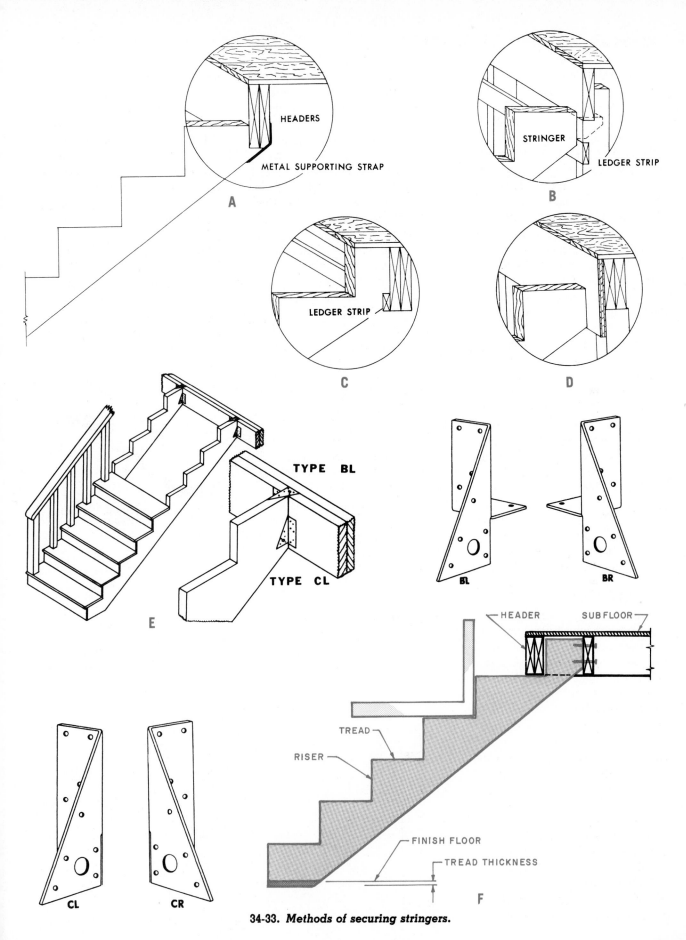

HEADERS

METAL SUPPORTING STRAP

A

STRINGER

LEDGER STRIP

B

LEDGER STRIP

C

D

TYPE BL

TYPE CL

E

BL **BR**

CL **CR**

HEADER **SUBFLOOR**

TREAD

RISER

FINISH FLOOR

TREAD THICKNESS

F

34-33. *Methods of securing stringers.*

method would apply most often at a platform where the headers are usually of less depth than the floor joists. It also affords full headroom underneath.

In E, Fig. 34-33, the stair carriages are framed to a header or trimmer with metal brackets. Another method of installing stringers is to cut the top stair tread deeper to permit the stringer to pass under the header. Then attach the stringer to a special framing member between the floor joists as shown in F, Fig. 34-33. Note also in this illustration that the stringer has been cut off at the bottom to allow for the thickness of the first tread.

Erecting the Stairway

The rough stringers are the first stairway members erected—except when a side of the stairway butts against a wall, in which case the wall (finish) stringer must be nailed on first. Temporarily nail the rough stringers in position. Check each stringer for plumb by holding the carpenter's level vertically against a riser cut. Then check the stringers for levelness with each other by setting the carpenter's level across the stringers on the tread cuts.

A stringer which lies against a trimmer joist should be spiked to the joist with at least three 16d nails. A stringer which is installed as shown at C in Fig. 34-33 and is not adjacent to a joist (a center stringer in a three-stringer stairway, for example) should be toenailed to the well header with 10d nails, three to each side of the stringer. The bottom of a stringer which is anchored on subflooring should be toenailed with 10d nails, four to each side if possible, driven into the subflooring and if possible into a joist below.

After the stringers are mounted to the wall and treads and risers cut to length, nail the

bottom riser to each stringer with two 6d, 8d, or 10d nails, depending on the thickness of the stock.

The first tread, if 1¹⁄₁₆″ thick, is then nailed to each stringer with two 10d finish nails and to the riser below with at least two 10d finish nails. Proceed up the stair in this same manner. If 1⅝″ thick treads are used, a 12d finish nail may be required. Use three nails at each stringer, but eliminate nailing to the riser below. All finish nails should be set.

RAILING

All stairways should have a handrail extending from one floor to the next. For closed stairways the rail is attached to the wall with suitable metal brackets. The rail should be set 30″ above the tread at the riser line and 34″ above the floor on a landing. Fig. 34-34. Handrails and balusters are used for open stairs and for open spaces around stairs. The handrails end against newel posts.

Stairs should be laid out so that stock parts may be used for newels, rails, balusters, and goose-

necks. Fig. 34-11. These parts may be very plain or elaborate, but they should be in keeping with the style of the house. The balusters are doweled or dovetailed into the treads and in some cases are covered by a return nosing. Fig. 34-35. For the dovetail method, a strip called a nosing return is cut off the end of the tread, as shown in the upper (plan) view of Fig. 34-35. Dovetails are shaped on the lower ends of the balusters, and dovetail recesses of corresponding size are cut in the end section of the tread. The dovetails on the balus-

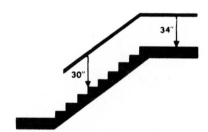

34-34. *Recommended heights for handrails.*

34-35. *Balusters are attached to the treads with either dowels or dovetails.*

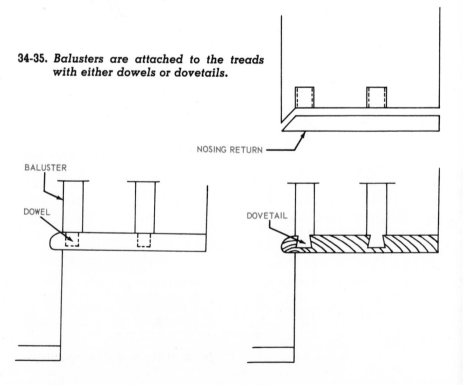

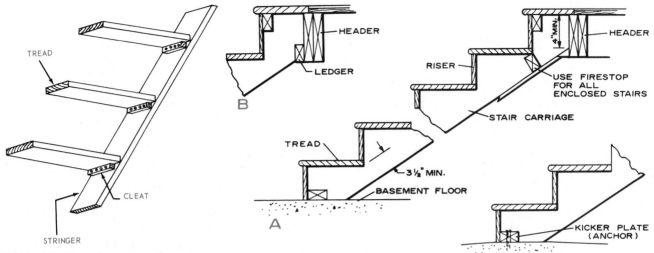

34-36. Cleat stairs are often installed in basements.

34-37. Basement stairs: A. Carriage details. B. Ledger for carriage. C. Kicker plate. The lower end of the stringer should be anchored against a kicker plate which has been bolted to the concrete floor.

ters are glued into the recesses in the tread, and the nosing return is then nailed back in place.

Newel posts should be firmly anchored. Where half-newels are attached to a wall, blocking should be provided at the time the wall is framed.

BASEMENT STAIRS

Basement stairs may be built either with or without riser boards. Cutout stringers are probably the most widely used supports for the treads, but in some cases the stringer is not cut out and the tread is fastened to the stringers by cleats. Fig. 34-36. The lower end of a basement stairway is usually anchored against a kicker plate which has been bolted to the concrete. Fig. 34-37.

Laying Out and Framing Cleat Stairway

A cleat stairway is inexpensive to build, it does not require risers, and the treads are usually made of softwood. First determine the total rise and run. Divide the rise by 7. If this does not result in even spacing adjust the divisor until equal spacings are

obtained. Try to keep this spacing between 6½″ and 7½″.

For the first tread position use the determined riser height *minus the thickness of the tread* and measure up the stringer at 90° from the bottom cut to establish point A. Fig. 34-38. Set the T bevel to the angle formed by the front edge of the stringer and the bottom cut as shown at B in Fig. 34-38. Slide the T bevel up the stringer until the tongue of the T bevel is at point A. Fig. 34-38. Mark a line across the stringer at this point.

Measure up from line A using the riser height and establish point C. Position the T bevel on the new point and mark another line across the stringer. Continue this operation until all tread positions are located. These lines locate the bottom of the tread and the top edge of the cleat.

Cut the cleats from 1″ × 2″ stock and nail the cleats in position below the line. Place the stringers in the stairwell and fasten them in place. Cut the treads to length. Starting with the bot-

tom tread, work up, placing each tread in position and nailing securely.

EXTERIOR STAIRS

Proportioning of risers and treads in laying out porch steps or approaches to terraces should be as carefully considered as the design of interior stairways. Similar riser-to-tread ratios can be used. However, the riser used in principal exterior steps should be between 6″ and 7″. The need for a good support or foundation for outside steps is often overlooked. If wood steps are used, the bottom step should be concrete. Fig. 34-39. If the steps are located over backfill or disturbed ground, the foundation should be carried down to undisturbed ground.

METAL SPIRAL STAIRWAYS

The standardized spiral stairway system provides unlimited versatility in design. These stairs are adaptable for use in all types of buildings, from a modest cottage to the most elegant residence, and are suitable for inte-

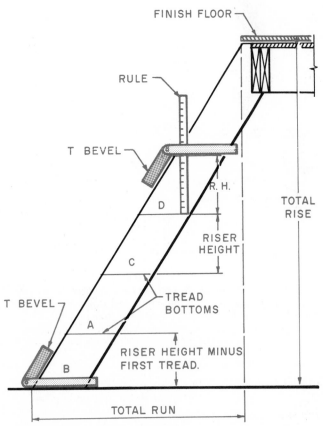

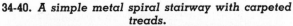

34-38. *Laying out a cleat stairway. Note that the distance from A to C is the same as the distance from C to D and is equal to the riser height. The distance between the floor and line A, however, is less than the riser height to allow for the thickness of the first tread.*

34-40. *A simple metal spiral stairway with carpeted treads.*

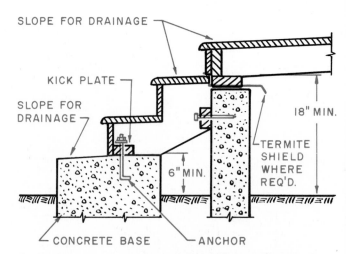

34-39. *Exterior steps of wood should have a bottom step of concrete. The stringer should be secured to the bottom step against a kick plate which has been bolted to the concrete.*

1. Place center column in position and slide on treads.

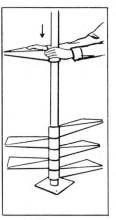

2. Attach rail standards.

3. Fasten treads to center column.

4. Connect handrail to rail standards with adjustable brackets.

34-41. *Installing a metal spiral stairway.*

34-42. *Disappearing stairs. When the stairs are in the stored position, the plywood door is barely noticeable.*

knocked down for easy handling at the job site. All of the parts are matched and marked, and the manufacturer furnishes complete shop drawings and installation instructions.

DISAPPEARING STAIRS

When attics are used primarily for storage and space for a fixed stairway is not available, hinged or disappearing stairs are often used. Disappearing stairways may be purchased ready to install. They operate through an opening in the ceiling and swing up into the attic space, out of the way, when not in use. Fig. 34-42. Where such stairs are installed, the attic floor should be designed for regular floor loading.

rior or exterior installation. Fig. 34-40.

Standardized construction provides a light, simple, strong, and durable stairway, meeting code requirements for design and carrying capacities. The basic stair structure is steel. To this can be added various types of treads, handrails, and railings.

The stairways are simple to install. Fig. 34-41. No welding is necessary. Handrails are attached to balusters with adjustable brackets. The stairs are shipped

QUESTIONS

1. What are the two general types of stairs?
2. What are the two main parts of a stairway?
3. What is the difference between a landing and a platform?
4. When designing a stairway, what are the three important considerations?
5. The total rise of a stairway is the vertical distance between what two points?

6. When laying out a stringer for a basement stairway, what is subtraced from the bottom of the stringer?
7. After the stair stringers are installed and the other parts are cut to length, what is the next piece to be installed?

Interior Trim

35

Wood or plastic molding and trim add individuality to home design. Figs. 35-1 & 35-2. Sometimes special patterns of millwork are required, but it is possible to create unique effects with standard patterns. Fig. 35-3. The standard patterns are readily available from local lumber dealers, allowing the development of many interesting design with maximum economy. Wood or plastic moldings can be used with wallpaper or fabric, as accent walls, or to create interesting shadow and highlight effects on flat surfaces. Figs. 35-4, 35-5, page 430, & 35-6. They may be painted to blend or contrast, or they may be stained to match or accent natural wood grains. Wood and plastic moldings in the home are most often used for trimming door and window frames, for base moldings, and, sometimes, for ceiling and wall moldings. Fig. 35-7, page 432.

Interior doors, trim, and other millwork may be painted or given a natural finish with stain, varnish, or other nonpigmented material. The paint or natural finish desired for the woodwork in various rooms often determines the species of wood to be used. Woodwork to be painted should be smooth, close-grained, and free from pitch streaks. Some species having these qualities in a high degree include ponderosa pine, northern white pine, redwood, and spruce. When hardness and resistance to hard usage are additional requirements, species such as birch or yellow poplar may be used.

35-1. Moldings may be installed on painted walls or with wallpaper. These moldings can be painted, stained, or even antiqued.

35-2. With the addition of molding, a plain room achieves definite character and depth. Moldings soften harsh corners, highlight room features, and, as can be seen in this stairway, emphasize direction.

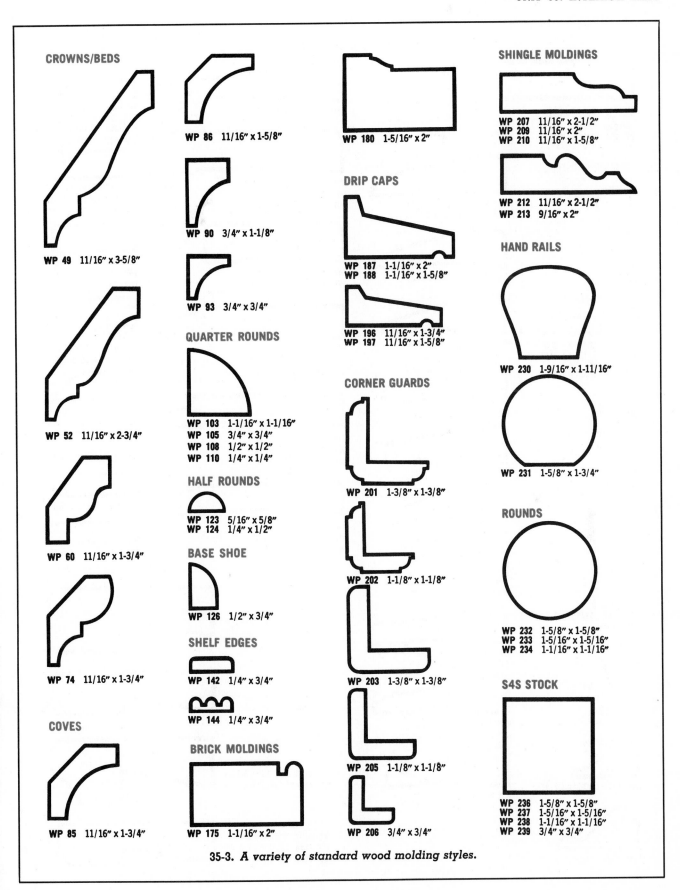

CROWNS/BEDS

WP 86   11/16" x 1-5/8"

WP 90   3/4" x 1-1/8"

WP 93   3/4" x 3/4"

WP 49   11/16" x 3-5/8"

QUARTER ROUNDS

WP 103   1-1/16" x 1-1/16"
WP 105   3/4" x 3/4"
WP 108   1/2" x 1/2"
WP 110   1/4" x 1/4"

WP 52   11/16" x 2-3/4"

HALF ROUNDS

WP 123   5/16" x 5/8"
WP 124   1/4" x 1/2"

BASE SHOE

WP 126   1/2" x 3/4"

WP 60   11/16" x 1-3/4"

SHELF EDGES

WP 142   1/4" x 3/4"

WP 144   1/4" x 3/4"

WP 74   11/16" x 1-3/4"

BRICK MOLDINGS

COVES

WP 175   1-1/16" x 2"

WP 85   11/16" x 1-3/4"

WP 180   1-5/16" x 2"

DRIP CAPS

WP 187   1-1/16" x 2"
WP 188   1-1/16" x 1-5/8"

WP 196   11/16" x 1-3/4"
WP 197   11/16" x 1-5/8"

CORNER GUARDS

WP 201   1-3/8" x 1-3/8"

WP 202   1-1/8" x 1-1/8"

WP 203   1-3/8" x 1-3/8"

WP 205   1-1/8" x 1-1/8"

WP 206   3/4" x 3/4"

SHINGLE MOLDINGS

WP 207   11/16" x 2-1/2"
WP 209   11/16" x 2"
WP 210   11/16" x 1-5/8"

WP 212   11/16" x 2-1/2"
WP 213   9/16" x 2"

HAND RAILS

WP 230   1-9/16" x 1-11/16"

WP 231   1-5/8" x 1-3/4"

ROUNDS

WP 232   1-5/8" x 1-5/8"
WP 233   1-5/16" x 1-5/16"
WP 234   1-1/16" x 1-1/16"

S4S STOCK

WP 236   1-5/8" x 1-5/8"
WP 237   1-5/16" x 1-5/16"
WP 238   1-1/16" x 1-1/16"
WP 239   3/4" x 3/4"

35-3. *A variety of standard wood molding styles.*

427

(Fig. 35-3 Continued from page 427)

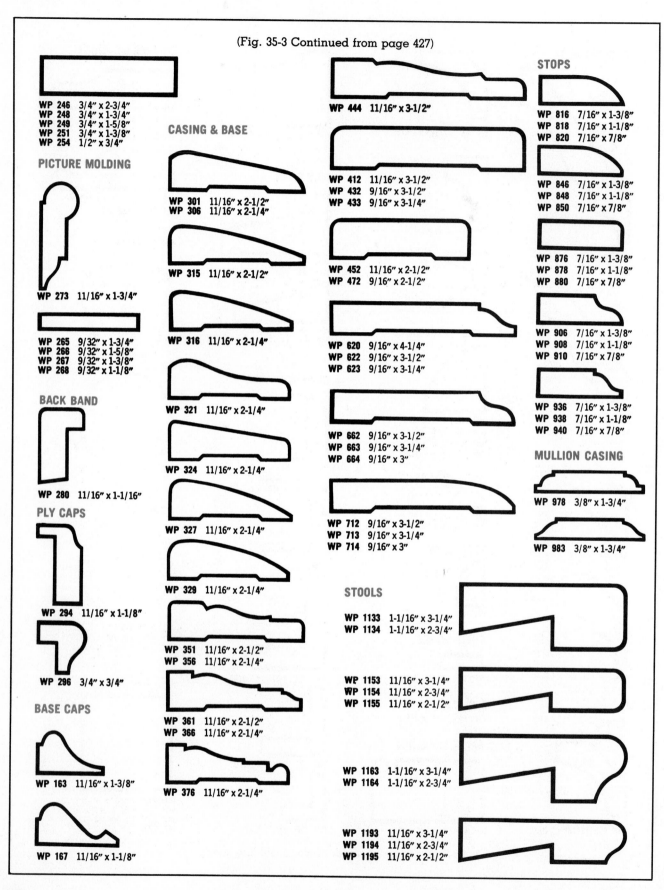

WP 246   3/4″ x 2-3/4″
WP 248   3/4″ x 1-3/4″
WP 249   3/4″ x 1-5/8″
WP 251   3/4″ x 1-3/8″
WP 254   1/2″ x 3/4″

PICTURE MOLDING

WP 273   11/16″ x 1-3/4″

WP 265   9/32″ x 1-3/4″
WP 266   9/32″ x 1-5/8″
WP 267   9/32″ x 1-3/8″
WP 268   9/32″ x 1-1/8″

BACK BAND

WP 280   11/16″ x 1-1/16″

PLY CAPS

WP 294   11/16″ x 1-1/8″

WP 296   3/4″ x 3/4″

BASE CAPS

WP 163   11/16″ x 1-3/8″

WP 167   11/16″ x 1-1/8″

CASING & BASE

WP 301   11/16″ x 2-1/2″
WP 306   11/16″ x 2-1/4″

WP 315   11/16″ x 2-1/2″

WP 316   11/16″ x 2-1/4″

WP 321   11/16″ x 2-1/4″

WP 324   11/16″ x 2-1/4″

WP 327   11/16″ x 2-1/4″

WP 329   11/16″ x 2-1/4″

WP 351   11/16″ x 2-1/2″
WP 356   11/16″ x 2-1/4″

WP 361   11/16″ x 2-1/2″
WP 366   11/16″ x 2-1/4″

WP 376   11/16″ x 2-1/4″

WP 444   11/16″ x 3-1/2″

WP 412   11/16″ x 3-1/2″
WP 432   9/16″ x 3-1/2″
WP 433   9/16″ x 3-1/4″

WP 452   11/16″ x 2-1/2″
WP 472   9/16″ x 2-1/2″

WP 620   9/16″ x 4-1/4″
WP 622   9/16″ x 3-1/2″
WP 623   9/16″ x 3-1/4″

WP 662   9/16″ x 3-1/2″
WP 663   9/16″ x 3-1/4″
WP 664   9/16″ x 3″

WP 712   9/16″ x 3-1/2″
WP 713   9/16″ x 3-1/4″
WP 714   9/16″ x 3″

STOOLS

WP 1133   1-1/16″ x 3-1/4″
WP 1134   1-1/16″ x 2-3/4″

WP 1153   11/16″ x 3-1/4″
WP 1154   11/16″ x 2-3/4″
WP 1155   11/16″ x 2-1/2″

WP 1163   1-1/16″ x 3-1/4″
WP 1164   1-1/16″ x 2-3/4″

WP 1193   11/16″ x 3-1/4″
WP 1194   11/16″ x 2-3/4″
WP 1195   11/16″ x 2-1/2″

STOPS

WP 816   7/16″ x 1-3/8″
WP 818   7/16″ x 1-1/8″
WP 820   7/16″ x 7/8″

WP 846   7/16″ x 1-3/8″
WP 848   7/16″ x 1-1/8″
WP 850   7/16″ x 7/8″

WP 876   7/16″ x 1-3/8″
WP 878   7/16″ x 1-1/8″
WP 880   7/16″ x 7/8″

WP 906   7/16″ x 1-3/8″
WP 908   7/16″ x 1-1/8″
WP 910   7/16″ x 7/8″

WP 936   7/16″ x 1-3/8″
WP 938   7/16″ x 1-1/8″
WP 940   7/16″ x 7/8″

MULLION CASING

WP 978   3/8″ x 1-3/4″

WP 983   3/8″ x 1-3/4″

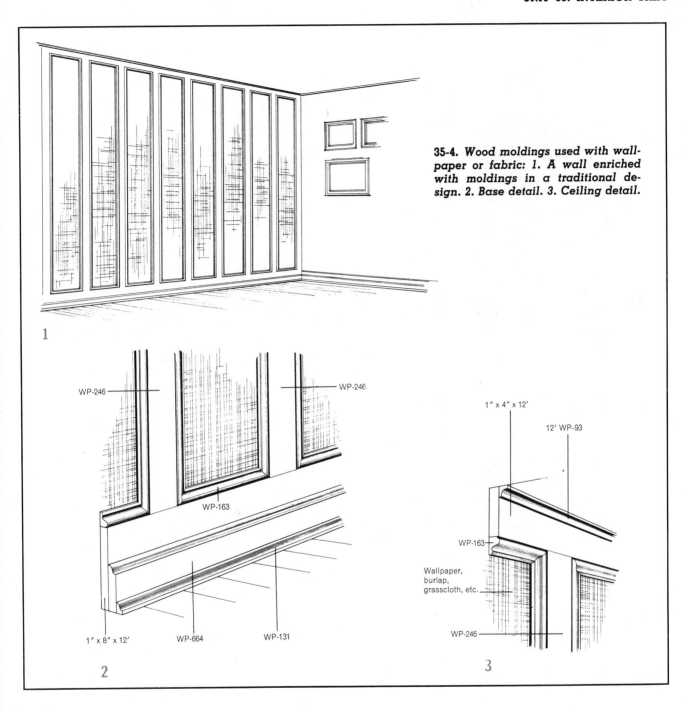

35-4. *Wood moldings used with wallpaper or fabric: 1. A wall enriched with moldings in a traditional design. 2. Base detail. 3. Ceiling detail.*

When the finish is to be natural, a pleasing figure, hardness, and uniform color are usually desirable. Species with these qualities include ash, birch, cherry, maple, oak, and walnut. Some require staining for best appearance.

The recommended moisture content for interior trim varies from 6 to 11 percent, depending on climate. The average moisture content for various parts of the United States is shown in Fig. 35-8, page 432.

INTERIOR DOOR AND WINDOW TRIM

After interior wall covering has been applied and the finish floor laid, all floor and wall surfaces should be scraped clean and free of any irregularities. Mark the location of all wall studs lightly on the floor or wall. Usually the marks are placed on the floor because the interior wall covering may be a finished surface.

Door and window frames are usually trimmed first to allow

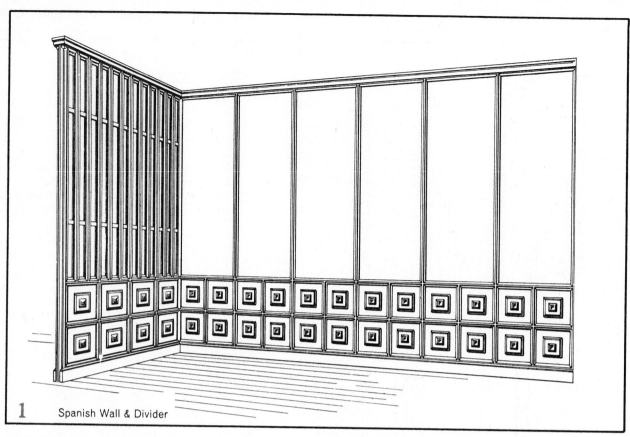

1 Spanish Wall & Divider

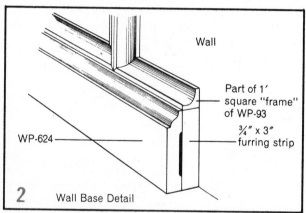

Wall

Part of 1'
square "frame"
of WP-93

¾" x 3"
furring strip

WP-624

2 Wall Base Detail

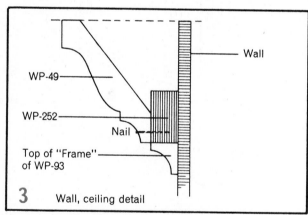

Wall

WP-49

WP-252

Nail

Top of "Frame"
of WP-93

3 Wall, ceiling detail

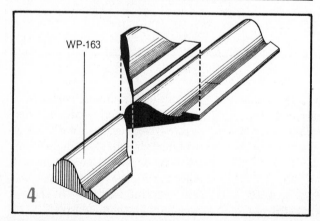

WP-163

4

Rosette Detail

2¾"

6½"

WP-163

WP-624

5

35-5. *Moldings and woodwork are used to give this wall a Spanish motif: 1. Spanish wall and divider. 2. Wall base detail. 3. Wall, ceiling detail. 4. Cutting miters in the molding to make the rosettes. Rosette detail.*

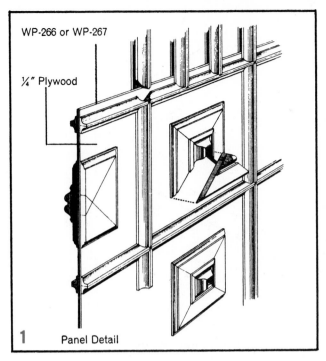

WP-266 or WP-267

¼" Plywood

1 Panel Detail

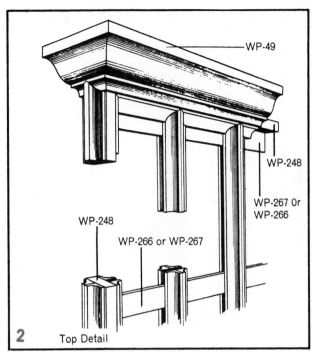

WP-49

WP-248

WP-267 Or WP-266

WP-248

WP-266 or WP-267

2 Top Detail

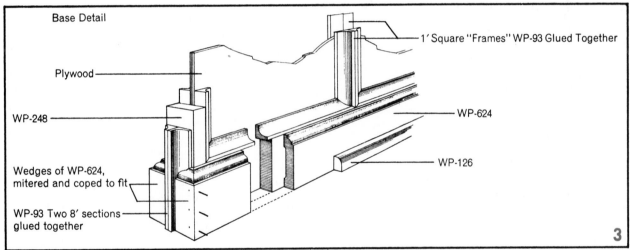

Base Detail

1' Square "Frames" WP-93 Glued Together

Plywood

WP-248

WP-624

Wedges of WP-624, mitered and coped to fit

WP-126

WP-93 Two 8' sections glued together

3

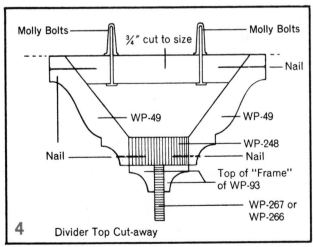

Molly Bolts

¾" cut to size

Molly Bolts

Nail

WP-49

WP-49

Nail

WP-248

Nail

Top of "Frame" of WP-93

WP-267 or WP-266

4 Divider Top Cut-away

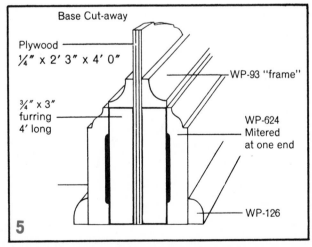

Base Cut-away

Plywood
¼" x 2' 3" x 4' 0"

WP-93 "frame"

¾" x 3" furring 4' long

WP-624 Mitered at one end

WP-126

5

35-6. *Details for the room divider shown in Fig. 35-5.: 1. Panel details. 2. Top detail. 3. Base detail. 4. Top cutaway. 5. Base cutaway.*

35-7. *Wood moldings added to an ordinary room divider will give a room a new focal point and added interest.*

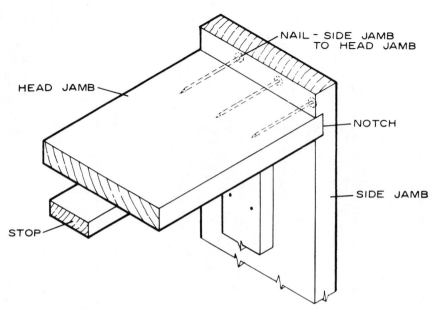

35-9. *Interior doorframe details.*

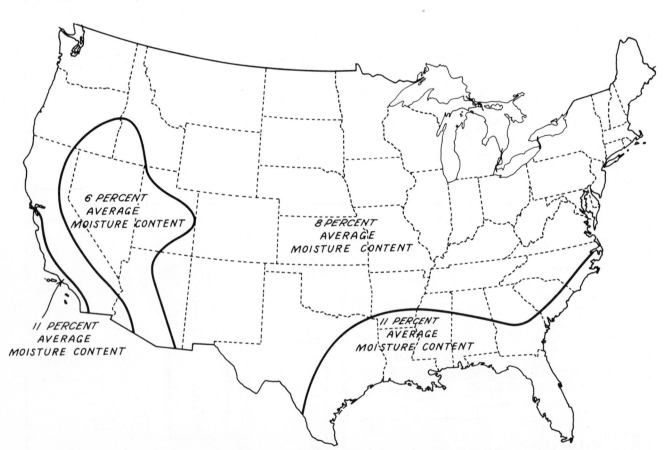

35-8. *Recommended average moisture content for interior wood trim in various parts of the United States. In Canada the recommended moisture contents are as follows: Vancouver, 11%; Saskatoon, 7%; Ottawa, 8%; Halifax, 9%. (These cities represent four major geographical areas.)*

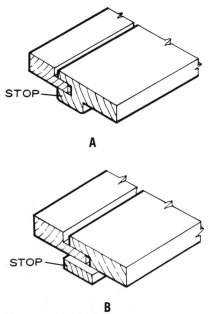

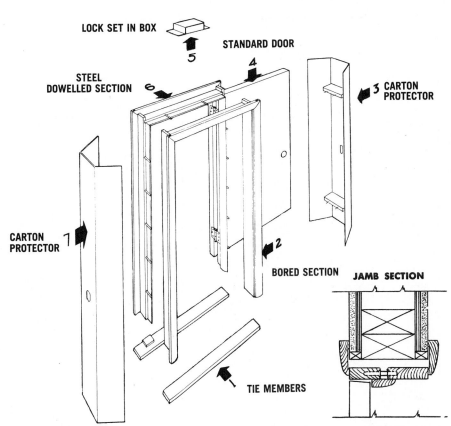

35-10. *Adjustable doorjambs: A. Two-piece adjustable jamb. B. Three-piece adjustable jamb. Two- and three-piece adjustable doorjambs adapt to various wall thicknesses.*

35-11a. *This prehung door has a two-piece jamb with pins for alignment behind the doorstop. The jamb can be adjusted to slight variations in wall thickness. However, when ordering a prehung door unit, be sure to specify the wall thickness. The manufacturer of this unit recommends a 2″ allowance in width and height over the nominal door size for the rough stud opening. The height is figured from the finished floor.*

other trim such as base moldings or wall moldings to be properly fitted between the door and window casings. Cabinets, built-in bookcases, fireplace mantels, and other millwork items are also installed at this time.

Doorframes

Rough openings in the stud walls for interior doors are usually framed out to be 3″ more than the door height and 2½″ more than the door width. This provides room for plumbing and leveling the frame in the opening. Interior doorframes are made up of two side jambs and a head jamb and include stop moldings which the door closes against. One-piece jambs are the most common. Fig. 35-9. They may be obtained in 5¼″ widths for plaster walls and 4½″ widths for walls with ½″ dry-wall finish. Two- and three-piece adjustable jambs are

also available. Fig. 35-10. Their chief advantage is in being adaptable to different wall thicknesses.

Some manufacturers produce interior doorframes with the door fitted and prehung, ready for installation. Application of the casing completes the job. When used with two- or three-piece jambs, casings are installed at the factory. Figs. 35-11 & 35-12.

DOORFRAME INSTALLATION

When the frames and doors are not assembled and prefitted, the side jambs should be nailed through the notch into the head jamb with three 7d or 8d coated nails. Fig. 35-9. Cut a spreader to a length exactly equal to the

distance between the jambs at the head jamb. Fig. 35-13.

The assembled frame is fastened in the rough opening by shingle wedges placed between the side jamb and the stud. Fig. 35-14. One jamb, usually the hinge jamb, is plumbed using four or five sets of shingle wedges for the height of the frame. Two 8d finishing nails are installed at each wedged area, one driven so the doorstop will cover it. Fig. 35-14.

Place the spreader in position at the floor line. Fig. 35-13. Fasten the opposite side jamb in place with shingle wedges and finishing nails, using the first jamb as a guide in keeping a uniform

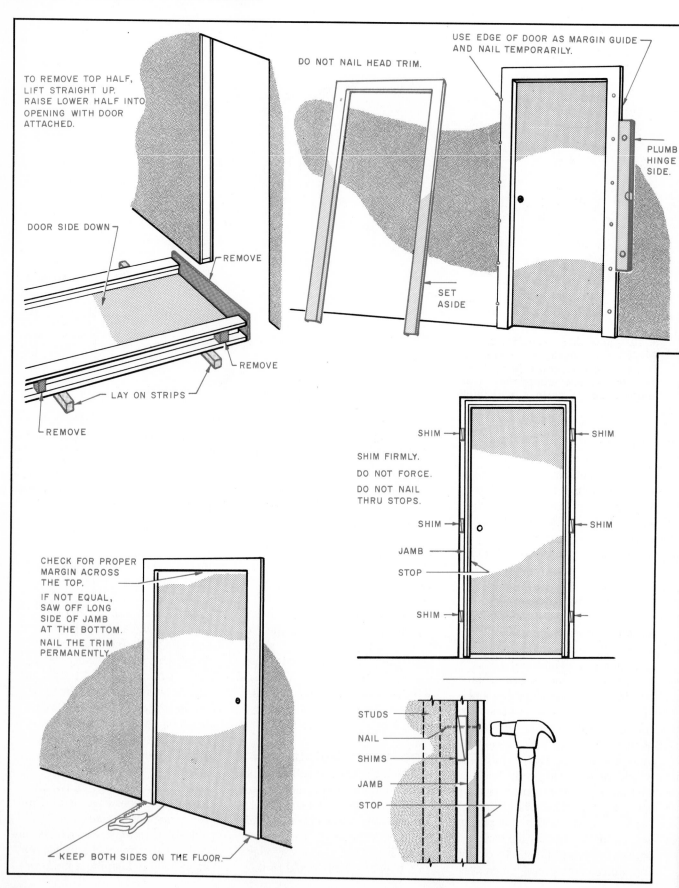

TO REMOVE TOP HALF, LIFT STRAIGHT UP. RAISE LOWER HALF INTO OPENING WITH DOOR ATTACHED.

DO NOT NAIL HEAD TRIM.

USE EDGE OF DOOR AS MARGIN GUIDE AND NAIL TEMPORARILY.

PLUMB HINGE SIDE.

DOOR SIDE DOWN

REMOVE

SET ASIDE

REMOVE

LAY ON STRIPS

REMOVE

SHIM

SHIM

SHIM FIRMLY.
DO NOT FORCE.
DO NOT NAIL THRU STOPS.

SHIM

SHIM

JAMB

STOP

SHIM

CHECK FOR PROPER MARGIN ACROSS THE TOP.

IF NOT EQUAL, SAW OFF LONG SIDE OF JAMB AT THE BOTTOM. NAIL THE TRIM PERMANENTLY.

STUDS

NAIL

SHIMS

JAMB

STOP

KEEP BOTH SIDES ON THE FLOOR.

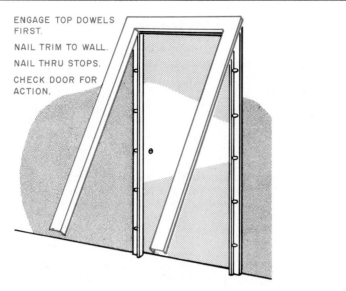

ENGAGE TOP DOWELS FIRST.

NAIL TRIM TO WALL.

NAIL THRU STOPS.

CHECK DOOR FOR ACTION.

35-11b. *Details for installing the door unit shown in Fig. 35-11a.*

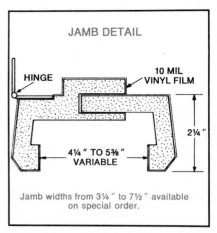

JAMB DETAIL

HINGE

10 MIL VINYL FILM

2¼ "

4¼ " TO 5⅜ " VARIABLE

Jamb widths from 3¼ " to 7½ " available on special order.

35-12. *The prefinished doors shown here have an adjustable two-piece split jamb which will accommodate walls from 4¼" to 5⅜" thick.*

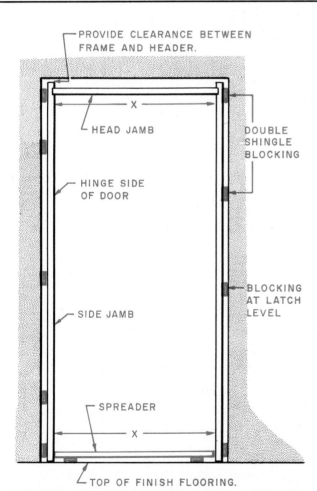

PROVIDE CLEARANCE BETWEEN FRAME AND HEADER.

X

HEAD JAMB

DOUBLE SHINGLE BLOCKING

HINGE SIDE OF DOOR

BLOCKING AT LATCH LEVEL

SIDE JAMB

SPREADER

X

TOP OF FINISH FLOORING.

35-13. *Cut a spreader equal to the distance (X) between the side jambs just below the head jamb. Place the spreader at the floorline to hold the side jambs parallel.*

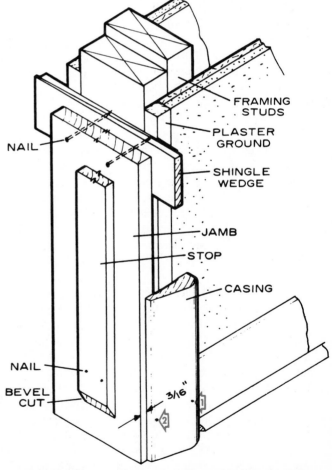

FRAMING STUDS

PLASTER GROUND

SHINGLE WEDGE

NAIL

JAMB

STOP

CASING

NAIL

BEVEL CUT

3/16

35-14. *Doorframe and trim installation details. Use a 6d or 7d finish nail at arrow 1 to nail through the casing into the wall stud. At arrow 2, use a 4d or 5d finish nail to fasten the casing to the jamb.*

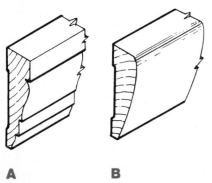

35-15. *Two common casings used for interior trim: A. Colonial. B. Ranch casing.*

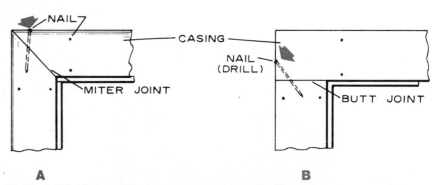

35-16. A. *Molded casing must have a mitered joint at the corner. B. Square-edge casing may be joined with a butt joint. In both cases, the joints may be reinforced by nailing at the arrows.*

width. This can be done by using a second precut spreader as a gauge, checking several points, or by carefully measuring at various points along the height of the doorframe between the side jambs.

Door Trim

Door trim, or *casing*, is nailed around interior door openings and is also used to finish the room side of windows and exterior doorframes. The most commonly used casings vary in width from 2¼″ to 3½″, depending on the style. Fig. 35-3. Thicknesses vary from ½″ to ¾″, although ¹¹⁄₁₆″ is standard in many of the narrow-line patterns. Two of the more common patterns are shown in Fig. 35-15.

DOOR TRIM INSTALLATION

Casings are nailed to both the jamb and the framing studs or header, allowing about a ³⁄₁₆″ edge distance from the face of the jamb. Fig. 35-14. Finish or casing nails either 6d or 7d, depending on the thickness of the casing, are used to nail into the stud. Arrow 1, Fig. 35-14. Fourpenny or 5d finishing nails or 1½″ brads are used to fasten the thinner edge of the casing to the jamb. Arrow 2, Fig. 35-14. With hardwood, it is advisable to predrill to prevent

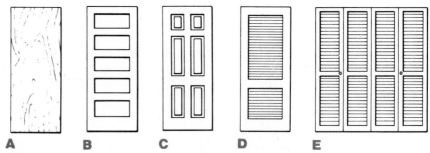

35-17. *Interior doors: A. Flush. B. Panel (5-cross). C. Panel (Colonial). D. Louvered. E. Folding (louvered).*

splitting. Nails in the casing are located in pairs and spaced about 16″ apart along the full height of the opening and at the head jamb. Fig. 35-14.

Casing with a molded shape must have mitered corner joints. A, Fig. 35-16. When casing is square-edged, a butt joint may be made at the junction of the side and head casing. B, Fig. 35-16. If the moisture content of the casing is well above the recommended amount, a mitered joint may open slightly at the outer edge as the material dries. This can be minimized by installing a small glued spline at the corner of the mitered joint. Actually, use of a spline joint under any moisture condition is considered good practice. Some prefitted jamb, door, and casing units are provided with splined joints. Nailing into

the joint after drilling will aid in retaining a close fit. Fig. 35-16.

The door opening is now complete except for fitting and securing the hardware and nailing the stops in proper position.

Interior Doors

As in exterior door styles, the two general interior types are the flush and the panel. Bifold and sliding door units might be flush or louvered. Most standard interior doors are 1⅜″ thick. Folding and sliding doors are usually 1⅛″ thick.

The flush interior door is usually made up of a hollow core of light framework faced with thin plywood or hardboard. A, Fig. 35-17. Plywood-faced flush doors may be obtained in gum, birch, oak, mahogany, and woods of other species, most of which are

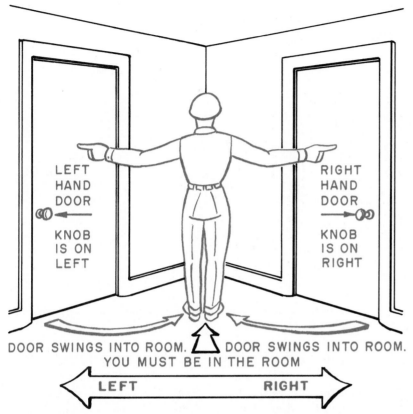

35-18. *Determining the hand of a door. When standing on the hinge side of the door, with the door closed, the knob is on the left for a left-hand door. For a right-hand door, the knob is on the right.*

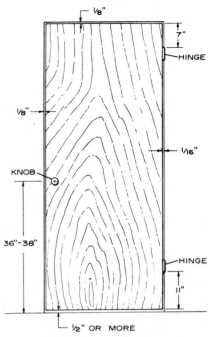

35-19. *Door clearance.*

suitable for natural finish. Nonselected grades are usually painted, as are hardboard-faced doors.

The panel door consists of solid stiles (vertical side members), rails (cross-pieces), and panels of various types. The five-cross panel and the Colonial panel doors are perhaps the most common of this style. B and C, Fig. 35-17. The louvered door is also popular and is commonly used for closets because it provides some ventilation. D, Fig. 35-17. Sliding or folding doors are installed in openings for wardrobes. These are usually flush or louvered. E, Fig. 35-17.

Common minimum widths for single interior doors are as follows:
● Bedrooms and other habitable rooms—2'6".
● Bathrooms—2'4".
● Small closets and linen closets—2'.

These sizes can vary a great deal. Sliding doors or folding door units, used for wardrobes, may be 6' or more in width. However, in most cases, the jamb, stop, and casing parts are still used to frame and finish the opening. The standard interior door height for first floors is 6'8". Doors on upper floors are sometimes 6'6".

Hinged doors should open or swing in the direction of natural entry, against a blank wall whenever possible. They should not be obstructed by other swinging doors. Doors should never be hinged to swing into a hallway. The door swing is designated as either right or left hand. A right-hand door is one in which the

latch is on the right when a person faces a closed door on the hinge side. A left-hand door is one in which the latch is on the left when a person faces a closed door on the hinge side. Fig. 35-18.

INTERIOR DOOR INSTALLATION

Interior doors are normally hung with two 3½" × 3½" loose-pin butt hinges. The door is fitted into the opening with the clearances shown in Fig. 35-19. The clearance and location of hinges, lock set, and doorknob may vary somewhat, but the dimensions in Fig. 35-19 are generally accepted and conform to most millwork standards. The edge of the lock stile should be beveled slightly to permit the door to clear the jamb when swung open. See "Fitting a Door" in Unit 27. If the door is to swing across heavy carpeting, the bottom clearance should be increased.

When fitting doors, the stops are usually temporarily nailed in

437

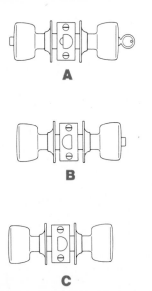

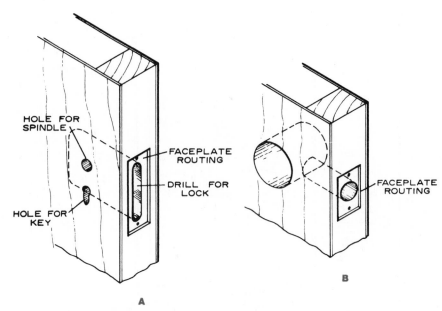

35-22. *Installation of lock set: A. Mortise lock. B. Bored lock.*

35-20. *Door sets: A. Exterior lock. Push button locks the outside knob, and a key is needed to unlock it. B. Bathroom or privacy door lock. Push button locks the outside knob. An emergency key may be used to unlock the door from the outside. C. Passage door lock. Both knobs are always free.*

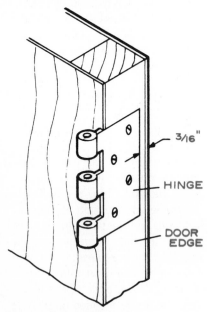

35-21. *Installation of the door hinge.*

place until the door has been hung. Stops for doors in single-piece jambs are generally 7/16" thick and may be 3/4" to 2 1/4" wide.

They are installed with mitered joints at the junction of the side and head jambs. A 45° bevel cut at the bottom of the stop, about 1" to 1 1/2" above the finished floor line, will eliminate a dirt pocket and make cleaning or refinishing the floor easier. Fig. 35-14. Review "Fitting a Door" and "Hanging a Door," Unit 27.

Door Hardware

Hardware for doors is made in a number of finishes, with brass, bronze, and nickel perhaps the most common. There are three kinds of door sets:

- Entry locks for exterior doors.
- Bathroom sets (inside lock control with safety slot for opening from the outside).
- Passage set (without lock). Fig. 35-20.

HINGES

Use two or three hinges for interior doors. Three hinges reduce the possibility of warpage and are useful on doors that lead to unheated attics or on wider and heavier doors.

Loose-pin butt hinges should be used. They must be the proper size for the door they will support. For 1 3/8" interior doors, choose 3 1/2" butts. After the door is fitted to the framed opening, with the proper clearances, hinge halves are fitted to the door. They are routed into the door edge with a back distance of about 3/16". Fig. 35-21.

LOCKS

Types of door locks differ with regard to installation, cost, and the amount of labor required to set them. Follow the installation instructions supplied with the lock set. Some types require drilling of the edge and face of the door and routing of the edge to accomodate the lock set and faceplate. A, Fig. 35-22. A more common bored type is much easier to install. It requires only one hole drilled in the edge and one in the face of the door. B, Fig. 35-22. Boring jigs and faceplate markers are available to provide accurate installation. Locks should be installed so that the doorknob is 36" to 38" above the floor line. Most sets come with paper templates

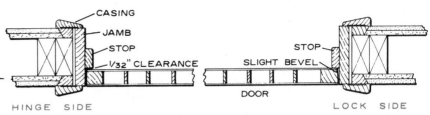

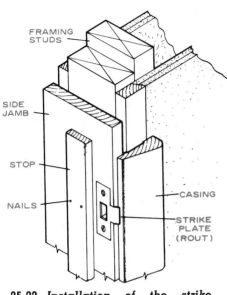

35-23. *Installation of the strike plate.*

35-24. *Doorstop installation details.*

marking the location of the lock and size of the holes to be drilled. See "Installing a Lock Set," Unit 27.

STRIKE PLATE

The strike plate, which is routed into the doorjamb, holds the door in place by contact with the latch. Fig. 35-23. When the door is latched, its face should be flush with the edge of the jamb. Review "Installing a Lock Set," Unit 27.

Doorstops

The stops which have been set temporarily during fitting of the door and installation of the hardware may now be permanently nailed in place. Finish nails or brads, 1½″ long, should be used. The stop at the lock side should be nailed first, setting it tight against the door face when the door is latched. Space the nails 16″ apart in pairs. Fig. 35-23.

The stop behind the hinge side is nailed next. A ¹⁄₃₂″ clearance from the door face should be allowed in order to prevent scraping as the door is opened. Fig. 35-24. The head-jamb stop is then

nailed in place. Remember that when door and trim are painted, some of the clearances will be taken up.

Sliding Doors

The bypass sliding door is designed for closets and storage walls. It requires no open swinging area, thus permitting a more effective and varied arrangement of furniture. No valuable floor space is lost due to door swing. Full access to the storage area is obtained by sliding the doors right or left. Fig. 35-25.

Sliding doors are usually installed in a standard doorframe. The track is mounted below the head jamb and then hidden from view with a piece of trim. Fig. 35-26. Standard size interior doors, 1⅜″ thick, 6′8″ or 7′0″ high, and any width, may be used. Most sliding door hardware will also adapt to ¾″ or 1⅛″ door thicknesses. Fig. 35-27. The door rollers are adjustable so that the door may be plumbed and aligned with the opening. Fig. 35-28. The doors are guided at the bottom by a small guide which is mounted on the floor where the doors overlap at the center of the opening. Fig. 35-29. Rough opening sizes differ slightly from one manufacturer to another. Be sure to consult the specifications provided with each particular door unit. Fig. 35-30.

Bifold doors

Bifold doors may be used to enclose a closet area, storage wall,

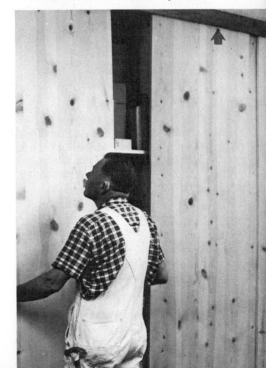

35-25. *Sliding bypass doors used on a storage wall.*

35-26. *Hanging the sliding door in the track. Note that the track is concealed behind a piece of trim mounted below the head jamb.*

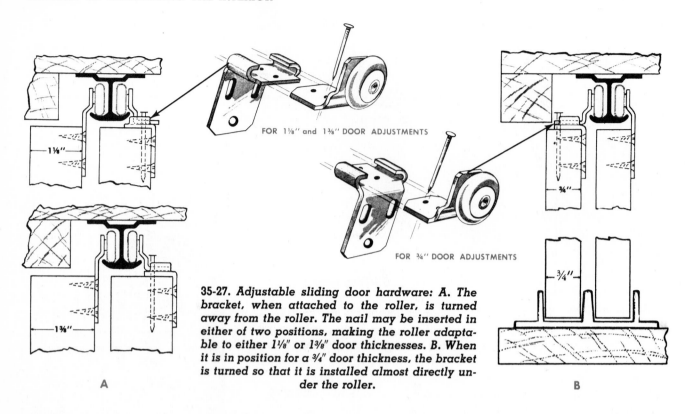

FOR 1⅛″ and 1⅜″ DOOR ADJUSTMENTS

FOR ¾″ DOOR ADJUSTMENTS

35-27. *Adjustable sliding door hardware: A. The bracket, when attached to the roller, is turned away from the roller. The nail may be inserted in either of two positions, making the roller adaptable to either 1⅛″ or 1⅜″ door thicknesses. B. When it is in position for a ¾″ door thickness, the bracket is turned so that it is installed almost directly under the roller.*

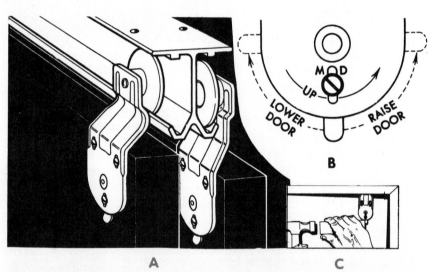

35-28. *Sliding door hardware is designed to permit aligning the doors with the opening without repositioning the hangers: A. The doors are mounted on the hangers and installed on the track. B. Details for adjusting the hangers when aligning the door. C. Tapping the adjustment with a block of wood and a hammer to make the alignment.*

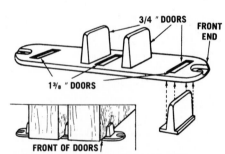

35-29. *The door guide on the floor is also adjustable. The plastic guides may be inserted in different slots of the metal plate to accomodate various door thicknesses.*

or laundry area. The doors may be wood, metal, or coated with plastic. They come in a large variety of styles to match the architecture of the home. Fig. 35-31.

The bifold unit has the advantage of opening up so that the entire opening is exposed at one time. With sliding bypass doors, the entire opening is accessible,

but only half is exposed at one time. Fig. 35-32.

The doors are available in 6′8″, 7′6″, and 8′0″ heights and in widths of 3′, 4′, 5′, and 6′. These are four-panel units and, if desired, the tracks may be cut in half and a two-panel unit installed. For example, two panels of the 3′0″ size could be used for a 1′6″ linen closet opening.

The bifold door is installed in a

Specifications

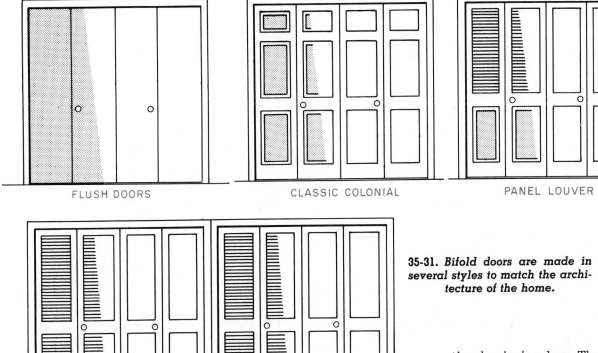

Front Opening Size	For 2 Doors	Rough Stud Opening
3'0" x 6'8"	1'6" x 6'8"	3'1½" x 7'0"
4'0" x 6'8"	2'0" x 6'8"	4'1½" x 7'0"
5'0" x 6'8"	2'6" x 6'8"	5'1½" x 7'0"
5'4" x 6'8"	2'8" x 6'8"	5'5½" x 7'0"
6'0" x 6'8"	3'0" x 6'8"	6'1½" x 7'0"

35-30. *A typical specification chart for sliding door units. Be sure to consult the manufacturer's specifications for each door to be installed.*

FLUSH DOORS

CLASSIC COLONIAL

PANEL LOUVER

FULL LOUVER

35-31. *Bifold doors are made in several styles to match the architecture of the home.*

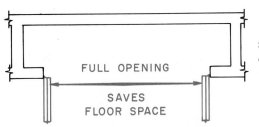

FULL OPENING

SAVES FLOOR SPACE

35-32. *A bifold door unit opens up to expose the entire opening.*

conventional doorframe. The frame may be trimmed with door casing to match the trim in the remainder of the house or, if desired, the jamb may be finished the same as the walls. Fig. 35-33. The rough opening is framed in the same way as for the conventional swinging door. The finish opening size, however, may vary with the manufacturer. Fig. 35-34.

To install a bifold door, install the top track first. Fasten the lower track to the floor, directly under the top track. Install the doors by inserting the bottom pivot into the bottom track socket. Insert the upper pivot into the top track socket. Adjust the panels to the opening by adjusting the track sockets. To make the tops of the panels even, raise or lower the panels by adjusting the lower pivot pin. Fig. 35-35.

Folding Doors

A folding door may be used as a room divider or to close off a laundry area, closet, or storage

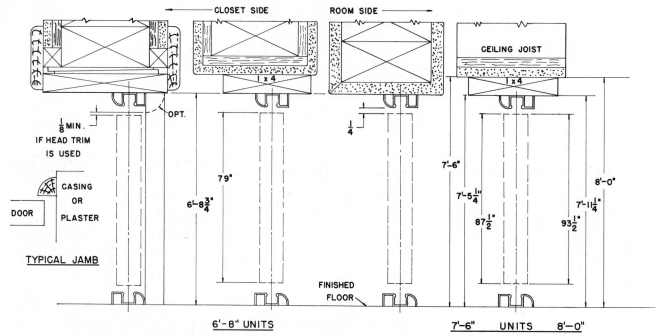

35-33. *Various header construction and trim details for a bifold door.*

Two-Panel Units*

Size	Door Width	Opening Width	Finished Heights		
			6'8" Units	7'6" Units	8' Units
1'6"	17½"	18½"	6'8¾"	7'5¼"	7'11¼"
2'0"	23½"	24½"	6'8¾"	7'5¼"	7'11¼"
2'6"	29½"	30½"	6'8¾"	7'5¼"	7'11¼"
3'0"	35½"	36½"	6'8¾"	7'5¼"	7'11¼"

Four-Panel Units

Size	Door Width	Opening Width	Finished Heights		
			6'8" Units	7'6" Units	8' Units
3'0"	35"	36"	6'8¾"	7'5¼"	7'11¼"
4'0"	47"	48"	6'8¾"	7'5¼"	7'11¼"
5'0"	59"	60"	6'8¾"	7'5¼"	7'11¼"
6'0"	71"	72"	6'8¾"	7'5¼"	7'11¼"

*Note: 2-panel units may be made on the job by cutting 4-panel unit tracks in half.

35-34. *A typical manufacturer's specifications for bifold doors.*

wall. Figs. 35-36 & 35-37. Folding doors are made from wood, reinforced vinyl, or plastic-coated wood. Wood folding doors, when closed, look like paneling. Folding doors made of a metal framework and covered with fabric or vinyl-coated materials are also available. These are called accordion-fold doors.

Folding doors are both convenient and attractive. They fold right inside their own doorway so that full advantage can be taken of every square foot of living space. The doors are hung on ny-lon rollers that glide smoothly in an aluminum track. The track at the top is concealed with beveled matching wood molding installed on each side. With such molding there is no exposed hardware to detract from the beauty of the door. Fig. 35-36.

Folding doors come from the factory already assembled. They are shipped complete in a package containing the door, hardware, latch fittings, and installation instructions. Standard or stock doors are available 6'8" high and 2'4", 2'8", 3', or 4' wide.

The folding door is installed in a standard doorframe. It may be trimmed in a conventional manner, or plaster jambs may be used. Fig. 35-38. A plaster channel is also available. This channel is installed before plastering, and the track is mounted after the plaster is applied, eliminating the need for the wood head molding normally used. Fig. 35-39. Folding doors may also be installed to fit into a wall cavity when

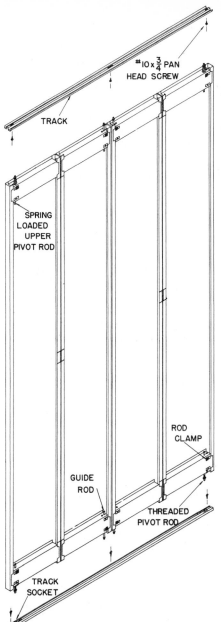

#10 x ¾ PAN HEAD SCREW

TRACK

SPRING LOADED UPPER PIVOT ROD

ROD CLAMP

GUIDE ROD

THREADED PIVOT ROD

TRACK SOCKET

35-35. Installation details for a bi-fold door unit. The track at the top is hidden by the matching wood molding.

35-36. Wood folding doors installed on a closet.

35-37. Wood folding doors used in an apartment to divide a large room.

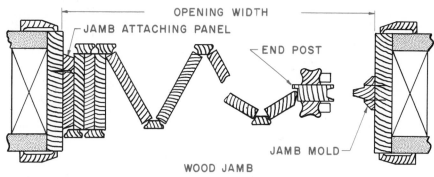

OPENING WIDTH

JAMB ATTACHING PANEL

END POST

JAMB MOLD

WOOD JAMB

OPENING WIDTH

JAMB ATTACHING PANEL

END POST

JAMB MOLD

PLASTER JAMB

35-38. Jamb sections showing the door installed with a wood jamb or a plaster jamb.

443

SURFACE MOUNTED ON WOOD

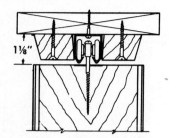

SURFACE MOUNTED ON PLASTER

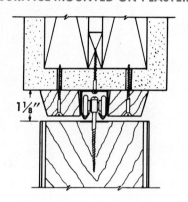

RECESSED IN PLASTER

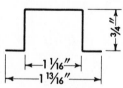

PLASTER CHANNEL

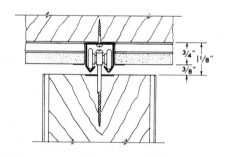

TRACK

35-39. *Head sections of folding doors showing the track installation details for a channel mounted on wood, plaster, or recessed in plaster.*

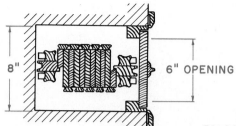

35-40. *A. Hinged recessed door. The door can be stacked into the recessed area and the panel closed to cover the end of the door. B. Sliding recessed panel. When the door is in use, the panel travels forward and covers the opening of the recessed area.*

FOLDED AWAY

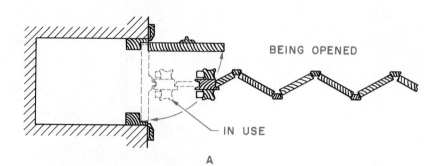

BEING OPENED

IN USE

A

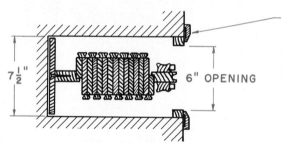

NOTE: INSTALL STOPS AND TRIM AFTER SLIDING RECESS PANEL IS IN PLACE.

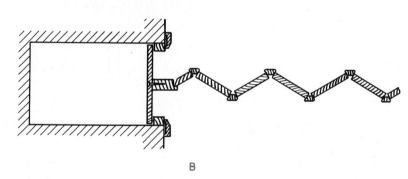

B

opened. The cavity must be at least 7½″ wide. Fig. 35-40.

Sliding Pocket Doors

The pocket type sliding door, in which the door slides into the wall, is often installed in places where the door is seldom closed. When the door stands open, it is out of the way. It is also convenient where there is minimum space for clearance of a swinging door. Fig. 35-41.

The unit can be bought complete with hardware, door, and trim. Universal sliding door hardware that will fit all door sizes can also be purchased. Fig. 35-42. Standard widths are 2′0″, 2′4″, 2′6″, 2′8″, and 3′0″. Any style of door with a thickness of 1⅜″ can be installed in the pocket to

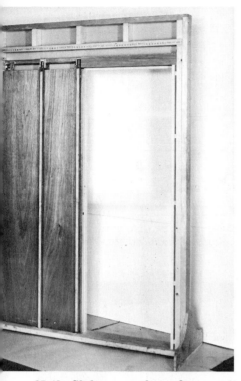

35-41. *Sliding pocket door unit without wall covering.*

35-42. *Metal-reinforced universal sliding door hardware. The hangers can be adjusted to align the door to the opening after the trim is installed.*

match the style of the other doors in the home.

When the opening for the door pocket unit is roughed in, the manufacturer should be consulted for specifications. The rough opening is usually 6'11½" or 7' high and twice the door width plus 2" or 2½". The wall header above the pocket must be adequate to support any weight on the wall and the structural frame of the building so that there is no weight on the pocket. Fig. 35-43.

The sliding door pocket frame comes complete and ready for installation into the rough wall opening. Install the pocket in the opening with 8d finish nails through shingle wedges at the head and the side jambs to level and plumb the unit. Fig. 35-44. After the wall covering has been applied, hang the door and install the stops. Fig. 35-45. Special care should be taken to use the correct

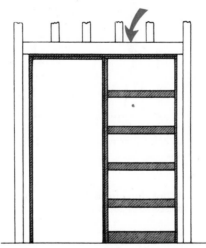

35-43. *A sliding pocket door frame installed in the rough opening. The wall header (arrow) must span the entire opening, which includes the pocket and the doorway.*

nails when applying the wall covering. If the nails are too long and project through the frame

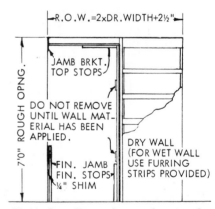

35-44. *Installing the pocket in the rough opening. Use shingle wedges at the head and side jambs to level and plumb the unit. Care should be taken to keep the jambs straight.*

slats into the opening, they may either scratch the surface of the door when it is slid into the pocket or prohibit it from entering the pocket. The same care in nail selection should be taken

445

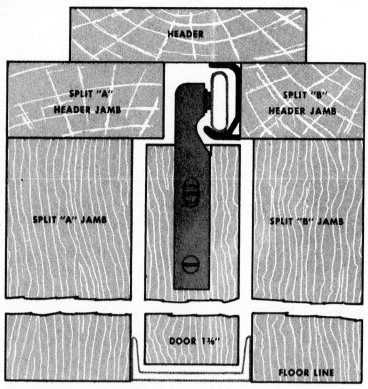

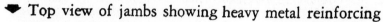

▲ End view of sliding door hardware assembly

▼ Top view of jambs showing heavy metal reinforcing

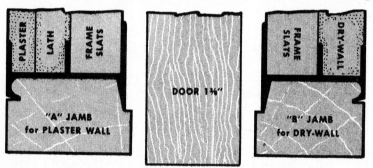

35-45. *Sliding door pocket frames are available for wet wall ("A" jamb) and dry wall ("B" jamb).*

35-46. *The hinges on a cafe door will permit the door to stand in the open position or freely return to the closed position.*

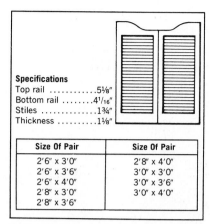

Specifications
Top rail ....5⅛″
Bottom rail ....4¹/₁₆″
Stiles ....1¾″
Thickness ....1⅛″

Size Of Pair	Size Of Pair
2′6″ x 3′0″	2′8″ x 4′0″
2′6″ x 3′6″	3′0″ x 3′0″
2′6″ x 4′0″	3′0″ x 3′6″
2′8″ x 3′0″	3′0″ x 4′0″
2′8″ x 3′6″	

35-47. *One manufacturer's specifications of standard cafe door sizes.*

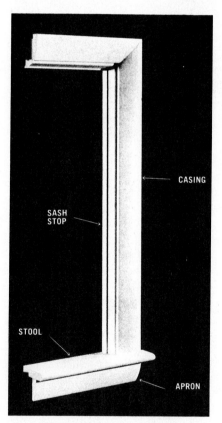

35-48. *Window trim parts.*

when installing the door casing and base molding.

Cafe Doors

The cafe door will add charm to the home and is adaptable to many uses. It may be installed in the kitchen, dining area, family room, or recreation room. Cafe doors swing from either direction and always freely return to the closed position. The hinges also permit the doors to be left open. Fig. 35-46. These doors are normally hung in pairs and come in a large variety of sizes. Fig. 35-47. They are installed with the door tops slightly below eye level to enable someone using the door to see anyone coming from the other side.

Window Trim

Casing for window trim should be of the same pattern as that selected for door casing. Other trim parts consist of sash stops, the stool, and the apron. Fig. 35-48. The stool is the horizontal trim member that laps the window sill and extends beyond the casing. The apron serves as a finish member below the stool. There are two common methods of installing wood window trim:

- With a stool and apron. Fig. 35-49.
- With complete casing trim. Fig. 35-50. Metal casing is sometimes used in place of wood trim around the window opening. Fig. 35-51.

WINDOW STOOL INSTALLATION

The stool is normally the first piece of window trim to be installed. It is notched out between the jambs so that the forward edge contacts the lower sash rail. Fig. 35-52.

The upper part of Fig. 35-52 shows a plan view of a stool in place. The lower view shows the stool laid out and cut, ready for installation. The distance A, the overall length of the stool, is equal to: the width of the finished opening, plus twice the width of the side casing, plus twice the amount that each end of the stool extends beyond the outer edge of the side casing. Distance B is equal to the width of the finished opening.

Distance C is equal to the horizontal distance measured along the face of the side jamb between

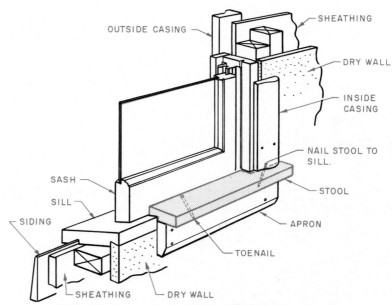

35-49. Window trim installed with a stool and apron.

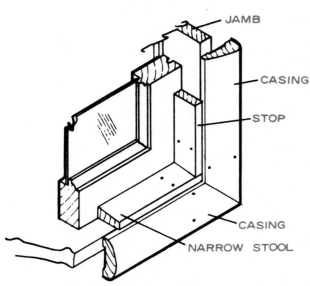

35-50. A window trimmed with casing at the bottom instead of a stool and apron.

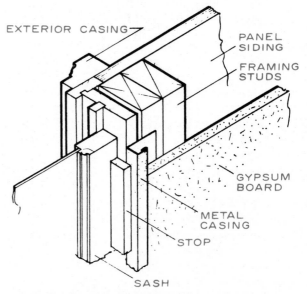

35-51. A window installed with metal casing.

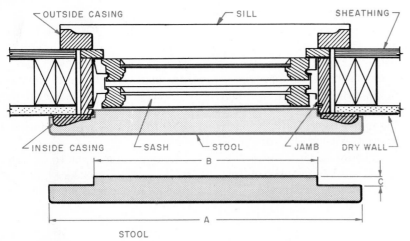

35-52. *Installation details for a window stool.*

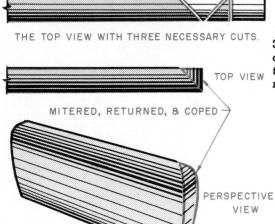

35-53. *The ends of the window apron should be coped to match the profile of the molding or mitered with a return nailed in place.*

WINDOW CASING INSTALLATION

The casing is applied after the stool is installed. It is nailed as described for the doorframes (Fig. 35-14) except that on some window types the inner edge is flush with the inner face of the jambs so that the stop covers the joint. Fig. 35-49. When stops are to be installed, they are fitted like the interior door stops and placed against the lower sash so that the sash can slide freely. A 4d casing nail or 1½" brad should be used. Place nails in pairs spaced about 12" apart. When the window is made with channel type weather stripping, it includes full-width metal subjambs in which the upper and lower sash slide, replacing the parting strip. Then, the stops are located against the subjambs instead of the sash to provide a small amount of pressure.

When metal casing is used as trim around window openings, it is applied to the sill as well as at the sides and head of the frame. Consequently, the jambs and sill of the frame are not as deep as when wood casing is used. The stops are also narrower by the thickness of the wall covering. The metal casing is installed flush with the inside edge of the window jamb. Fig. 35-51. This type of trim is installed at the same time as the wall covering.

WINDOW APRON INSTALLATION

Cut the apron to a length equal to the outer width of the casing line. The ends of the apron should be cut with a coping saw to appear as though they had been returned, or a return may be cut and nailed in place. Fig. 35-53. The apron is attached to the framing sill below with 8d finish nails. Fig. 35-49.

When casing is used instead of a stool and apron to finish the

the inside edge of the side jamb and the inside face of the lower sash. An allowance of about 1/32" should be deducted for clearance between the sash and the stool. Lay out this width from the outside edge of the stool. If the stool is held in place (centered) against the inside edge of the window jamb, the distance between the front edge of the stool and the lower sash, dimension C, can be set on a scriber. The distance along the wall can then be scribed

onto the top of the stool. Cut out the notch at each corner of the stool on the layout lines.

The stool is blind-nailed at the ends with 8d finish nails so that the casing at the sides will cover the nailheads. With hardwood, predrilling is usually required to prevent splitting. The stool should also be nailed at the center to the sill, and to the apron when it is installed. Toenailing may be substituted for face-nailing to the sill. Fig. 35-49.

BATHROOM

KITCHEN

BEDROOM

ARCHED WINDOW

LIVING ROOM

BAR

35-54. *Louvered shutters may be used throughout the home. They allow light and air to pass through, yet assure complete privacy and freedom from drafts.*

bottom of the window frame, a narrow stool or stop may be needed on some window types. Miter the side casings at the bottom corners and apply the bottom casing in the same way as the side and head casings of the window. Fig. 35-50. When this method is used to trim a window, the casings can be cut to finished length with miters on each end. The four pieces of mitered casing are then laid face down on a clean, smooth surface and fastened together with $\frac{1}{4}'' \times 5''$ corrugated fasteners from the back. The assembled casing, much like a picture frame, is then nailed as a unit to the window jambs and studs as described earlier.

Other types of windows, such as the awning, hopper, or casement, are trimmed about the same as the double-hung window. Casings of the same types are used for trimming these units.

Interior Shutters

Movable interior shutters were popular in the great mansions of New Orleans and in many of the finer homes of America from 1700 to the early part of the 19th century. The use of shutters has once again become popular. Fig. 35-54. They are found in the traditional interior, the provincial (country style) setting, and in the modern home, studio, or office. Shutters offer better control of light, air,

visibility, and privacy than most other types of window treatment. They also make ideal foldaway

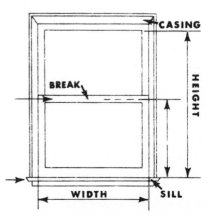

35-55. *Measuring a window for shutters.*

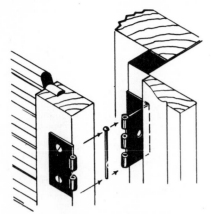

35-56. *A hinge strip installed on a double-hung window.*

35-58. *The hinge is sometimes reversed on a flat casing so that the shutter will not cover the casing.*

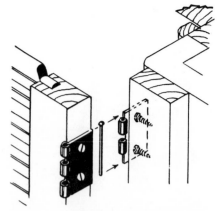

35-60. *A wood hinge strip is applied to the wall if the window jamb is not wood.*

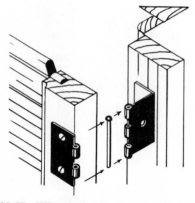

35-57. *When the window casings are flat, the hinge may be mounted directly on the casing.*

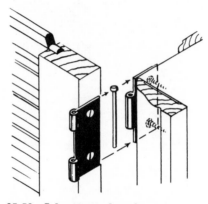

35-59. *A loose-pin butt hinge is used on a wide jamb.*

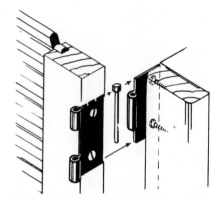

35-61. *On a thick casing, the hinge may be applied directly to the edge.*

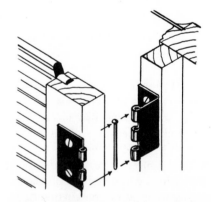

35-62. *Louvers applied to French doors or casement windows must be furred out to provide sufficient operating clearance for hardware.*

partitions or room separators that distinguish one living area from another.

To determine the size of the louver to be installed in a window, measure the width between the inside edges of the jamb or casing, depending on where you wish to hinge the shutters. Measure the height of the opening from the top of the sill to the inside edge of the top jamb or casing. If a cross rail is desired, it will be located at the approximate center of the shutter, referred to as the break. Fig. 35-55.

INSTALLING MOVABLE INTERIOR SHUTTERS

When installing movable shutters on a double-hung window, a hinge strip replaces the sash stop. Fig. 35-56. The strip should be ¾" wide and have a depth equal to the distance from the front of the jamb to the sash. Mount the shutter to these strips with ¾" flush door hinges.

On casings that are reasonably flat, the shutter may be applied directly with a ¾" flush door hinge. The hinge should be set no more than 1" back from the inside edge of the casing. Fig. 35-57.

Another method of hanging shutters to casings that are reasonably flat is shown in Fig. 35-58. In this installation the hinges are applied in reverse so that the shutter will not cover the casing.

When there are no obstructions in back of the shutter, loose-pin butt hinges may be applied directly to the jamb. Fig. 35-59.

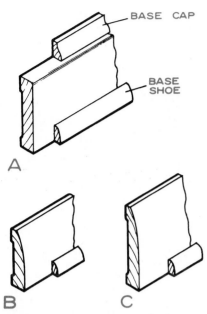

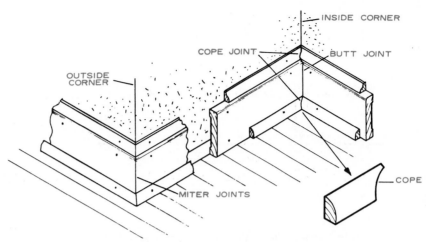

35-64. *Base molding installation details.*

35-63. *Base moldings: A. Two-piece baseboard with a square-edged base and a base cap. B. Narrow ranch base. C. Wide ranch base.*

35-65. *When several pieces of molding are needed to provide sufficient length, they should be joined with a lap miter joint on a wall stud. Use two 8d finish nails. The bottom nail should be close enough to the floorline to be covered by the base shoe molding.*

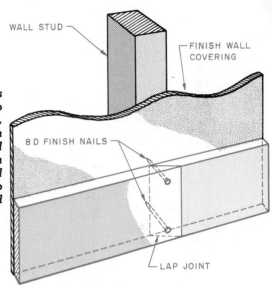

If the window jamb is other than wood and the wall is not cased, or if the casing has been removed, apply a hanging strip to the wall and use either loose-pin butt hinges or ¾″ flush door hinges. Fig. 35-60.

When a casing is at least ¾″ thick, the shutter may be applied directly to the casing with either type hinge. Fig. 35-61.

On French doors or casement windows, install a framing strip, ¾″ wide × 1″ deep, to the sash. Fig. 35-62. The shutter may then be installed with ¾″ flush door hinges. Shutter and frame should overlap each side and top and bottom outside the glass area by ¾″ to leave room for the hinges and operating clearance for louvers.

BASE, CEILING, AND WALL MOLDING

Installing Base Molding

Base molding is usually about the last trim to be installed. It must be installed after all the doors are trimmed and the cabinets are in place. Also, it usually butts wall openings such as warm and cold air registers. The base molding serves as a finish between the finished wall and floor. It is made in a number of sizes and shapes. Fig. 35-3.

Base molding may have several parts. Two-piece base consists of a baseboard topped with a small base cap. A, Fig. 35-63. When the wall covering is not straight and true, the small base cap molding will conform more closely to the variations than will a wider base alone. A common size for two-piece base is ⅝″ × 3¼″ or wider. One-piece base varies in size from ⁷⁄₁₆″ × 2¼″ to ½″ × 3¼″ and wider.

Most baseboards are finished with a base shoe, ½″ × ¾″ in size. B and C, Fig. 35-63. Base molding without the shoe is sometimes placed at the wall-floor junction, especially when carpeting is installed. Although a baseboard is desirable at the wall-floor junction to serve as a protective "bumper," wood trim is sometimes eliminated entirely.

Square-edged baseboards should be installed with a butt

451

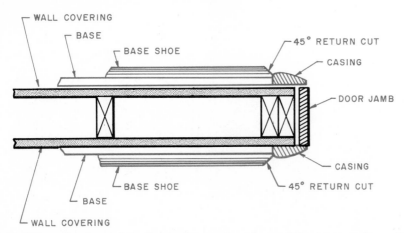

35-66. *When the face of the base shoe projects beyond the face of the molding which it abuts, a 45° return cut should be made on the base shoe.*

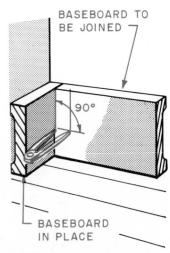

35-68. *When an inside corner is butt-jointed, it must be scribed to insure a good tight joint.*

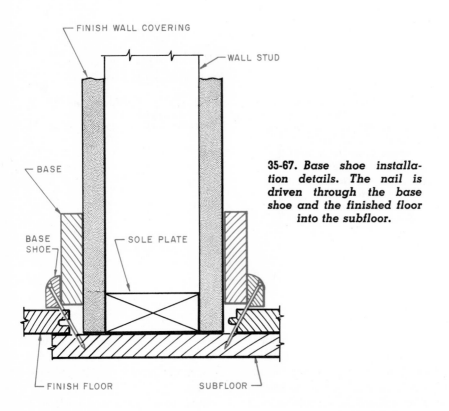

35-67. *Base shoe installation details. The nail is driven through the base shoe and the finished floor into the subfloor.*

molding to which it butts, the end of the base shoe should be returned onto itself. The return is cut at 45° on the end of the shoe molding. This return will eliminate dirt pockets and give a better appearance. Fig. 35-66. The base shoe should be nailed into the subfloor and not into the baseboard itself. Thus, if there is a small amount of joint shrinkage, no opening will occur under the shoe. Fig. 35-67.

SCRIBING A BUTT JOINT

To butt join a piece of square-edged baseboard to another piece already in place at an inside corner, set the piece to be joined in position on the floor. Bring the end against or near the face of the piece already installed and scribe a line parallel to the installed piece on the face of the piece to be joined. Fig. 35-68. Be careful to hold the legs of the scriber at right angles to the reference surface (the baseboard which has already been installed). This will insure a parallel line. Follow the same procedure when putting ends of moldings against the side of door casings or wall registers.

joint at inside corners and a miter joint at outside coners. Molded baseboards are also mitered at outside corners but they are coped at inside corners. Fig. 35-64. These methods of joining are necessary to provide tight joints because the walls at corner baseboard locations may not be perfectly vertical. When cutting molding to fit between walls, always cut it a little long so that the molding can be bowed slightly and sprung into place. When it is necessary to use more than one length of molding along a wall, join the pieces on a wall stud with a lap miter joint. Fig. 35-65. The baseboard is secured to each stud with two 8d finishing nails.

When the face of the base shoe projects beyond the face of the

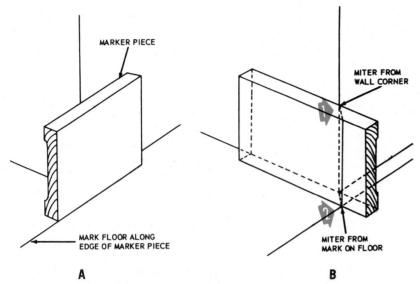

35-69. *Laying out a miter joint at an outside corner.*

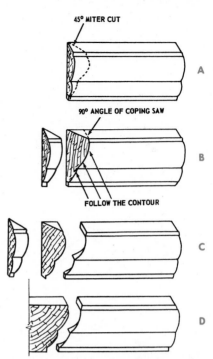

35-70. Coping a joint: A. Make a 45° miter cut. B. Set the coping saw at 90° to the back edge. C. Make the cut, following the contour line created by the 45° cut. D. The coped end of the molding will fit tightly against the face of the other member.

MITERING A JOINT

When miter joining an outside corner, set a marker piece of baseboard across the wall corner. A, Fig. 35-69. Mark a line on the floor along the edge of the piece. Set the piece to be mitered in place. Mark the place where the wall corner intersects the top edge and the point where the mark on the floor intersects the bottom edge. B, Fig. 35-69. Draw a 45° line across the top edge from the point marked at arrow 1 and draw a 45° line across the bottom edge from the point shown at arrow 2 in Fig. 35-69. Connect these lines with a line drawn across the face of the board and cut along the line.

COPING A JOINT

Inside corner joints between molding trim members are usually made by cutting the end of one member to fit against the face of the other. Shaping the end of the butting member to fit the face of the other member is called coping. Fig. 35-64. To cope a molding, miter the end at 45° the same way as if the molding were to have a plain mitered inside corner joint. A, Fig. 35-70. Then set the coping saw on the line at the top of the miter cut. Hold the saw at 90° to the back of the molding and saw along the face contour line created by the 45° miter cut. B and C, Fig. 35-70. The end profile of the coped member will match the face of the other member. D, Fig. 35-70. The result is a good, tight joint. It will not open up when the molding is nailed in place, and it is not likely to open up as the wood shrinks after installation.

BASE MOLDING INSTALLATION SEQUENCE

When applying base moldings in a room, the installation sequence should be carefully planned before starting the job to save time in making necessary and difficult cuts. The drawing in Fig. 35-71 shows the outline of a room with one door. Here is one suggested sequence for installing the molding: Cut and fit a piece of molding to go along the wall marked 1, scribing each end to fit walls 2 and 3. Install molding on the walls marked 2 and 3. The molding for these walls should have one end coped to fit the molding on wall 1. Cope the joint

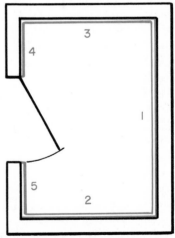

35-71. *An outline of a room with one door. Can you figure the best sequence for cutting and installing base molding in this room?*

first; then measure and mark the piece and cut it to finish length. Cope one end of the moldings for

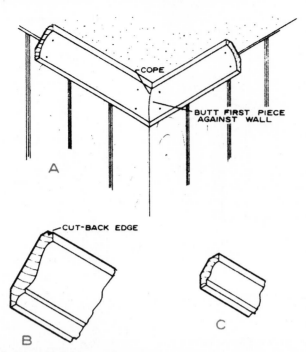

35-72. *Ceiling moldings: A. Installation details at an inside corner. B. Crown molding with a cut-back edge. C. A small crown molding.*

walls 4 and 5 to fit against the face of the moldings for walls 2 and 3. Cut the other end to fit against the door casing.

Another approach to trimming this same room (Fig. 35-71) is to begin at the right and work around the room in a counterclockwise direction. The first piece is cut to fit along wall 5 between the door casing and the end wall. The piece of molding on wall 2 is coped on one end to fit against the molding on wall 5. It is then measured, marked, and cut to length so that the other end fits against wall 1. The first end of wall molding 1 is coped to fit against piece 2, then measured, marked, and cut to length. This procedure is continued for walls 3 and 4.

A right-handed person will find in working around the room counterclockwise that the coping of the molding will be much easier. This is true for most molding shapes because the sawing is

started at the top of the molding, where it is lighter and where most moldings are shaped and have a narrow edge. This narrow edge is weak and, if the cut is started at the bottom, the molding may break as the cut is finished.

Ceiling Molding

Moldings are sometimes used at the junction of wall and ceiling for architectural effect or to terminate dry wall or wood paneling. Fig. 35-72. A cut-back edge at the top of the molding will partially conceal any unevenness of the plaster and make painting easier when molding and ceiling are different colors. B, Fig. 35-72. For gypsum dry wall construction a small simple molding might be preferred. C, Fig. 35-72.

Cut and fit ceiling molding in the same way as described for base moldings to insure tight joints and retain a good fit if there are minor moisture

changes. To secure the molding, a finish nail should be driven through the molding and into the upper wall plates. For large moldings, if possible, also drive a nail through the molding into each ceiling joist.

Wall Molding

Wall molding, sometimes called dado molding or chair rail, consists of strips of molding which run along the walls at 3' to 4' above the floor. This type of molding serves to protect the finished wall from the backs of chairs. It may also serve as trim between two different wall finishes. For example, a wall may be painted below the molding and wallpapered above. Regular casing, base, band, and cap moldings may be used as wall moldings, allowing a wide range of choices with an almost unlimited selection of thicknesses and widths. Wall moldings are installed in the same manner as base moldings.

Applying Molding to a Flush Door

The style of a flush door can be varied by the application of different moldings. Fig. 35-73. A Spanish style door is shown in Part 1 of Fig. 35-73. This door is trimmed with two molding patterns used in conjunction with a screen stock.

Other moldings applied to a flush door can give it a contemporary or a traditional style. Parts 2 and 3, Fig. 35-73. Silhouettes of the various molding styles and their pattern numbers are shown in Fig. 35-74. Review the sections on cutting and coping moldings in this unit before cutting the moldings to size for the applications shown here or applications of your own design.

Trimming a Clothes Closet

Base and shoe moldings of the

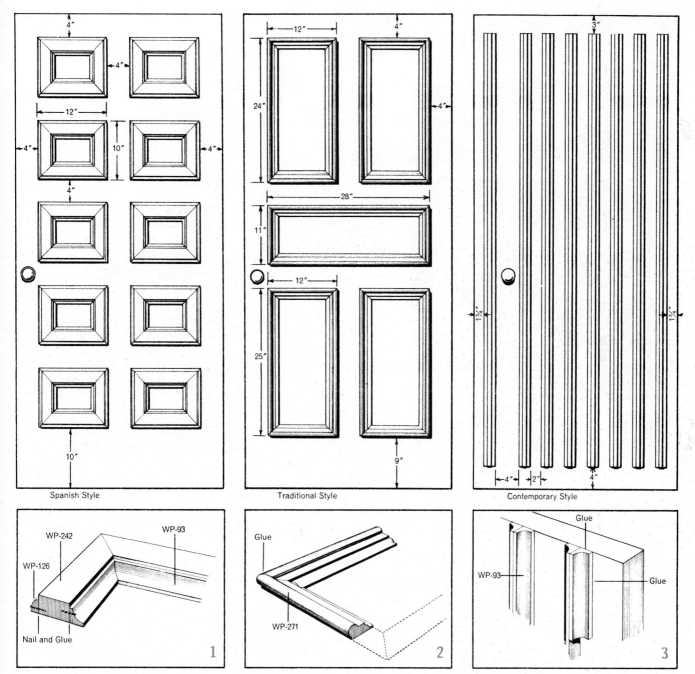

35-73. *Details for applying molding to a flush door: 1. Spanish style. 2. Traditional. 3. Contemporary.*

same pattern used in the adjoining room are installed at the floor-line of clothes closets. These moldings are cut, fit, and installed in the same manner as the base moldings in the room.

A piece of wood molding or a piece of 1″ × 3″ clear stock may be used as a hook strip and a shelf support in which the closet pole is installed. Many times a piece of wood base turned upside down (with the square edge up) is used as a hook strip. Locate the height of the hook strip, usually about 66″ above the finished floor line. Fig. 35-75. Level a line at this point on each wall of the closet. Cut and fit the molding in the same manner as the base molding.

Lay out and bore holes for the clothes pole. On one side, make a cut down from the top to remove

455

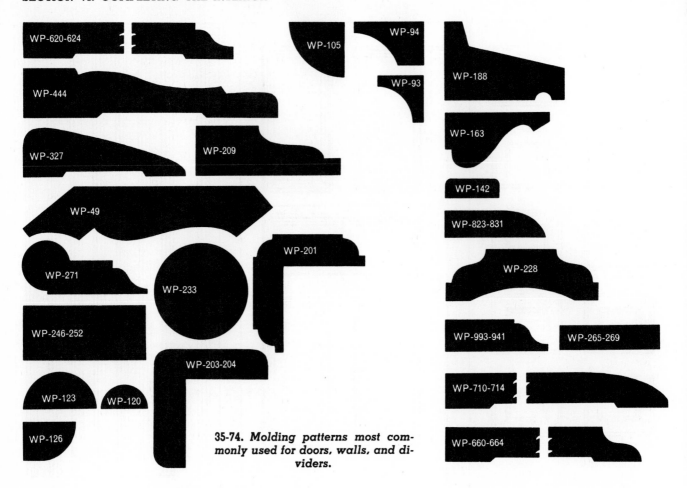

35-74. *Molding patterns most commonly used for doors, walls, and dividers.*

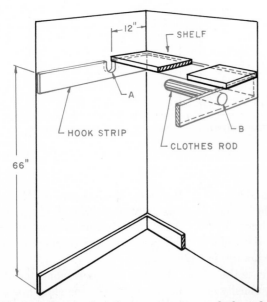

35-75. *Installation details for trimming a clothes closet.*

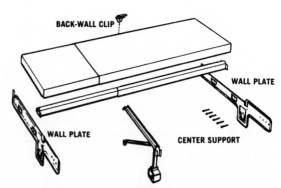

35-76. *Prefinished adjustable shelving, wall plates, brackets, and clothes rods made from metal are available for closet installations.*

the stock above the hole. Point A, Fig. 35-75. This will allow the pole to be inserted into the other hole and then slipped into the

35-77. *A metal clothes hanger bar which attaches to the front edge of the closet shelf.*

slot. Install the molding as you would base molding.

Cut the shelf to length and set it on top of the hook strip. The shelf is usually not nailed in order to make cleaning and painting easier.

Metal adjustable shelving is available prefinished in a variety of colors and wood grains. Also available are wall plates and brackets for mounting and a rod which adjusts to various lengths and fits into the wall plates. Fig. 35-76. Another accessory for clothes closets is a clothes hanger bar which attaches to the front edge of a wood shelf. This unit, which will reinforce and trim the front edge of the shelf, comes in varying lengths up to 14'. Fig. 35-77.

QUESTIONS

1. What are the most common uses of wood or plastic moldings in the home?

2. What work must be completed before trimming interior doors and windows in a home?

3. What is the main advantage of a two- or three-piece doorjamb?

4. What is the standard thickness of interior doors?

5. What is the standard height of an interior door?

6. What is the advantage of a bifold door over a sliding door for a closet installation?

7. What are the two common methods of installing window trim?

8. What is the first piece of window trim to be installed?

9. How is the length of the window apron determined?

10. Why is base molding about the last trim to be installed?

11. What is the standard height for a clothes closet shelf?

SECTION VII

Methods for Conserving Energy

Energy Conservation 36

ENERGY USE IN THE HOME

About one-fifth of our total national use of energy is for climate conditioning of houses, apartments, condominiums, mobile homes, and other living units. Energy is needed for heating, air conditioning, humidifying, dehumidifying, and air cleaning. As the cost of energy rises and the supply diminishes, it becomes increasingly important to make all types of living units more energy efficient. Anyone planning to build or remodel a living unit must consider every possible way of reducing energy needs. Also, this must be done without adding too much to the cost of construction. There are energy-saving materials and devices available that are not too expensive and save enough energy to justify their cost.

Energy is used in daily living in a wide variety of ways. Precious energy is consumed in almost everything we do, such as cooking a meal, switching on a light, watching television, taking a shower, or washing clothes. When the weather turns cold or hot, the wind blows, or we have rain, energy is consumed to compensate for these conditions.

Of the many factors that influence internal temperature, the most important is outside weather. If it is cold outside, heat is lost through the living unit in many ways:

• By *exfiltration*, the conduction of heat through the exterior surfaces and leakage of warm air to the outside through cracks in windows, walls, and doors.
• By *infiltration*, the leakage of cold outside air into the house.
• By *radiation*, in which warmth is lost to the sky and the surroundings. (A warm body tends to lose its heat by infra-red radiation.)
• By *heat loss through water drains*. Energy is used to heat water. As the hot water goes down the drain, this energy is lost from the home. Fig. 36-1.

It has been estimated that as much as 40% of heat loss is due to exfiltration and infiltration.

Heat is generated and gained in living quarters by appliances, electric lights, and people. Some heat is also gained through solar energy, even in the standard home. In cold weather, the difference between heat loss and heat gain must be made up by some type of heating unit to maintain a comfortable temperature range of 68° to 78° F. (20 to 26°C). In addition, energy is used to clean the air and to add humidity for comfort.

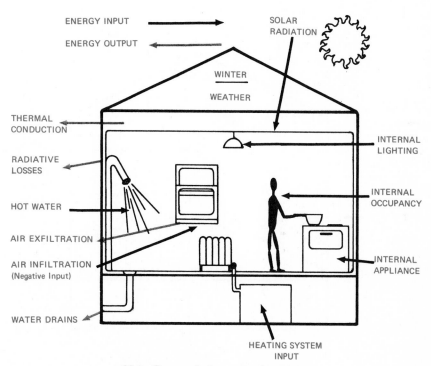

36-1. *Energy balance in the winter.*

In warm weather, many of the same factors are at work. When the outside temperature is greater than the indoor temperature, heat tends to flow into the house (infiltration). To maintain comfort, an air conditioner is often used to remove this excess heat. Excess moisture is removed from the air by a dehumidifier. Fig. 36-2.

There are three ways construction can aid in energy conservation:

1. Standard homes can be made more energy efficient with such things as proper insulation, energy-saving sheathing, storm doors and windows, proper calking, and a light color of roofing.

2. Energy-saving (conservation) homes can be designed using different materials and methods to increase energy conservation.

3. Solar energy can be used for heating water and/or for a large part of the space heating and cooling. However, even homes planned for solar energy must also have some kind of supplementary heating and cooling units.

In planning any structure to include energy-saving devices, the following must be given careful consideration:

• Will the device or material save sufficient energy to pay for itself within a ten- to fifteen-year period? At current costs, active solar energy units cannot meet this requirement.

• Will the structure be within current construction capabilities and local building codes?

• Will the structure be acceptable from an architectural standpoint? For example, will tiny windows sparsely placed be acceptable?

• Will the design drastically change the lifestyle of the residents? In the years ahead, most living units must be smaller and more space efficient. For example, it may not be possible to have both a living room and a family room. Many current house plans call for a *great room* as an all-purpose living area. Today, fewer homes have separate dining rooms, and bedrooms are smaller.

BUILDING MATERIALS

Plastics

The use of plastics in building construction is increasing rapidly. It is estimated that by the twenty-first century, plastics may be the major building material, surpassing even wood. Much of this increase is due to demand for energy-efficient materials in insulation, sheathing, and siding. Plastic materials offer characteristics and properties different from many natural materials. Some advantages of plastics are:

• Light weight.
• Good electrical insulating properties.
• Good heat insulating properties.
• Resistance to atmospheric corrosion.
• Attractive appearance.
• Lower cost.
• Variety in design and styling.

Plastics may be defined as non-metalic materials that are capable of being formed or molded with the aid of heat, pressure, chemical reaction, or a combination of these. There are many different types of plastics used in building construction. The following nine types are the most important:

PVC (polyvinyl chloride) is by far the most important plastic material for building construction. It is made in both rigid and flexible form. PVC is used widely for siding, sash, roofing, and interior trim. It can be used for all types of wiring insulation and in home plumbing systems for cold water lines and for sanitation.

ABS (acrylonitrile-butadiene-styrene) is a tough, colorful material that has many uses in building construction for drain-waste-ventilating pipe and fittings. Fig. 36-3. Solid ABS is used for sewer

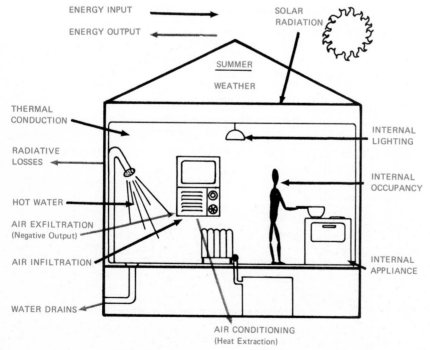

36-2. *Energy balance in the summer.*

36-3. *Pipe for drainage, waste, or ventilation can be made of ABS or PVC plastics.*

pipe, conduit, and air and water pressure pipe. It can be chrome-plated for such products as faucet handles and soap dispensers. ABS can be formed by heat and pressure (thermoformed) into bath and shower units. There is also an ABS foam-core pipe that is somewhat less expensive than solid ABS pipe and which has wide usage in sewer, conduit, and duct work.

Acrylic is popular for glass replacement since it will not break. Acrylic is thermoplastic, which means that it can be easily formed for such items as skylights, covers for basement windows, and lighting fixtures. Acrylics make excellent sealants and calks that adhere to most construction materials, such as metal, glass, ceramics, plastics, wood, masonry, and concrete.

Rigid plastic foams play an important part in energy conserva-tion. The materials are superior insulators against heat loss and are being used more and more for sheathing. However, plastic foams have no structural strength, and great care must be taken in installing them. The most common kinds of rigid plastic foams are Styrofoam, polysty-rene, urea-formaldehyde, and polyurethane. Another foam that is used for sheathing is isocyanur-ate.

High-pressure laminates, such as the decorative laminates, are important interior design materi-als. They are commonly used for counter tops, walls, and many other interior surfaces.

Fiber-reinforced plastics are made primarily of spun glass and polyester. These materials are widely used for bathtubs, show-ers, and vanity units. In fact, fi-ber-reinforced plastic bathtub and shower units account for a major-ity of the market, surpassing steel and ceramic units by a wide margin.

Epoxy resins are used in var-ious ways in construction, such as for finishings, for assembling parts, and for adhesives. Epoxies can be used in assembling units of wood, steel, or concrete.

Acetal plastics have high im-pact strength and chemical resis-tance and have been widely used in furniture hardware for such pieces as drawers, swivel compo-nents in chairs, and casters. Two types of acetal are now extremely important in the plumbing indus-try. Acetal homopolymer is re-placing copper and brass for plumbing parts. Acetal copolymer is used for plumbing fittings for-merly made of brass. Its uses are endless for such products as fau-cets, faucet underbodies, lavatory basins, valves, fittings, shower heads, faucet sprayers, pipe cou-plings, and many other items.

Polycarbonate resins and sheets can be used in place of glass to make nonbreakable win-dows. Extruded polycarbonates are used for window and door frame components.

Adhesives, Sealants, and Calks

Many different types of adhe-sives are needed in the manufac-ture of building products, particu-larly such materials as plywood, hardboard, and particleboard. On the job, the builder uses several different kinds, the primary ones being contact adhesives (cement), elastomeric construction adhe-sives (mastics), and calks. Other types of adhesives are used by tile setters and floor covering spe-cialists.

Contact adhesives, or cements, are used in installing plastic lami-nates for counter tops and walls. Elastomeric construction adhe-sives (mastics) have the consis-

Sheathing Materials

	3/4" foil-faced kraft-fiber ply	1" polystyrene (beadboard)	1" Styrofoam	3/4" foil-faced urethane	1/2" plywood	1/2" fiberboard (standard grade)	1/2" gypsum
Can it be overlapped to stop air infiltration?	yes	no	no	no	no	no	no
Can it be used without corner bracing?	yes	no	no	no	yes	no	yes
Is it damage resistant, not vulnerable to vandalism, chipping of edges, breakage, or puncture?	yes	no	no	no	yes	no	no
Can it increase R-value of wall beyond traditional 1/2" wood fiber?	yes	yes	yes	yes	no	no	no
Lightweight, easy to lift?	yes	yes	yes	yes	no	no	no
Cuts easily with a knife?	yes	yes	yes	yes	no	no	no
Provides strong base for siding application?	yes	no	no	no	yes	yes	yes
Low burn rate?	yes	no	no	no	yes	no	yes
Requires no jamb modification for traditional windows or door sizes?	no	no	no	no	yes	yes	yes
Adds structural strength beyond traditional 1/2" wood fiber?	yes	no	no	no	yes	no	yes
Does not require any special interior wall materials such as drywall for safe installation?	yes	no	no	no	yes	yes	yes

36-4. *Advantages and disadvantages of common sheathing materials.*

36-5. *Styrofoam applied horizontally as sheathing.*

tency of putty and are used for fastening plywood, hardboard, and particleboard to floor joists, ceiling joists, and studs. Mastics usually come in cylindrical containers that fit into calking guns or in cans. They are also available as ribbons of material that can be pressed into place. Mastics have good gap-filling properties and may be used with or without supplementary nailing or stapling. There are many kinds of mastics, some synthetic and others of natural rubber. Most of the modern mastics are of plastic origin, such as acrylic products.

Many different types of calks and sealants are available. Polysulfide, polyurethane, and silicone types are preferred. Calking may be semi- or slow-drying. Most calking comes in a sealed cartridge for use in a calking gun. Calking is also available in a can for use with a putty knife.

It is important that the builder select the correct kind of adhesive and/or calk for each particular purpose and to follow the instructions of the manufacturer. For example, acetate calks are used primarily by builders to replace earlier calking made of oils, artificial rubber, and polyvinyl acetate.

Wall Sheathing

In Unit 15 the four most common types of sheathing—wood, plywood, fiberboard, and gypsum board—were discussed. While these materials are still used, more and more builders are choosing sheathings designed specifically for energy conservation.

Each sheathing material has certain advantages and disadvantages that must be considered in making the best selection for each job. Fig. 36-4. Some of the most common of the energy-conserving sheathing materials include the following:

Styrofoam insulation can be

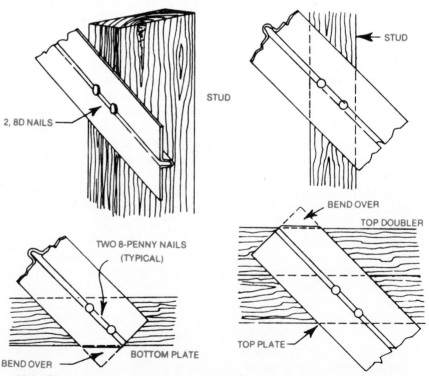

36-6. *Sheet metal let-in braces must be used with Styrofoam sheathing.*

36-7. *Installing foil-faced kraft fiber plies as sheathing. Note that the material is being stapled to the studs.*

used for both wall sheathing and the exterior of basement walls. Fig. 36-5. Styrofoam for walls can reduce energy use by as much as 14 percent. Applying the insulation to the outside of the foundation adds another 10 percent savings, giving a total of 24 percent.

Styrofoam sheathing adds no structural strength to the wall; so sheet metal let-in (wind) braces are needed. Fig. 36-6. These heavy T-shaped sheet metal braces are similar to the wood type described on pages 124 to 126. The sheet metal (wind) braces can be installed diagonally across either the outside or the inside of a wall. A shallow, narrow kerf (gain) must be cut into each stud to insert the sheet

metal brace. The brace is nailed in place with the cross section of the T nailed to the outside of the studs. Siding must be nailed directly to the studs using long, thin siding nails because the porous Styrofoam cannot serve as a nailing base.

Foil-faced kraft fiber ply sheathing is made of high-quality long-fibered plies that are pressure-laminated together using a special water-resistant adhesive. Reflective aluminum foil is fastened to each side. Fig. 36-7. The sheathing is a highly efficient energy-saving material. It has good structural strength and therefore can be used without corner bracing.

Polystyrene is another type of

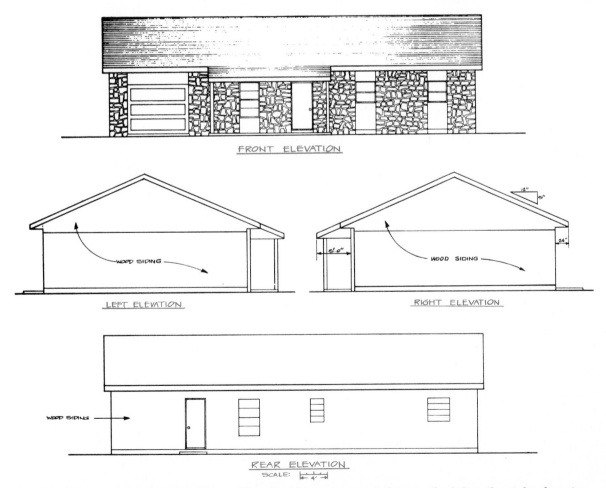

36-8. *Elevations of the Arkansas Home. Note that there are no windows in the left or the right elevation.*

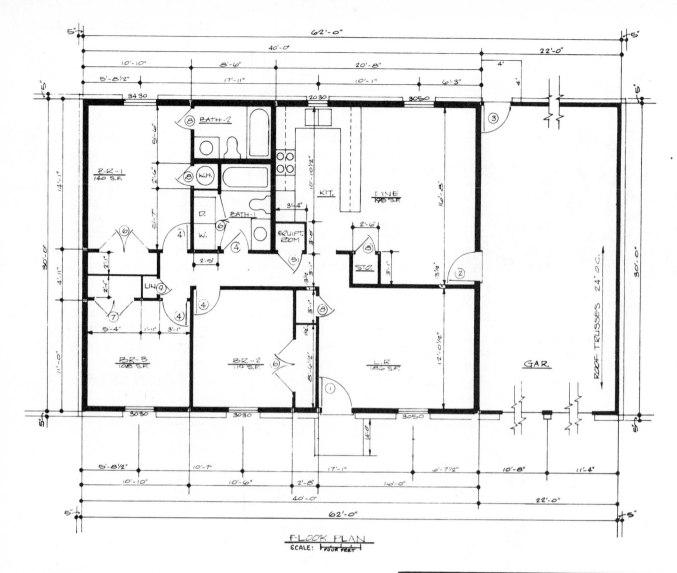

FLOOR PLAN
SCALE: 1" = FOUR FEET

insulating board. It is made from expanded polystyrene. It has many of the same advantages and disadvantages as Styrofoam.

Foil-faced urethane insulating board is a panel made of a core of plastic (urethane) covered on both sides by aluminum foil.

STANDARD HOME CONSTRUCTION

Many things can be done to improve the energy efficiency of the standard home. Some of these things can be done at small costs; others are expensive.

● Limit the size of the living unit (in square feet) by redesigning the living quarters. If each area can be made more flexible (have several uses), the original cost of

36-9. Floor plan. Most rooms have one small window. The bathrooms have no windows.

construction can be reduced and energy efficiency improved.

● Use the correct kind of sheathing. Installation of energy-efficient sheathing instead of standard materials provides great fuel savings.

● Add proper insulation in walls, ceilings, and floors and around the perimeter of the building.

● Add proper, well-fitted storm

windows and doors and double-glazed windows.

● Install the correct type of weather stripping around all doors and windows.

● Add calking around all windows and doors where the frame meets the siding and around all other possible openings. Calking should be applied around an outside faucet, between the basement and

DOOR SCHEDULE		
1.	3-0 x 6-8 x 1¾"	THERMA-TRU
2.	2-8 x 6-8 x 1¾"	THERMA-TRU
3.	2-8 x 6-8 x 1¾"	H.C. EXT.
4.	2-6 x 6-8 x 1⅜"	H.C. INT.
5.	2-6 x 6-8 x 1⅜"	H.C. INT. W/LOUVER
6.	2-6 x 6-8 x 1⅜"	H.C. INT. PAIR
7.	2-0 x 6-8 x 1⅜"	H.C. INT. PAIR
8.	2-0 x 6-8 x 1⅜"	H.C. INT.
9.	1-6 x 6-8 x 1⅜"	H.C. INT.
10.	9-0 x 7-0	O.H. GAR. DOOR

464

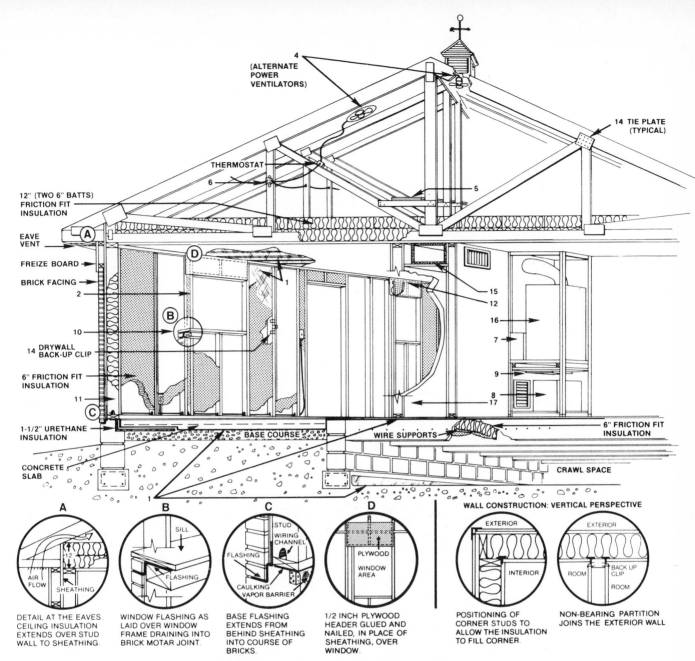

36-10. *Design features of the Arkansas Home.*

the floor framing, the drip cap and siding, the corners formed by siding, where pipes and wires penetrate the ceiling, below an unheated attic or chiminey, and where chimney and masonry meet siding.

● Add attic ventilation. Attics with a ceiling vapor barrier must be ventilated with one square foot of vent area for each 300 square feet of ceiling. Attics without a ceiling vapor barrier must be ventilated with one square foot of vent area for each 150 square feet of ceiling. The vents may be in

the roof, soffit, or gable. One of the most efficient venting systems is that installed in the ridge of the roof.

● Insulate duct work. Duct work may be insulated by adding insulating tape around each joint and furnace opening. Also, batt insulation completely covering all exposed duct work can be installed.

ENERGY-SAVING (CONSERVATION) HOME

An energy-saving home was sponsored by the U.S. Department of Housing and Urban De-

velopment and built in Little Rock, Arkansas. It has become known as the *Arkansas Plan* or *Arkansas Home*. Figs. 36-8 & 36-9. Basically, this home involved the redesigning of wall and ceiling construction to allow for 6″ of insulation in walls and 12″ in ceilings. The design also called for smaller windows equipped with storm windows, metal exterior doors with insulation cores and magnetic weather stripping, power attic ventilation, humidifier, dehumidifier, and air filtration equipment. Figs. 36-10, 11, &

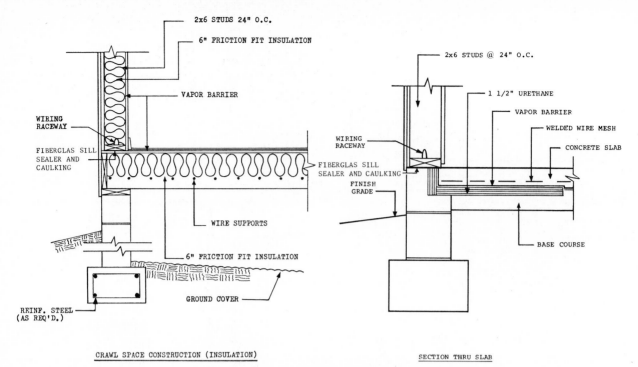

2x6 STUDS 24" O.C.

6" FRICTION FIT INSULATION

2x6 STUDS @ 24" O.C.

VAPOR BARRIER

1 1/2" URETHANE

VAPOR BARRIER

WELDED WIRE MESH

CONCRETE SLAB

WIRING RACEWAY

WIRING RACEWAY

FIBERGLAS SILL SEALER AND CAULKING

FIBERGLAS SILL SEALER AND CAULKING

FINISH GRADE

WIRE SUPPORTS

BASE COURSE

6" FRICTION FIT INSULATION

GROUND COVER

REINF. STEEL (AS REQ'D.)

CRAWL SPACE CONSTRUCTION (INSULATION)

SECTION THRU SLAB

36-11. *Section of crawl space, floor, and wall.*

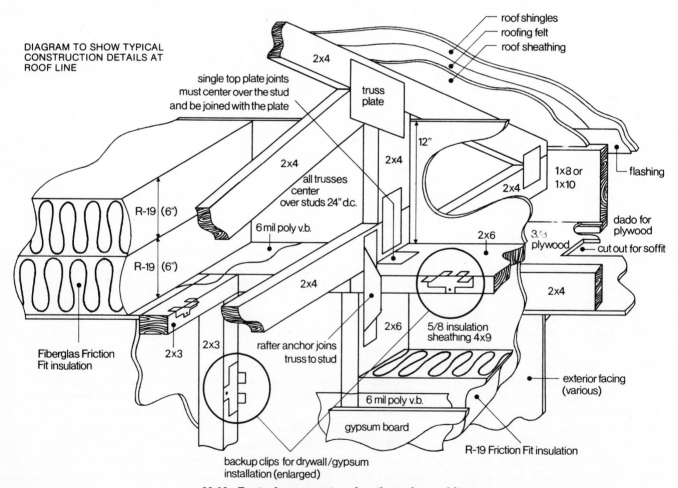

DIAGRAM TO SHOW TYPICAL CONSTRUCTION DETAILS AT ROOF LINE

roof shingles
roofing felt
roof sheathing

2x4

single top plate joints must center over the stud and be joined with the plate

truss plate

12"

2x4

2x4

flashing

1x8 or 1x10

2x4

R-19 (6")

2x4

all trusses center over studs 24" d.c.

2x4

dado for plywood

3/8 plywood

cut out for soffit

6 mil poly v.b.

2x6

R-19 (6")

2x4

2x4

Fiberglas Friction Fit insulation

2x3

2x3

rafter anchor joins truss to stud

2x6

5/8 insulation sheathing 4x9

6 mil poly v.b.

exterior facing (various)

gypsum board

R-19 Friction Fit insulation

backup clips for drywall/gypsum installation (enlarged)

36-12. *Typical construction details at the roof line.*

12. Specifications include the following, most of which are shown on Fig. 36-10.

1. Vapor barriers covering walls, ceilings, and floors.

2. Windows that have storm windows and are calked.

3. Metal exterior doors 1¾" thick with urethane core.

4. Attic space with power roof ventilators and eave vents.

5. Inspection catwalk.

6. Wiring and piping installed to permit correct placement of insulation.

7. Humidifier.

8. Dehumidifier.

9. Air filtering device.

10. Sill and window flashing.

11. Wall studs (2" × 6") spaced 24" on center.

12. Window headers.

13. Proper structural support.

14. Tie plates and drywall back-up clips.

15. Insulated ducts.

16. Centrally located climate conditioning equipment.

17. Partition walls of 2" × 3" studs.

18. Construction strength that is soundly engineered.

The design of the house is unique in many ways. The home has 1200 square feet of living space. It was designed with windows and doors on the front and rear walls only, with no windows or doors on either the left or the right side of the house. Fig. 36-8. As a result, there is only one small window in each major room of the house. Fig. 36-9. This necessitates the use of air conditioning in almost every location in the country in which the house is built.

This home reduced the heat loss 66 percent over a typical standard home of exactly the same size (1200 square feet) built to FHA standards. Both homes were built without a basement, but with a crawl space for heating ducts.

Conservation in the *energy-saving house* was achieved as follows:

• Windows/doors—32.2 percent. Achieved by using metal doors with urethane cores, magnetic weather stripping, and double-glazed windows. The windows were limited to an area of not more than 8 percent of the floor area.

• Flooring—18.2 percent. Achieved with floor insulation over crawl space and around the slab perimeters.

• Duct loss—18.4 percent. Achieved by using special insulation on the heating and air-conditioning ducts.

• Walls—7.7 percent. Achieved by using 2 × 6 studs on 24" centers, instead of the standard construction two 2 × 4 studs on 16" or 24" centers. Six inches of insulation (R-19) were put into the walls.

• Ceiling—7.5 percent. Achieved by using 12" (R-38) of insulation in the ceiling and by adding attic fans.

• Infiltration—15 percent. Achieved by using friction fit batts of polyurethane vapor barriers for walls, floors, and ceilings instead of the batts with integral vapor barriers used in standard construction.

SOLAR HOMES

The energy provided by the sun is enormous. For example, on a global scale, two weeks' worth of solar energy is equal to the fossil energy stored in all of the earth's known reserves of coal, oil, and

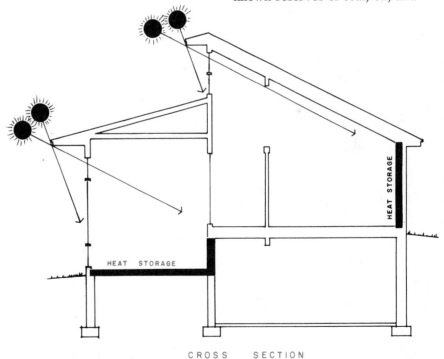

HEAT STORAGE

HEAT STORAGE

CROSS SECTION

36-13a. A large percentage of a home's heating needs can be supplied with a passive solar system. A furnace is needed to provide heat when solar energy isn't available. Passive solar heat is obtained when the sun comes through the south-facing windows. The rays are absorbed by the heavy walls in the house and the concrete (or stone) floor in the greenhouse and/or entryway. When the sun goes down, the insulated drapes are closed. During the evening, the heat that was absorbed by the mass walls and floors radiates into the rooms to help keep them warm. During the very cold weather and on dark days, additional heat must be supplied by a furnace. A passive solar energy system is the most practical solution to using solar energy.

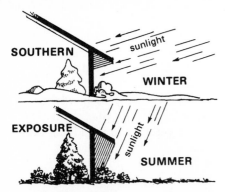

36-13b. *The angle of the sun's rays is very different in the summer and winter. Southern exposure glass areas can provide considerable passive solar energy in the winter. With a long overhanging roof, these glass areas are protected from the sun's rays in the summer.*

natural gas. Solar energy can be applied in climate conditioning living units.

There are two main types of solar energy systems: *passive* and *active*. In the passive solar system, the basic design includes south-facing windows and/or glass walls that invite in the winter sun and a large overhang to block off unwanted heat in the summer. Often the north-facing wall is designed with no windows. This kind of energy-efficient home must be properly designed and carefully situated to take full advantage of natural solar energy. Fig. 36-13.

In the active solar system, there is a means of collecting and storing solar energy. Some homes use solar energy for water heating only. Fig. 36-14. It is theoretically possible to achieve close to 100 percent solar heating and cooling in some parts of the country. However, in most areas, present technology can achieve only about 70 percent solar space heating and about 90 percent solar water heating under ideal conditions. There are few sections of the country that can achieve even this kind of efficiency. Therefore, living units must include supplementary heating and cooling units.

The design and building of active solar energy houses and other living units is a complicated subject on which many books are available. The house owner must decide how an active solar system can be incorporated in the house design. A major consideration is the cost-benefit ratio of such a system. The added cost of the heat collectors, heat storage, and servicing and the additional construction costs are too high to offset the energy saved over a ten- to fifteen-year period.

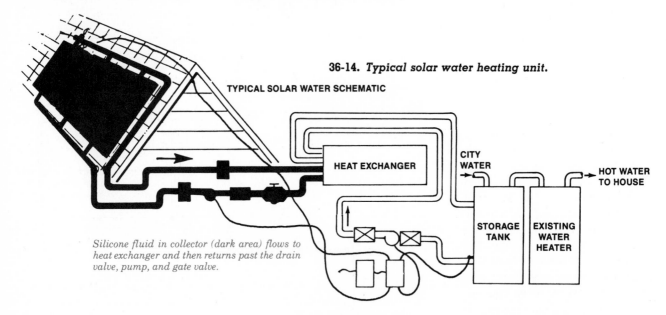

36-14. *Typical solar water heating unit.*

TYPICAL SOLAR WATER SCHEMATIC

HEAT EXCHANGER

CITY WATER

HOT WATER TO HOUSE

STORAGE TANK

EXISTING WATER HEATER

Silicone fluid in collector (dark area) flows to heat exchanger and then returns past the drain valve, pump, and gate valve.

QUESTIONS

1. What is the most important factor influencing internal temperature?

2. Name three considerations in planning an energy-efficient house.

3. What type of plastic is primarily used for bathtubs and showers?

4. Name three types of energy-efficient wall sheathing.

5. Name four things that can improve the energy efficiency of a standard house.

6. How does the framing in the Arkansas Home differ from that in a standard house?

7. What are the two main types of solar energy systems?

Index